# Lecture Notes in Civil Engineering

Volume 260

**Series Editors**

Marco di Prisco, Politecnico di Milano, Milano, Italy

Sheng-Hong Chen, School of Water Resources and Hydropower Engineering, Wuhan University, Wuhan, China

Ioannis Vayas, Institute of Steel Structures, National Technical University of Athens, Athens, Greece

Sanjay Kumar Shukla, School of Engineering, Edith Cowan University, Joondalup, WA, Australia

Anuj Sharma, Iowa State University, Ames, IA, USA

Nagesh Kumar, Department of Civil Engineering, Indian Institute of Science Bangalore, Bengaluru, Karnataka, India

Chien Ming Wang, School of Civil Engineering, The University of Queensland, Brisbane, QLD, Australia

**Lecture Notes in Civil Engineering** (LNCE) publishes the latest developments in Civil Engineering—quickly, informally and in top quality. Though original research reported in proceedings and post-proceedings represents the core of LNCE, edited volumes of exceptionally high quality and interest may also be considered for publication. Volumes published in LNCE embrace all aspects and subfields of, as well as new challenges in, Civil Engineering. Topics in the series include:

- Construction and Structural Mechanics
- Building Materials
- Concrete, Steel and Timber Structures
- Geotechnical Engineering
- Earthquake Engineering
- Coastal Engineering
- Ocean and Offshore Engineering; Ships and Floating Structures
- Hydraulics, Hydrology and Water Resources Engineering
- Environmental Engineering and Sustainability
- Structural Health and Monitoring
- Surveying and Geographical Information Systems
- Indoor Environments
- Transportation and Traffic
- Risk Analysis
- Safety and Security

To submit a proposal or request further information, please contact the appropriate Springer Editor:

- Pierpaolo Riva at pierpaolo.riva@springer.com (Europe and Americas);
- Swati Meherishi at swati.meherishi@springer.com (Asia—except China, and Australia, New Zealand);
- Wayne Hu at wayne.hu@springer.com (China).

**All books in the series now indexed by Scopus and EI Compendex database!**

M. S. Ranadive · Bibhuti Bhusan Das ·
Yusuf A. Mehta · Rishi Gupta
Editors

# Recent Trends in Construction Technology and Management

Select Proceedings of ACTM 2021

Volume 1

*Editors*
M. S. Ranadive
Department of Civil Engineering
College of Engineering Pune
Pune, Maharashtra, India

Yusuf A. Mehta
Department of Civil and Environmental
Engineering
Rowan University
Glassboro, NJ, USA

Bibhuti Bhusan Das
Department of Civil Engineering
National Institute of Technology Karnataka
Mangalore, Karnataka, India

Rishi Gupta
Department of Civil Engineering
University of Victoria
Victoria, BC, Canada

ISSN 2366-2557  ISSN 2366-2565 (electronic)
Lecture Notes in Civil Engineering
ISBN 978-981-19-2147-6  ISBN 978-981-19-2145-2 (eBook)
https://doi.org/10.1007/978-981-19-2145-2

© The Editor(s) (if applicable) and The Author(s), under exclusive license to Springer Nature Singapore Pte Ltd. 2023
This work is subject to copyright. All rights are solely and exclusively licensed by the Publisher, whether the whole or part of the material is concerned, specifically the rights of translation, reprinting, reuse of illustrations, recitation, broadcasting, reproduction on microfilms or in any other physical way, and transmission or information storage and retrieval, electronic adaptation, computer software, or by similar or dissimilar methodology now known or hereafter developed.
The use of general descriptive names, registered names, trademarks, service marks, etc. in this publication does not imply, even in the absence of a specific statement, that such names are exempt from the relevant protective laws and regulations and therefore free for general use.
The publisher, the authors, and the editors are safe to assume that the advice and information in this book are believed to be true and accurate at the date of publication. Neither the publisher nor the authors or the editors give a warranty, expressed or implied, with respect to the material contained herein or for any errors or omissions that may have been made. The publisher remains neutral with regard to jurisdictional claims in published maps and institutional affiliations.

This Springer imprint is published by the registered company Springer Nature Singapore Pte Ltd.
The registered company address is: 152 Beach Road, #21-01/04 Gateway East, Singapore 189721, Singapore

# Contents

## Recent Trends in Concrete Technology

**Self-healing Behavior of Microcapsule-Based Concrete** .............. 3
B. S. Shashank and P. S. Nagaraj

**Durability Properties of Fibre-Reinforced Reactive Powder Concrete** ........................................................... 15
Abbas Ali Dhundasi, R. B. Khadiranaikar, and Kashinath Motagi

**Performance of Geopolymer Concrete Developed Using Waste Tire Rubber and Other Industrial Wastes: A Critical Review** ......... 29
Dhiraj Agrawal, U. P. Waghe, M. D. Goel, S. P. Raut, and Ruchika Patil

**Thermal Behaviour of Mortar Specimens Embedded with Steel and Glass Fibres Using *KD2* Pro Thermal Analyser** .................. 43
P. Harsha Praneeth

**Modulus of Elasticity of High-Performance Concrete Beams Under Flexure-Experimental Approach** ........................... 57
Asif Iqbal A. Momin, R. B. Khadiranaikar, and Aijaz Ahmad Zende

**Impact of Phase Change Materials on the Durability Properties of Cementitious Composites—A Review** ........................... 71
K. Vismaya, K. Snehal, and Bibhuti Bhusan Das

**Comparative Study of Performance of Curing Methods for High-Performance Concrete (HPC)** ............................ 83
Sandesh S. Barbole, Bantiraj D. Madane, and Sariput M. Nawghare

**Development of a Mathematical Relationship Between Compressive Strength of Different Grades of PPC Concrete with Stone Dust as Fine Aggregate by Accelerated Curing and Normal Curing** .................................................. 101
N A G K Manikanta Kopuri, S. Anitha Priyadharshani, and P. Ravi Prakash

**Using Recycled Aggregate from Demolished Concrete to Produce Lightweight Concrete** .................................. 111
Abd Alrahman Ghali, Bahaa Eddin Ghrewati, and Moteb Marei

**Recent Trends in Construction Materials**

**A Study in Design, Analysis and Prediction of Behaviour of a Footbridge Manufactured Using Laminate Composites—Static Load Testing and Analysis of a Glass Fibre Laminate Composite Truss Footbridge** ........................... 125
Col Amit R. Goray and C. H. Vinaykumar

**State-of-the-Art of Grouting in Semi-flexible Pavement: Materials and Design** ............................................. 135
Hemanth Kumar Doma and A. U. Ravi Shankar

**Utilization of Agro-Industrial Waste in Production of Sustainable Building Blocks** ................................................ 149
S. S. Meshram, S. P. Raut, and M. V. Madurwar

**Effect of Exposure Condition, Free Water–cement Ratio on Quantities, Rheological and Mechanical Properties of Concrete** ....................................................... 161
Mahesh Navnath Patil and Shailendra Kumar Damodar Dubey

**Development of Sustainable Brick Using Textile Effluent Treatment Plant Sludge** ......................................... 185
Uday Singh Patil, S. P. Raut, and Mangesh V. Madurwar

**Utilization of Pozzolanic Material and Waste Glass Powder in Concrete** ........................................................ 201
Lomesh S. Mahajan and Sariputt R. Bhagat

**Recent Trends in Construction Technology and Management**

**Integrating BIM with ERP Systems Towards an Integrated Multi-user Interactive Database: Reverse-BIM Approach** ............ 209
M. Arsalan Khan

**Application of Game Theory to Manage Project Risks Resulting from Weather Conditions** ......................................... 221
Abd Alrahman Ghali and Vaishali M. Patankar

## Contents

**Environmental Impact Analysis of Building Material Using Building Information Modelling and Life Cycle Assessment Tool** ...... 233
Kunal S. Bonde and Gayatri S. Vyas

**Enhancing the Building's Energy Performance through Building Information Modelling—A Review** ...... 247
Dhruvi Shah, Helly Kathiriya, Hima Suthar, Prakhar Pandya, and Jaykumar Soni

**Analysis of Clashes and Their Impact on Construction Project Using Building Information Modelling** ...... 255
Samkit V. Gandhi and Namdeo A. Hedaoo

**Predicting the Performance of Highway Project Using Gray Numbers** ...... 271
Supriya Jha, Manas Bhoi, and Uma Chaduvula

**COVID-19—Assessment of Economic and Schedule Delay Impact in Indian Construction Industry Using Regression Method** ...... 283
Soniya D. Mahind and Dipali Patil

**Comparison of Afghanistan's Construction and Engineering Contract with International Contracts of FIDIC RED BOOK (2017) and NEC4—ECC** ...... 299
Mohammad Ajmal and C. Rajasekaran

**Comparative Analysis of Various Walling Materials for Finding Sustainable Solutions Using Building Information Modeling** ...... 315
Amey A. Bagul and Vasudha D. Katare

**Studies on Energy Efficient Design of Buildings for Warm and Humid Climate Zones in India** ...... 327
Santhosh M. Malkapur, Sudarshan D. Shetty, Kishor S. Kulkarni, and Arun Gaji

### Recent Trends in Environmental and Water Resources Engineering

**Spatial Variability of Organic Carbon and Soil pH by Geostatistical Approach in Deccan Plateau of India** ...... 351
N. T. Vinod, Amba Shetty, and S. Shrihari

**Hydro-geo Chemical Analysis of Groundwater and Surface Water Near Bhima River Basin Jewargi Taluka Kalburgi, Karnataka** ...... 361
Prema and Shivasharanappa Patil

**Conjunctive Use Modeling Using SWAT and GMS for Sustainable Irrigation in Khatav, India** .................................................. 373
Ranjeet Sabale and Mathew K. Jose

**Modelling and Simulation of Pollutant Transport in Porous Media—A Simulation and Validation Study** .................................................. 387
M. R. Dhanraj and A. Ganesha

**Adsorptive Removal of Malachite Green Using Water Hyacinth from Aqueous Solution** .................................................. 401
Sayali S. Udakwar, Moni U. Khobragade, and Chirag Y. Chaware

**Sediment Yield Assessment of a Watershed Area Using SWAT** ........ 415
Prachi A. Bagul and Nitin M. Mohite

**"NDVI: Vegetation Performance Evaluation Using RS and GIS"** ...... 425
A. Khillare and K. A. Patil

**Comparison of Suspended Growth and IFAS Process for Textile Wastewater Treatment** .................................................. 437
Sharon Sudhakar, Nandini Moondra, and R. A. Christian

**Innovative Arch Type Bridge Cum Bandhara for Economical and Quick Implementation of Jal Jeevan Mission** .................................................. 449
P. L. Bongirwar and Sanjay Dahasahasra

**Green Synthesis of Zinc Oxide Nanoparticles and Study of Its Adsorptive Property in Azo Dye Removal** .................................................. 467
C. Anupama and S. Shrihari

**Anthropogenic Impacts on Forest Ecosystems Using Remotely Sensed Data** .................................................. 481
Gaurav G. Gandhi and Kailas A. Patil

**Seasonal and Lockdown Effects on Air Quality in Metro Cities in India** .................................................. 497
K. Krishna Raj and S. Shrihari

**Inter-Basin Pipeline Water Grid for Maharashtra** .................................................. 511
Raibbhann Sarnobbat, Pritam Bhadane, Vaibhav Markad, and R. K. Suryawanshi

**Removal of Heavy Metals from Water Using Low-cost Bioadsorbent: A Review** .................................................. 527
Praveda Paranjape and Parag Sadgir

**Adsorptive Removal of Acridine Orange Dye from Industrial Wastewater Using the Hybrid Material** .................................................. 547
Vibha Agrawal and M. U. Khobragade

**Basin Delineation and Land Use Classification for a Storm Water Drainage Network Model Using GIS** .......................... 561
Kunal Chandale and K. A. Patil

**Prediction of BOD from Wastewater Characteristics and Their Interactions Using Regression Neural Network: A Case Study of Naidu Wastewater Treatment Plant, Pune, India** .................. 571
Sanket Gunjal, Moni Khobragade, and Chirag Chaware

**Floodplain Mapping of Pawana River Using HEC-RAS** .............. 579
Tejas R. Bhagwat and Aruna D. Thube

**Performance Evaluation of Varying OLR and HRT on Two Stage Anaerobic Digestion Process of Hybrid Reactor (HUASB) for Blended Industrial Wastewater as Substrate** ..................... 597
Rajani Saranadagoudar, Shashikant R. Mise, and B. B. Kori

**Analysis of Morphometric Parameters of Watershed Using GIS** ....... 603
Bhairavi Pawar and K. A. Patil

**Mapping Ground Water Potential Zone of Fractured Layers by Integrating Electric Resistivity Method and GIS Techniques** ....... 613
R. Chandramohan and B. Kesava Rao

**Sustainable Development in Circular Economy: A Review** ............ 629
Mohnish Waikar and Parag Sadgir

**Biogas Generation Through Anaerobic Digestion of Organic Waste: A Review** .................................................... 641
Vaishali D. Jaysingpure and Moni U. Khobragade

**Recent Trends in Geotechnical Engineering**

**A Cost Comparison Study of Use of Cased and Uncased Stone Column in Marshy Soil** ........................................... 653
Starina J. Dias and Wilma Fernandes

**Review on Field Direct Shear Test Methodologies** ................... 665
Kakasaheb D. Waghmare and K. K. Tripathi

**Numerical Prediction of Tunneling Induced Surface Settlement of a Pile Group** ................................................... 673
B. Swetha, S. Sangeetha, and P. Hari Krishna

**Numerical Study of Pile Supported Embankment Resting on Layered Soft Soil** ............................................... 685
Uzma Azim and Siddhartha Sengupta

**A Review on Application of NATM to Design of Underground Stations of Indian Metro Rail** ...................................... 715
Sandesh S. Barbole, M. S. Ranadive, and Apurva R. Kharat

**Slope Stability Analysis of Artificial Embankment of Fly Ash and Plastic Recycled Polymer Using Midas GTS-NX** .................. 729
Prajakta S. Chavan, Kalyani G. Patil, and Sariput M. Nawghare

**Lateral Capacity of Step-Tapered Piles in Sand Deposits** ............. 739
K. V. S. B. Raju, K. S. Rajesh, L. Dhanraj, and H. C. Muddaraju

**Strength and Dilatancy Behaviour of Granular Slag Sand** ............ 753
K. V. S. B. Raju and Chidanand G. Naik

**Parametric Study of the Slope Stability by Limit Equilibrium Finite Element Analysis** ........................................... 767
Prashant Sudani, K. A. Patil, and Y. A. Kolekar

**Model Testing on PET Bottle Mattress with Aggregate Infill as Reinforcement Overlaying on Fly Ash Under Circular Loading** .... 779
Shahbaz Dandin, Mrudula Kulkarni, and Maheboobsab Nadaf

**Recent Trends in Structural Engineering**

**Sustainable Project Planning of Road Infrastructure in India: A Review** ............................................................. 799
Appa M. Kale and Sunil S. Pimplikar

**Behaviour of Space Frame with Innovative Connector** ............... 805
Pravin S. Patil and I. P. Sonar

**Seismic Analysis of Base Isolated Building Frames with Experimentation Using Shake Table** ......................... 819
Mahesh Kalyanshetti, Ramankumar Bolli, and Shashikant Halkude

**Critical Study of Wind Effect on RC Structure with Different Permeability** ...................................................... 839
Ankush Asati and Uday Singh Patil

**Performance Assessment of SMA-LRB Isolated Building Structure Due to Underground Blast-Induced Ground Motion** ........ 861
Sonali Upadhyaya, Narendranath Gogineni, and Sourav Gur

**Research Progress on the Torsion Behavior of Externally Bonded RC Beams: Review** ......................................... 875
Rajesh S. Rajguru and Manish Patkar

**Study on Static Analysis and Design of Reinforced Concrete Exterior Beam-Column Joint** ...................................... 887
Yogesh Narayan Sonawane and Shailendrakumar Damodar Dubey

**Refined Methodology in Design of Reinforced Concrete Shore Pile: A Design Aid** ................................................. 897
Mahesh Navnath Patil and Shailendra Kumar Damodar Dubey

**Investigating the Efficacy of the Hybrid Damping System for Two-Dimensional Multistory Building Frame Using Time History Analysis** .................................................... 919
A. P. Kote and R. R. Joshi

**Thermal Buckling Analysis of Stiffened Composite Cutout Panels** ..... 935
K. S. Subash Chandra, T. Rajanna, and K. Venkata Rao

**Effect of Isolated Wind Incidence on Local Peak Pressure** ............ 949
Supriya Pal, Ritu Raj, and S. Anbukumar

**Investigation of Performance of Perforated Core Steel Buckling Restrained Brace** ................................................. 961
Prajakta Shete, Suhasini Madhekar, and Ahmad Fayeq Ghowsi

**A Method for Evaluating Maximum Response in Multi-storied Buildings Due to Bi-directional Ground Motion** .................... 973
P. B. Kote, S. N. Madhekar, and I. D. Gupta

**Finite Element Analysis of Piled Raft Foundation Using Plaxis 3D** .... 991
Anupam Verma and Sunil K. Ahirwar

**FRP Strengthened Reinforced Concrete Beams Under Impact Loading: A State of Art** .......................................... 1001
Swapnil B. Gorade, Deepa A. Joshi, and Radhika Menon

**Effect of Lateral Stiffness on Structural Framing Systems of Tall Buildings with Different Heights** ................................. 1015
A. U. Rao, Sradha Remakanth, and Aditya Karve

**Free Vibration Response of Functionally Graded Cylindrical Shells Using a Four-Node Flat Shell Element** ...................... 1031
R. B. Dahale, S. D. Kulkarni, and V. A. Dagade

**The Behaviour of Transmission Towers Subjected to Different Combinations of Loads Due to Natural Phenomenon** ................. 1047
Devashri N. Varhade and R. R. Joshi

**Fragility Assessment of Mid-Rise Flat Slab Structures** ............... 1061
B. P. Dhumal and V. B. Dawari

**Seismic Response of RC Elevated Liquid Storage Tanks Using Semi-active Magneto-rheological Dampers** ........................ 1073
Manisha V. Waghmare, Suhasini N. Madhekar, and Vasant A. Matsagar

**Virtual Testing of Prototypes Using Test Frame Designed for Lateral Load** .................................................. 1089
Suyog Nikam and I. P. Sonar

**Crack Simulation and Monitoring of Beam-Column Joint by EMI Technique Using ANSYS** .................................... 1101
Tejas Shelgaonkar and Suraj Khante

**The Impact of Perforation Orientation on Buckling Behaviour of Storage Rack Uprights** ............................................... 1115
Kadeeja Sensy, Ashish Gupta, K. Swaminathan, and J. Vijaya Vengadesh Kumar

**Modelling Interfacial Behaviour of Cement Stabilized Rammed Earth Using Cohesive Contact Approach** .......................... 1127
T. Pavan Kumar Reddy and G. S. Pavan

**Four Node Flat Shell Quadrilateral Finite Element for Analysis of Composite Cylindrical Shells** .................................. 1135
V. A. Dagade and S. D. Kulkarni

**Exact Elasticity Analysis of Sandwich Beam with Orthotropic Core** ................................................................. 1149
Ganesh B. Irkar and Y. T. LomtePatil

**Flexural Fatigue Analysis of Cross Ply and Angle Ply Laminates** ...... 1171
Sammed Patil and Y. T. LomtePatil

**Recent Trends in Transportation and Traffic Engineering**

**Review on Mechanisms of Bitumen Modification: Process and Variables** .................................................... 1185
N. T. Bhagat and M. S. Ranadive

**Alkali Activated Black Cotton Soil with Partial Replacement of Class F Fly Ash and Areca Nut Fiber Reinforcement** .............. 1193
B. A. Chethan, A. U. Ravi Shankar, Raghuram K. Chinnabhandar, and Doma Hemanth Kumar

**Development of Road Safety Models by Using Linear and Logistic Regression Modeling Techniques** ...................... 1205
Krantikumar V. Mhetre and Aruna D. Thube

**Finite Element Analysis for Parametric Study of Mega Tunnels** ....... 1227
Shilpa Kulkarni and M. S. Ranadive

**Development of Financial Model for Hybrid Annuity Model Road Project** .................................................... 1245
Pratiksha B. Gilbile and Gayatri S. Vyas

**Analysis of Perpetual Pavement Design Considering Subgrade CBR, Life-Cycle Cost, and $CO_2$ Emissions** ........................ 1257
Saurabh Kulkarni and M. S. Ranadive

**Laboratory Investigation of Lateritic Soil Stabilized with Arecanut Coir Along with Cement and Its Suitability as a Modified Subgrade** .......................................... 1273
B. A. Chethan, B. M. Lekha, and A. U. Ravi Shankar

**Pavement Analysis and Measurement of Distress on Concrete and Bituminous Roads Using Mobile LiDAR Technology** ............. 1287
Prashant S. Alatgi and Sunil S. Pimplikar

**Laboratory Study on New Type of Self-consolidating Concrete Using Fly Ash as a Pavement Material** ............................. 1295
Bhupati Kannur and Hemant Chore

**FTIR Analysis for Ageing of HDPE Pyro-oil Modified Bitumen** ....... 1311
H. P. Hadole and M. S. Ranadive

**Utilization of Aluminium Refinery Residue (ARR), GGBS and Alkali Solution Mixes in Road Construction** .................... 1329
Nityanand S. Kudachimath, Raviraj H. Mulangi, and Bhibuti Bhusan Das

# About the Editors

**Prof. M. S. Ranadive** is currently working as Professor and Head, Department of Civil Engineering, College of Engineering Pune (COEP), Pune, Maharashtra. He obtained his B.E. (Civil) from the University of Pune, Maharashtra; M.E. (Civil) From Shivaji University, Kolhapur and Ph.D. from the University of Pune. His major areas of research interests include quality monitoring of pavements, use of anti-stripping agents in bituminous concrete, effective use of bio-oil obtained by pyrolysis of municipal solid waste in flexible pavement, continuous pavement monitoring through dynamic responses by instrumentation. He has published 22 papers in leading international journals, 20 papers in national journals and in all there are more than 100 papers on his account. Professor M. S. Ranadive was a Guest Editor for the Journal of Performance of Constructed Facilities, ASCE. He is a member of American Society of Civil Engineers, Member of Indian Roads Congress, the Life Member of the Indian Geotechnical Society, Member of Indian Concrete Institute, and many more. He is a reviewer for various journals like *Journal of Materials in Civil Engineering*, ASCE, *International Journal of Construction Management and Economics*, Germany; *Journal of Transportation Engineering*; *International Journal of Pavement Engineering*, Taylor and Francis; *Journal of Building Engineering*, Elsevier Publication; *International Journal of Innovative Infrastructure Solutions*, Springer and many more.

**Dr. Bibhuti Bhusan Das** is currently working as an Associate Professor at the Department of Civil Engineering, National Institute of Technology Karnataka (NITK), Surathkal, Mangalore, India. He obtained his B.Tech. (Civil) from Orissa University of Agriculture and Technology, Orissa; M.Tech. from Indian Institute of Technology (IIT) Delhi in Construction Engineering and Management; Ph.D. from Indian Institute of Technology (IIT) Bombay and Post-doctoral from Lawrence Technological University, Southfield, Michigan, USA. His major areas of research interests include concrete technology, sustainable construction, building materials. He has published more than 40 research papers in and SCI Scopus indexed journals. Dr. Bibhuti Bhusan Das has edited four books namely *Sustainable Construction and*

*Building Materials, Recent Developments in Sustainable Infrastructure, Smart Techniques for Sustainable Development,* and *Recent Trends in Civil Engineering* which were published by Springer Nature Singapore. Dr. Bibhuti Bhusan Das is a member of the Indian Concrete Institute, Chennai, India; American Society for Testing and Materials; Prestressed Concrete Institute, Chicago; American Society of Civil Engineers and various other reputed Societies. He is also a reviewer for reputed journals such as *Journal of Materials in Civil Engineering*, ASCE, *Canadian Journal of Civil Engineering, Construction and Building Materials*, Elsevier Publications, and many others.

**Prof. Yusuf A. Mehta** is currently a Professor at the Department of Civil and Environmental Engineering, Rowan University, USA; and Director, Centre of Research and Education in Advanced Transportation Engineering Systems. He obtained his B.S. from the University of Bombay, India; M.S. from the University of Oklahoma, Norman; and Ph.D. from Pennsylvania State University. Since coming to Rowan, Dr. Mehta, has received approximately $25 million dollars of external funding in pavements and materials. He has extensive experience working on several research projects with New Jersey, Florida, Wisconsin and Rhode Island departments of transportations, Federal Highway Administration and Federal Aviation Administration, and Department of Defense. He has also led the effort to acquire the Heavy Vehicle Simulator (HVS) that can simulate 10-20 years of traffic in a few years. Dr. Mehta has received several teaching awards, such as ASCE-NJ Educator of the Year Award, May 2014 and the 2012 Louis J. Pignataro Memorial Transportation Engineering Education Award. The award was for outstanding record of achievement in transportation engineering research, and undergraduate and graduate engineering education. He has received the Mid-Atlantic American Society of Engineering Education Section Distinguished Teaching Award, West Point, 2008. He has also received the faculty research achievement award in 2014. Under the direction of Dr. Mehta, CREATEs has expanded its capabilities to integrate research in Intelligent Transportation Systems, transportation safety, geotechnical engineering, cementitious materials, and Structural engineering by collaborating with faculty members at Rowan University. These collaborations will allow CREATEs to seek research funding in the above mentioned areas. The CREATEs award is approximately $37 M since its inception. CREATEs provides hands-on experience to sixty undergraduate and graduate students in various fields of transportation. This expansion has allowed CREATEs to conduct research, education, and outreach in all the above mentioned fields.

**Dr. Rishi Gupta** is a professor at the Department of Civil Engineering, University of Victoria, Canada where he also leads the Facility for Innovative Materials and Infrastructure Monitoring (FIMIM). He obtained his Diploma from Bombay Technical Board (India); B.E. (Civil) from Pune University (India); M.A.Sc. in Civil Engineering from University of British Columbia and Ph.D. in Civil Engineering (Materials) from University of British Columbia. His major areas of research interests include shrinkage and self-sealing characteristics of concrete and development of 'crack-free' cement composites, structural health monitoring (SHM) of infrastructure, durability and corrosion studies of reinforced concrete, sustainable construction technologies, and advanced materials for structures. Dr. Rishi Gupta holds several patents to his name. He has published more than 70 papers in refereed international journal publications and more than 50 refereed conference publications. Dr. Rishi Gupta has been awarded UVic's REACH award for excellence in Undergraduate Research Enriched teaching, June 2020; Drishti award for Innovation in Science and Technology, 2020; Nominated for Medal for distinction in Engineering Education, Engineers Canada award, 2020; Recipient of EGBC's President's Awards for Teaching Excellence in Engineering and Geoscience Education, 2019; Awarded fellowship with Engineers Canada or Geoscientists Canada, February 2017. Dr. Rishi Gupta is a Member of the Board of Examiners at Engineers and Geoscientists of British Columbia (since July 2018); Editorial board member (2010–2017), EGBC Burnaby/NW Chair (2011-2012); Vice-Chair (2009–2011), Volunteer (2007–to date), EGBC-Burnaby/New Westminster Branch; UVic faculty liaison for EGBC's Victoria Branch (Since May 2015); Member of The Centre for Advanced Materials and Related Technology (CAMTEC); University of Victoria, (Since October 2013) and Board member of Habitat for Humanity Victoria and many other respected boards. He is a reviewer for *Construction and Building Materials*, Elsevier Publications, ASCE *Journal of Materials in Civil Engineering, Journal of American Society of Testing and Materials, Canadian Journal of Civil Engineering, Construction Materials*, Institute of Civil Engineers, UK, and many more.

# Recent Trends in Concrete Technology

# Self-healing Behavior of Microcapsule-Based Concrete

**B. S. Shashank and P. S. Nagaraj**

**Abstract** In general, it is a well-known fact that it is very difficult to detect the cracks at the very initial stage of concrete, and the same cracks later create problems for structure. Further adding to this always the physical intervention is essential for periodic inspection and repair of these cracks. To increase the durability of concrete structures, many self-healing mechanisms are used, out of which microcapsule-based healing mechanisms appear to be feasible. The microcapsule-based self-curative process is an effective process for sealing the cracks in the concrete. Microcapsule-based self-curative starts with the crack occurrence and progresses once it gets in contact with the microcapsule, then self-healing agents are released into cracks, which are contained in capsules. In this study, an effort has been made to synthesize self-healing materials that are synthetic or artificially produced polymer-based substances that have the incorporated ability to naturally repair damage to themselves without any physical contact or human interference. The sodium silicate agent is used as a self-healing material and in situ polymerization method is used for the manufacturing of capsules. It is found that adding 2 and 3% amount of these capsules will heal the concrete after cracking and better the durability.

**Keywords** Self-healing · In situ polymerization · Microcapsule · Strength

## 1 Introduction

In the current scenario, concrete is the most widely consumed material other than water. Due to its wonderful behavior of showing defects, durability is a big concern; hence different modern tools are being experimented with to meet the standards of the construction industry to new heights [1–4]. Different types of processes, methods,

---

B. S. Shashank (✉)
School of Construction Management, NICMAR University, Pune, India
e-mail: shashankbsbhat.bs@gmail.com

P. S. Nagaraj
Department of Civil, UVCE Bangalore University, Bengaluru, India

© The Author(s), under exclusive license to Springer Nature Singapore Pte Ltd. 2023
M. S. Ranadive et al. (eds.), *Recent Trends in Construction Technology and Management*, Lecture Notes in Civil Engineering 260,
https://doi.org/10.1007/978-981-19-2145-2_1

and raw materials are used to attain sustainable, very good, and more economic concrete. But due to human errors, unskilled labor, and incorrect handling [5], an efficient building is difficult to sustain life for a longer duration. Many issues like weathering, micro-cracks, leakages, excessive bending, etc., arise after the construction. To solve these types of problems, many remedial methods are used before and after the construction. One of the remedial techniques is microcapsule-based self-healing concrete [6].

The concept of microcapsules healing is carried out by the healing agent which is coated by a shell material when it is placed in concrete. After the crack appears in concrete and reaches the microcapsule, the microcapsule breaks and the healing material is circulated into the crack to heal by sodium silicate [7]. The microcapsule consists of a healing agent as sodium silicate and the encapsulating material as urea–formaldehyde. And an additional material is added along with this, i.e., polypropylene fiber to advance the strength and property of the concrete at a certain percentage of the cement content [8].

## 1.1 Objectives

- Manufacturing of microcapsule with suitable process
- To heal the cracks and to increase the durability of the structures using microcapsules.

## 2 Microcapsule

Microcapsule is a polymer-based compound that fulfills the criteria that it can incorporate the core material with an encapsulation material. The strength of the microcapsules mainly depends on the shell material and its thickness.

## 2.1 Morphology of Microcapsules

The structure of the microcapsule is mainly dependent on the healing agent called core material and coating material called encapsulation material. The core material is evenly distributed within the shell material to form a microcapsule as shown in Fig. 1 [9].

**Fig. 1** Microcapsule with core healing agent

## 2.2 Core Material

The core material is a polymer matrix to be encapsulated by coating material and it may be in a solid or liquid state depending on the functional requirements. In the present study, sodium silicate is used as a core material that will act as a healing agent in concrete [10].

## 2.3 Coating Material

The physical and chemical properties of microcapsule are mainly dependent on the encapsulation material which protects the core material from external pressure and extend the sustainability of core material [9].

## 2.4 Release Mechanisms

The main purpose of microencapsulation is to release the core material which is coated by encapsulation material at the time of usage of the microcapsule. The shell material should break at the time of usage and react with the surroundings. Whenever a crack appears in concrete and that gets into contact with a microcapsule, it will break and the healing agent will be released into the crack to heal the concrete [10, 11]. The same can be observed in Fig. 2.

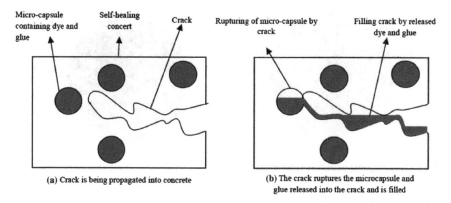

(a) Crack is being propagated into concrete

(b) The crack ruptures the microcapsule and glue released into the crack and is filled

**Fig. 2** Healing agent release mechanisms

**Table 1** List of chemicals required for the synthesis process

| Chemical | Quantity |
| --- | --- |
| Sodium silicate | 60 ml |
| Urea | 5.00 g |
| Formaldehyde | 12.7 g |
| Ammonium chloride | 0.50 g |
| Resorcinol | 0.50 g |
| Sodium hydroxide | 20 ml (dropwise) |
| Hydrochloric acid | 20 ml (dropwise) |

## 3 Materials and Methods

### 3.1 Synthesis of Microcapsules

The production process of sodium silicate microcapsule was based on the in situ polymerization method and the process is mainly based on some of the parameters such as agitation, temperature, and pH range [12, 13]. Table 1 and Fig. 3 show the materials for the preparation of microcapsule and prepared sample, respectively.

### 3.2 Concrete Mix Design

Mix design of concrete was made according to IS 10262-2019 and two different grades of concrete were designed M20 and M40 with testing all basic materials required for the test as cement sand and aggregate, and the mix design for M20 and M40 is shown in Table 2.

**Fig. 3** Preparation of microcapsule and prepared sample

**Table 2** Mix design

| Grade of concrete | Cement kg/m$^3$ | Fine aggregate kg/m$^3$ | Coarse aggregate kg/m$^3$ | Polypropylene fibre kg/m$^3$ | Water l/m$^3$ |
|---|---|---|---|---|---|
| M20 | 394 | 750 | 1037 | 0.25% | 197 |
| M40 | 420 | 625 | 1085 | 0.25% | 180 |

**Table 3** Mix design designation

| S. No. | Grade designation | | | | Grade of concrete |
|---|---|---|---|---|---|
| | CC | 0.25% fibers | 2% microcapsule | 3% microcapsule | |
| 1 | CC-1 | CCF-1 | CCFM-1 | CCFM-3 | M20 |
| 2 | CC-2 | CCF-2 | CCFM-2 | CCFM-4 | M40 |

## 3.3 Mix Design Designation

As there are many mixes to understand better grade designation mentioned, it is shown in Table 3.

# 4 Results and Discussion

## 4.1 SEM Analysis

The microcapsules were scanned using scanning electron microscopy (SEM) in the range of 400–6000 cm$^{-1}$ to get a clear picture of the microcapsules which have

**Fig. 4** SEM images of the microcapsule

been synthesized, by focusing on the particular range that can say that the prepared microcapsules as a diameter of 5–10 μm, as shown in Fig. 4. Some of the capsules are irregular in shape, so average diameter has been taken to predict the size of the capsules.

## 4.2 FTIR Analysis

Fourier-transform infrared spectroscopy (FTIR) examines the peak range of urea–formaldehyde and sodium silicate which was used and the results showed that they were almost matched with the standard range of peak value. So, by this test, we can confirm the presence of the core and encapsulate, and the same can be noted from the graph shown in Fig. 5 and Table 4.

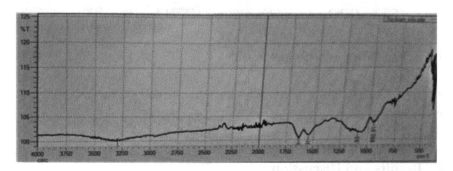

**Fig. 5** FTIR analysis graph

**Table 4** Chemicals present in the core material

| S. No. | Peak | Content | Standard peak range |
|---|---|---|---|
| 1 | 3296.35 | Sodium silicate | 3000–3500 |
| 2 | 1097.50 | Urea–formaldehyde | 1000–1250 |

**Table 5** Variation in a slump across the different mix

| Mix | Conventional concrete | 0.25% fibers | 2% microcapsule | 3% microcapsule | Grade of concrete |
|---|---|---|---|---|---|
| Slump in mm | CC-1 | CCF-1 | CCFM-1 | CCFM-3 | M20 |
|  | 95 | 89 | 85 | 83 |  |
| Slump in mm | CC-2 | CCF-2 | CCFM-2 | CCFM-4 | M40 |
|  | 92 | 82 | 79 | 79 |  |

## 4.3 Slump Flow Test

The workability of concrete was tested according to IS 7320:1974 and it was found that workability will reduce as the percentage of microcapsule increases in both M20 and M40 grade concrete as shown in Table 5. Further with the addition of fibers also contributed to the reduction in the slump of concrete. The effect of both can be solved by proper admixture usage.

## 4.4 Compressive Strength

A compressive strength test was performed based on the IS 516:1959. Concrete with microcapsule is initially tested by applying 50% of the load for inducing minor cracks, then sufficient time was provided to heal. After healing the concrete was once again tested for its compressive strength. It was observed that for M20 grade concrete it took 22 days and for M40 it took 26 days to heal up to a considerable extent. Testing samples revealed that after healing of concrete the compressive strength decreased by 14% and 12% for 2% and 3% microcapsule, respectively, compared to conventional concrete and original samples for M20 grade concrete, and the compressive strength decreased by 11% and 8% for 2% and 3% microcapsule, respectively, for M40 grade concrete compared to conventional concrete. The same has been shown in Fig. 6. But the decrease in the compressive strength of concrete has not affected the grade of concrete.

## 4.5 Split Tensile Strength

A split tensile strength test was performed according to IS 5816:1999. Similar to the compressive strength of concrete, samples with microcapsule are showing a decrease in split tensile strength as the percentage of microcapsule are increased compared to original concrete but the strength of these concretes is better than that of conventional concrete. Further, the healing was observed in the specimens after 22 and 26 days

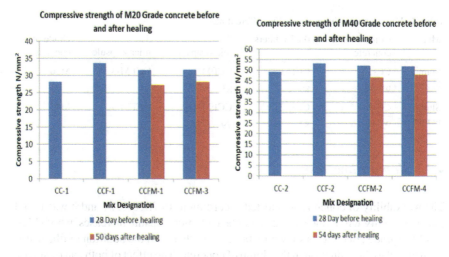

**Fig. 6** Variation in compressive strength of M20 and M40 grade concrete

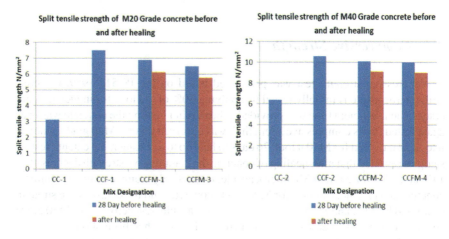

**Fig. 7** Variation in split tensile strength

for M20 and M40 grade concrete, respectively. The same has been shown in Fig. 7, and also Fig. 8 shows the healing of the concrete cylinder.

## 4.6 Flexural Strength

A flexural strength test was performed according to IS 516:1959 with a beam size of 150 × 150 × 500 mm. By observing variation in flexural strength, it was found that as the percentage of microcapsule increases, the flexural strength of concrete was

# Self-healing Behavior of Microcapsule-Based Concrete

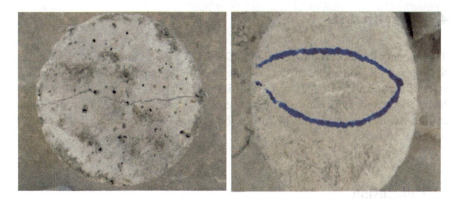

**Fig. 8** Healing of concrete cylinder due to microcapsule

decreased for both M20 and M40 grade concrete compared to conventional concrete. But after healing the flexural strength reduced by 15 and 29% for M20 grade concrete and 11 and 16% for M40 grade concrete for 2% and 3% microcapsule, respectively, compared to its original samples. Variation in flexural strength is shown in Fig. 9.

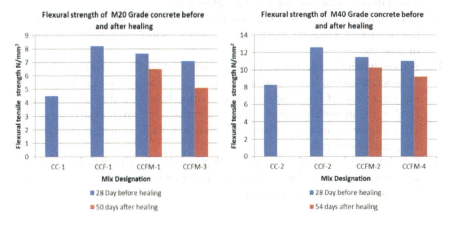

**Fig. 9** Variation in flexural strength

## 4.7 Cost Analysis

The amount of microcapsules required for 2% and 3% for 1 cum meter of concrete is 8.82 kg and 13.23 kg, respectively. And the cost for the preparation of the microcapsules of 2% and 3% is Rs 4410 and Rs 6615, respectively. The cost of microcapsules-based concrete is more compared to conventional concrete, but taking into consideration the maintenance and the life of the self-healing concrete the cost can be justified.

## 5 Conclusion

1. By using the in situ polymerization process microcapsules are prepared and these are having a size of less than 1 micron which is vital for any material to blend with cement and also FTIR analysis and FEM analysis are validating the presence of a healing agent in the microcapsule which helps to self-heal the concrete during the crisis.
2. Incorporation of the microcapsule showed excellent healing of concrete within 22–28 days, which is a significant development.
3. The compressive strength before and after cracking showed that sodium silicate can be used as a healing agent, which contributes to the strength development of concrete.
4. Although there is a decrease in compressive, split tensile, and flexural strength of concrete and an increase in microcapsule percentage before healing, but as the percentage of microcapsule increases, the healing ability also increases. Further, the strength of concrete is more compared to conventional concrete but less compared to its original value, which is acceptable as there is no change in the grade of concrete.
5. Cost analysis concludes that self-healing concrete cost is slightly more than conventional concrete, but compared to its durability and maintenance aspects, it is a worthy investment in the end.

## References

1. Bonilla L et al (2018) Dual self-healing mechanisms with microcapsules and shape memory alloys in reinforced concrete. J Mater Civ Eng 30(2):04017277
2. Sahoo S, Das BB, Mustakim S (2017) Acid, alkali, and chloride resistance of concrete composed of low-carbonated fly ash. J Mater Civ Eng 29(3):4016242
3. Snehal K, Das BB, Akanksha M (2020) Early age, hydration, mechanical and microstructure properties of nano-silica blended cementitious composites. Constr Build Mater 233:117212
4. Snehal K, Das BB (2021) Acid, alkali and chloride resistance of binary, ternary and quaternary blended cementitious mortar integrated with nano-silica particles. Cement Concr Compos 123:104214

5. Sangadji S (2017) Can self-healing mechanism helps concrete structures sustainable? Procedia Eng 171:238–249
6. Beglarigale A et al (2018) Sodium silicate/polyurethane microcapsules used for self-healing in cementitious materials: monomer optimization, characterization, and fracture behavior. Constr Build Mater 162:57–64
7. Restuccia L et al (2017) New self-healing techniques for cement-based materials. Procedia Struct Integr 3:253–260
8. Bashir J et al (2016) Bio concrete-the self-healing concrete. Indian J Sci Technol 9(47)
9. Li W et al (2016) Preparation and properties of melamine urea-formaldehyde microcapsules for self-healing of cementitious materials. Materials 9(3):152
10. Kanellopoulos A et al (2017) Polymeric microcapsules with switchable mechanical properties for self-healing concrete: synthesis, characterisation and proof of concept. Smart Mater Struct 26(4):045025
11. Ahn E et al (2017) Principles and applications of ultrasonic-based nondestructive methods for self-healing in cementitious materials. Materials 10(3):278
12. Gruyaert E et al (2016) Capsules with evolving brittleness to resist the preparation of self-healing concrete. Mater Constr 66(323):e092
13. Šavija B et al (2017) Simulation-aided design of tubular polymeric capsules for self-healing concrete. Materials 10(1):10

# Durability Properties of Fibre-Reinforced Reactive Powder Concrete

**Abbas Ali Dhundasi, R. B. Khadiranaikar, and Kashinath Motagi**

**Abstract** Concrete used for sewer lines, piers and deck slabs of bridges, highways and nuclear power plants are prone to acid attacks. Hence it is required to study the mechanical properties along with durability aspects to assign its service life. In the present study, fibre-reinforced reactive powder concrete (FRPC) with varying silica fume content, superplasticizer and dosages of steel fibres are produced. Strengths under compression, flexure and tension have been obtained. Durability studies have been carried out by acid tests, salt crystallization test and permeability tests. Strengths of 140, 150 and 160 MPa have been achieved. To study the resistance against acid attacks, concrete is subjected to $H_2SO_4$ solution with 0.5, 1, 1.5 and 2% concentrations. Crystallization of salts is studied by immersing specimens in $Na_2SO_4$ solutions for 10–50 cycles. The concrete shows greater resistance towards the permeability of chloride ions. A considerable amount of loss in strength and mass is observed when subjected to a higher dosage of acid environment. An optimum dosage of 2.5% of fibres is suggested. Addition of fibres improved the strength and durability properties.

**Keywords** Reactive powder concrete · Steel fibres · Acid test · Salt crystallization test · Rapid chloride ion penetration test

## 1 Introduction

The performance of any concrete is governed by its mechanical properties and durability aspects under adverse environmental conditions throughout its service life. Based on strength parameters concrete is categorized as normal strength (20–40 MPa), high strength (40–60 MPa) and ultrahigh-strength (>80 MPa) concretes [1]. Reactive powder concrete is a new-generation ultrahigh-strength concrete with strengths in the range of 100–800 MPa. Its applications include the construction of storage bunkers for nuclear waste, sea harbour structures, bridge piers and pathways, highways, airways, sewage canals etc.

---

A. A. Dhundasi (✉) · R. B. Khadiranaikar · K. Motagi
Department of Civil Engineering, Basaveshwara Engineering College, Bagalkot, Karnataka, India
e-mail: abbasdhundasi@gmail.com

P. Richard and M. Cheyrezy from France are the pioneers who produced reactive powder concrete in the early 1990s. Coarse aggregates were eliminated to enhance the homogeneity. Silica fume was added as pozzolanic material. W/B ratio was reduced to 0.15–0.19 and a workable mix was obtained by adding superplasticizers. The microstructure was enhanced by the application of pressure and post heat treatment at 20 and 90 °C. The ductility was improved by the addition of steel fibres. RPC with strengths of 200–800 MPa were produced [2]. A study on various types of superplasticizers and their effect on workability at a low w/c ratio was carried out by Collepardi et al. [3]. RPC with 285 MPa strength was produced with low-pressure steam curing and 90°C hot water curing [4]. Applying 50 MPa presetting pressure and a very high dosage of 1900 kg/m$^3$ of cement yielded RPC with 500 MPa compressive strength [5]. RPC with strengths 120–300 MPa is produced with autoclave curing, hot water, hot air and steam curing [6–10]. The researchers concluded that heat curing has a profound influence on strength development and hence higher strength RPCs can be produced.

The ductility and tensile strengths of concrete can be improved by the addition of fibres [11–14]. Various types of fibres that can be incorporated in concrete are steel, polypropylene, nylon, glass, basalt fibres etc. The fibre-reinforced concretes (FRC) can be used for the construction of bridge deck slabs, highway slabs as well as in runways of airports. A study on the dosage of polypropylene fibres with different lengths was carried out on self-compacting concrete. Mechanical properties and spalling tendency were evaluated. The concrete was cured under elevated temperatures [15]. A model using ANSYS was proposed to determine the tensile strength of concrete with fibres varying in aspect ratio (length/depth) [16].

An extensive study on the mechanical properties of FRC was done by Zuzana et al. The fibres were made of steel, nylon, polypropylene and glass materials with cross-sections of round, rectangular and irregular in nature. Shapes considered were (a) straight, (b) hooked, (c) wave and (d) deformed. The ends of fibres were (a) hooked, (b) button and (c) deformed. The results concluded that straight fibres with hooked ends improved density, tensile strength, fracture energy and shear resistance [17]. RPC with strength 200–315 MPa was produced with a w/c ratio of 0.2 and 20% silica fume. Steel fibres with 16 mm dia. and length 6 mm were added in 0.5–4% by volume. Properties and microstructure were evaluated. It was concluded that the addition of fibres increases both compressive and flexural strengths. It was also suggested that the volume fraction of fibres should not exceed 3% to get optimum results [18].

The concrete is susceptible to chemical attacks from an acidic environment and dissolved acid which precipitate in the form of acidic rain. Many researchers have shown that the durability of concrete can be expressed in terms of loss in weight/mass and compressive strength. Amongst the acids, sulphuric acid ($H_2SO_4$), hydrochloric acid (HCl), nitric acids ($HNO_3$) etc. cause massive deterioration of concrete. The concrete used in sewer pipes is severely exposed to sulphuric acid attacks. These cause corrosion in concrete. The chemical reactions by several bacterial activities under low pH values and anaerobic conditions result in the generation of sulphuric acid around concrete. Commonly found bacteria are desulphovibrio, *Thiobacillus plumbophilus, Halothiobacillus neapolitanus,* and *Acidithiobacillus thiooxidans.* An investigation on corrosion of concrete caused by such bacteria has been carried out [19]. Acid resistance of normal strength blended concrete (20–40 MPa) was carried out by immersing specimens in 5% $H_2SO_4$ and 5% HCl solution for 28 and 90 days. It was observed that 40% weight loss had taken place [16]. Effect of sulphuric acid on fly ash-based geopolymer concrete was done for 7, 28 and 56 days. The strengths were 53 and 62 MPa. The specimens were immersed in 10% solution. The results indicated that GPC shows higher resistance to sulphuric acid attack [20]. Sulphuric acid has a significant effect on the weight loss and the compressive strength of concrete [21–27]. Durability tests on RPC with 180 MPa strength were done. Acid tests, accelerated corrosion and RCPT tests were performed. Mass was reduced up to 20% and compressive strength up to 60% [28]. In chemical industries producing medicines, artificial manure etc., nitric acid is formed in the presence of water by the reaction of compounds and radicals of nitrates. It is then released into the atmosphere. The product of cement hydration ($Ca(OH)_2$) reacts with such acid and transforms into highly soluble nitro aluminate calcium hydrates and salts of calcium nitrate. This leads to the deterioration of concrete [29]. A comparative study on various ageing tests to evaluate salt crystallization damage was also done to assess the durability characteristics of concrete [30].

Many researches have been carried out on assessing the mechanical properties and durability studies on normal, high strength, reinforced, self-compacting and geopolymer concrete etc. Many studies have been carried out only on the production of RPC by varying constituent materials and curing conditions. However, very few investigations are done on durability properties. Hence the present research focuses on producing steel fibre-reinforced reactive powder concrete with constituent materials available in the southern part of India and studying its mechanical properties and durability properties. The optimum dosage of fibres along with the variation of silica fume is determined. The durability properties of fibre-reinforced RPC are assigned by performing various tests which will help the constructure engineers to utilize the material efficiently for many applications.

## 2 Materials and Mix Proportions

The reactive powder concrete mix proportions are prepared with OPC 53 Grade cement complying with IS:12269(1987). Superfine silica fume 920D is added as pozzolanic material procured from Elkem India Pvt. Ltd. Quartz powder (QP) of 300–600 μm size is used to improve microstructure. It has high silica content and acts as filler material unless the curing temperature exceeds 200 °C, at which it produces silicates and contributes to an increase in strength. Potable water, free from any impurities, is used with variable ratios. A BASF product Glanium-8233 is added as a superplasticizer. The aggregates constitute high-purity silica sand. The details of mix proportions are shown in Table 1.

**Table 1** Design mix proportions for RPC

| Mix | Cement | Sand | W/B ratio | Silica fume | | QP | Superplasticizer | | Fibre |
|---|---|---|---|---|---|---|---|---|---|
| | kg/m$^3$ | kg/m$^3$ | | kg/m$^3$ | % | kg/m$^3$ | kg/m$^3$ | % | % |
| R1-A | 1000 | 959.51 | 0.21 | 50 | 5 | 100 | 15 | 1.5 | 1.5 |
| R1-B | 1000 | 947.69 | | 50 | | | 15 | | 2.0 |
| R1-C | 1000 | 935.88 | | 50 | | | 15 | | 25 |
| R2-A | 1000 | 924.06 | 0.20 | 100 | 10 | 100 | 20 | 2.0 | 1.5 |
| R2-B | 1000 | 912.24 | | 100 | | | 20 | | 2.0 |
| R2-C | 1000 | 900.42 | | 100 | | | 20 | | 2.5 |
| R3-A | 1000 | 897.67 | 0.19 | 150 | 15 | 100 | 25 | 2.5 | 1.5 |
| R3-B | 1000 | 885.88 | | 150 | | | 25 | | 2.0 |
| R3-C | 1000 | 874.06 | | 150 | | | 25 | | 2.5 |

## 3 Experimental Programme

### 3.1 RPC Production

Three grades of fibre-reinforced ultrahigh-strength concrete, i.e. R1, R2 and R3, with compressive strengths of 140, 150 and 160 MPa are produced with a varying dosage of silica fume and fibre content. The silica fume is varied in 5, 10 and 15% and fibres are added in 1.5, 2.0 and 2.5% of the dosage. Dry mix and wet mix are done in a 200 l capacity pan mixer. The mixer rotates at a speed of 140–280 RPM [7]. To get a workable mix superplasticizer dosage is varied w.r.t. w/c ratio. Care is taken to disperse the steel fibres evenly throughout the mix to avoid flocculation. It takes 18 ± 2 min to get a workable mix. Wet mix is then compacted in three layers on a vibrating table to remove air voids. The demoulded specimens are kept for hot water steam curing at 90°C for 48 h and then for normal water curing for up to 28 days. Cube specimens of 100 × 100 × 100 mm, cylinder specimens of 100 mm dia. and 150 mm height and beams of 100 × 100 × 500 mm are cast and the mechanical properties are noted down for uniaxial monotonic loads.

### 3.2 Durability Tests

RPC samples are immersed in acid-resistant trays of 100 × 150 × 150 cm, with a calculated amount of predetermined concentration of acid solutions. Separate specimens are used for each durability study.

a. **Acid Test**

The concentration of $H_2SO_4$ solution is varied in the range of 0.5, 1.0, 1.5 and 2%. The observations were taken at intervals of 7, 28, 60 and 90 days. At the end of each period, samples are taken out of solutions and washed thoroughly with clean water. To remove absorbed water, samples are oven-dried. The samples exposed to different solutions of $H_2SO_4$ show deterioration. Due to the chemical reactions leading to the degradation of concrete, the pH of the solution is observed to change. Hence at each 24-h interval, a constant pH is maintained carefully. The deterioration of concrete is measured in terms of weight loss and compressive strength loss. The percentage loss is calculated from the following equations:

Reduction in compressive strength (%):

$$\frac{f'_i - f'_f}{f'_i} \times 100 \qquad (1)$$

where $f'_i$, $f'_f$ are initial and final compressive strengths at 28 days and after degradation, respectively.

Percentage weight/mass loss is calculated as follows:

$$\frac{W_1 - W_2}{W_1} \times 100 \qquad (2)$$

where $W_1$ and $W_2$ are initial and final weights after deterioration.

b. **Salt Crystallization Test**

The 28 days cured, clean and oven-dried samples are immersed in 14% $Na_2SO_4$ solution. The steps for the salt crystallization test of each cycle are as follows: (1) Samples are immersed in an acidic solution for 18 h at room temperature. (2) The removed samples are drained for 30 min. (3) The drained samples are oven-dried for 4 h at $105 \pm 5$ °C and later cooled down to room temperature. This completes one cycle of 24 h. For the present study, the number of cycles is varied from 0, 10, 20, 30, 40 to 50 cycles.

c. **Rapid Chloride Penetration Test**

Cylindrical specimens are cut using a diamond saw cutter to get samples of 50 mm thickness and 100 mm dia. RCPT tests were performed on these samples [22]. A DC voltage of 60 V is applied between two cells of 0.3 N NaOH and 3% NaCl solutions. The chloride ion penetration is governed by the measure of the total amount of charges passed after 360 min in each specimen. The resistance towards chloride ion penetration is calculated for the grades of FRPC (Figs. 1 and 2).

**Fig. 1** Specimens in $H_2SO_4$ solution

**Fig. 2** Specimens in $Na_2SO_4$ solution

## 4 Results and Discussion

### 4.1 Production of Fibre-Reinforced RPC

The steel fibres are added to improve the mechanical properties of all grades of RPC designed for 1.5, 2 and 2.5% dosage. The fibres are of round shape with hooked ends. The aspect ratio ($l/d$) is 60, with a length of 12 mm and a diameter of 0.2 mm. The 28 days strengths obtained for each mix are shown in Table 2.

It is observed that a higher dosage of fibres contributes to an increase in strength. An optimum of 2.5% dosage is suggested. The flexural strength of FRPC is observed to be 20–25 MPa, i.e., 15% of compressive strength. An average of 15–20 MPa tensile strength was achieved. These strengths are much greater than normal strength and high strength reinforced concretes [6–10]. Thus, in fibre-reinforced ultrahigh strength

**Table 2** Mechanical proportions of UHSC

| Mix | Compressive strength (MPa) | | Flexural strength (MPa) | Tensile strength (MPa) |
|---|---|---|---|---|
| | Designed | Achieved | | |
| R1-A | 140 | 138.84 | 14.12 | 8.89 |
| R1-B | | 141.21 | 16.94 | 12.11 |
| R1-C | | 142.85 | 20.89 | 15.31 |
| R2-A | 150 | 147.95 | 16.34 | 11.33 |
| R2-B | | 150.19 | 18.67 | 14.02 |
| R2-C | | 154.54 | 22.28 | 16.94 |
| R3-A | 160 | 156.71 | 18.45 | 13.86 |
| R3-B | | 161.08 | 21.72 | 16.74 |
| R3-C | | 165.47 | 24.88 | 20.16 |

concrete, steel bars as reinforcement to resist tensile stresses can be omitted, which helps in the conservation of natural resources for future generations.

## 4.2 Durability Tests

### 4.2.1 Acid Tests

The effect of different concentrations of $H_2SO_4$ is discussed as follows:

a. **0.5% $H_2SO_4$**: The visual observations show that there is no change in shape or structure of the specimen exposed for up to 7 days. However, a small amount (2%) of weight loss is observed in R1 samples. R2 and R3 samples show better resistance at the initial stage and up to 28 days exposure period. Samples show white deposits as a result of surface erosion for 60 days of exposure. A noticeable amount of weight loss and compressive strength loss is observed. Scaling of the top surface layer is observed for samples kept for 90 days, shown in Fig. 3a. The weight loss is 10.49%, 9.63% and 7.23% for R1, R2 and R3, respectively. The strength loss is 26.95%, 20.59% and 19.24%, respectively. These are shown in Figs. 4a and 5a.

b. **1.0% $H_2SO_4$**: Surface erosion is initiated even at 7 days of exposure. Discolouration of samples from greenish-grey to whitish-grey is observed at 28 days of exposure. No signs of corrosion are observed. Scaling of the concrete takes place at a higher exposure period, which increases progressively and peeling of the surface takes place at 90 days of exposure as shown in Fig. 3b. The weight is reduced by 13.63%, 12.82% and 10.60% and correspondingly compressive strength reduction is 38.45%, 29.27% and 24.55% for R1, R2 and R3, respectively. These are shown in Figs. 4b and 5b.

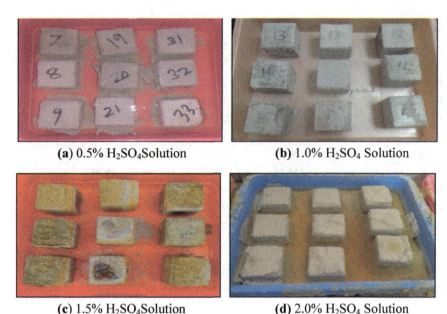

Fig. 3 FRPC samples exposed to H$_2$SO$_4$ solutions

c. **1.5% H$_2$SO$_4$**: It is observed that the scaling of concrete is predominant even on lower days of exposure. Capillary pores are formed. Corrosion of steel is initiated. Reddish-brown deposits are seen on the surface as shown in Fig. 3c. These are the results of steel fibre corrosion. However, concrete shows resistance towards spalling. As shown in Figs. 4c and 5c, 16.91%, 15.82% and 14.01% of mass reduction and 56.07%, 46.78% and 40.22% of loss in compressive strength are observed for R1, R2 and R3.

d. **2.0% H$_2$SO$_4$**: This acidic solution induces severe damage to all the specimens. Corrosion of steel fibres is predominant. The solution percolates through capillary pores and oxidation of Fe$^{2+}$ ions results in the formation of ferric oxides and ferric hydroxides in the form of rust. The volume of this layer reaches up to seven times its original value inducing bursting pressure. Thus, spalling of concrete is observed as shown in Fig. 3d. The embedded fibres are exposed and the rate of corrosion increases with an increase in the exposure period. Samples are disintegrated to a greater extent. A huge amount of weight loss is observed, i.e. 23.98%, 20.09% and 18.38% with strength loss of 76.05%, 67.63% and 59.72% for all the grades of concrete. It is shown in Figs. 4d and 5d.

### 4.2.2 Salt Crystallization Test

Ordinary portland cement is susceptible to damage caused by various acids. This can be overcome by adding superfine pozzolanic material which improves microstructure

Durability Properties of Fibre-Reinforced Reactive ...

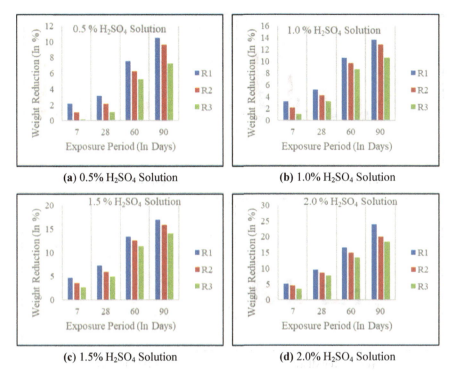

**Fig. 4** Reduction in weight for FRPC samples

and density by reducing porosity and permeability. The addition of fibres improves bond strength between aggregate phase and paste which further improves density. The 5% of silica fume in R1 and w/c ratio of 0.21 produced a homogeneous mix and design strength is achieved. However, it is observed that more air voids and pores are generated compared to R2 and R3 when subjected to the salt crystallization test. Thus, from Figs. 8 and 9, it is noted that the percentage of weight loss and reduction in compressive strength in R1 is relatively more than other mixes of FRPC. From Fig. 9 it is seen that all the mixes of FRPC show better resistance towards salt crystallization up to 20 cycles. Thereafter significant damage is induced with an increase in the number of cycles. The damage increases progressively with an increase in each cycle ultimately leading to the failure of the specimen. The results are tabulated in Table 3. The maximum weight loss is up to 6–10% of its original weight corresponding to a 50% loss of strength in compression at 50 cycles. It can be stated that the higher the grade of concrete, the higher the resistance towards salt crystallization (Figs. 6 and 7).

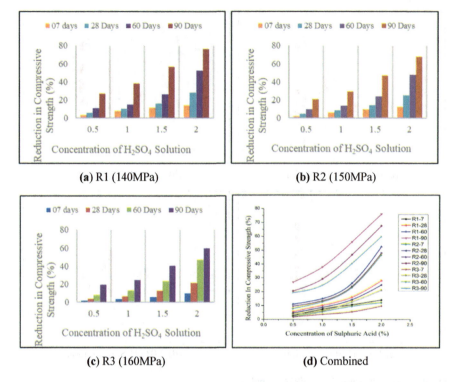

**Fig. 5** Reduction in compressive strength for FRPC samples

**Table 3** Variation of weight loss and compressive strength with durability cycles

| Cycles | Weight reduction (%) | | | Compressive strength (MPa) | | |
|---|---|---|---|---|---|---|
| | M1 | M2 | M3 | M1 | M2 | M3 |
| 0 | 0.00 | 0.00 | 0 | 140 | 150 | 160 |
| 10 | 1.75 | 1.24 | 0.89 | 134.36 | 143.26 | 152.62 |
| 20 | 2.67 | 2.03 | 1.76 | 128.49 | 135.42 | 144.17 |
| 30 | 6.23 | 5.09 | 4.01 | 116.92 | 121.65 | 130.82 |
| 40 | 7.92 | 6.48 | 5.24 | 103.14 | 108.62 | 117.54 |
| 50 | 10.92 | 8.21 | 6.45 | 87.92 | 97.05 | 108.36 |

### 4.2.3 Rapid Chloride Ion Penetration Test (RCPT)

Three samples of each grade of concrete were cast with a w/c ratio varying between 0.19, 0.20 and 0.21. The purpose of this test is to determine the impact of the w/c on chloride permeability. Re-recorded values of average charges passed through each specimen are shown in Table 4. It is observed that higher charges are passed through

Durability Properties of Fibre-Reinforced Reactive … 25

**Fig. 6** Salt crystals in FRPC samples

**Fig. 7** Delamination of surfaces in R1, R2 and R3

**Fig. 8** Weight loss in FRPC samples

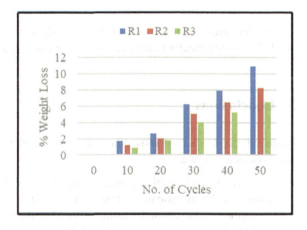

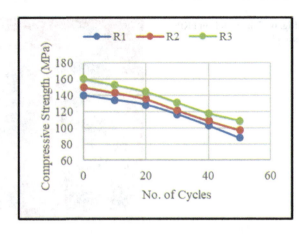

Fig. 9 Strength loss in FRPC samples

Table 4 RCPT test results

| FRPC | W/C ratio | Charges passed (in Coulomb) | Average charges (in Coulombs) | Chloride ion permeability |
|---|---|---|---|---|
| R1-A | 0.21 | 153 | 158 | Very low |
| R1-B |  | 157 |  |  |
| R1-C |  | 163 |  |  |
| R2-A | 2.0 | 119 | 125 | Very low |
| R2-B |  | 125 |  |  |
| R2-C |  | 131 |  |  |
| R3-A | 0.19 | 103 | 109 | Very low |
| R3-B |  | 109 |  |  |
| R3-C |  | 115 |  |  |

R1 samples compared to R2 and R3; however, these are very low. Hence it is stated that FRPC shows greater resistance to chloride ion penetration.

## 5 Conclusions

Based on the laboratory tests conducted on the RPC specimen, the following conclusions are drawn:

1. Fibre-reinforced reactive powder concretes of 140, 150 and 160 MPa are produced. An increase in the dosage of silica fume and steel fibres improves the strengths of concrete under compression, flexure and tension, and an optimum of 2.5% is obtained for the dosage of fibres.

2. Concrete properties can be assessed with visual observations of discolouration, scaling, peeling, corrosion of fibres and spalling of concrete.
3. The highest form of deterioration is observed for samples exposed to acid solutions of $H_2SO_4$. The higher concentration of solutions resulted in weight loss of 20–25% and strength loss of 60–75%.
4. FRPC with high density shows better resistance towards the crystallization of soluble salts with an average weight loss of 6–10% at 50% strength reduction.
5. Higher grade FRPC (R3) shows better resistance to acid attacks compared with other grades of concrete (R1 and R2).
6. A very low amount of charges is passed through all FRPC specimens, which indicates that it has high resistance towards the permeability of chloride ions.

# References

1. Hiremath PN, Yaragal SC (2017) Effect of different curing regimes and durations on early strength development of reactive powder concrete. Constr Build Mater 154:72–87. https://doi.org/10.1016/j.conbuildmat.2017.07.181
2. Richard P, Cheyrezy M (1994) Reactive powder concretes with high ductility and 200–800 MPa compressive strength. ACI SP 144(4):507–518
3. Coppola L, Cerulli T, Troli R, Collepardi M (1996) The influence of raw materials on performance of reactive powder concrete. In: International conference on high-performance concrete and performance and quality of concrete structures, Florianopolis, pp 502–513
4. Bonneau O, Lachemi M, Dallaire E, Dugat J, Aitcin P-C (1997) Mechanical properties and durability of two industrial reactive powder concretes. ACI Mater J 94:286–290
5. Teichmann T, Schmidt M (2004) Influence of the packing density of fine particles on structure, strength and durability of UHPC. In: First international symposium on ultra high perform concrete, Die Duetsche Bibliothek, Kassel, 2004, pp 313–325
6. Dhundasi AA, Khadiranaikar RB (2019) Effect of curing conditions on mechanical properties of reactive powder concrete with different dosage of quartz powder. In: Das B, Neithalath N (eds) Sustainable construction and building materials. Lecture notes in civil engineering, vol 25. Springer, Singapore. https://doi.org/10.1007/978-981-13-3317-0_33
7. Hiremath PN, Yaragal SC (2018) Performance evaluation of reactive powder concrete with polypropylene fibers at elevated temperatures. Constr Build Mater 169:499–512. ISSN 0950-0618. https://doi.org/10.1016/j.conbuildmat.2018.03.020
8. Hiremath P, Yaragal SC (2017) Investigation on mechanical properties of reactive powder concrete under different curing regimes. Mater Today: Proc 4(9):9758–9762. https://doi.org/10.1016/j.matpr.2017.06.262
9. Ipek M, Yilmaz K, Sümer M, Saribiyik M (2011) Effect of pre-setting pressure applied to mechanical behaviours of reactive powder concrete during setting phase. Constr Build Mater 25:61–68. https://doi.org/10.1016/j.conbuildmat.2010.06.056
10. Courtial M, De Noirfontaine MN, Dunstetter F, Signes-Frehel M, Mounanga P, Cherkaoui K, Khelidj A (2013) Effect of polycarboxylate and crushed quartz in UHPC: microstructural investigation. Constr Build Mater 44:699–705. https://doi.org/10.1016/j.conbuildmat.2013.03.077
11. George RM, Das BB, Goudar SK (2019) Durability studies on glass fiber reinforced concrete. Sustain Constr Build Mater:ials, 2019, 747–756
12. Yadav S, Das BB, Goudar SK (2019) Durability studies of steel fibre reinforced concrete. Sustain Constr Build Mater 737–745

13. Srikumar R, Das BB, Goudar SK (2019) Durability studies of poly propylene fiber reinforced concrete. Sustain Constr Build Mater 727–736
14. Snehal K, Das BB (2018) Mechanical and permeability properties of hybrid fibre reinforced porous concrete. Indian Concr J 98:54–59
15. Influence of length and dosage of polypropylene fibres on the spalling tendency and the residual properties of self-compacting concrete after heated at elevated temperatures
16. Sideris KK, Manita P (2013) Influence of length and dosage of polypropylene fibres on the spalling tendency and the residual properties of self-compacting concrete after heated at elevated temperatures. MATEC Web of Conferences 6.https://doi.org/10.1051/matecconf/20130602004
17. Vasily R, Belyakov V, Moskovsky S (2016) Properties and design characteristics of the fiber concrete. Procedia Eng 150:1536–1540. https://doi.org/10.1016/j.proeng.2016.07.107
18. deb T, Jacek Ś (2011) The influence of selected material and technological factors on mechanical properties and microstructure of reactive powder concrete (RPC). Arch Civ Eng 57:227–246. https://doi.org/10.2478/v.10169-011-0017-1
19. Okabe S, Odagiri M, Ito T, Satoh H (2007) Succession of sulfur-oxidizing bacteria in the microbial community on corroding concrete in sewer systems. Appl Environ Microbiol 73(3):971–980. https://doi.org/10.1128/AEM.02054-06
20. Murthi P, Venkatachalam S (2008). Studies on acid resistance of ternary blended concrete. Asian J Civ Eng 9
21. Song XJ, Marosszekya M, Brungs M, Munn R (2005) Durability of fly ash based geopolymer concrete against sulphuric acid attack. In: International conference on durability of building materials and components, Lyon, France, 17–20 Apr 2005
22. Rendell F, Jauberthie R (1999) The deterioration of mortar in sulphate environments. Constr Build Mater 13(6):321–327
23. Aydin S, Yazici Yigiter H, Baradan B (2007) Sulfuric acid resistance of high-volume fly ash concrete. Build Environ 42(2):717–721
24. Ariffin MAM, Bhutta MAR, Hussin MW, Tahir MM, Aziah N (2013) Sulphuric acid resistance to blended ash geopolymer concrete. Constr Build Mater 43:80–86
25. Khatri RP, Sirivivatnnon V, Yang JL (1997) Role of permeability in sulphate attack. Cem Concr Res 27:1179e1189
26. Snehal K, Das BB (2021) Acid, alkali and chloride resistance of binary, ternary and quaternary blended cementitious mortar integrated with nano-silica particles. Cem Concr Compos 128:104214
27. Sahoo S, Das BB, Mustakim S (2017) Acid, alkali, and chloride resistance of concrete composed of low-carbonated fly ash. J Mater Civ Eng 29(3):04016242
28. Muranal SM, Khadiranaikar RB (2014) Study on the durability characteristics of reactive powder concrete. Int J Struct Civ Eng Res 3(2):45–56
29. Olusola KO, Joshua O (2012) Effect of nitric acid concentration on the compressive strength of laterized concrete. Civ Environ Res 2(10):48–57. http://www.iiste.org/Journals/index.php/CER/article/view/3539/3587
30. Lubelli B, Van Hees RPJ, Nijland TG (2014) Salt crystallization damage: how realistic are existing ageing tests? In: AMS'14 proceedings of the international conference on aging of materials and structures, 2014, pp 103–111. http://resolver.tudelft.nl/uuid:363f94aa-21c6-40ca-933e-93271b29497b

# Performance of Geopolymer Concrete Developed Using Waste Tire Rubber and Other Industrial Wastes: A Critical Review

Dhiraj Agrawal, U. P. Waghe, M. D. Goel, S. P. Raut, and Ruchika Patil

**Abstract** The use of concrete as a construction material is one of the highest among other materials used throughout the world. Due to this, enormous demand for concrete exists and its constituents became very vital economically and technologically with respect to the growth of any nation. The primary binder material used in the concrete is cement, and the process of making cement plays a vital role in infrastructural growth. Cement production releases a vast amount of carbon dioxide into the atmosphere, causing severe environmental health hazards like global warming and other issues allied to it. To control the cost of cement and to keep a tab on problems arising from its manufacturing, there is a need to find a substitute of cement so that this substitute can be used as a partial/complete replacement of cement in the manufacturing of concrete. Geopolymer concrete is one way to tackle this problem. Further, industrialization is established very rapidly in the last three decades, particularly in developing countries. Distinct industrial wastes are being generated from these industries leading to the problem of their disposal and various health and environmental concerns. The proper utilization of such wastes is the need of the present time. Extensive research has been carried out on developing the geopolymer concrete along with the use of different industrial wastes. Most of the studies on geopolymer concrete have given promising results for strengths and durability in comparison to conventional concrete. This work presents a detailed review of the studies based on the use of various industrial wastes, like fly ash, ground granulated blast furnace slag, metakaolin etc. along with the various alkaline activators for developing the geopolymer concrete to reduce cement footprint. A detailed review is reported considering the performance of the geopolymer concrete for different mechanical properties, its strengths and durability in comparison with the conventional concrete. It has been found that the enhancement in the strengths is observed for the geopolymer concrete as compared to the

D. Agrawal (✉) · U. P. Waghe · S. P. Raut · R. Patil
Department of Civil Engineering, Yeshwantrao Chavan College of Engineering (YCCE), Nagpur 441110, India
e-mail: dgagrawal@ycce.edu

M. D. Goel
Department of Applied Mechanics, Visvesvaraya National Institute of Technology (VNIT), Nagpur 440010, India

© The Author(s), under exclusive license to Springer Nature Singapore Pte Ltd. 2023
M. S. Ranadive et al. (eds.), *Recent Trends in Construction Technology and Management*, Lecture Notes in Civil Engineering 260,
https://doi.org/10.1007/978-981-19-2145-2_3

concrete prepared using cement. Furthermore, limitations in the manufacturing of the geopolymer concrete using these wastes are also discussed.

**Keywords** Industrial wastes · Geopolymer concrete · Cement · Alkaline activators

## 1 Introduction

The rapid growth of industrialization is observed in the last three decades in all developing countries of the world. Infrastructural development is considered to be the most important part of the growth of a nation. Instant infrastructural expansion is being carried out using concrete structures in several nations. The use of concrete leads to the demand for cement as a binder material, hence the manufacturing of cement increased very rapidly during this tenure. The production of cement releases almost an equal amount of carbon dioxide ($CO_2$) into the atmosphere [1–4]. Owing to this, cement consumption has become a foremost challenge for sustainable progression. 5–8% of anthropogenic $CO_2$ emission was caused by concrete and 95% of it was due to cement manufacturing [5–7]. The worldwide cement production from 2015 to 2019 in various fastest-growing countries is displayed in Fig. 1, which can be considered as the base for the amount of $CO_2$ emissions and its effect on the environment [8]. From Fig. 1 it can be observed that China, India and other Southeast Asian kingdoms produce almost 65–68% of the total production of cement, and ultimately they also

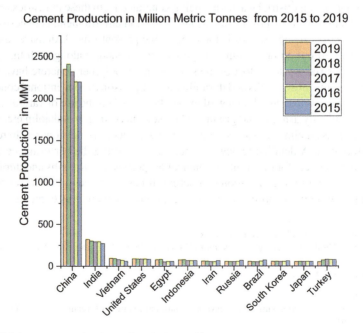

**Fig. 1** Major cement production nations in the world

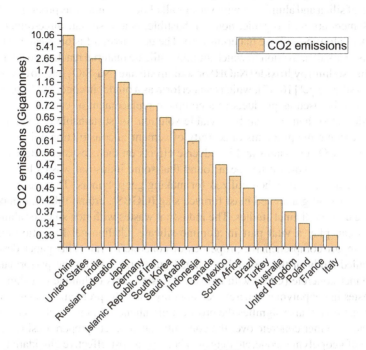

**Fig. 2** Carbon dioxide emission for top 20 countries for 2018

contribute to the same amount of $CO_2$ emissions. Figure 2 presents the scenario of $CO_2$ emission in the world by the top 20 $CO_2$ emitting countries in 2018 [9]. Almost 7% of total greenhouse gas emission is caused due to production of cement [10]. So for developing countries like India, it becomes a serious concern in aspects of a good and healthy environment and sustainable development to develop an alternative to the current conventional construction culture. At the same time, while considering the infrastructural growth around the globe, production of cement, its acute shortage due to gigantic demand and the effect of cement production on the environment, industrial evolution is also at its peak in the entire biosphere with the production of enormous industrial wastes, creating the problem of disposal, land, water and air pollution. This thing also needs strong attention. Extended research was carried out around the world for finding an alternative for binding material in concrete using numerous agro-industrial wastes [11–13]. These investigations have concluded the possible use of some industrial wastes like fly ash (FA), rice husk ash (RHA), ground granulated blast furnace slag (GGBS), metakaolin as a partial replacement for cement but it has some technical precincts and it also does not provide the widespread solution on the concern. Along with this, these wastes also contributed to the generation of $CO_2$ during electricity generation or in iron industries.

The concept of geopolymer was first announced by Joseph Davidovits in which the reaction of silica and aluminum with various alkaline activators was pronounced [14]. Geopolymers are an inorganic, non-combustible, heat-resistant, three-dimensional network of alumino-silicate materials [15]. The geopolymer fusion is established on the basis of alkali activation in which alumino-silicate material reacts with alkali activators like sodium hydroxide (NaOH) or sodium silicate ($Na_2SiO_3$) etc. to develop the alumino-silicate gel [16, 17], which can perform as a binder in concrete. Geopolymer binders can be used as products as a complete replacement of cement in concrete production [18], hence it can be a viable solution for sustainable development as it can overcome the problems of scarcity of cement in constructional development of the state, $CO_2$ emissions and its adverse effects on the ecosystem. Based on the earlier research studies it has been found that some industrial wastes have shown very promising results when utilized for making geopolymers. The wastes like fly ash (FA), ground granulated blast furnace slag (GGBS), crumb rubber, metakaolin etc. were used in distinct studies. The industrial wastes with rich silica or aluminum contents can play a vital part in alumino-silicate alkaline gel formation and are able to impart comprehensive properties of concrete/mortar. This paper describes the detailed study of the researches carried out in the past on the performance of geopolymer concrete (GPC) with industrial wastes. This technique of using industrial wastes in geopolymer concrete can not only solve the problem of environmental contamination but also signifies the prospect of enhancement in mechanical properties of geopolymer concrete over the conventional concrete prepared using cement. The use of geopolymer concrete may also become a cost-effective elucidation in the infrastructure industry.

## 2 Development of Geopolymer Concrete Using Industrial Wastes

### 2.1 Methodology to Prepare Geopolymer Concrete

The key constituent of geopolymer concrete is the alkaline activator solution (AAS), and its dosages are fixed in the mix. As the focus of this study is on the use of industrial wastes in geopolymer concrete, the response of these industrial wastes with AAS must be studied well in advance before testing the other parameters of concrete. Sodium or potassium-based alkaline activators are generally used in the preparation of geopolymers in the concrete [19]. NaOH and $Na_2SiO_3$ are mainly used. The simplified reaction mechanism of geopolymerization consists of the procedure as solid alumino-silicates are dissolved due to alkaline hydrolysis by the inclusion of water. At high pH values dissolution of alumino-silicates is rapid. Due to condensation, gel formation is quicker [20]. In the process of preparing GPC, the mix design should be calculated using relevant standards. Later on, proper mixing of dry constituents of concrete should be carried out before adding some amount of

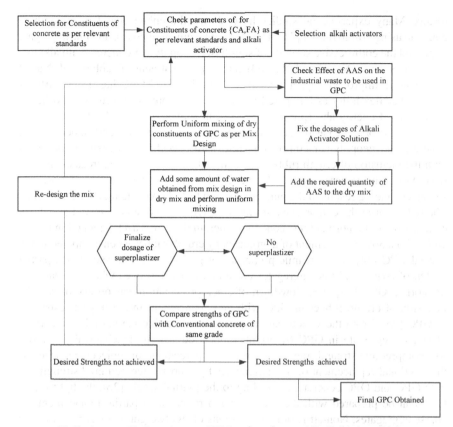

**Fig. 3** Methodology to prepare the mix of geopolymer concrete

water out of the total quantity of water obtained from the mix design. Polymer activator and catalyst activator are to be added subsequently and uniform mixing should be performed. After the preparation of geopolymer concrete (GPC) the mechanical properties of newly prepared GPC should be compared with conventional concrete for the desired grade. Figure 3 illustrates the methodology to prepare the mix of geopolymer concrete using industrial wastes.

## 2.2 Use of Various Industrial Wastes in Geopolymer Concrete

### 2.2.1 Use of Crumb Rubber in Geopolymer Concrete

As per past studies, the use of crumb rubber as a partial substitution of coarse aggregates and fine aggregates in concrete and geopolymer concrete has shown promising results. Many experimental studies have shown the enhancement of strength in concrete using crumb rubber [21]. The 10% replacement of coarse and fine aggregate improved the compressive strength of slag-based geopolymer concrete with alkaline activators such as NaOH and $Na_2SiO_3$ due to pre-treatment of rubber with NaOH [22]. Due to improved impact resistance, the rubberized GPC has a low stiffness with higher flexibility as compared to the GPC without crumb rubber and hence the enhanced energy absorption is observed in rubberized GPC [22]. The density of concrete is observed as the same for convention ordinary portland cement (OPC) concrete and geopolymer control concrete with a decline in density is observed with the inclusion of crumb rubber as fractional substitution of aggregates in GPC [23]. As per the research study [24], the tensile strengths were increased. GPC has shown enhancing results in the pull-off strength test of concrete than OPC concrete. The reduction in the compressive strength for OPC concrete is observed at 52% in comparison to the normal OPC concrete, wherein there was a 41% reduction for the GPC with a 30% replacement of aggregates by crumb rubber when compared to the normal GPC [24]. The dynamic properties of geopolymer concrete after a partial switch of coarse and fine aggregates are found to be improved when compared to the normal GPC [25]. The slowdown of crack propagation was observed with the inclusion of crumb rubber in GPC with an enhancement in the impact resistance of GPC [25]. From the various studies, it is observed that the partial replacement of coarse aggregate in GPC by crumb rubber reduces the mechanical properties like compressive strength rapidly, even at less percentage of replacement whereas the fractional replacement of fine aggregates by rubber offers required strengths to both GPC and OPC concrete. According to the journal article [26], the lightweight GPC can be prepared with the use of crumb rubber as a partial replacement for coarse aggregates. Non-structural applications of lightweight geopolymer concrete can be promising using rubber waste. Reduction of the slump was noted with the replacement of more than 30% of aggregates by rubber [27]. The size of aggregates has shown its effect on the compressive strengths. The activator solution provides the pre-treatment to the rubber, so the bonding of rubber with other constituents of concrete is enriched. The clear zones of phenolphthalein indicators were not seen for assessing the carbonation depth [27]. The modulus of elasticity is found to be greatly reduced by implementing the partial replacement of aggregates using crumb rubber while the difference between the densities of normal GPC to rubberized GPC is found to be reduced [28]. Due to the weak bonding between rubber crumbs and other ingredients of GPC, the porosity is found to be increased, which in turn resulted in the reduction of compressive strengths. Along with this, the density of normal GPC and rubberized GPC is found to be increased due to the immersion of specimens in seawater [29]. The maximum decrease in compressive strength is detected up to 60% as related to the normal GPC and rubberized GPC with 15% replacement of aggregates by crumb rubber [30]. Table 1 indicates the compressive strengths obtained for distinct grades of GPC with varying replacement ratios of coarse and fine aggregates by crumb rubber.

**Table 1** Compressive strengths for geopolymer concrete in MPa for research papers studied

| % replacement of fine/coarse aggregates by crumb rubber | Source of information |||||||
|---|---|---|---|---|---|---|---|
| | [23] | [24] | [25] | [26] | [27] | [28] | [29] | [30] |
| 0 | 37.40 | 56.23 | 54.00 | 42.10 | 57.90 | 46.04 | 50.00 | 65.00 |
| 5 | – | – | – | – | – | 41.55 | 40.00 | 48.00 |
| 10 | 40.00 | 49.67 | – | 29.10 | – | 37.01 | 35.00 | 30.00 |
| 15 | – | – | 26.20 | – | 19.20 | 34.06 | 24.00 | 20.60 |
| 20 | 28.30 | 42.11 | – | 28.90 | – | – | 20.00 | 15.80 |
| 25 | – | – | – | – | – | – | – | – |
| 30 | 24.80 | 33.00 | 11.70 | – | 6.20 | – | – | – |

### 2.2.2 Use of Siliceous Material in Geopolymer Concrete

Rashad [31] used the combination of GGBS and quartz powder in GPC. The addition of quartz powder has shown the enhancement of compressive strengths and other mechanical properties of GPC. According to the detailed study by Rashad [32], the setting time and workability were found to be increased with the inclusion of fly ash. In further research, he concluded that the use of silica fume showed positive results as the production of hydrates of calcium and silicates intensified with densifying the pore structure. With the use of quartz powder, the flowability of concrete is increased with a reduction in porosity. At the same time, the quartz powder also showed resistance to the cyclic thermal loadings [31]. Rendering to the studies by Pasupathy et al., it is observed that carbonation is increased with maximum use of fly ash, whereas the proper mixing of FA and GGBS gives similar trends of carbonation of OPC concrete [33]. The effect of palm oil fuel ash as a replacement of GGBS in GPC has been studied, and it is concluded that there is a noteworthy decrement in the setting time with an increment in the replacement of GGBS by palm oil fuel ash. In continuation with this, it is also noted that the 100 palm oil fuel ash-based GPC has 35% reduced strength as compared to the GPC with 100% GGBS [34]. In comparison of the alkali-activated concrete FA and GGBS with Portland cement concrete, it is detected that the tensile strength for alkali-activated concrete is more than Portland cement concrete. On the other hand, the comparison of alkali-activated concrete made using FA with the alkali-activated concrete prepared using GGBS has shown similar mechanical properties [35]. The use of metakaolin in GPC has displayed upgrading for the reduction in shrinkage of exposed specimens. The fire resistance of GPC is also improved due to the inclusion of metakaolin [36]. In continuation to the use of metakaolin, it is concluded that the enhancement of compressive strength by 18% is seen when 20% aluminum oxide is mixed with 80% metakaolin in GPC and matched to the GPC with 100% metakaolin [35]. As per the study conducted by Lee et al. [36], on alkali-activated controlled low-strength materials (CLSMs) with the use of fly ash, slag and bottom ash, it is observed that the increase in the amount of bottom ash has shown a lessening in flowability of alkali-activated CLSM. In addition, they also commented that the bleeding rate is reduced with an increase in the quantity of slag. At the same instant, the compressive strength is also increased due to the inclusion of slag as it imparts later strength to CLSM. Chemical analysis of alkali-activated mortars is carried out by Winnefeld et al. [37] with high and low calcium fly ashes obtained from the combustion of hard coal. They commented that the low calcium fly ash is more reactive than the high calcium lignite coal fly ashes in alkali-activated systems. The specimens with low calcium fly ashes also showed sufficient compressive strengths [37]. According to the study on porous concrete [38], it is seen that the compressive strength of porous concrete made up of coal-bottom ash using geopolymeric binder was less than that of porous concrete with gravel and cement, whereas the porous concrete with coal bottom ash has provided immobility to the heavy metals leached from bottom ash. A research on the Australian brown coal fly ash collected from three different sources was carried out, and it is observed that the brown coal fly ash is insufficient to achieve satisfactory results. Blending these brown coal ashes with class F fly ash and blast furnace slag is required to get promising outcomes [39]. An experimental study on pervious concrete is completed by Zaetang [40], the ordinary portland cement is replaced with FA and bottom ash was used as coarse aggregate in this study. Limited substitution of cement by FA is carried out and the obtained results show that sodium hydroxide plays a vital role in augmenting strengths. It is recommended that 15% replacement of cement by FA gives encouraging results in aspect of strengths. The study on the combined utilization of crumb rubber as a partial replacement for sand and FA as a binder [42] was conceded and it is concluded that there is a reduction in compressive strength of GPC due to the inclusion of rubber irrespective of the type of fly ash used. The fineness of fly ash reduces the rate of reduction of compressive strength. The use of FA and GGBS is done in the study [41, 42], and it is found that the replacement of FA with GGBS boosts the modulus of elasticity and compressive strength of GPC. The compressive strength is amplified by 1.48 times when the strengths of GPC with 100% FA are harmonized with the GPC with 30% FA and 70% GGBS on the 28th day of testing of specimens. For the same comparison, the modulus of elasticity is increased by 42.85%. As per an experiment by Olivia and Nikraz [43], it was concluded that the GPC was made with the use of FA as a binder material and a total of nine distinct specimens were cast with one OPC sample with 55 MPa strength. The almost same or high strength is achieved by the samples of GPC. The higher tensile and flexural strengths were observed for GPC as compared to OPC concrete. Tawatchai Tho-in presented experimental results on the pervious GPC using FA as the binder material, the ratio of coarse aggregate to FA was 1:8. The compressive strength and splitting tensile strengths were determined, and it is found that these strengths for pervious alkali-activated GPC are slightly higher than that of normal pervious concrete prepared using cement as a binder. The density of pervious GPC is 30% less than the orthodox pervious concrete [43]. A combined study was conceded on the use of cement and FA as a binder material in concrete. The conventional concrete was compared to fractional and integral replacement of cement by FA and it is perceived that the mechanical properties like compressive, flexural and split tensile strengths were improved by 43.6%, 36.07% and 51.13%, respectively, at 50% replacement of cement by FA and compared with the specimens having 100% cement as binder material [44]. Table 2 presents the chemical oxide composition of distinct industrial wastes used in the formation of GPC in referred papers. Figure 4 presents the amount of oxides of silica, aluminum and iron present in different industrial wastes, which are the main components of cement.

## 3 Discussion

The widespread literature revisions were accomplished concerning the development of sustainable concrete without using cement and other conventional constituents of concrete like sand. Figures 1 and 2 depict how harmful activity cement production is. The generation of greenhouse gases and $CO_2$ emissions will be the long-lasting glitches as the development of nations cannot be compromised. So it is the need and duty of each nation to think about sustainable development. As reviewed from the studies cement is the main matter to emit $CO_2$, so GPC became the most viable elucidation as an alternative to cement that can be generated using alkaline activators and other siliceous by-products disposed of by industries as binder material. It is marked in Table 1 that the researchers have used crumb rubber as a replacement for fine aggregate in concrete and in GPC by overcoming the environmental intimidations generated by the production of cement due to stockpiled tires. The generation of discarded tires is a never-ending problem for our ecosystem as the automobile industry grows rapidly day by day. It is observed from [20–22] that the crumb rubber gives promising results in view of the strengths of GPC while comparing them with orthodox concrete, and the dynamic behavior of rubberized GPC is improved to that of normal concrete. At the same time, it is also seen that the molarity ratio, blending of NaOH and $Na_2SiO_3$ also affects the performance of crumb rubber in GPC and in normal concrete. From Table 2 it is realized that many siliceous materials recycled from industrial wastes can be used as binder materials in GPC as a replacement for cement. The use of quartz powder [31] has shown enhancement in the strengths of concrete. Figure 4 proves the efficiency of industrial wastes in the replacement of cement in concrete. GGBS and FA are the most used binders in GPC. More focus is needed on experimental studies on GPC as more encouraging results are predictable for practical implementation of GPC over conventional concrete. The source/collection of industrial wastes should be studied as it affects the properties of concrete.

## 4 Conclusion

The use of divergent industrial wastes concerning to prepare geopolymer concrete has been reviewed. In accordance with the studied literature on experimental lessons, the following conclusions have been drawn:

Table 2 Chemical oxide composition for siliceous material used in GPC

| Source | IW | SiO2 | Al2O3 | Fe2O3 | CaO | MgO | Na2O | K2O | SO3 | TiO2 | P2O5 | MnO2 | Cl | L.O.I |
|---|---|---|---|---|---|---|---|---|---|---|---|---|---|---|
| [19] | GGBS | 36.95 | 10.01 | 1.48 | 33.07 | 6.43 | 1.39 | 0.74 | 3.52 | 0.52 | 0.10 | 0.52 | 0.05 | 0.00 |
| | QP | 98.81 | 0.14 | 0.05 | 0.51 | 0.02 | 0.04 | 0.03 | 0.03 | 0.02 | 0.32 | 0.00 | 0.00 | 0.32 |
| [21] | GGBS | 36.95 | 10.01 | 1.48 | 33.07 | 6.43 | 1.39 | 0.74 | 3.52 | 0.52 | 0.10 | 0.52 | 0.05 | 0.00 |
| | QP | 98.81 | 0.14 | 0.05 | 0.51 | 0.02 | 0.04 | 0.03 | 0.03 | 0.02 | 0.32 | 0.00 | 0.00 | 0.32 |
| [22] | FA | 80.40 | 14.00 | 0.04 | 0.31 | 3.57 | 0.10 | 0.09 | 0.85 | 0.04 | 0.08 | 0.49 | – | 0.54 |
| | GGBS | 34.20 | 13.80 | 43.10 | 5.40 | 0.40 | 0.10 | – | 0.40 | – | 0.80 | – | – | 1.80 |
| [23] | POFA | 47.37 | 3.53 | 6.19 | 11.83 | 4.19 | – | – | – | – | – | – | 0.88 | 1.84 |
| | GGBS | 34.10 | 13.50 | 0.36 | 42.70 | 4.50 | – | – | – | 0.24 | – | 0.20 | 0.01 | 1.40 |
| [24] | GGBS | 36.00 | 10.50 | 0.70 | 39.80 | 7.90 | 0.30 | 0.20 | 2.10 | – | – | – | – | – |
| | FC | 37.70 | 20.00 | 5.60 | 23.40 | 4.30 | 1.70 | 0.60 | 2.40 | – | – | – | – | – |
| [26] | FA | 42.10 | 28.60 | 14.40 | 6.26 | 2.60 | – | 2.40 | 0.61 | – | – | – | – | – |
| | BFS | 35.17 | 13.93 | 0.58 | 42.50 | 4.12 | – | 0.46 | 2.03 | – | – | – | – | – |
| [27] | LCFA | 56.00 | 25.00 | 8.70 | 1.90 | 3.10 | 0.20 | 3.30 | 0.45 | 1.00 | 0.29 | – | 0.05 | 2.19 |
| | HCFA | 44.00 | 7.30 | 3.80 | 29.00 | 4.10 | 1.10 | 0.70 | 10.00 | 0.46 | 0.02 | – | 8.90 | 2.24 |
| [28] | FA | 42.10 | 28.60 | 14.40 | 6.26 | 2.60 | – | 2.40 | 0.61 | – | – | – | – | – |
| | BFS | 35.17 | 13.93 | 0.58 | 42.47 | 4.12 | 0.15 | 0.46 | 2.03 | – | – | – | – | – |
| [29] | LYDP | 48.90 | 15.16 | 9.68 | 2.44 | 5.38 | 4.86 | 0.41 | 7.48 | 1.58 | – | 0.05 | – | 4.04 |
| | GGBS | 32.38 | 12.24 | 0.49 | 44.04 | 5.13 | 0.22 | 0.33 | 4.21 | 0.51 | – | 0.37 | – | 0.08 |
| | FC | 51.88 | 25.92 | 12.66 | 4.35 | 1.54 | 0.78 | 0.71 | 0.24 | 1.3 | – | 0.15 | – | 0.55 |
| [30] | OPC | 14.40 | 2.70 | 3.40 | 70.40 | 0.90 | 0.20 | 0.60 | 4.30 | 0.30 | – | – | – | 2.40 |
| | FA | 35.90 | 15.10 | 17.30 | 17.20 | 2.30 | 0.90 | 3.20 | 5.90 | 0.90 | – | – | – | 1.10 |

(continued)

Table 2 (continued)

| Source | IW | SiO₂ | Al₂O₃ | Fe₂O₃ | CaO | MgO | Na₂O | K₂O | SO₃ | TiO₂ | P₂O₅ | MnO₂ | Cl | L.O.I |
|---|---|---|---|---|---|---|---|---|---|---|---|---|---|---|
|  | BA | 31.80 | 12.10 | 18.00 | 25.30 | 2.40 | 0.90 | 2.50 | 3.70 | 0.50 | – | – | – | 2.60 |
| [31] | T-1 FC | 50.67 | 18.96 | 6.35 | 14.14 | 3.12 | – | 0.69 | – | 0.74 | – | – | – | – |
|  | T-2 FC | 58.05 | 21.59 | 5.10 | 9.42 | 1.86 | – | 0.92 | – | 0.39 | – | – | – | – |
|  | T-3 FC | 54.70 | 29.00 | 6.74 | 1.29 | 0.80 | – | 1.88 | – | 0.10 | – | – | – | – |
| [32] | FA | 58.27 | 26.10 | 3.60 | 2.89 | 1.31 | 0.51 | 0.90 | 1.06 | 0.56 | 0.30 | 2.90 | – | 1.60 |
|  | GGBS | 33.20 | 13.20 | 1.05 | 45.77 | 3.25 | 0.88 | – | 0.82 | – | – | 1.97 | – | 1.95 |
| [34] | FA | 50.50 | 26.57 | 13.77 | 2.13 | 1.54 | 0.45 | 0.77 | 0.41 | – | 1.00 | – | – | 0.60 |
|  | Cement | 21.10 | 4.70 | 2.80 | 63.80 | 2.00 | 0.50 | – | 2.50 | – | – | – | 0.01 | 2.10 |
|  | FA | 57.90 | 31.11 | 5.07 | 1.29 | 0.97 | 0.09 | 1.00 | 0.05 | – | – | – | 0.04 | 0.80 |

*IW* Industrial waste, *GGBS* Ground granulated blast furnace slag, *FA* Fly ash, *QP* Quartz powder, *POFA* Palm oil fuel ash, *FC* Class F fly ash, *BFS* Blast furnace slag, *LCFA* Low calcium fly ash, *HCFA* High calcium fly ash, *LYDP* Loy young dry precipitator ash, *OPC* Ordinary portland cement, *BA* Bottom ash, *T-1 FC* Type 1 class C fly ash, *T-2 FC* Type 2 ultra-fine class C fly ash, *T-3 FC* Type 3 class f fly ash

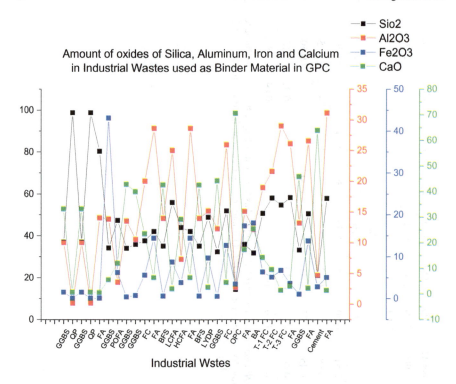

**Fig. 4** Amount of oxides of silica, aluminum, iron, and calcium available in industrial wastes used as binder materials in GPC

- The alkali-based geopolymers are the finest alternative to the cement as sufficient strength can be achieved.
- Limited use of crumb rubber as partial substitution of fine aggregate can be acceptable as it gives desired strengths in both normal concrete and GPC. 30% optimum replacement of fine aggregate by crumb rubber is possible, beyond this the compressive strengths of specimens may be reduced.
- Dynamic properties of concrete have been improved in some studies after the incorporation of crumb rubber. The effect of the size of rubber particles on the strengths of GPC should be studied.
- GGBS and FA have shown very interesting results when reacted with an alkali-activator solution and achieved higher strengths than conservative concrete.
- GPC can be used as an environment-friendly and economical solution for infrastructural development as it involves the use of recycled industrial by-products and minimizes environmental pollution.

# References

1. Mehta A, Siddique R (2016) An overview of geopolymers derived from industrial by-products. Constr Build Mater 127:183–198
2. Snehal K, Das BB (2021) Acid, alkali and chloride resistance of binary, ternary and quaternary blended cementitious mortar integrated with nano-silica particles. Cem Concr Compos 128:104214
3. Sahoo S, Das BB, Mustakim S (2017) Acid, alkali, and chloride resistance of concrete composed of low-carbonated fly ash. J Mater Civ Eng 29(3):04016242
4. Snehal K, Das BB, Akanksha M (2020) Early age, hydration, mechanical and microstructure properties of nano-silica blended cementitious composites. Constr Build Mater 233:117212
5. Shivaprasad KN, Das BB (2018) Determination of optimized geopolymerization factors on the properties of pelletized fly ash aggregates. Constr Build Mater 163:428–437
6. Farsana C, Das BB, Snehal K (2021) Influence of fineness of mineral admixtures on the degree of atmospheric mineral carbonation. Smart Technol Sustain Dev 117–136
7. Ouellet-Plamondon C, Habert G (2015) Life cycle assessment (LCA) of alkali-activated cements and concretes. Handbook of alkali-activated cements, mortars and concretes, pp 663–686
8. Datis Export Group (2019) Worldwide cement production from 2015 to 2019, 12 July 2020 [online]. Available at: https://datis-inc.com/blog/worldwide-cement-production-from-2015-to-2019/. Accessed 15 Dec 2020
9. Union of Concerned Scientists (2020) "https://www.ucsusa.org," Union of Concerned Scientists, 12 Aug 2020 [Online]. Available at: https://www.ucsusa.org/resources/each-countrys-share-co2-emissions. Accessed 20 Dec 2020
10. Oyebisi S, Ede A, Olutoge F, Omole D (2020) Geopolymer concrete incorporating agro-industrial wastes: effects on mechanical properties, microstructural behaviour and mineralogical phases. Constr Build Mater 256
11. Agrawal D, Hinge P, Waghe UP, Raut SP (2014) Utilization of industrial waste in construction material—a review. Int J Innov Res Sci Eng Technol (IJIRSET) 3(1):8390–8397
12. Dalinaidu A, Das BB, Singh DN (2007) Methodology for rapid determination of pozzolanic activity of materials. J ASTM Int 4(6):1–11
13. Shivaprasad KN, Das BB (2018) Effect of duration of heat curing on the artificially produced fly ash aggregates. IOP Conf Ser: Mater Sci Eng 431(9):092013
14. Grant Norton M, Provis JL (2020) 1000 at 1000: geopolymer technology—the current. Springer
15. Abbas Mohajerani A, Suter D, Jeffrey-Bailey T, Song T, Arulrajah A, Horpibulsuk S, Law D (2018) Recycling waste materials in geopolymer concrete. Clean Technol Environ Policy
16. Nuruddin MF, Malkawi AB, Fauzi A, Mohammed BS, Almattarneh HM (2016) Geopolymer concrete for structural use: recent findings and limitations. In: International conference on innovative research 2016—ICIR Euroinvent. IOP Publishing
17. Sharath BP, Shivaprasad KN, Athikkal MM, Das BB (2018) Some studies on sustainable utilization of iron ore tailing (IOT) as fine aggregates in fly ash based geopolymer mortar. IOP Conf Ser: Mater Sci Eng 431(9):092010
18. Pavithra P, Srinivasula Reddy M, Dinakar P, Hanumantha Rao B, Satpathy BK, Mohanty AN (2016) A mix design procedure for geopolymer concrete with fly ash. J Clean Prod 117–125
19. McKenzie W (2020) Geopolymer concrete—playing in the laboratory, 25 June 2014. [Online] Available at: https://www.youtube.com/watch?v=-86Grly1E-c&feature=youtu.be. Accessed 12 Dec 2020
20. Agrawal D, Waghe UP, Raut SP (2021) Performance evaluation of rubberized concrete with the use of steel fibers. In: Advances in civil engineering and infrastructure development. Springer, Singapore, pp 709–717
21. Aly AM, El-Feky MS, Kohail M, Nasr E-SAR (2019) Performance of geopolymer concrete containing recycled rubber. Constr Build Mater 207:136–144
22. Luhar S, Chaudhary S, Luhar I (2019) Development of rubberized geopolymer concrete: Strength and durability studies. Constr Build Mater 204:740–753

23. Pham TM, Liu J, Tran P, Pang V-L, Shi F, Chen W, Hao H, Tran TM (2020) Dynamic compressive properties of lightweight rubberized geopolymer concrete. Constr Build Mater
24. Aslani F, Deghani A, Asif Z (2020) Development of lightweight rubberized geopolymer concrete by using polystyrene and recycled crumb-rubber aggregates. J Mater Civ Eng
25. Dong M, Elchalakani M, Karrech A, Yang B (2020) Strength and durability of geopolymer concrete with high volume rubber replacement. Constr Build Mater
26. Taylor SJ (2018) Mechanical properties of crumb rubber geopolymer concrete. UNSW Canberra at ADFA, Canberra
27. Yahya Z, Abdullah MMAB, Ramli SNH, Minciuna MG, Abd Razak R (2018) Durability of fly ash based geopolymer concrete infilled with rubber crumb in seawater exposure. In: IOP conference series: materials science and engineering, Euroinvent ICIR 2018
28. Azmi AA, Al Bakri Abdullah MM, Ghazali CMR, Sandu AV, Hussin K (2016) Effect of crumb rubber on compressive strength of fly ash based geopolymer concrete. In: MATEC web of conferences IConGDM 2016
29. Rashad AM, Zeedan SR, Hassan HA (2012) A preliminary study of autoclaved alkali-activated slag blended with quartz powder. Constr Build Mater 70–77
30. Rashad AM (2013) A comprehensive overview about the influence of different additives on the properties of alkali-activated slag—a guide for civil engineer. Constr Build Mater 47:29–55
31. Rashad AM (2014) An exploratory study on alkali-activated slag blended with quartz powder under the effect of thermal cyclic loads and thermal shock cycles. Constr Build Mater 70:165–174
32. Rashad AM (2013) Alkali-activated metakaolin: a short guide for civil engineer—an overview. Constr Build Mater 41:751–765
33. Pasupathy K, Berndt M, Castel A, Sanjayan J, Pathmanathan R (2016) Carbonation of a blended slag-fly ash geopolymer concrete in field conditions after 8 years. Constr Build Mater 125:661–669
34. Salih MA, Farzadnia N, Ali AAA, Demirboga R (2015) Development of high strength alkali activated binder using palm oil fuel ash and GGBS at ambient temperature. Constr Build Mater 93:289–300
35. Thomas RJ, Peethamparan S (2015) Alkali-activated concrete: engineering properties and stress–strain behavior. Constr Build Mater 93:49–56
36. Lee NK, Kim HK, Park IS, Lee HK (2013) Alkali-activated, cementless, controlled low-strength materials (CLSM) utilizing industrial by-products. Constr Build Mater 49:738–746
37. Winnefeld F, Leemann A, Lucuk M, Svoboda P, Neuroth M (2010) Assessment of phase formation in alkali activated low and high calcium fly ashes in building materials. Constr Build Mater 24:1086–1093
38. Jang JG, Ahn YB, Souri H, Lee HK (2015) A novel eco-friendly porous concrete fabricated with coal ash and geopolymeric binder: heavy metal leaching characteristics and compressive strength. Constr Build Mater 79:173–181
39. Tennakoon C, Sagoe-Crentsil K, San Nicolas R, Sanjayan JG (2015) Characteristics of Australian brown coal fly ash blended geopolymers. Constr Build Mater 101:396–409
40. Zaetang Y, Wongsa A, Sata V, Chindaprasirt P (2015) Use of coal ash as geopolymer binder and coarse aggregate in pervious concrete. Constr Build Mater 96:289–295
41. Park Y, Abolmaali A, Kim YH, Ghahremannejad M (2016) Compressive strength of fly ash-based geopolymer concrete with crumb rubber partially replacing sand. Constr Build Mater 118:43–51
42. Olivia M, Nikraz H (2012) Properties of fly ash geopolymer concrete designed by Taguchi method. Mater Des 36:191–198
43. Tho-in T, Sata V, Chindaprasirt P, Jaturapitakkul C (2012) Pervious high-calcium fly ash geopolymer concrete. Constr Build Mater 30:366–371
44. Shehab HK, Eisa AS, Wahba AM (2016) Mechanical properties of fly ash based geopolymer concrete with full and partial cement replacement. Constr Build Mater 126:560–565

# Thermal Behaviour of Mortar Specimens Embedded with Steel and Glass Fibres Using *KD2* Pro Thermal Analyser

**P. Harsha Praneeth**

**Abstract** Understanding the thermal behaviour aspects of cement mortar has a significant role in the applications of sustainable buildings, chimneys, cooling towers, bridges, etc. as mortar is the main constituent in all concrete structures and masonry walls. It is often desirable to have lower thermal conductivity ($K$) values for the sake of thermal insulation; higher values are needed to optimize the thermal stresses in the structure. Hence, a need exists in understanding these behaviours, so that the desired material can be tailored and used in the places of need. The current study aims at investigating the thermal properties of mortar specimens, i.e. thermal conductivity, specific heat capacity ($C$), thermal diffusivity ($D$), and thermal resistivity ($\rho$) of reinforced mortar specimens embedded with steel and glass fibres. Thermal properties are determined using a *KD2* pro thermal property analyser on five mortar specimens at temperature ranges of room temperatures (27) 50, and 100 °C and hydration periods of 3, 14, and 28 days. The results indicated that in specimens reinforced with 1% glass fibres, thermal conductivity decreased significantly with the increase in the hydration period and temperatures, while the effect of the hydration period of 1%steel fibre specimens on thermal conductivity is relatively low. It is concluded that mortar specimens reinforced with steel fibres have higher $K$; relatively glass fibres have K on the lower side. Mortar specimens with 0.5 and 1% of glass fibre have shown the highest resistance to heat flow. While the inclusion of 1% steel fibre is most favourable for effective heat transfer across the mortar specimens.

**Keywords** Thermal conductivity · Thermal behaviour · Thermal diffusivity · Mortar

---

P. Harsha Praneeth (✉)
Department of Civil Engineering, Geethanjali College of Engineering and Technology, Hyderabad, India
e-mail: iharshaent@gmail.com

# 1 Introduction

The energy efficiency of buildings, nuclear power plants, rocket launching facilities, large furnaces, gravity dams, etc. can be improved, by engineering a material, which can mimic the properties of an insulating material, where the ability of a material to conduct heat is at its optimum. In order to achieve this, the inclusion of minerals, chemicals, and fibres in the concrete or mortar mix has been the trend of the day. By incorporating such supplements in the preparation of concrete mixes, the mechanical and thermal properties of concrete will be significantly affected [7, 16, 22, 21]. The inclusion of various fibre dosages and their dispersion, during the concrete manufacturing process, can affect the thermal properties of concrete enormously [8, 20, 27] (Arikumar et al., 2019). Especially the effects of fibres on the values of thermal conductivity ($K$) can be very significant. As $K$ is not the only critical aspect in thermal behaviour studies, the applications focusing on this aspect are of significance in the design of certain aspects of construction material deployed in buildings.

Studies pertaining to the changes taking place in mechanical properties and thermal properties with the inclusion of limestone [1], conductive fibres [5], rubber particles [10], mineral admixtures [6], etc. were extensively examined. Keeping such material in mind, numerous studies carried out by the researchers indicated variations of moisture content in the mortar specimens were not observed up to 50 °C, while declination is linear and steep until 100 °C [26]. The inclusion of sand has decreased the specific heat ($C$) while $K$ has increased, a complete opposite phenomenon was observed due to the inclusion of silica fumes $C$ [26]. $K$ is usually determined by three methods: two linear parallel probe methods (TLPP), lane heat source (PHS), and hot guarded plate (HGP). PHS and HGP methods are similar to the TLPP method (Kim et al.). Mortar specimens were prepared with partial replacement of mineral admixtures, such as Fly ash (FA), Silica fume (SF), and Blast furnace slag (BFS) in Portland cement (PC). Compressive strength, $K$ values are compared with plane mortar specimens, in relation to the specimens replaced partially with 10, 20, and 30% of FA, SF, and BF. Comparing all the results, lower $K$ values are reported for specimens prepared with SF, with respect to other mineral admixtures [6]. Mortar specimens with Palm oil fuel ash (POFA) as a mineral admixture and partially replaced with Ordinary Portland cement (OPC), along with Acrylic and Polypropylene (PP) fibre. The casted specimens were exposed to thermal loads of 300 and 600 °C, and compressive strength, water adsorption, and $K$ are investigated. A drop of 50% in $K$ was achieved in the mixes replaced with 50% POFA, due to the presence of higher internal air voids, and similar behaviour was observed with the specimens incorporated with acrylic fibres [11]. Cork fibre and paper waste are embedded into the gypsum matrix during the manufacturing of composite panels. The composite panels are investigated for water adsorption, compression, flexural strength, sound propagation, and $K$. Inclusion of 60% of cork fibre into the composites has decreased the $K$ by 300% [17]. Mendes et al. [9] proved the relation between variations in the $K$ is dependent on the porosity of the specimens, a relation was stated between the testing methods of Ultrasonic pulse velocity and heat flow meter. Contrafatto et al. [4] indicated that a significant correlation does not exist between $K$ and open porosity of the specimens prepared using pyroclasts that are porous in nature and that are embedded into the mortar mix. Hot wire parallel technique was adopted to determine the effects of moisture content and porosity on $K$, $D$, and $C$ of the aluminium refractory concrete specimens in the temperature ranges of ambient temperature $-1000$ °C [18]. $K$, $D$, and density decrease with the increase in temperature. While moisture content will directly affect the $K$ [19], conductivity and $D$ are determined for hempcrete at higher temperatures using a hot box test [15]. Methods such as the guarded hot plate method, hot wire method, heat flow method, and probe method are adopted in common. For the current experimental investigations, probe methods were adopted. Studies indicated that with the increasing temperatures, the thermal diffusivity decreases; as a result of increasing temperatures, moisture content tends to reduce [28]. Thermal analysers such as Thermogravimetric Analysis, Differential Thermal Analysis, and Differential Scanning Calorimetry are used to understand the mass loss and heat flow of cement specimens at elevated temperatures [14–14, 23, 24].

The primary focus of this research article is to determine the effects of steel and glass fibre concentration on the thermal properties of mortar specimens. Thermal properties such as $K$, $C$, $D$, and $\rho$ are determined using *KD2* Pro thermal analyser for the first time. The variations in the thermal properties are determined for the mortar specimen prepared with 0.5 and 1% of steel and glass fibre concentrations, and the results are compared to plane mortar specimens subjected to 50 and 100 °C. The suitability of *KD2* Pro thermal analyser in determining the thermal properties was investigated and discussed.

## 2 Experimental Groundwork

### 2.1 Materials

Cement binder of OPC, i.e. OPC-53 grade of cement (GoC) was chosen for the current experimental study, due to the ease in availability and its frequent use. The standard consistency of the cement paste is 30%, determined using IS:4031 (Part-4)-1988. While specific gravity of cement is 3.15, determined using IS:2720(Part-3). The fine aggregate (FA) used for the present investigation is Ennore sand, confining to the code specifications of IS: 650-1991. FA confined to three groups of particle size distribution, i.e. <2 mm and >1 mm, <1 mm and >50 μm, and <500 μm and >90 μm; while the ratios of each grade of *FA* are 33.33%. Two types of commercially available fibres are incorporated in to the mortar mix.

For the mortar mix, corrugated crimped stainless steel fibres that are cold drawn from mild steel wire are used. The mechanical properties of crimped fibres and their samples are shown in Table 1 and Fig. 1a. Along with steel fibre, Alkali resistance (Ar) glass fibre, where the integral part of the glass fibre is zirconium, constitutes 19%

**Table 1** Mechanical properties of steel fibres

| Type of fibre | Steel | Glass |
|---|---|---|
| Diameter (∅) | 0.55 mm | 13.5–14 μm |
| Length (L) | 50 mm | 6 mm |
| Tensile strength | 1168 MPa | 3500 MPa |
| L/∅ | 90.90 | 444.44 |
| K (W/m K) | 16 | 1.05 |
| Specific gravity | 7.85 | 2.6 |

**Fig. 1** Appearance of **a** corrugated crimped steel fibre and **b** Ar glass fibre

of the total fibre content. Ar glass fibres are highly abrasive resistant and noncombustible. The mechanical properties and the fibre samples are shown in Table 1 and Fig. 1b.

## 2.2 Mix Proportions

To determine the compressive strength of mortar specimens, i.e. plane mortar specimens are cast using *OPC-53* (200 g), and Ennore sand (Zone-I, II, and III—200 g each), a total of 600 g, is mixed with a water content of (*P*/4) + 3% (*P*—water percentage required for producing a paste of standard consistency as per *IS*:4031(Part-1)-1988), i.e. 84 g. The mortar specimens are cast, as per the code specifications of *IS*: 4031(Part-6)–1988. Five types of specimens are cast for the current investigation, i.e. specimens with Plane mortar, 0.5%, 1% concentrations of corrugated crimped steel fibres, and Ar fibres incorporated into the plane mortar mix.

## 2.3 Testing

Mortar mix specimens are placed in 70.7 × 70.7 × 70.7 mm$^3$ molds. While thermal properties are determined on cylindrical specimens, for which a set of three cylindrical specimens were cast into the molds 50 mm in diameter (Ø) and length (L). The mortar mix was poured into the desired mold, as three layers, while each layer was compacted by means of a mechanical vibrator. In order to place the sensors in Fig. 2b into the mortar specimens, two grooves are provided on top of the specimen, before the hardening of the mortar mix Fig. 2a, and placed in the curing tank for the desired curing periods. Once the desired curing periods are achieved, the specimens are taken out of the curing tank and subjected to desired temperature loading using an oven, since the heating temperatures are very low.

In order to acquire the parameters under investigation, from KD2 Pro thermal analyser, the following procedure was adopted for all the specimens. $K$, $C$, $D$, and $\rho$ of the cast mortar specimens are determined using *Decagon KD2 Pro* setup. The setup uses a method of transient line heat source for determining the thermal properties, while the dual needle *SH-1* sensor is used to obtain the parameters under investigation. A read time of 2 min was adopted, from which the data is acquired to predict the thermal properties. Out of the considered 2 min of read time taken for the SH-1 sensor, half of the time is spent for temperature equilibrium prior to initiation of heat. While the rest is the time utilized for heating; however, the measurements are taken for the entire duration. Depending on the type of material used for the study, different sensors can be used, and the read time to be adopted will be altered, as per the manufacturer's specifications.

Care is taken while inserting the dual needle sensor setup, and needles should remain parallel during the process of insertion into the specimens. To achieve this, a red tab with sample holes was inserted into the specimens at the time of cast, mortar specimens with dual needle holes Fig. 2a. Since the sensor releases a heat pulse, a

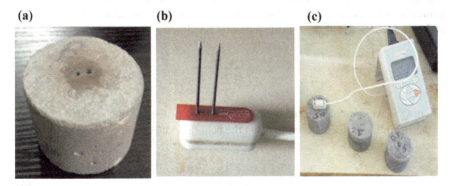

**Fig. 2** a Specimen with pre-inserted whole, b dual needle SH1 sensor, and c thermal properties test setup

minimum extra spacing of 1.5 cm is left parallel to the sensor in all the directions, to avoid errors while the collection of the readings.

In order for the *KD2 Pro* analyser to predict the results accurately, a good thermal contact between sensor and mortar specimens needs to be achieved, for which the thermal grease has been applied to the sensors, before placing the needles into the provided slots.

K is the ability of a material to conduct heat, the unit being $\frac{W}{m-K}$. The quantity of heat flow through a material is directly proportional to the cross-sectional area and the varying temperature gradient, along the surface through which the heat flows, while inversely proportional to the material thickness. It is calculated by Eq. 1:

$$Q = \left(\frac{k * A * (t_2 - t_1)}{L}\right) \tag{1}$$

where $Q$ = Quantity of heat; $A$ = Area of the cross section; $(t_2 - t_1)$ = Temperature difference; L = Distance of flow (Thickness).

D is the rate at which the temperature can be spread across the material. Higher diffusivity indicates the rate of heat transfer will be high. D is dependent on K, C, and density of the material. D is measured using SH-1 dual needle sensor, derived by Eq. 2:

$$D = \frac{K}{(d * C)} \tag{2}$$

where $K$ = Thermal conductivity; $d$ = Density; $C$ = Specific heat capacity.

In solid members, the C is determined by low heat exchange. C is the amount of heat required per unit mass to raise the temperature by a °C. C is very much influenced by the type of aggregates, and the effects of embedded steel fibre reinforcement are negligible, in the high-strength concrete (HSC) at elevated temperatures [25]. C values increase up to a temperature of 500 °C, then decrease from 700 to –900 °C and later increases [19]. 'C' is calculated by Eq. 3:

$$C = \frac{Q}{m(t_2 - t_1)} \tag{3}$$

where $Q$ = Amount of heat transfer (J); $m$ = Mass of the substance (kg).

$\rho$ is the ability of a material to resist the heat flow; it is the reciprocal of '$K$'. It is calculated by Eq. 4:

$$\rho = \frac{1}{k} \tag{4}$$

## 3 Results and Discussions

To understand the effects of hydration periods on the values of $K$, for specimens prepared with plane mortar, 0.5% steel and glass fibre, 1% steel and glass fibre at curing periods of 3, 14, and 28 days are plotted and discussed. These fibre concentrations are selected, as an increase in steel fibre concentrations causes the balling effect, which resulted in an improper compaction of mortar specimens. Along with these results, $K$, $C$, $D$, and $\rho$ effects of temperature, i.e. at 27, 50, and 100 °C on all the five mixes were determined. The values plotted in the table are the average of three readings of KD2 pro thermal analyser for each specimen, and the averages of the three specimen's values are plotted. Such a method was adopted to achieve consistency in the readings, and the discussion for the achieved results is as follows:

### 3.1 Thermal Conductivity (K)

The final values showing the variations in $K$ values of all the specimens at 3, 14, and 28 days of hydration, for all the five mortar mixes, are shown in Fig. 3. The mortar specimens of 0.5% glass fibre, 1% steel, and glass fibre have shown a reasonable drop in the $K$ values, with the increase in hydration, while in the plane mortar and

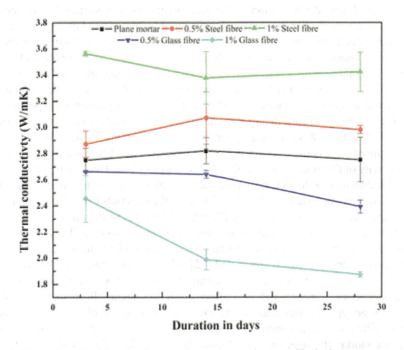

**Fig. 3** Thermal conductivity of various specimens at several stages of hydration

0.5% steel fibre specimens, variations in the $K$ are negligible. From the data obtained in the current investigation and mentioned by the other researchers, the effects of $K$, with varying hydration periods, are not observed in the specimens of plane mortar and 0.5% steel fibre.

The specimens with 1% glass fibre replacement in the mortar have the lowest $K$ values, while 1% steel fibre has the highest. Among the specimens under investigation, for specimens replaced with 1% glass fibre, the drop in $K$ values was 23.67% from the day of the cast to 28 days of hydration. This indicates a dispersion of 1% glass fibre is well spread across the specimen, with respect to 1% steel fibre. From Table 1, one can observe the lower values of $K$ for glass fibre, i.e. 1.05, while for steel fibre is 16, which has a significant effect on the overall $K$ values when embedded in the specimens. An average $K$ value of 0.5% steel fibre is 2.87 and 2.98 at 3 and 28 days of hydration, while 3.56 and 3.42 for 1% steel fibre. This indicates with the increasing steel fibre dosages, the $K$ values increases, thereby making the material highly conductive. While 0.5% glass fibre has $K$ value 2.66 and 2.39, for 1% glass fibre, it is 2.45 and 1.87 at 3 and 28 days of hydration. This phenomenon can be attributed due to the inclusion of fibres in the mortar specimens with different dosages of weight percentages, increasing the water adsorption capacity of the specimens. The higher the percentage inclusion of fibre content into the specimens, the greater will be the distribution of pores across the specimens. This behaviour was observed during the preparation of the fibre-reinforced mixes, which affected the consistency of the mix. As the clustering of fibres in the specimens leads to entrapment of air during the hydration of the mix, the pastes with standard consistency couldn't penetrate the fibres. This results in a lot of interconnected pores in the fibre specimens, due to the clustering of fibres. As specific gravity of glass fibres is three times lower than that of steel fibres. In the case of glass fibre, the distribution of fibre content is higher with respect to steel fibre, which increases the entrapped air content. From Fig. 3, we can say that 1% steel fibre specimens will have lesser pore content, since the density of specimens with fibre reinforcement is greater, and specimens with higher densities result in larger $K$ values [5]. However, due to the higher aspect ratios of glass fibre, $K$ values are fifteen times lower than that of steel fibres, while the density of the specimens is low in relation to steel fibre. This resulted in the lower conductivity of the mortar mixes embedded with glass fibres. One can infer from the available data that the higher the concentration of glass fibre dosages in the mix, the lower $K$ values can be achieved, and in doing so, the materials can act as insulators.

Temperature effects on the '$K$' at 27, 50, and 100 °C are shown in Fig. 4b, and the effects on the $K$ due to changes in temperature are significant. The optimum $K$ values of concrete specimens should be greater than 1.74 (W/m K), else the flux that can be attenuated will be restricted [15]. The porosity present in the concrete specimens plays a crucial role in governing the values of $K$. Types of pores, i.e. filled with air or moisture, will also affect the $K$ [5]. Figure 4b indicates that the effects of the types of fibre and their ratios have a significant role in governing the thermal efficiency of the concrete as a whole. The average $K$ values determined for the plane mortar specimens at 27, 50, and 100 °C are 2.75, 2.67, and 2.35. The $K$ value at 27 and 100 °C has dropped by 15%, while in 0.5 and 1% of steel fibre

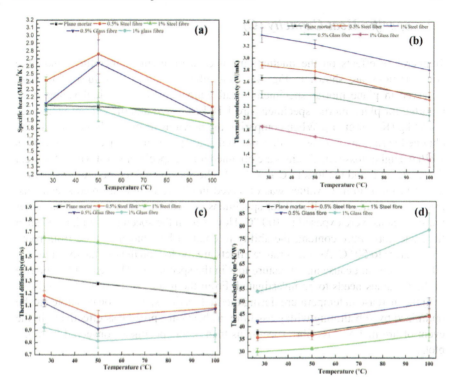

**Fig. 4** Relation between **a** specific heat and temperature, **b** thermal conductivity and temperature, **c** thermal diffusivity and temperature, and **d** thermal resistivity and temperature

specimens, it is 22.82 and 18.12%. The effects of 0.5% steel fibre incorporation into plane mortar specimens are negligible, as the values are very much similar, i.e. 2.35 and 2.30 at 50 and 100 °C. The inclusion of the steel fibre into the specimens has increased the K values, in comparison to the rest of the specimens. As $K$ values are higher for steel fibre concentration, among the specimens with different fibre concentrations, in order for the concrete to conduct temperature effectively across the concrete members, during thermal loading, higher steel fibre concentrations are desirable. As the maximum permitted steel reinforcement is 6% (Column) and 4% (slab and beam) in concrete members provided with steel reinforcement. Hence, $K$ transfer during fire accidents will be more in the column, with respect to other structural members. The reported $K$ for the mortar specimens embedded with 0.5 and 1% glass fibre at 100 °C is 2.40 and 1.30. While the loss in $K$ values due to 0.5 and 1% glass fibre incorporation into the mortar is 14.65 and 30.48%. In specimens with 1% glass and steel fibre, the $K$ values at 100 °C are 1.30 and 2.80, indicating that the steel fibre specimens can conduct heat two times more effectively than the specimens with 1% glass fibre. The 1% glass fibres specimens at 100 °C has lower $K$ values, which indicates the glass fibre replacement can restrict the attenuation heat flux more effectively, with respect to any other specimens in the current study.

## 3.2 Specific Heat Capacity (C)

Temperature effects on the fluctuations in $C$ at different temperatures and fibre concentrations are shown in Fig. 4a. Negligible variations in values of $C$ were observed in plane mortar specimens at all temperatures. The initial values of $C$ reported for plane mortar specimens are 1.18 kJ/kg [25] and the current values are 2.1 kJ/kg (Kodur et al.). When comparing the previous results with the current, the obtained variations in $C$ are within the limits, at specimens subjected to 100 °C is 2 kJ/kg. Higher losses in $C$ values are obtained for the specimens cast with 0.5% steel fibre, i.e. 2.42, and the loss in $C$ is 14%. As the specimens' temperature increased, a significant drop in $C$ values was observed in all the fibre-reinforced specimens. Drop in $C$ values is significant for specimens embedded with 1% glass fibre, after the specimens were exposed to 100 °C. However, in the specimens with 0.5% steel fibre and glass fibre content, the ability to conduct heat has increased from room temperature to 50 °C. As temperature effects are too low on the mortar specimens at 50 °C, it does not cause any moisture loss in the specimens. The reason for the rise in the $C$ values needs to be investigated. From the results of $C$, it is observed that with the increase in temperature during thermal loadings, effects on concrete can be optimized by substituting with glass fibre, as heat can spread less effectively in the event of fire in 1% glass fibre specimens than other specimens, observed from the plots.

## 3.3 Thermal Diffusivity (D)

Temperature effects on $D$ are shown in Fig. 4c, plotted for all plane and fibre-reinforced specimens. As $D$ values are dependent on the mass of the specimen, fluctuations in the density of the specimen are evident from the results. The specimens with higher glass and steel fibre content, i.e. 1% have set the upper and lower threshold values of '$D$'. As diffusivity is derived from $K$ divided by the product of specific heat and density of the material [2], the $D$ values obtained for the plane mortar specimens are 1.34, 1.28, and 1.18 $m^2/s$ at 27, 50, and 100 °C. The drop in the values of $D$ is very less under the temperatures of 100 °C, as the effects of changes taking place in moisture content and density are very optimum [3, 15]. The $D$ values for 1% glass fibre are 0.92, 0.81, and 0.86 $m^2/s$, while for 1% steel fibre, the values are 1.65, 1.61, and 1.48 $m^2/s$ at the above-mentioned temperatures.

## 3.4 Thermal Resistivity (ρ)

Variations in the values of $\rho$ with different types of fibre concentrations at three temperatures are shown in Fig. 4d. $\rho$ is the ability of a material to resist the ability to conduct heat, i.e. higher values of ρ indicate greater will be the ability to resist heat. One can observe from Fig. 4d that the specimens with 1% glass fibre have the highest resistance to heat, i.e. 53.89, and as the temperature increased to 100 °C, the resistance increased by 45%. Among all the specimens, the mortar specimens with 0.5 and 1% of glass fibre have shown the highest resistance to heat flow. As the concentration of steel fibre is 1%, the least ρ is shown by the mortar specimens. The specimens with plane mortar and 0.5% steel fibre are very much similar at 100 °C, i.e. 44.45 and 44.13. Hence, the inclusion of 0.5% steel fibre is insignificant in controlling the ability of the heat to flow across the specimens. The inclusion of 1% steel fibre is most favourable for effective heat transfer across the mortar specimens, while 1% inclusion of glass fibre optimizes the heat flow.

## 4 Conclusion

An experimental investigation is conducted to understand the thermal properties such as thermal conductivity, thermal diffusivity, thermal resistivity, and specific heat of mortar specimens embedded with 0.5 and 1% of glass and steel fibre into the plane mortar specimens. While variations in the thermal properties were observed with respect to the fibre embedded and plane mortar specimens, the following research findings are made from the study.

- The higher the concentration of glass fibre dosages in the mix, the lower $K$ values can be achieved, and in doing so, the mortar insulation properties can be enhanced.
- Higher doses of steel fibre are effective in conducting the temperature across the mortar during thermal loading.
- An increase in temperature effects on mortars during thermal loadings can be optimized by substituting with glass fibre, as heat can spread less effectively in the event of fire in 1% glass fibre specimens, rather than plane and steel fibre specimens, observed from the plots of $C$.
- Mortar specimens with 0.5 and 1% of glass fibre have shown the highest resistance to heat flow. While the inclusion of 1% steel fibre is most favourable for effective heat transfer across the mortar specimens.
- Kd2 Pro thermal analyser is a suitable test method for determining the thermal properties of mortar specimens at elevated temperatures; however, the determination of thermal properties is limited to 150 °C.

# References

1. Bosiljkov VB (2003) SCC mixes with poorly graded aggregate and high volume of limestone filler. Cem Concr Res 33(9):1279–1286. https://doi.org/10.1016/S0008-8846(03)00013-9
2. Carman AP, Nelson RA (2007) The thermal conductivity and diffusitity of concrete. Univ Illinois Bull 18(34):1–46
3. Choktaweekarn P et al (2009) A model for predicting thermal conductivity of concrete. Mag Concr Res 61(4):271–280. https://doi.org/10.1680/macr.2008.00049
4. Contrafatto L et al (2020) Physical, mechanical and thermal properties of lightweight insulating mortar with recycled etna volcanic aggregates. Constr Build Mater 240:117917. https://doi.org/10.1016/j.conbuildmat.2019.117917
5. Cook DJ (1974) The thermal conductivity of fibre-reinforced concrete. Cem Concr Res 4:497–509
6. Demirboğa R (2003) Influence of mineral admixtures on thermal conductivity and compressive strength of mortar. Energy Build 35(2):189–192. https://doi.org/10.1016/S0378-7788(02)00052-X
7. Farsana C, Snehal K, Das BB (2020) Influence of fineness of mineral admixtures on the degree of atmospheric mineral carbonation. Smart technologies for sustainable development. Lecture notes in civil engineering, vol 78, pp 117–136
8. George RM, Das BB, Goudar SK (2019) Durability studies on glass fiber reinforced concrete. Sustain Constr Build Mater 747–756
9. Mendes JC et al (2020) Correlation between ultrasonic pulse velocity and thermal conductivity of cement-based composites. J Nondestruct Eval 39(2). https://doi.org/10.1007/s10921-020-00680-7
10. Meshgin P et al (2012) Utilization of phase change materials and rubber particles to improve thermal and mechanical properties of mortar. Constr Build Mater 28(1):713–721. https://doi.org/10.1016/j.conbuildmat.2011.10.039
11. Mo KH et al (2017) Thermal conductivity, compressive and residual strength evaluation of polymer fibre-reinforced high volume palm oil fuel ash blended mortar. Constr Build Mater 130:113–121. https://doi.org/10.1016/j.conbuildmat.2016.11.005
12. Nari V et al (2020) A comparative study on the thermal behaviour of PPC and OPC cement. Mater Today: Proc 2–7. https://doi.org/10.1016/j.matpr.2020.05.708
13. Pavani HP et al (2020) Thermal behaviour of PPC and OPC-53 when exposed to extreme temperatures. Adv Cem Res 32(8):358–370. https://doi.org/10.1680/jadcr.18.00066
14. Praneeth H et al (2020) Characterisation of micro- and mesoporosity in portland cement at elevated temperatures. Mag Concr Res 72(6):304–313. https://doi.org/10.1680/jmacr.18.00321
15. Reilly A et al (2019) The thermal diffusivity of hemplime, and a method of direct measurement. Constr Build Mater 212:707–715. https://doi.org/10.1016/j.conbuildmat.2019.03.264
16. Sahoo S, Das BB, Mustakim S (2017) Acid, alkali, and chloride resistance of concrete composed of low-carbonated fly ash. J Mater Civ Eng 29(3):4016242
17. Sair S et al (2019) Development of a new eco-friendly composite material based on gypsum reinforced with a mixture of cork fibre and cardboard waste for building thermal insulation. Compos Commun 16:20–24. https://doi.org/10.1016/j.coco.2019.08.010
18. dos Santos WN (2003) Effect of moisture and porosity on the thermal propertiers of a conventional refractory concrete. J Eur Ceram Soc 23(5):745–755. https://doi.org/10.1016/S0955-2219(02)00158-9
19. Shin K-Y, Kim S-B (1999) Thermophysical properties and transient heat transfer of concrete at elevated temperature, pp 233–241
20. Snehal K, Das BB (2018) Mechanical and permeability properties of hybrid fibre reinforced porous concrete. Indian Concr J 98:54–59
21. Snehal K, Das BB (2021) Acid, alkali and chloride resistance of binary, ternary and quaternary blended cementitious mortar integrated with nano-silica particles. Cem Concr Compos 123:104214

22. Snehal K, Das BB, Akanksha M (2020) Early age, hydration, mechanical and microstructure properties of nano-silica blended cementitious composites. Constr Build Mater 233:117212
23. Snehal K, Das BB (2020) Effect of phase-change materials on the hydration and mineralogy of cement mortar. Proc Inst Civ Eng-Constr Mater. ISSN 1747*650X. https://doi.org/10.1680/jcoma.20.00045
24. Snehal K, Das BB, Sumit K (2020) Influence of integration of phase change materials on hydration and microstructure properties of nanosilica admixed cementitious mortar. J Mater Civ Eng 32(6):04020108. https://doi.org/10.1061/(ASCE)MT.1943-5533.0003178
25. Taerwe L, De Schutter G (1995) Specific heat and thermal diffusivity of hardening concrete. Mag Concr Res 47(172):203–208
26. Xu Y, Chung DDL (2000) Effect of sand addition on the specific heat and thermal conductivity of cement. Cem Concr Res 30(1):59–61. https://doi.org/10.1016/S0008-8846(99)00206-9
27. Yadav S, Das BB, Goudar SK (2019) Durability studies of steel fibre reinforced concrete. Sustain Constr Build Mater 737–745
28. Zhang Z-X (2016) Effect of temperature on rock fracture. Rock Fract Blast 111–133. https://doi.org/10.1016/b978-0-12-802688-5.00005-1

# Modulus of Elasticity of High-Performance Concrete Beams Under Flexure-Experimental Approach

Asif Iqbal A. Momin, R. B. Khadiranaikar, and Aijaz Ahmad Zende

**Abstract** The modulus of elasticity of High-Performance Concrete (HPC) is one of the characteristics found to be greater than conventional concrete. This enhanced elastic property of HPC makes it suitable for most of the structures with heavy loads and long spans. Also, the elastic property of reinforced HPC differs as compared to pure HPC. This research work aims at determining the modulus of elasticity of reinforced HPC beam under flexure using an experimental approach. The beam models of size 150 mm × 300 mm × 2000 mm with varying percentages of tension reinforcement for $M_{60}$, $M_{80}$ and $M_{100}$ grade HPC beams are studied experimentally. The modulus of elasticity of reinforced HPC beam under flexure is determined using experimental stress–strain curves and bending equations from the experimental data. The equation for modulus of elasticity is proposed in terms of longitudinal tension reinforcement ratio and grade of HPC. The modulus of elasticity of reinforced HPC beam under flexure increased with the increase in longitudinal tension reinforcement ratio of the concrete section.

**Keywords** High-performance concrete · Modulus of elasticity · Conventional concrete · Longitudinal tension reinforcement ratio

## 1 Introduction

The elastic property of concrete, i.e. the modulus of elasticity measures its rigidity. This property of concrete is defined as the ratio between the stress applied and the strain obtained within the defined limit of proportionality. The limit of proportionality

---

A. I. A. Momin (✉) · A. A. Zende
Department of Civil Engineering, BLDEA's Vachana Pitamaha Dr. P.G Halakatti College of Engineering and Technology Vijayapur, Affiliated to VTU, Belagavi, Karnataka, India
e-mail: asifarzanmomin@gmail.com

R. B. Khadiranaikar
Department of Civil Engineering, Basaveshwar Engineering College, Bagalkot, Affiliated to VTU, Belagavi, Karnataka, India

© The Author(s), under exclusive license to Springer Nature Singapore Pte Ltd. 2023
M. S. Ranadive et al. (eds.), *Recent Trends in Construction Technology and Management*, Lecture Notes in Civil Engineering 260,
https://doi.org/10.1007/978-981-19-2145-2_5

is "the maximum stress the material tolerates without deviating from the stress–strain proportionality (Hooke's law)." The applied stress may be static or dynamic; however, this study is limited to static stress. It is generally related to the compressive strength of concrete; it increases with an increase in compressive strength. Hence, the modulus of elasticity of concrete may be articulated as a function of the compressive strength of concrete [1]. But the increase in the rate of modulus of elasticity is comparatively less than that of the compressive strength of concrete [2]. Modulus of elasticity generally depends upon many parameters like aggregate type, the mix proportions, curing conditions, rate of loading, etc. The larger the amount of coarse aggregate with a high elastic modulus, the higher would be the modulus of elasticity of concrete. Researchers have also studied that the addition of mineral admixtures enhances the strength, modulus of elasticity and durability of concrete [2–9]. The concrete specimens tested in wet conditions show about 15% higher elastic modulus than those tested in dry conditions [10]. This property is evaluated by drawing the slope for the stress–strain curve. As most of the part of the stress–strain curve is non-linear, different methods for computing the modulus of elasticity are shown in Fig. 1. The secant modulus is the commonly used method for evaluating this property by different codes and researchers [11–14]. Now the modulus of elasticity of reinforced concrete members is very indifferent to the modulus of elasticity of concrete alone [15, 16]; hence, the use of the modulus of elasticity of concrete overestimates the design values in the design of structural members. Also, in the design of structural members, the total cross-sectional area of the section is considered neglecting the effect of confining steel reinforcements. The high-strength concrete and the high-performance concrete differ largely from the conventional cement concrete because of use of the cementitious material like silica fume, fly ash, etc. and hence because of which the mechanical properties of HSC/ HPC also differs [17–19] and the proposed equation for modulus of elasticity of concrete or reinforced concrete may not be applied either for HPC or reinforced HPC.

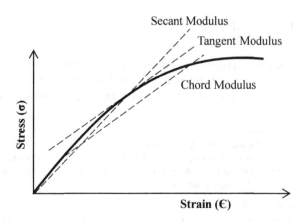

**Fig. 1** Stress–strain curve for concrete

**Table 1** Modulus of elasticity of concrete

| Sl. No. | Code/researcher(s) | Equation for modulus of elasticity of concrete (MPa) |
|---|---|---|
| 1 | IS-456:2000 | $E_c = 5000\sqrt{f_{ck}}$, applicable for concrete grades < $M_{55}$ |
| 2 | ACI-318-14 | $E_c = 4730\sqrt{f'_c}$, applicable for $f'_c < 41.4$ MPa |
| 3 | ACI-363-10 [20] | $E_c = 3320\sqrt{f'_c} + 6900$, applicable for $f'_c < 83$ MPa |
| 4 | NS 3473 [21] | $E_c = 9500(f'_c)^{0.3}$, applicable for $25 \le f'_c \le 85$ MPa |
| 5 | CSA A23.3-M84 [22] | $E_c = 5000\sqrt{f'_c}$ |
| 6 | EN 1992-1-1:2004 | $E_c = 22000\left[\frac{(f_{ck}+8)}{10}\right]^{0.3}$ |
| 7 | CSA A23.3 | $E_c = 4500\sqrt{f'_c}$ |

**Table 2** Modulus of elasticity of reinforced concrete

| Sl. No. | Code/researcher(s) | Equation for modulus of elasticity of reinforced concrete (MPa) |
|---|---|---|
| 1 | Kulkarni et al., 2014 | $E_c = 3774.23 p_t^2 + 2789.67 p_t + 5000\sqrt{f_{ck}}$ |

The database for modulus of elasticity of concrete, reinforced concrete and high-strength concrete/high-performance concrete as proposed by different codes and researchers is presented in Tables 1, 2 and 3.

**Table 3** Modulus of elasticity of HSC/ HPC

| Sl. No. | Code/researcher(s) | Equation for modulus of elasticity of HSC/HPC (MPa) |
|---|---|---|
| 1 | D. Mostofinejad and M. Nozhati [23] | $E_c = 10.25(f'_c)^{0.316}$, ($R^2 = 0.87$) for limestone aggregate<br>$E_c = 8(f'_c)^{0.352}$, ($R^2 = 0.85$) for andesite aggregate<br>$E_c = 10.75(f'_c)^{0.312}$, ($R^2 = 0.88$) for quartzite aggregate |
| 2 | M. A. Rashid, M. A. Mansur and Paramasivam [24] | $E_c = 8900\beta(f'_c)^{0.33}$, $\beta$ = coarse aggregate coefficient, applicable for $20 \le f'_c \le 130$ MPa |
| 3 | Hani H. Nassif et al. | $E_c = 0.036(w_c)^{1.5}(f'_c)^{0.5}$, $w_c$ = unit weight of concrete in kg/m$^3$ |
| 4 | Andrew Logan et al. [25] | $E_c = 0.000035 k_1 (w_c)^{2.5}(f'_c)^{0.33}$, applicable up to 124 MPa<br>$k_1$ is the correction factor to account for the source of aggregates |

**Table 4** Details of beam models

| Beam designation | Longitudinal reinforcement | % of reinforcement | | $\rho_b$ | $\rho/\rho_b$ |
|---|---|---|---|---|---|
| | | Ast (mm$^2$) | $\rho$ | | |
| 60SB1 | 2 # 12 mm | 226.19 | 0.789 | 4.762 | 0.166 |
| 60SB2 | 2 # 10 + 1# 12 mm | 270.16 | 0.938 | 4.850 | 0.193 |
| 60SB3 | 2 # 16 + 1# 10 mm | 480.66 | 1.695 | 4.800 | 0.353 |
| 60SB4 | 2 # 16 + 2# 10 mm | 559.20 | 1.972 | 4.849 | 0.407 |
| 80SB1 | 2–12 mm and 1–10 mm | 304.73 | 0.778 | 5.104 | 0.152 |
| 80SB2 | 3–12 mm | 339.29 | 0.866 | 4.981 | 0.174 |
| 80SB3 | 2–10 mm and 1–16 mm | 358.13 | 0.911 | 5.069 | 0.180 |
| 80SB4 | 2–12 mm and 1–16 mm | 427.25 | 1.091 | 5.052 | 0.216 |
| 100SB1 | 2 # 12 mm | 226.19 | 0.789 | 5.997 | 0.132 |
| 100SB2 | 2 # 10 + 1# 12 mm | 270.16 | 0.938 | 6.087 | 0.154 |
| 100SB3 | 2 # 16 + 1# 10 mm | 480.66 | 1.695 | 6.137 | 0.276 |
| 100SB4 | 2 # 16 + 2# 10 mm | 559.20 | 1.972 | 5.922 | 0.333 |

The applied Moment '$M$' under two-point loading is determined using Eq. (3).

$$f_c = \frac{M \, y_t}{I} \qquad (2)$$

$$M = \frac{wl}{6} \qquad (3)$$

The stresses-strain curves obtained for varying longitudinal reinforcement ratios and HPC grades for single-span HPC beams are shown in Figs. 4, 5 and 6.

## 5 Evaluation of Modulus of Elasticity of HPC Beam

(a) **Using experimental stress–strain curves**

The commonly used method of secant modulus is employed for evaluating the modulus of elasticity of the HPC beam from the stress–strain curves. The modulus of elasticity is determined from the slope of the line joining origin and a point on the curve representing 40% stress at failure. Thus, the modulus of elasticity obtained using experimental stress–strain curves is shown in Table 5.

(b) **Using bending equation from the experimental data**

Experimental evaluation of the secant modulus of elasticity is also done for the same beam specimens by using the bending equation. The single-span HPC beam being subjected to two-point loads, and the maximum deflection due to

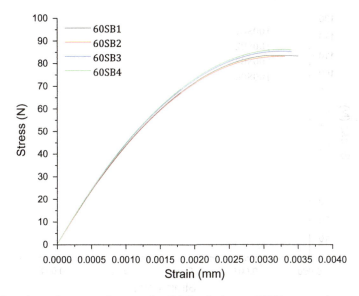

**Fig. 4** Experimental stress–strain curves for 60 MPa single-span HPC beam specimens

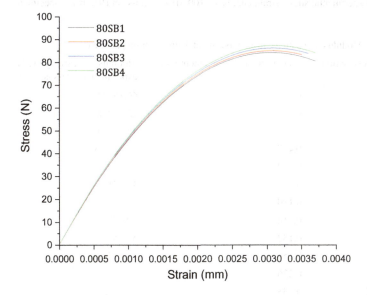

**Fig. 5** Experimental stress–strain curves for 80 MPa Single-span HPC beam specimens

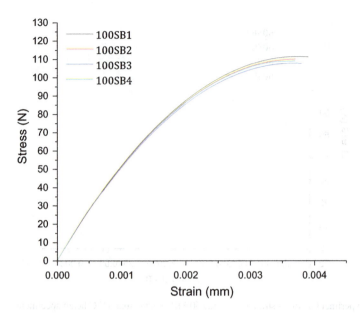

**Fig. 6** Experimental stress–strain curves for 100 MPa single-span HPC beam specimens

**Table 5** Modulus of elasticity from experimental stress–strain parameters

| Beam designation | Longitudinal reinforcement ratio $\rho/\rho_b$ | Modulus of elasticity (MPa) from experimental stress–strain curves |
|---|---|---|
| 60SB1 | 0.166 | 46,188.00 |
| 60SB2 | 0.193 | 45,662.60 |
| 60SB3 | 0.353 | 46,296.70 |
| 60SB4 | 0.407 | 46,486.88 |
| 80SB1 | 0.152 | 48,849.50 |
| 80SB2 | 0.174 | 49,031.82 |
| 80SB3 | 0.180 | 49,366.58 |
| 80SB4 | 0.216 | 49,672.86 |
| 100SB1 | 0.132 | 52,959.17 |
| 100SB2 | 0.154 | 52,612.30 |
| 100SB3 | 0.276 | 52,061.04 |
| 100SB4 | 0.333 | 52,496.29 |

the applied load '$W$' is given by Eq. 4.

$$\delta_l = \frac{23\,Wl^3}{648\,E_c\,I} \qquad (4)$$

The deflection due to the self-weight of the HPC beam is given by Eq. 5 and the total deflection considering the self-weight of the HPC beam is given by Eq. 6.

$$\delta_s = \frac{5wl^4}{384E_c I} \tag{5}$$

$$\delta = \delta_s + \delta_l \tag{6}$$

The applied load '$W$' is taken as half the load at the first visible crack and the corresponding deflection as '$\delta$'. The values of load at the first visible crack and its corresponding deflection for different beam specimens tested with varying strength of the HPC and longitudinal reinforcement ratio are shown in Table 6. By substituting these values, the static modulus of elasticity is determined using Eq. 7 for HPC beam specimens.

$$E_c = \frac{1}{\delta I}\left[\frac{5wl^4}{384} + \frac{23\,Wl^3}{648}\right] \tag{7}$$

**Table 6** Modulus of elasticity from experimental data using bending equation

| Beam designation | Longitudinal reinforcement ratio $\rho/\rho_b$ | Load corresponding to first visible crack (kN) | Deflection corresponding to first visible crack (mm) | Modulus of elasticity (MPa) |
|---|---|---|---|---|
| 60SB1 | 0.166 | 66.75 | 5.00 | 50,160.65 |
| 60SB2 | 0.193 | 71.45 | 4.75 | 53,090.57 |
| 60SB3 | 0.353 | 109.05 | 5.70 | 51,739.64 |
| 60SB4 | 0.407 | 117.52 | 5.60 | 52,546.43 |
| 80SB1 | 0.152 | 105.00 | 3.16 | 51,867.06 |
| 80SB2 | 0.174 | 117.00 | 3.30 | 53,461.88 |
| 80SB3 | 0.180 | 95.32 | 2.50 | 53,781.80 |
| 80SB4 | 0.216 | 122.30 | 3.10 | 54,236.83 |
| 100SB1 | 0.132 | 70.35 | 5.00 | 57,748.24 |
| 100SB2 | 0.154 | 94.05 | 6.22 | 58,875.64 |
| 100SB3 | 0.276 | 124.00 | 6.60 | 59,289.27 |
| 100SB4 | 0.333 | 128.32 | 6.23 | 59,326.57 |

## 6 Results and Discussion

The static modulus of elasticity of HPC beam specimens determined using experimental stress–strain curves and bending equation from experimental data are compared in Table 7. The average static modulus of elasticity is also presented in Table 7 for different longitudinal reinforcement ratios.

The stiffness of the HPC beam increases with an increase in longitudinal reinforcement ratio as such, and the trend line of the average variation of modulus of elasticity from different methods shows that it increases with an increase in the grade of HPC and longitudinal reinforcement ratio (Fig. 7). The values obtained from different methods are almost the same for the same longitudinal reinforcement ratio with a variation of 9.61%. However, from Fig. 7, the variation for $M_{80}$ grade HPC beams was found to be lesser compared to $M_{60}$ and $M_{100}$ grade because the variation of percentage of tension reinforcement within the beams was kept smaller in the range of 0.778–1.091, while the range for $M_{60}$ and $M_{100}$ grade was higher and of same ratios (0.789–1.972). It is also observed that the variation of modulus of elasticity within the same grade does not have higher variation as compared to variation within the different grades of HPC.

Taking the data of Modulus of Elasticity from bending equation, a regression test is performed, and following generalized Eq. 8 is derived, for reinforced single-span HPC beams.

$$E_{\text{Ref,HPC}} = 5000\sqrt{1.075 f_{ck}} + \left(300\sqrt{1.36 f_{ck}}\right) \rho \qquad (8)$$

**Table 7** Comparison of modulus of elasticity from different methods

| Beam designation | Modulus of elasticity (MPa) | | Average modulus of elasticity (MPa) |
|---|---|---|---|
| | Using experimental stress–strain curves | Using bending equation from the experimental data | |
| 60SB1 | 46,188.00 | 50,160.65 | 47,465.27 |
| 60SB2 | 45,662.60 | 53,090.57 | 48,193.54 |
| 60SB3 | 46,296.70 | 51,739.64 | 48,161.48 |
| 60SB4 | 46,486.88 | 52,546.43 | 48,602.50 |
| 80SB1 | 48,849.50 | 51,867.06 | 49,762.66 |
| 80SB2 | 49,031.82 | 53,461.88 | 50,782.92 |
| 80SB3 | 49,366.58 | 53,781.80 | 50,931.92 |
| 80SB4 | 49,672.86 | 54,236.83 | 51,280.11 |
| 100SB1 | 52,959.17 | 57,748.24 | 54,618.09 |
| 100SB2 | 52,612.30 | 58,875.64 | 54,760.74 |
| 100SB3 | 52,061.04 | 59,289.27 | 54,529.10 |
| 100SB4 | 52,496.29 | 59,326.57 | 54,831.22 |

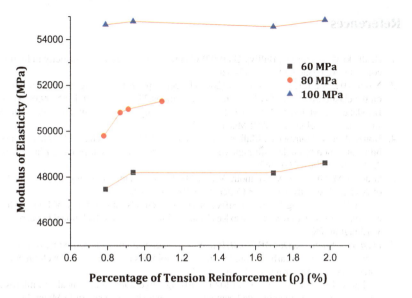

**Fig. 7** Average variation of modulus of elasticity with longitudinal tension reinforcement

Equation 8 predicts the value of Modulus of Elasticity for varying grades of HPC and percentage of tension reinforcement. The first part of the equation represents stiffness due to the HPC element, while the second part represents stiffness due to reinforcement. The second part of the equation is itself the additional stiffness because of the reinforcement (Fig. 7).

## 7 Conclusions

The present work evaluates the modulus of elasticity of HPC beam under flexure for 60–80 MPa strength of concrete. The following are the conclusions drawn based on the present experimental and analytical study.

1. The modulus of elasticity of reinforced HPC beam under flexure increases with the increase in longitudinal reinforcement ratio of the concrete section.
2. The modulus of elasticity of reinforced HPC beam for 60–80 MPa can be determined for different longitudinal reinforcement ratios by the proposed Eq. (2).
3. The maximum variation obtained was 11% from the different methods of evaluation of modulus of elasticity of HPC beam (Table 7).

## References

1. Baalbaki W, Aitcin PC, Ballivy G (1992) On predicting modulus of elasticity in high-strength concrete. ACI Mater J 89(5):517–520
2. Nassif HH, Najm H, Suksawang N (2005) Effect of pozzolanic materials and curing methods on the elastic modulus of HPC. Cem Concr Compos 27(6):661–670. ISSN 0958–9465
3. Bilodeau A, Malhotra VM (1992) High-volume fly ash system: concrete solutions for sustainable development. ACI Mater J 97(1):41–48
4. Bouzoubaa N, Fournier B, Malhotra VM, Golden D (2002) Mechanical properties and durability of concrete made with high volume fly ash blended cement produced in cement plant. ACI Mater J 99(6):560–567
5. Graciela M, Giaccio GM, Malhotra VM (1988) Concretes incorporating high volume fractions of ASTM class F fly ash. Cem Concr Aggr 10(12):88–95
6. Nassif H, Suksawang N (2002) Effect of curing methods on durability of high-performance concrete. Transp Res Rec: J Transp Res Board (1798):31–38. TRB, National Research Council, Washington, DC
7. Farsana C, Snehal K, Das BB (2020) Influence of fineness of mineral admixtures on the degree of atmospheric mineral carbonation. In: Smart technologies for sustainable development. Lecture notes in civil engineering, vol 78, pp 117–136
8. Snehal K, Das BB, Akanksha M (2020) Early age, hydration, mechanical and microstructure properties of nano-silica blended cementitious composites. Constr Build Mater 233:117212
9. Snehal K, Das BB (2021) Acid, alkali and chloride resistance of binary, ternary and quaternary blended cementitious mortar integrated with nano-silica particles. Cem Concr Compos 233:104214
10. Zia P, Ahmad S, Leming M (1989–1994) High-performance concretes a state-of-art report. Federal Highway Administration Research and Technology, FHWA Publications
11. Building Code Requirements for Structural Concrete (ACI 318–14) Commentary on Building Code Requirements for Structural Concrete (ACI 318R-14). Farmington Hills, MI, 317 p
12. Kankam CK, Meisuh BK, Sossou G, Buabin TK (2017) Stress-strain characteristics of concrete containing quarry rock dust as partial replacement of sand. Case Stud Constr Mater 7(2017):66–72. https://doi.org/10.1016/j.cscm.2017.06.004
13. Eurocode 2: Design of concrete structures - Part 1–1: General rules and rules for buildings, Dec-2004
14. Domagała L (2017) A study on the influence of concrete type and strength on the relationship between initial and stabilized secant moduli of elasticity. Solid State Phenom 258:566–569
15. Kulkarni S, Shiyekar MR, Shiyekar SM (2017) Confinement effect on material properties of RC beams under flexure. J Inst Eng India Ser A 98:413. https://doi.org/10.1007/s40030-017-0221-3
16. Kulkarni SK, Shiyekar MR, Shiyekar SM et al (2014) Elastic properties of RCC under flexural loading-experimental and analytical approach. Sadhana (39):677–697. Indian Academy of Sciences. https://doi.org/10.1007/s12046-014-0245-6
17. Iravani S (1996) Mechanical properties of high-performance concrete. ACI Mater J 93(5):416–426
18. Das BB, Pandey SP (2011) Influence of fineness of fly ash on the carbonation and electrical conductivity of concrete. J Mater Civ Eng 23(9):1365–1368
19. Goudar SK, Das BB, Arya SB (2019) Microstructural study of steel-concrete interface and its influence on bond strength of reinforced concrete. Adv Civ Eng Mater 8(1):171–189
20. ACI Committee 363, Report on High-Strength Concrete (ACI 363R – 10), ACI, Farmington Hills, MI, p 27
21. Norwegian Code NS 3473, Design of Concrete Structures, Norwegian Council for Standardization, Oslo, Norway, 1992
22. CSA A23.3-94 (1995) Design of concrete structures. Canadian Standard Association, Rexdale, Ontario, Canada

23. Mostoufinezhad D, Nozhati M (2005) Prediction of the modulus of elasticity of high strength concrete. Iranian J Sci Tech Trans B Eng 29(B3)
24. Rashid MA, Mansur MA, Paramasivam P (2002) Correlations between mechanical properties of high-strength concrete. J Mater Civ Eng 203–238
25. Logan A, Choi W, Mirmiran A, Rizkalla S, Zia P (2009) Short-term mechanical properties of high-strength concrete. ACI Mater J 106(5)
26. Momin AA, Khadiranaikar RB (2019) Experimental and finite element analysis of 80 MPa two-span high-performance concrete beam under flexure. In: Sustainable construction and building materials, vol 25, Chapter 35, pp 381–396
27. Kumar PS, Mannan MA, Kurian VJ, Achuytha H (2007) Investigation on the flexural behaviour of high-performance reinforced concrete beams using sandstone aggregates. Build Environ 42(7):2622–2629

# Impact of Phase Change Materials on the Durability Properties of Cementitious Composites—A Review

K. Vismaya, K. Snehal, and Bibhuti Bhusan Das

**Abstract** Phase change materials (PCMs) are the novel thermal storage materials which have an ability to engross and dispel heat during the process of phase transition from solid to liquid and vice versa. Utilization of PCMs in cementitious composites has gained a lot of attention from the research fraternity to minimize the energy loadings used for space conditioning and heating in building. Impact of PCM's presence in cementitious composites on the durability parameters is the need for its better usage. This paper gives the state of review on the influence of inclusion of phase change materials in the cementitious system on its various durability aspects. Durability properties such as porosity, water absorption, shrinkage, chloride ingression, and chemical attacks are compiled in this article. It is stated that the integration of PCM in cement composites enhances the porosity of cementitious system. Major hindrance described by the researchers is the interruption of hydration activity of cementitious system by the addition of PCM. Literature also signified that the micro/nano encapsulates PCMs and the use of highly reactive Pozzolans such as silica fume or nano-silica in conjunction with PCMs has the ability to lock up the limitations of PCMs.

**Keywords** Phase change material · Durability · Shrinkage · Porosity

## 1 Introduction

Rapid growth of industrialization and urbanization has imposed many environmental threats including the rise in global temperature, increase in the emission of greenhouse gases, and over exploitation of natural resources. The present scenario forces us to significantly target sustainability when dealing with innovations. The construction industry is one of the main consumers of material resources and energy. As the energy sources are of diminishing and expensive in nature, submissive methods of controlling the temperature inside the building become more significant. Mostly in the absence of heating and cooling systems the thermal conditions inside the

---

K. Vismaya · K. Snehal · B. B. Das (✉)
N. I. T. K., Mangalore, India
e-mail: bibhutibhusan@gmail.com

building will be often outside the acceptable range of human comfort [21]. Hence the development of energy efficient buildings should be our priority to contribute toward increased energy security. Many researchers engrossed to improve the heat storing ability of structures by incorporating phase change materials (PCMs) [4]. PCMs have received substantial attention due to the ability to store latent heat in building envelopes. Various studies report the extensive application of PCMs in plasters and wall boards to create more effectiveness in HVAC systems [1, 16]. The studies also reported that PCMs are helpful in mitigating thermal cracking in cementitious composites by controlling thermal variations [24]. The principle of PCMs in optimizing the thermal variations is quite simple. When the marginal temperature increases the material changes from solid phase to liquid by absorbing heat. Correspondingly, the moment when temperature drops to certain level material changes its state from liquid to solid by desorbing stored heat through an exothermic reaction [15].

Based on the chemical composition phase change materials are categorized into three groups (i) Organic compounds (ii) Inorganic compounds (iii) Eutectic mixtures [3, 12, 18]. The group of organic PCMs can be divided into paraffins and non-paraffins. Paraffins are advantageous by the fact that their melting point matches the human coziness temperature ranges and additionally they have high heating ability, compatible melting without segregation, and hardly show the tendency to supercool during phase change [5]. But Paraffins exhibit low thermal conductivity and high-volume changes [18]. Fatty acids, esters, alcohols, and glycols can be classified as non-paraffins. They have eminent melting and freezing properties, low flammability and renewable nature but they are expensive than paraffins [4]. Inorganic PCMs include hydrated salts. They are cheap and non-flammable. These PCMs have high heat of fusion and good thermal conductivity but they have a corrosive effect on metals and undergo supercooling [1]. Based on the composition material Eutectics are divided into three groups (i) organic-organic, (ii) inorganic-inorganic, and (iii) inorganic–organic. They have sharp melting points and relatively high storage density than that of organic compounds [3]. PCMs can be incorporated into a number of building materials including gypsum boards, masonry wall with bricks, concrete or natural stone, asphalt, etc. Since cement-based concrete/mortar is a porous and widely used construction material, it acts as an ideal media for PCMs. Mainly they can be incorporated by three methods (i) direct incorporation, (ii) immersion, and (iii) encapsulation [8, 22, 31]. As the name signifies in direct incorporation PCM is directly added into the mortar mix. In immersion technique PCM is immersed into a porous construction material like gypsum, brick, or concrete blocks. PCM gets easily adsorbed into the pores of the material due to the capillary forces [30]. In the encapsulation technique PCM is encapsulated by a polymeric shell in the form of a capsule [25].

Previous studies reported that the method of incorporation of PCMs influences the properties of cement composites [1, 4, 7, 26–29]. Cunha et al. studied the effect of direct incorporation of PCM and stated that it is the most economical and promising method as compared to microencapsulation technique. Sharma et al. reported that direct incorporation method altered the micropore structure of the cement matrix by increasing the intruded pore volume due to lesser hydration. It was found that the unhydrated cement particles were coated with PCM which made it unable to hydrate further. The microstructural studies of Snehal and Das [26, 27] described that direct addition of PCM resulted in a porous media with pronounced traces of calcium hydroxide and lesser formation of Calcium silicate hydrate (CSH). Jayalath et al. studied the influence of microencapsulated PCMs in cement composites and reported that the fineness of PCM imparted a filler effect and thereby increased the hydration of cement. Studies reported that microencapsulated PCMs are vulnerable to damage or rupture of the shell during the mixing stage and the leaked PCM interferes with the hydration products causing strength reduction. Microencapsulated PCMs tend to reduce the density of the concrete as it replaces aggregates which possess comparatively a higher density [9, 13]. Paksoy et al. reported that the addition of PCM in bulk and microencapsulated forms reduces the compressive strength of the concrete. PCMs can be impregnated in concrete products for latent heat storage but the stability of PCM is the primary factor of consideration for this method. It is reported that the use of modified concrete can successfully incorporate PCMs through immersion method especially those which are unstable in alkaline media [12, 25]. Hence it can be understood that different methods of PCM incorporation have both benefits and drawbacks which can influence the strength and durability parameters of the composite.

This paper discusses the comprehensive findings of various researches on the influence of phase change materials (PCMs) on the durability parameters such as porosity, water absorption, density, and shrinkage characteristics of cementitious composites.

## 2 Durability Properties of PCM—Cementitious Composites

The durability properties of concrete influence the overall performance as it stipulates the resistance of the structures to the severe environmental conditions to which it may get exposed to. Various researchers made attempts to assess the effect of incorporation of PCMs on the durability parameters of cementitious composites.

### 2.1 Porosity

Various studies conducted by researchers [8, 9, 13] reported that the integration of PCM in cement composites increased the percentage of porosity matrix which is collated and represented in Fig. 2. The influence of PCM content on the critical pore diameter which is one of the controlling factors of durability of concrete is investigated by various researchers [2, 8] which is represented in Fig. 3 (Fig. 1).

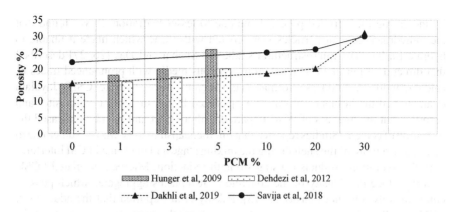

**Fig. 1** Variation of porosity with percentage of PCM

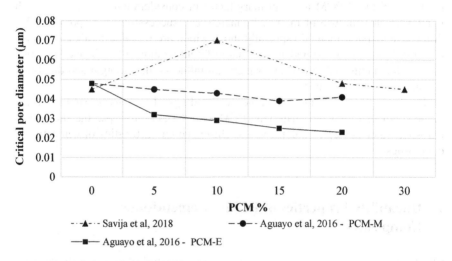

**Fig. 2** Representation of critical pore diameter with different percentages of phase change materials

Hunger et al. studied the porosity of self-compacting concrete with different PCM contents directly mixed in microencapsulated form into the concrete. A mixture of paraffins commercially named as Micronal DS 5008 X having melting point at 23 °C was used for the study. The results obtained showed that porosity increased with increase in PCM content which is due to the structural change of packing density of concrete and lower density of PCM than the other ingredients of concrete.

Dakhli et al. studied the porosity of cement passively integrated with microencapsulated PCM composed of vegetal wax in a powder form. The PCM used for the study is industrially referred as INERTEK 23 P. It was observed that porosity increased with the percentage of PCM which could be attributed to the presence of vegetable waxes. From the graph plotted, the sample with 30% PCM was found to

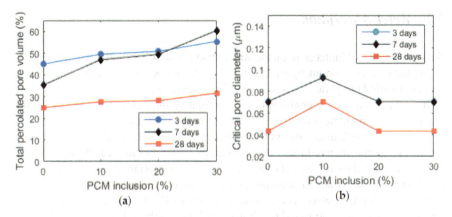

**Fig. 3** Plot representing **a** total pore volume **b** critical pore size for cement paste with different PCM dosages at varied ages [24]

be more porous than that of 0, 10, and 20% PCM. Additionally, it can be also noticed that the relationship between porosity and percentage of PCM is non-linear.

Dehdezi et al. studied the porosity of PCM incorporated cementitious system and stated that Porosity increased with PCM content due to its lower particle density (0.9 kg/m$^3$) as compared to that of the sand replaced (2.64 kg/m$^3$). The PCM used for the study was a microencapsulated paraffin wax in powdered form with an acrylic outer shell having a melting point at around 26 °C.

Savija et al. studied the critical pore diameter and porosity of PCM incorporated cement paste. Microencapsulated PCM composed of paraffin core with melamine formaldehyde (MF) shell was used for the study. It was found that the critical pore volume remained constant for different PCM dosages except for 10% PCM mix which showed comparatively a large pore volume. The study reported that the critical pore size was non-dependent on PCM content but dependent on the hydration degree of the composite. The study also reported that the percentage of percolated porosity increased with increase in percentage of PCM.

Aguayo et al. investigated the critical pore diameter of a cementitious system incorporating microencapsulated, paraffinic PCMs of size 10 μm and 7 μm, respectively. A significant reduction was obtained in the parameter owing to the enhanced reactivity of the PCMs due to their smaller particle size. Due to the smaller particle size, the PCM act as nucleation sites for the increased formation of hydrated calcium hydroxide.

## 2.2 Water Absorption

Data presented by Cunha et al. on the water absorption by capillarity and immersion by the inclusion of non-encapsulated PCM in a developed mortar and a fiber-based mortar at varied temperatures is plotted in Fig. 4 (Fig. 5).

It can be noted that water absorption by capillarity and immersion decreased with the addition of non-encapsulated PCM. It can be verified that the water absorption is not varying much under varied temperatures which indicates that PCM remains within the mortar matrix in its different phases (solid and liquid). With reference to the obtained data researchers have observed that mortar mixes with no PCM proliferated the water absorption rate. This was due to the availability of larger pores in the cementitious matrix. The mortars exposed to 10 °C temperatures exhibited low water absorption and found that it was attributed to the solid state of PCM. The mortars submitted to higher temperatures (25 and 40 °C) exhibited a higher rate of

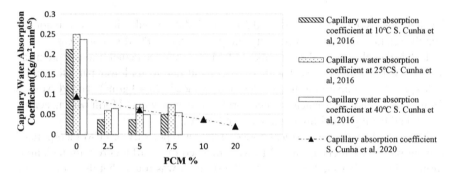

**Fig. 4** Representation of capillary water absorption coefficient with different percentages of PCMs

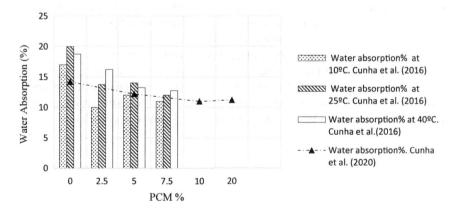

**Fig. 5** Plot representing percentage of water absorption with different percentages of PCM inclusion

capillary water absorption, due to the phase transition (solid to liquid) of PCM as it occupies in a relatively smaller volume of the pores.

## 2.3 Density

The influence of PCM incorporation on the density of cementitious matrix investigated by various researchers [9, 11, 13, 15] is represented in Fig. 6. From the results of the study it can be observed that the integration of PCM tends to lower the density of the matrix as the PCM normally substitutes the materials with greater density. It was also reported that the reduction in density can be owed to the increased in porosity of the mixes with the incorporation of PCMs.

Hunger et al. studied the influence of direct mixing of microencapsulated PCM in self-compacting concrete on its material properties. The PCM used was a mixture of paraffin waxes in powder form, encapsulated in polymethyl methacrylate microcapsules named Micronal DS 5008 X with a melting point of 23 °C. Dehdezi et al. investigated the thermal, mechanical, and microstructural properties of concrete incorporating different amounts of microencapsulated phase change materials (PCMs). Paraffin wax core in the form of dry powder encapsulated with acrylic outer shell having a melting temperature of 26 °C was used for the study. Both the studies reported that the concrete density decreased with increase in PCM content which can be attributed to the low density of the PCM (0.90–0.915 g/cm$^3$) and a physical change of concrete filling density which can be shown from the increased porosity of the matrix.

Kheradmand et al. experimentally investigated the thermal and physical properties of mortar integrated with phase change materials (PCMs) having peak melting point of 34 °C. The PCM was dispersed in grated pristine form into mixtures as macro capsule core. It was reported that density reduced by 87 and 83% of the reference mortar with the addition of 10 and 20% of PCM in grated pristine form. The reduction

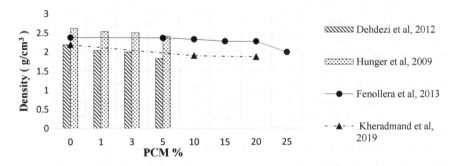

**Fig. 6** Plot representing the variation of density with different percentages of PCM

of density for PCM mixtures can be ascribed to the lesser density of the PCM than the sand particles.

Fenollera et al. analyzed the mechanical behavior of a self-compacting concrete incorporated with microencapsulated phase change material. Microencapsulated PCM of Micronal DS 5007 X in aqueous dispersion having a melting point at 23 °C was used for the study. The density and the difference of average densities of the samples were tested for the fresh state as well as after 24 h for the same volume. It was reported that density reduced proportionately with the increase in PCM content. This decrease was more than for the 25% mix due to the total replacement of the filler by PCM, which leads to an increase in water/(cement + filler) ratio. Density in the fresh state decreased by 1.1% for each 5% of PCM inclusion.

Niall et al. studied the thermal behavior and properties of PCM—concrete composite incorporated with microencapsulated and impregnated forms of PCM. Microencapsulated paraffin was added to the fresh concrete during the mixing process and the light weight aggregate used in the concrete mix was impregnated with butyl stearate. The study also investigated the influence of 50% GGBS cement replacement in the mixes. The density of both composites was lesser than control concrete due to the lesser density of PCM material. The panels containing GGBS showed a insignificant rise in density which may be attributed to the lowering of porosity due to its filler effect.

## 2.4 Shrinkage

Yang et al. investigated the mechanical properties of PCM admixed concrete by adopting mixes with PCM addition as well as replacement of total volume under two temperature state of, i.e., at 23 °C (PCM is in solid state) and 40 °C (PCM is in liquid state). It was reported that in PCM—modified concrete the drying shrinkage was found to be amplified PCM content regardless of temperature conditions. The lack of strength/stiffness for the PCM and its replacement with materials possessing comparatively high strength/stiffness minimized the ability to act against volumetric changes which attributed to higher drying shrinkage of the composite. In the mixes with total replacement increase in PCM content increased the w/c ratio and decreased the cement which caused a rise in drying shrinkage characteristics. The study also evaluated the change in drying shrinkage value during sat 23 and 40 °C. Drying shrinkage at 40 °C was found to be lesser than that at 23 °C for mixes without PCM. But for the mix with 10% and 20% replacement of fine aggregates with PCM the drying shrinkage at 40 °C was reported as higher than that at 23 °C by 6.3% and 4.7%, respectively.

N. P. Sharifi et al. investigated the enhancement in thermal performance of buildings and pavements by incorporating Phase Change materials. Three different PCMs of paraffin blend with melting points of −10, 6, and 28 °C with a specific gravity less than one were used for the study. The applicability of Light weight aggregate and Rice husk ash as a PCM carrier was examined. It was reported that the mix with light weight aggregate pre-soaked in PCM underwent autogenous expansion. The observations of the study suggested that a portion of the pre-soaked PCM was released to the media by light weight aggregate (LWA) or some amount of PCM from the surface of the light weight aggregate entered the mix. Similarly, an autogenous expansion was reported for the mixes with pre-soaked Rice Husk Ash (RHA) which can be implied to the leakage or adhering on the surface of RHA which may further enter the bulk cement paste. However, it was reported that the total shrinkage increased in the mixes with incorporation of LWA and RHA pre-soaked with PCM regardless of the amount of PCM pre-soaked (Fig. 7).

Wei et al. examined the influence of PCM addition on drying shrinkage in cementitious composites containing microencapsulated PCMs which is represented in Fig. 8. The study was conducted in mortars prepared at different proportions of microencapsulated PCM and/or quartz inclusions. It was reported that for the mixes containing both PCM and quartz, the shrinkage reduced with increasing quartz inclusion as the stiff inclusion impeded the shrinkage of the paste. Furthermore, it was also reported that the dosage of PCM inclusion did not influence the drying shrinkage of the composites as the soft inclusion does not have the ability to resist the shrinkage upon drying.

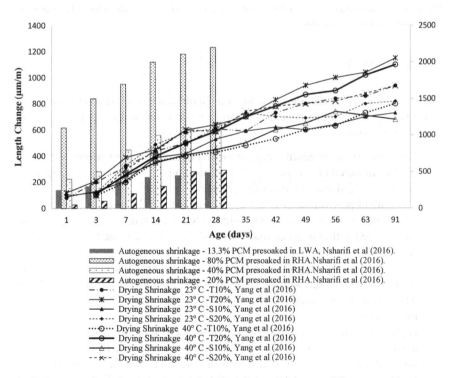

**Fig. 7** Representation of autogenous shrinkage and drying shrinkage at different ages (days)

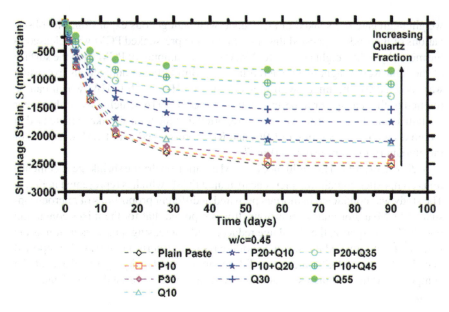

**Fig. 8** Plot representing the rate of drying shrinkage for PCM and/or quartz admixed cementitious composites with respect to time

## 3 Conclusion

The review paper gives an understanding on the influence of PCMs on the durability properties of cementitious system. The following conclusions can be made from the referred literatures.

1. The porosity of cementitious composite increased with PCM content which can be attributed to structural change of packing density of concrete and lower density of PCM than the other ingredients of concrete.
2. The critical pore diameter is not dependent of the PCM content but depended on the hydration rate of the composite. A significant reduction in the parameter can be owed to the enhanced reactivity of the PCMs due to their smaller particle size.
3. Water absorption caused due to the facts of capillarity and immersion for the composites decreased with the addition of PCM without polymeric capsulation which can be attributed to the reduction in the volume of free pores in the microstructure.
4. Higher ambient temperature conditions enhanced the capacity of capillary water absorption which can be attributed to the transformation of PCM from solid to liquid phase as it occupies relatively a smaller volume in the mortar pores.
5. The density of the matrix decreased with the incorporation of PCMs as it replaces filler materials having relatively higher density. It can be also attributed to the increased porosity of the matrix.

6. Drying shrinkage in PCM—modified concrete increased with PCM inclusion. This can be owed to the reduced capacity of the matrix to resist volume changes due to the replacement of the materials with higher stiffness or strength.
7. The composites with PCM impregnated into light weight aggregate and rice husk ash as carriers underwent autogenous expansion which implies the release of PCM into the media by leakage or adhering to the surface of the carrier.

# References

1. Adesina A (2019) Use of phase change materials in concrete: current challenges. Renew Energy Environ Sustain 4:9
2. Aguayo M, Das S, Maroli A, Kabay N, Mertens JC, Rajan SD, Sant G, Chawla N, Neithalath N (2016) The influence of microencapsulated phase change material (PCM) characteristics on the microstructure and strength of cementitious composites: experiments and finite element simulations. Cem Concr Compos 73:29–41
3. Baetens R, Jelle BP, Gustavsen A (2010) Phase change materials for building applications: a state-of-the-art review. Energy Build 42:1361–1368
4. Berardi U, Gallardo A (2019) Properties of concretes enhanced with phase change materials for building applications. Energy Build 199:402–414
5. Cedeño FO, Prieto MM, Espina A, García JR (2001) Measurements of temperature and melting heat of some pure fatty acids and their binary and ternary mixtures by differential scanning calorimetry. Thermochim Acta 369(1–2):39–50
6. Cunha S, Leite P, Aguiar J (2020) Characterization of innovative mortars with direct incorporation of phase change materials. J Energy Stor 30:101439
7. Cunha S, Aguiar J, Ferreira V, Tadeu A (2013) Influence of adding encapsulated phase change materials in aerial lime-based mortars. Adv Mater Res 687:255–261
8. Dakhli Z, Chaffar K, Lafhaj Z (2019) The effect of phase change materials on the physical, thermal and mechanical properties of cement. Sci 1(2019):27
9. Dehdezi PK, Hall MR, Dawson AR, Casey SP (2012) Thermal, mechanical and microstructural analysis of concrete containing microencapsulated phase change materials. Int J Pave Eng 1–14
10. Fabiani C, Anna Laura P, D'Alessandro A, Ubertini F, Luisa FC, Franco C (2018) Effect of PCM on the hydration process of cement-based mixtures: a novel thermo-mechanical investigation. Materials 11(6):871
11. Fenollera M, Míguez JL, Goicoechea I, Lorenzo J, Ángel Álvarez A (2013) The influence of phase change materials on the properties of self-compacting concrete. Materials 6:3530–3546
12. Hawes DW, Banu D, Feldman D (1991) The stability of phase change materials in concrete. Sol Energy Mater Sol Cells 27:103–118
13. Hunger M, Entrop AG, Mandilaras I, Brouwers HJH, Founti M (2009) The behavior of self-compacting concrete containing micro-encapsulated Phase Change Materials. Cem Concr Compos 31:731–743
14. Jayalath A, Nicolas RS, Sofi M, Shanks R, Ngo T, Ayea L, Mendis P (2016) Properties of cementitious mortar and concrete containing micro-encapsulated phase change materials. Constr Build Mater 120:408–417
15. Kheradmand M, Vicente R, Azenha M, Aguiar JL (2019) Influence of the incorporation of phase change materials on temperature development in mortar at early ages: experiments and numerical simulation. Constr Build Mater 225:1036–1051
16. Kuznik F, David D, Johannes K, Roux JJ (2011) A review on phase change materials integrated in building walls. Renew Sustain Energy Rev 15(1):379–391

17. Lecompte T, Le Bideau P, Glouannec P, Nortershauser D, Le Masson S (2015) Mechanical and thermo-physical behaviour of concretes and mortars containing Phase Change Material. Energy Build 94(2015):52–60
18. Ling TC, Poon CS (2013) Use of phase change materials for thermal energy storage in concrete: an overview. Constr Build Mater 46:55–62
19. Sharifi NP (2016) Application of phase change materials to improve the thermal performance of buildings and pavements. Doctoral Thesis, Worcester Polytechnic Institute
20. Niall D, West R, McCormack S, Kinnane O (2017) Mechanical and thermal evaluation of different types of PCM–concrete composite panels. J Struct Integr Maint 2(2017):100–108
21. Norvell C, Sailor DJ, Dusicka P (2013) The effect of microencapsulated phase-change material on the compressive strength of structural concrete. J Green Build 8(3):16–124
22. Paksoy H, Kardas G, Konuklu Y, Cellat K, Tezcan F (2017) Characterization of concrete mixes containing phase change materials. IOP Conf Ser: Mater Sci Eng 251:012118
23. Cunha S, Lima M, Aguiar JB (2016) Influence of adding phase change materials on the physical and mechanical properties of cement mortars. Constr Build Mater 127:1–10
24. Savija B (2018) Smart crack control in concrete through use of phase change materials (PCMs): a review. Materials 11:654
25. Sharma B (2013) Incorporation of phase change materials into cementitious systems. Thesis, Arizona State University, USA
26. Snehal K, Das BB, Kumar S (2020) Influence of integration of phase change materials on hydration and microstructure properties of nanosilica admixed cementitious mortar. J Mater Civ Eng 32(6)
27. Snehal K, Das BB (2020) Effect of phase-change materials on the hydration and mineralogy of cement mortar. Proc Inst Civ Eng—Constr Mater. ISSN: 1747-650X
28. Snehal K, Das BB (2020) Influence of incorporating phase change materials on cementitious system—a review. Recent trends in civil engineering. Lecture notes in civil engineering, vol 105, pp 33–63
29. Snehal K, Das BB (2020) Experimental set-up for thermal performance study of phase change material admixed cement composites—a review. Smart technologies for Sustainable development. Lecture notes in civil engineering, vol 78
30. Snehal K, Archana Dinesh T, Das BB Experimental investigation on the influence of phase change material (PCM) on the properties of cement mortar. In: 4th UKIERI concrete congress—concrete: the global builder, 5–8 March 2019, Dr B R Ambedkar National Institute of Technology Jalandhar, India
31. Souayfane F, Fardoun F, Biwole PH (2016) Phase Change Materials (PCM) for cooling applications in buildings: a review. Energy Build 129:396–431
32. Wei Z, Falzone G, Wang B, Thiele A, Puerta-Falla G, Pilon L, Neithalath N, Sant G (2017) The durability of cementitious composites containing microencapsulated phase change materials. Cem Concr Compos 81:66–76
33. Yang HB, Liu TC, Chern JC, Lee MH (2016) Mechanical properties of concrete containing phase-change material. J Chin Inst Eng 39(5):521–530

# Comparative Study of Performance of Curing Methods for High-Performance Concrete (HPC)

Sandesh S. Barbole, Bantiraj D. Madane, and Sariput M. Nawghare

**Abstract** In this paper, the results statistical and graphical analysis of a comparative study of curing methods for high-performance concrete is discussed. The results are obtained from an experimental investigation carried out on concrete cubes. Mainly, two experiments were conducted: concrete cube strength test using a Compressive Testing Machine (CTM) and a permeability test. Concrete cubes of M35, M45, M50, M60, and M70 were cast and they were cured for 8 h, 3 days, 7 days, 14 days, 28 days, 56 days, and 128 days by three curing methods, namely, steam curing, water curing and sealing compound curing. By using results obtained from this experimental investigation is compared to find out the strength of concrete at the age of all curing periods mentioned above for individual grades of concrete. In addition, a comparison of the compressive strength of concrete with the early strength of concrete across all grades is investigated. Furthermore, a study on the percentage change in strength w. r. t. water is also conducted.

**Keywords** Steam curing · Water curing · Compound curing · Compressive testing machine (CTM) · Permeability test

## 1 Introduction

The quality, strength, and durability of concrete completely depend on the efficient continuous curing method. The method of curing depends on the nature of concrete,

---

S. S. Barbole (✉)
P.G. Diploma in Rail & Metro Technology (PGDRMT), College of Engineering Pune (COEP), Pune, India
e-mail: barboless19.pgdrmt@coep.ac.in

B. D. Madane
Maharashtra Metro Rail Corporation Limited (Maha-Metro), Pune, India
e-mail: bantiraj.madane@mahametro.org

S. M. Nawghare
College of Engineering Pune (COEP), Pune, India
e-mail: smn.civil@coep.ac.in

© The Author(s), under exclusive license to Springer Nature Singapore Pte Ltd. 2023
M. S. Ranadive et al. (eds.), *Recent Trends in Construction Technology and Management*, Lecture Notes in Civil Engineering 260,
https://doi.org/10.1007/978-981-19-2145-2_7

properties of concrete, application of concrete, the surrounding temperature, and relative humidity. It is essential to keep concrete moist and stop the loss of moisture from it, while it is gaining strength through the hydration process. There are many methods of curing, which are depending on site conditions and the requirement of construction. There are many arguments on which method of curing concrete gives good strength, therefore this topic has gained wide scope for research and application in the construction industry. There was much research conducted to find effective curing and the effect of type of curing on curing properties. Common properties of concrete, which are measures, are strength and permeability. Very few studies had been conducted on curing oh High-Performance Concrete (HPC). Site constraints restrict the curing by a conventional method, hence there is a need to explore and establish modern curing for high strength and performance of concrete.

The application of steam curing at atmospheric pressure to the curing chamber is one of the oldest and most widely used methods of accelerated curing of concrete. Under ideal conditions, the curing of concrete by steam at atmospheric pressure (low-pressure accelerated) has the advantage over other methods of accelerated curing in that the curing environment of the concrete is near saturation in regards to moisture. Evaporation of water from the products is minimized, which is especially important where products like block, pipe, etc. are involved. Although the steam curing of masonry units, pipe, and precast or prestressed concrete products follows the same basic rules, curing procedures are different for each. Heat transfer and evaporation will be fast. Pipe and precast products may be cured in a form where evaporation is minimum or may be of such large mass that heat transfer is slow, and large temperature gradients and resultant stresses may exist between the centre and the outside of the mass [1]. Steam curing has been used for many years. It is used to speed up the strength development of concrete products. With an increase in curing temperature, the rate of hydration of cement also increases. Therefore, the speed of strength gain of concrete can be increase with the help of steam curing [2]. The curing temperature has a huge influence on the strength development rate of concrete.

One of the major challenges is the lack of availability of Indian standard codes suitable for Indian climatic conditions for steam curing and compound curing. In order to study across all three methods, i.e., water curing, steam curing, compound curing we have referred IS 456(2000) for water curing, ASTM C 309 for compound curing, and ACI SP-32 for steam curing. Synchronization and simulation of practical conditions was the biggest challenge faced by us. This research will help in the standardization of Indian codes for steam and compound curing.

## 2 Literature Review

The following are the previous research reviews based on the effect of various curing methods on concrete properties.

Tan and Zhu [3] have studied the influence of curing methods and mineral admixtures on the strength and chloride permeability of concrete is discussed. Analysis was conducted by using results obtained from the experimental analysis. The chloride permeability of concrete was measured using rapid chloride ion penetration tests (ASTM C 1202). The results obtained by tests show that the compressive strength of plain concrete is decreased by 11% by steam curing. The reason for this decreased in strength was the use of slag, fly ash, and silica fume. However, after using autoclave, the strength of concrete surpasses 80 MPa. Compared with normal curing methods, steam and autoclave curing increases the electric charge passed through plain cement concrete by 110 and 224%, respectively.

Yazıcı and Arel [4] have investigated the effect of steam curing on mortar. Mortar includes mineral admixtures with 3, 1, and 0.5 of aggregate, binders, and water ratios are adopted. In mortars 0, 10, 20, and 30% of fly ash or 0, 5, 10, and 15% of silica fumes were used. 24, 48, and 72 h strength of concrete cubes were measured. This discussed accelerated curing temperature and early strength time is discussed according to compressive strength test results.

Mahmet Gesog Lu has discussed the influence of steam curing on the compressive strength, ultrasonic pulse velocity, water sortivity, chloride ion permeability, and electrical resistivity of metakaolin and silica fume blended concrete. Various combinations of Portland cement, silica fumes, and metakaolin are studied with constant water or binder ration. The use of silica fumes and metakaolin shows a decrease in water sorptivity and chloride ion permeability of concrete.

## 3 Methods of Curing

### 3.1 Water Curing

Water curing is the oldest and most common method of curing. In this method, concrete cubes are submerged in water for a designed period. We have done water curing has been done as per IS 456:2000. This method is commonly used in a laboratory. The main purpose of water curing is to continuously keep concrete in a moist environment. Another purpose of water curing is to maintain water temperature. Water curing plays a key role in the promotion of hydration, elimination of shrinkage, and absorption of heat of hydration. The temperature of the water should not decrease by 5 °C than the temperature of concrete. In water curing, cold water should not be used as it may give a thermal shock, which might lead to cracking. On-site, water curing is done by various methods like ponding, immersion, water spraying, wet covering, etc. Precast concrete structures are commonly immersed in water is adopted. The curing of vertical structures is usually conducted by spread curing. Wet curing is done by using gunny bags, hessian bags, jute matting, etc.

## 3.2 Compound Curing

Compound curing involves the application of the liquid sealing compound to the hardened concrete. It restricts penetration of liquids and gases, which might cause reinforcement corrosion, acid attack, and damages to concrete. There is a variety of curing compounds available on market. With the curing compound, conventional curing is also important. It forms a protective layer of moisture-retentive film over the concrete. Curing cannot be avoided with curing compounds. Curing compounds increase the durability of concrete. This curing compound must not be used over paint, additional concrete layer, or tiles. This method of curing is used at the site where there is a shortage of water. As Indian standards for compound curing are not yet been thoroughly defined, we have done compound curing as per ASTM C 309.

## 3.3 Steam Curing

Currently, the use of steam curing in the construction industry is increasing with the demand for rapid and speedy construction. The need for early and high-strength concrete is completed by this method of curing. It is mostly adopted for precast members where early high strength is required. It is executed by using water vapor at atmospheric or high pressure. Pressure is approximately between 40 and 70 °C (100–160 °F). In the steam curing procedure, it is suggested that to start the curing procedure a few hours after casting. This period is called as pre-steaming period, which can be from 2 to 6 h. Initial steam curing temperature starts with 10–20 °C and the maximum curing temperature permitted is 85–90 °C. Steam curing is very advantageous in a cold-weather environment. It is done with help of canvas covering or sheets to cover the structure and inside the covering, steam curing is taken place. As Indian standards for compound curing are not yet been thoroughly defined, we have done compound curing as per ACI SP 32 (Figs. 1 and 2).

## 4 Material Used for High-Performance Concrete (HPC)

The details of concrete mix design are given in Table 1.
  Table 2 represents the properties of Fine Aggregates.
  Table 3 represents the properties of Fine Aggregates.
  Table 4 represents the properties of Cement.

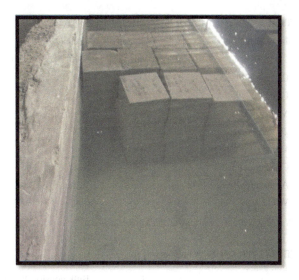

Fig. 1 Water curing by ponding

Fig. 2 Compound curing

## 5 Methodology for Concrete Mix of HPC

1. Identification of site for raw materials—various sites in the vicinity of the Pune area of Maharashtra, India are visited for suitable raw materials. While deciding on the suitability of raw materials following points are mainly considered.

   i. Availability of enough requirement of quantity required for research.
   ii. Materials having required properties as per IS2386.

**Table 1** Material used for high-performance concrete (HPC)

| Ingredients | Grades of concrete (N/mm$^2$) | | | | |
|---|---|---|---|---|---|
| | M35 | M45 | M50 | M60 | M70 |
| Cement OPC53 | 215 | 250 | 430 | 440 | 465 |
| GGBS | 215 | 230 | 70 | 35 | NA |
| Microsilica | NA | NA | NA | 25 | 35 |
| 20 mm down aggregate | 814 | 758 | 782 | 788 | 778 |
| 10 mm down aggregate | 349 | 357 | 368 | 371 | 366 |
| Crusher sand | 1006 | 832 | 832 | 832 | 827 |
| Admixture dosage (%) | 1.2 | 1.2 | 0.8 | 1.1 | 1.1 |
| W/C ratio | 0.40 | 0.36 | 0.29 | 0.28 | 0.29 |

*Note* Ingredient's weights are in kilogram (kg/m$^3$)

**Table 2** Properties of fine aggregates

| Sr. No. | Properties | Values |
|---|---|---|
| 1 | Bulk density (loose) (kg/m$^3$) | 1650 |
| 2 | Bulk density (compacted) (kg/m$^3$) | 1770 |
| 3 | Specific gravity | 2.87 |
| 4 | Water absorption (percentage) | 3.12 |

**Table 3** Properties of course aggregates

| Sr. No. | Properties | Values |
|---|---|---|
| 1 | Bulk density (loose) | 1678 |
| 2 | Bulk density (compacted) (kg/m$^3$) | 1800 |
| 3 | Specific gravity | 2.98 |
| 4 | Water absorption | 1.06 |

**Table 4** Properties of cement (PPC)

| Sr. No. | Properties | Values |
|---|---|---|
| 1 | Consistency (percentage) | 28 |
| 2 | Initial setting time (min) | 135 |
| 3 | Final setting time (min) | 195 |
| 4 | Compressive strength (MPa) | 33 |
| 5 | Fineness (m$^2$/kg) | 290 |
| 6 | Soundness (mm) | 0.8 |
| 7 | Density (g/cc) | 3.14 |

iii. Materials and water are free from impurities as well as located close to each other in order to reduce transportation costs to a lab
2. Based on the above factors site at Talegaon, Pune, Maharashtra India has been selected.
3. Raw materials, i.e., coarse aggregate, fine aggregate collected and tested as per IS 2386 for properties such as flakiness index, elongation index, sieve analysis for gradation of aggregates similarly cement is tested for soundness, Initial setting time, final setting time, etc.
4. Proportioning of aggregates on maximum density approach as per IS 2386. Admixtures and other materials are tested for the properties at the time of procurement from the vendor.
5. Now, for finalization of design mix, design is done according to IS10262 for grades M35, M45, M50, M60, and M70 at the same time relevant international codes such as ACI are referred for design. Trial mixes are carried out at the lab according to the design accordingly, mix design is finalized as per getting satisfactory results.
6. Both W/C ratio and Plasticizer dosage are finalized on the basis of mix design and trail mixes carried out consequently to get desired strength, i.e., M35, M45, M50, M60, and M70. Also, the study has been done on industrial practice which is currently being used for W/C ratio and Plasticizer dosage in the heavy civil industry to make the study practically viable.
7. Standard sized cubes are casted, i.e., 1 M × 1 M × 1 M. 80 no's of cubes are cast for each grade i.e., a total of 500 cubes are cast for grades M35, M45, M50, M60, and M70. The intention behind huge sampling was to reduce possible errors due to site handling and bring uniformity to the data.
8. Casting of samples is done at three different curing conditions, i.e., Steam curing, curing by using curing compound, and water curing.
9. Calibration of testing apparatus, i.e., Compression Testing Machine (CTM) is done before going for cube tests.
10. Testing of samples at age of 8 h, 3 days, 14 days, 28 days, 56 days, 90 days, and 180 days is performed as shown in the figure on CTM. Strengths got across different concrete grades and different curing conditions are will be discussed and interpreted further in this paper.
11. It is to be noted here that weather conditions are jot down at the time of casting of the cube, testing of cube whenever felt necessary to eliminate variations on the readings.

**Fig. 3** Cube casting

## 6 Statistically and Graphical Analysis

### 6.1 Methodology

1. Compressive strength after 7, 28 days is plotted against M50, M60 design mix, respectively.
2. Results are compared with the target strength equation given by IS 10262:2019.

$$f_{ck}'' = f_{ck}' + 1.65 \times S$$

where $f'_{ck}$ = target mean compressive strength at 28 days, in N/mm$^2$;
$f_{ck}$ = characteristic compressive strength at 28 days, in N/mm$^2$;
$S$ = standard deviation (5 N/mm$^2$).

3. Finally, the graphical analysis is done (Figs. 3 and 4).

### 6.2 Results and Interpretations

#### 6.2.1 Compressive Strength of Steam Curing, Water Curing, and Compound Curing

The compressive strength of M35, M45, M50, M60, and M70 of steam curing, water curing, and compound curing are given in Table 5.

**Fig. 4** Cube testing

Comparative Interpretation Across the Different Grades

Table 5 and Graphs 1, 2, 3, 4, and 5 demonstrate strengths of concrete subjected to different methods of curings, i.e., water curing, steam curing, and curing using curing compound.

Horizontal Interpretation for All Grades for Early Strength

Early Strength (i.e., 8 h) achieved by steam curing varies between 32.12 and 19.61% (grades M35–M70). As grade of concrete is increased from M35 to M70 variation in the achievement of the early strength achievement decreases, i.e., difference between hydration of concrete decreases as the grade increases. When same conditions and grades are repeated for comparison of compound curing with water curing is found that early strength (8 h) found in the experimental test varies between 12.25 and 10.59% between grades M35 and M70. Here it can be deduced that early strength achieved by using steam curing is higher for all grades (Table 6) after which compound curing is found effective and water curing at last. This might be due to tricalcium aluminate (it has potential capacity to yield a high early strength.) formation reaction speeds up by steam curing due to steam temperature and ability to penetrate the concrete pores and carry out early hydration. In case of curing compound, reasons might be attributed to chemical properties of curing compound penetrate the concrete and speed up hydration of reaction forming tricalcium aluminate which in turn ends up adding to early strength of concrete [1–5].

**Table 5** Compressive strength of concrete

| Grade of concrete | Curing in days | Compressive strength (MPa) | | |
|---|---|---|---|---|
| | | Steam curing | Water curing | Compound curing |
| M35 | 0.33 (8 h) | 03.50 | 03.02 | 03.06 |
| | 3 | 23.27 | 24.40 | 23.58 |
| | 7 | 33.00 | 30.89 | 31.94 |
| | 14 | 44.25 | 40.10 | 43.53 |
| | 28 | 50.37 | 53.89 | 53.97 |
| | 56 | 53.30 | 58.20 | 55.89 |
| | 90 | 55.11 | 59.85 | 57.09 |
| | 180 | 56.46 | 61.20 | 57.74 |
| M45 | 0.33 (8 h) | 03.77 | 03.27 | 03.59 |
| | 3 | 31.88 | 33.30 | 20.77 |
| | 7 | 45.20 | 49.91 | 28.13 |
| | 14 | 60.61 | 59.60 | 38.33 |
| | 28 | 69.00 | 72.67 | 47.53 |
| | 56 | 73.01 | 80.22 | 49.23 |
| | 90 | 75.48 | 81.99 | 50.28 |
| | 180 | 77.34 | 83.91 | 50.85 |
| M50 | 0.33 (8 h) | 04.47 | 03.90 | 04.21 |
| | 3 | 33.15 | 38.5 | 33.10 |
| | 7 | 47.01 | 52.88 | 44.84 |
| | 14 | 63.03 | 69.35 | 61.10 |
| | 28 | 71.76 | 79.80 | 75.77 |
| | 56 | 75.93 | 83.90 | 78.46 |
| | 90 | 78.50 | 85.57 | 80.14 |
| | 180 | 80.43 | 87.19 | 81.05 |
| M60 | 0.33 (8 h) | 05.25 | 04.30 | 04.89 |
| | 3 | 34.13 | 42.81 | 36.77 |
| | 7 | 55.13 | 59.16 | 49.81 |
| | 14 | 76.13 | 78.16 | 67.87 |
| | 28 | 81.67 | 86.28 | 84.16 |
| | 56 | 83.87 | 93.40 | 87.15 |
| | 90 | 87.16 | 95.04 | 89.02 |
| | 180 | 88.88 | 96.43 | 90.03 |
| M70 | 0.33 (8 h) | 06.10 | 05.10 | 05.64 |
| | 3 | 42.70 | 45.14 | 44.60 |

(continued)

**Table 5** (continued)

| Grade of concrete | Curing in days | Compressive strength (MPa) | | |
|---|---|---|---|---|
| | | Steam curing | Water curing | Compound curing |
| | 7 | 61.01 | 70.36 | 60.42 |
| | 14 | 81.80 | 84.93 | 82.33 |
| | 28 | 96.13 | 99.10 | 102.10 |
| | 56 | 103.70 | 114.40 | 105.70 |
| | 90 | 105.20 | 116.10 | 108.00 |
| | 180 | 107.61 | 117.40 | 109.20 |

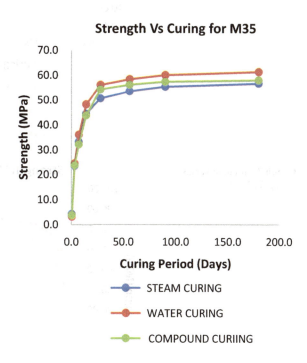

**Graph 1** Strength versus curing for M35

Hence, from above it can be deduced that early strength of concrete varies mentioned in the following order: **Steam curing > Compound curing > Water curing**.

Horizontal Interpretation for All Grades for 28 and 180 days Strength

Cubes casted as mentioned in methodology, tested in parts according to requirement. 28 days strength of concrete by method of steam curing varies from −9.87 to −3% (−ve sign indicates % strength lower than water) lower than strength achieved by

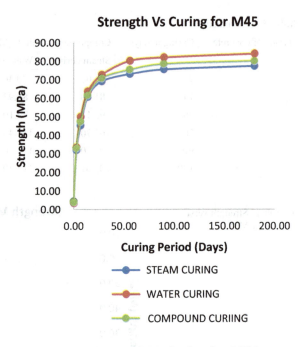

**Graph 2** Strength versus curing for M45

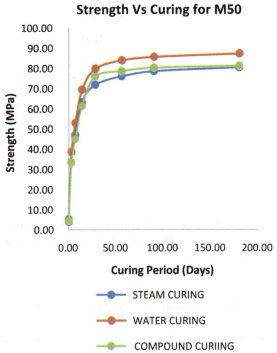

**Graph 3** Strength versus curing for M50

Comparative Study of Performance of Curing Methods ... 95

**Graph 4** Strength versus curing for M60

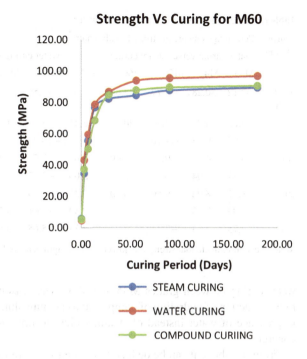

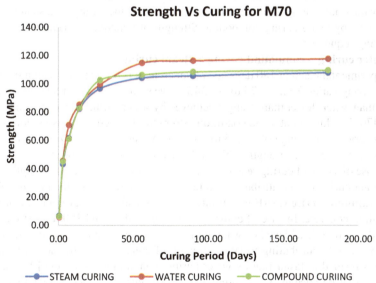

**Graph 5** Strength versus curing for M70

**Table 6** Percentage change in strength with water

| Curing period | Percentage change in strength with water ||||||||||
|---|---|---|---|---|---|---|---|---|---|---|
| | Water curing versus steam curing ||||| Water curing versus compound curing |||||
| | M35 | M45 | M50 | M60 | M70 | M35 | M45 | M50 | M60 | M70 |
| 0.33 (8 h) | 32.12 | 30.28 | 30.77 | 22.09 | 19.61 | 16.56 | 15.29 | 14.54 | 13.72 | 10.59 |
| 3 | −4.63 | −4.27 | −13.90 | −20.29 | −5.41 | −3.36 | −1.60 | −14.03 | −14.12 | −1.20 |
| 7 | −8.05 | −9.43 | −11.10 | −6.81 | −13.29 | −10.99 | −5.57 | −15.20 | −15.81 | −14.13 |
| 14 | −8.01 | −4.70 | −9.11 | −2.60 | −3.69 | −9.51 | −3.56 | −11.90 | −13.17 | −3.06 |
| 28 | −9.87 | −5.04 | −10.08 | −5.34 | −3.00 | −3.43 | −2.93 | −5.06 | −2.46 | 3.02 |
| 56 | −8.42 | −8.98 | −9.50 | −10.20 | −9.35 | −3.96 | −6.23 | −6.48 | −6.69 | −7.59 |
| 90 | −7.93 | −7.93 | −8.26 | −8.29 | −9.39 | −4.61 | −4.52 | −6.35 | −6.34 | −6.99 |
| 180 | −7.75 | −7.83 | −7.75 | −7.83 | −8.34 | −5.65 | −4.84 | −7.04 | −6.64 | −6.98 |

*Note* −ve sign indicate that strength is lower than strength achieved by water curing

water curing between grades M35 and M70. The reasons can be attributed to the rate of reaction of hydration of compound tricalcium alumina ferrite C3AF is higher in presence of water instead of steam, which in turn adds to later strength of the concrete.

From the above it can be deduced that long-term strength, i.e., 28 days is always greater for water curing than the other two methods of the curing, i.e., steam curing and curing by using curing compound. Strength of concrete 28 days varies in the following sequence:

**Water curing > compound curing > steam curing**

Experimental result at 180 days age of concrete, it is found that strength by using steam curing varies from −7.75 to −8.34% (−ve sign indicates percentage strength lower than water) lower than strength achieved by water curing between grades M35 and M70. In addition, at same conditions strength of concrete by using compound curing varies in the range of −5.65 to −6.98 (−ve sign indicates % strength lower than water) lower than strength achieved by using water curing [6–8].

In case of steam of curing reason for getting 28 days and 180 days strength lower than water curing can be attributed to rate of hydration reaction which forms **tricalcium alumino ferrite** C4AF which intern is responsible for development of later strength in concrete. In case of compound curing 28 days and 180 days strength is lower than strength achieved by water curing under same conditions can be deduced that compound contributing to latest strength of concrete, i.e., tricalcium alumino ferrite is formed at faster rate by water curing than compound curing [9, 10]. The reason might be attributed to continuous contact with water accelerates hydration reaction of formation of all 4 compounds C3S C2S C4A C4AF but steam curing and compound curing affect these rates differently. Compound responsible for early strength of concrete is formed at a faster rate in case steam curing, compound curing, and water curing respectively but as strength achieved over longer period is cumulative of all 4 compounds, which are better formed by water curing (Table 5).

Another notable observation is that strength achieved through experiment for same curing period, i.e., 180 days but across different grades of concrete, i.e., M35, M45, M50, M60, M60, and M70 variation from later strength (180 days) achieved by water curing later strength (180 days) of concrete increases as compared with respective grades of concrete in water curing (Table 5). As exact reason behind this behavior cannot be deduced from available research, it could be a potential gray area of the research.

Summary of Interpretation of All the Results

Water curing always beats the other two methods of curing when it comes to the later strength of the concrete (Tables 5 and 6). Compound contributing to early strength of concrete (**tricalcium alumino ferrite** C4AF) (8 h strength) is hydrated faster in the sequence of steam curing, compound curing, and water curing, respectively. But it is found that trend reverses in case of strength achieved in case of later strength (28, 180 days) in which sequence found to be water curing, compound curing, and stream curing, respectively. Hence, condition where water curing is feasible then the method should be preferred over the other two, i.e., steam curing and compound curing.

### 6.2.2 Permeability Test or Water Penetration Test

See Table 7 and Graph 6.

Permeability is directly proportional to the durability of concrete. For low-grade concrete, more water penetration results were obtained and for higher grades almost negligible water penetration shows. This shows that higher grade HPC has more resistance against hydrostatic pressure [11].

Table 7 Permeability test/water penetration text

| Sr. No. | Grade of concrete | Penetration in mm | |
|---|---|---|---|
| | | Average | Maximum |
| 5 | M35 | 11.33 | 14 |
| 6 | M45 | 9.33 | 12 |
| 7 | M50 | 4.67 | 6 |
| 8 | M60 | 1.17 | 4 |
| 9 | M70 | 0.67 | 1 |

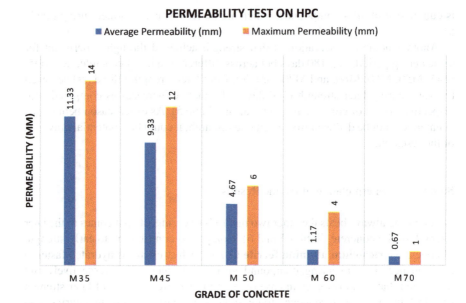

**Graph 6** Permeability test/water penetration text

## 7 Conclusion

It is found that early strength (8 h) by steam curing is in the range of 32–19% higher than water curing and early strength (8 h) achieved by curing compound is in the range of 17–11% higher than water curing also. Further, it is observed that 3, 28, and 180 days strength trend said above reverses, i.e., here strengths decrease in the range of 7.75–8.34% for steam curing and −5.65 to −6.98% for compound curing which implies that water curing is best as compared to other two in long term.

Horizontal comparison, i.e., across the grades for same age (M35, M45, M50, M60, M70) showed that achievement of strength varies inversely with the grades, i.e., strength achieved decreases w.r.t water curing as grade of concrete increases irrespective of age of concrete.

It is found that there is a direct correlation between permeability and grade of concrete, i.e., as grade of concrete increases permeability decreases which can be interpreted as concrete which is cured by different methods of curing will have permeability lower for higher grade of concrete which is achieved as per sequence mentioned in results point this paper.

## 7.1 Potential Applications of the Research Outcomes

1. The industries where achievement of the early lifting strength is important steam curing of the concrete could be done but concrete of higher grade than mentioned should be used in order to avoid effect on ultimate strength or if same grade is being used then it should be made sure that minimum required grade is being achieved for long term according to relevant standard codes at that place.
2. The situations where curing by using steam is not possible compound curing could be done by taking care of things said above.

## References

1. Reported by ACI Committee 5 17. Accelerated curing of concrete at atmospheric pressure-state of the art. Revised 1992
2. Gesoğlu M (2010) Influence of steam curing on the properties of concretes incorporating metakaolin and silica fume. Mater Struct
3. Tan K, Zhu J (2016) Influences of steam and autoclave curing on the strength and chloride permeability of high strength concrete. Springer
4. Yazıcı S, Arel HS (2016) The influence of steam curing on early-age compressive strength of pozzolanic mortars. Arab J Sci Eng
5. ACI Committee 305R-99 "Hot Weather Concreting" Reported by ACI Committee 305. ACI Manual of Concrete Practice (2009)
6. Goel A, Narwal J, Verma V, Sharma D, Singh B (2013) A comparative study on the effect of curing on the strength of concrete. Int J Eng Adv Technol (IJEAT)
7. Lee MG. Effect of steam-curing on the strength of precast concrete. In: Proceedings of the seventh international conference on composites engineering, ICCE, Denver, USA, pp 513–514
8. Singh DN, Pandey SP, Das BB (2014) Influence of initial curing humidity on compressive strength and ultrasonic properties of concrete. Indian Concr J 88(2):48–56
9. Williams RP, Van Riessen A (2010) Determination of the reactive component of fly ashes for geopolymer production using XRF and XRD. Fuel 89(12):3683–3692
10. Olivia M, Nikraz H (2012) Properties of fly ash geopolymer concrete designed by Taguchi method. Mater Des (1980–2015) 36:191–198
11. Lü J, Guan H, Zhao W, and Ba H (2011) Compressive strength and permeability of high-performance concrete. J Wuhan Univ Technol Mater Sci Ed 26

# Development of a Mathematical Relationship Between Compressive Strength of Different Grades of PPC Concrete with Stone Dust as Fine Aggregate by Accelerated Curing and Normal Curing

**N A G K Manikanta Kopuri, S. Anitha Priyadharshani, and P. Ravi Prakash**

**Abstract** Concrete is a composite material made from cement, water, fine and coarse aggregates. In recent times, a lot of studies are conducted to find a new fine aggregate material from industrial wastes. In this study, full replacement of fine aggregate with stone dust is carried out. Concrete specimens are cast by replacing fine aggregate with stone dust. The quality of concrete is an important factor in assimilating the strength of a structure. Curing is a significant factor that influences the quality of concrete. In this study, normal water curing and accelerated curing by boiling water method are considered. The important aspect of construction is the time constraint that leads to a high impact on the economy. The strength and the curing of concrete can be expedited by rising the temperature and thereby increasing the rate of hydration. This study aims at investigating the behavior of the strength of $M_{20}$, $M_{25}$, $M_{30}$, $M_{35}$, and $M_{40}$ grades of concrete with **PPC** when cured under both conditions. The mix design is prepared with reference to IS10262:2019. The specimens are prepared and tested according to IS: 516-1959. From the study, it is concluded that the strength of specimens with normal and accelerated curing is more than the target mean strengths of all grades of concrete.

**Keywords** Portland Pozzolana cement · Stone dust · Normal curing · Accelerated curing

---

N A G K Manikanta Kopuri (✉) · S. Anitha Priyadharshani · P. Ravi Prakash
Department of Civil Engineering, NIT Warangal, Warangal, India
e-mail: na720012@student.nitw.ac.in

S. Anitha Priyadharshani
e-mail: priyadharshanianitha@nitw.ac.in

P. Ravi Prakash
e-mail: rprakash@nitw.ac.in

© The Author(s), under exclusive license to Springer Nature Singapore Pte Ltd. 2023
M. S. Ranadive et al. (eds.), *Recent Trends in Construction Technology and Management*, Lecture Notes in Civil Engineering 260,
https://doi.org/10.1007/978-981-19-2145-2_8

# 1 Introduction

Concrete is a versatile material used as a principal element in the construction industry. The use of industrial waste materials as fine aggregates in concrete is a good alternative to conventionally used fine aggregates. Stone dust is one of the best alternative materials for fine aggregate. Accelerated curing method is used to get early high compressive strength in concrete [7]. This method is also used to find out 28 days' compressive strength of concrete in 28 h. (As per IS: 9013-1978-Method of making, curing, and determining compressive strength of accelerated cured concrete test specimens.)

# 2 Literature Review

The review of literature on the replacement of fine aggregate with different materials and the comparison of normal and accelerated curing is presented below.

Pooravshah and Bhavanashah [1] developed a mathematical model to predict early age strength for blended cement through accelerated curing. They proposed the mathematical model of 28 and 56 days compressive strength of cubes for OPC cement and blended cement individually which gives the confidence level of around 95%. This mathematical model was also helpful for precast manufacturers.

Gholap [2] studied the behavior of concrete made of blended cement with accelerated curing. They also studied the formation of the mathematical model. It is concluded that the compressive strength of concrete with accelerated curing is increased significantly and higher than the target strength of concrete.

Pawar et al. [3] compared the 28 days' compressive strength of concrete under accelerated curing and normal moist curing. It is noticed that the warm water method gives quite comparable results and is hence acceptable for obtaining early strength.

Patel et al. [4] studied the effect of traditional and accelerated curing methods on the compressive strength of concrete prepared with industrial waste. The compressive strength of concrete subjected to the accelerated curing method was found to be higher than membrane curing and saturated wet covering.

Das and Gattu [5] conducted studies to understand the performance of quarry dust as fine aggregate in concrete. The results showed that with the increased proportion of quarry dust, maximum strength was observed at 40% proportion, followed by a subsequent drop in strength and decreased workability.

Jyothi and Rao [6] investigated the compressive strength of high-strength concrete made with fly ash by accelerated curing. A slight increase in compressive strength was observed from the accelerated curing test.

## 3 Experimental Procedure

In this study, five sets of ($M_{20}$, $M_{25}$, $M_{30}$, $M_{35}$, and $M_{40}$) specimens are cast and tested. Each set comprises 3 cubes for determining the compressive strength of concrete at 28 days of ordinary curing and accelerated curing, respectively.

## 4 Materials Used

PPC (Portland Pozzolana cement), fine aggregates (Zone II stone dust), coarse aggregates, and water are used in this study.

The properties of cement, quarry dust, and coarse aggregate obtained from the tests are shown in Tables 1, 2 and 3. The quantity of materials required by weight

Table 1 Cement properties

| S. No. | Parameter | Result | Limits as per IS 1489 (Part 1): 1991 |
|---|---|---|---|
| 1 | Consistency | 31% | 31% |
| 2 | (a) IST | 35 min | Not less than 30 min |
|   | (b) FST | 7 h | Not more than 600 min |
| 3 | Fineness | 3.1% | Not more than 10% |
| 4 | Specific gravity | 2.68 | Not more than 3 |

Table 2 Quarry dust properties

| S. No. | Parameter | Result | Limits as per IS:383-1970 |
|---|---|---|---|
| 1 | Specific gravity | 2.79 | Should be between 2.4 and 2.8 |
| 2 | $p^H$ | 8 | Should be between 6 and 8 |
| 3 | Fineness modulus | 2.44 | Should be between 2.2 and 3.2 |
| 4 | Zone | II | – |

Table 3 Coarse aggregate properties

| S. No. | Parameter | Result | Limits as per IS:2386(PART IV)-1963 |
|---|---|---|---|
| 1 | Aggregate abrasion value | 25% | Not more than 30% |
| 2 | Specific gravity | 2.85 | Should be between 2.5 and 2.9 |
| 3 | Fineness modulus | 6.70 | Should be between 6 and 10 |
| 4 | Aggregate crushing value | 21.25% | Not more than 30% |
| 5 | Aggregate impact value | 21.58% | Not more than 30% |

**Table 4** Materials required by weight for concrete of grade M 20

| Cement | Fine aggregate | Coarse aggregate | W/C ratio |
|---|---|---|---|
| 357.7 | 710.89 | 1188.98 | 186 |
| 1 | 1.99 | 3.32 | 0.52 |

**Table 5** Materials required by weight for concrete of grade M 25

| Cement | Fine aggregate | Coarse aggregate | W/C ratio |
|---|---|---|---|
| 395.74 | 695.85 | 1163.81 | 186 |
| 1 | 1.76 | 2.94 | 0.47 |

**Table 6** Materials required by weight for concrete of grade M 30

| Cement | Fine aggregate | Coarse aggregate | W/C ratio |
|---|---|---|---|
| 432.55 | 681.28 | 1139.46 | 186 |
| 1 | 1.58 | 2.63 | 0.43 |

**Table 7** Materials required by weight for concrete of grade M 35

| Cement | Fine aggregate | Coarse aggregate | W/C ratio |
|---|---|---|---|
| 442.86 | 677.21 | 1132.64 | 186 |
| 1 | 1.53 | 2.56 | 0.42 |

**Table 8** Materials required by weight for concrete of grade M 40

| Cement | Fine aggregate | Coarse aggregate | W/C ratio |
|---|---|---|---|
| 450 | 683.07 | 1146.59 | 180 |
| 1 | 1.517 | 2.55 | 0.4 |

for various grades of concrete ($M_{20}$, $M_{25}$, $M_{30}$, $M_{35}$, and $M_{40}$) is shown in Tables 4, 5, 6, 7 and 8.

## 5 Preparation of Testing Specimens

- **Mixing**

  Mixing of ingredients is done by the method of hand mixing.
- Casting of specimens;
- Curing of the specimens (normal curing and accelerated curing).

# Development of a Mathematical Relationship Between ...

## 6 Test Results

### Compressive Strength values PPC $M_{20}$ at 28 days of Normal curing and Accelerated curing

Table 9 presents compressive strength values and their mathematical relationship to $M_{20}$ concrete mix at 28 days of normal curing and accelerated curing.

### Compressive Strength values PPC $M_{25}$ at 28 days of Normal curing and Accelerated curing

Table 10 presents compressive strength values and their mathematical relationship to $M_{25}$ concrete mix at 28 days of normal curing and accelerated curing.

**Table 9** Compressive strength values and mathematical relationship for normal and accelerated curing of $M_{20}$ concrete

| S. No. | Normal curing compressive strength (MPa) | Accelerated curing compressive strength (MPa) |
|---|---|---|
| 1 | 31.95 | 41.69 |
| 2 | 31.51 | 44.89 |
| 3 | 30.71 | 41.18 |

**Table 10** Compressive strength values and mathematical relationship for normal and accelerated curing of $M_{25}$ concrete

| S. No. | Normal curing compressive strength (MPa) | Accelerated curing compressive strength (MPa) |
|---|---|---|
| 1 | 36.62 | 48.47 |
| 2 | 36.13 | 44.02 |
| 3 | 36.71 | 46.35 |

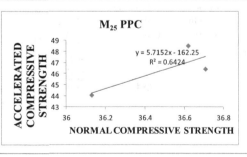

## Compressive Strength values PPC $M_{30}$ at 28 days of Normal curing and Accelerated curing

Table 11 presents compressive strength values and their mathematical relationship to $M_{30}$ concrete mix at 28 days of normal curing and accelerated curing.

## Compressive Strength values PPC $M_{35}$ at 28 days Normal curing and Accelerated curing

Table 12 presents compressive strength values and their mathematical relationship to $M_{35}$ concrete mix at 28 days of normal curing and accelerated curing.

**Table 11** Compressive strength values and mathematical relationship for normal and accelerated curing of $M_{30}$ concrete

| S. No. | Normal curing compressive strength (MPa) | Accelerated curing compressive strength (MPa) |
|---|---|---|
| 1 | 40.84 | 50.00 |
| 2 | 40.17 | 49.13 |
| 3 | 41.06 | 50.43 |

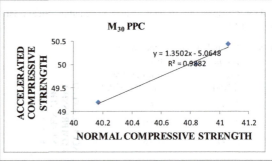

**Table 12** Compressive strength values and mathematical relationship for normal and accelerated curing of $M_{35}$ concrete

| S. No. | Normal curing compressive strength (MPa) | Accelerated curing compressive strength (MPa) |
|---|---|---|
| 1 | 42.66 | 52.04 |
| 2 | 42.35 | 53.57 |
| 3 | 44.57 | 52.84 |

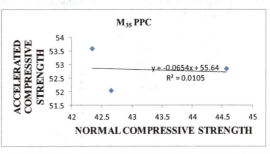

Development of a Mathematical Relationship Between ... 107

**Table 13** Compressive strength values and mathematical relationship for normal and accelerated curing of $M_{40}$ concrete

| S. No. | Normal curing compressive strength (MPa) | Accelerated curing compressive strength (MPa) |
|---|---|---|
| 1 | 46.26 | 56.48 |
| 2 | 45.46 | 58.31 |
| 3 | 45.02 | 55.75 |

$M_{40}$ PPC

$y = 0.2394x + 45.936$
$R^2 = 0.013$

### Compressive Strength values PPC $M_{40}$ at 28 days of Normal curing and Accelerated curing

Table 13 presents compressive strength values and their mathematical relationship to $M_{40}$ concrete mix at 28 days of normal curing and accelerated curing.

Table 14 gives regression equation for different grades of concrete.

**Table 14** Regression equation for normal curing and accelerated curing

| Grade of concrete | Regression equation |
|---|---|
| M20 (PPC) | $Y = 0.9246x + 13.564$ |
| M25 (PPC) | $Y = 5.7152x - 162.25$ |
| M30 (PPC) | $Y = 1.3502x - 5.0648$ |
| M35 (PPC) | $Y = 0.0654x + 55.64$ |
| M40 (PPC) | $Y = 0.2394x + 45.936$ |

## 7 Reason for Enhanced Compressive Strength Under Accelerated Curing

Accelerated curing uses heat or a combination of heat and moisture, in the early stages of the curing process to increase the rate of cement hydration. Heat causes cement hydration reactions to occur at an expedited rate, which causes concrete to develop strength at a faster rate. The increase in strength with increased curing temperature is due to the speeding up of chemical reactions of hydration. This increase affects only the early strengths without any harmful effects on the ultimate strength and cracking of the concrete due to thermal shock. Hence, the curing of concrete and its gain of strength can be speeded up by raising the temperature of curing in which the curing period is reduced.

## 8 Conclusions

1. Based on the study, it is concluded that accelerated curing values are more than normal curing values for all grades of concrete.
2. Based on the study, it is concluded that the average strength of accelerated curing specimens is 35.64% more than the average strength of normal curing specimens for M20 grade of concrete.
3. Based on the study, it is concluded that the average strength of accelerated curing specimens is 26.86% more than the average strength of normal curing specimens for M25 grade of concrete.
4. Based on the study, it is concluded that the average strength of accelerated curing specimens is 22.51% more than the average strength of normal curing specimens for M30 grade of concrete.
5. Based on the study, it is concluded that the average strength of accelerated curing specimens is 22.27% more than the average strength of normal curing specimens for M35 grade of concrete.
6. Based on the study, it is concluded that the average strength of accelerated curing specimens is 24.70% more than the average strength of normal curing specimens for M40 grade of concrete.
7. Based on the study, it is recommended that accelerated curing can be adopted as a time-consuming method than normal curing.
8. The replacement of fine aggregate with stone dust is more economical.

## References

1. Pooravshah, Bhavanashah (2011) Development of mathematical model to predict early age strength for blended cement through accelerated curing. In: National conference on recent trends in engineering & technology
2. Gholap MS(2014) Experimental study of concrete with blended cement with accelerated curing and formation of mathematical model. IOSR J Mech Civ Eng (ISOR-JMCE) 11(2):42–51. e-ISSN: 2278-1684, p-ISSN: 2320-334x
3. Pawar AJ, Nikam JS, Dhake PD (2015) Comparison of 28 days concrete compressive strength by accelerated curing and normal moist curing. Int J Res Eng Technol (IJRET) 04(13):219–222. CISSN: 2319-1169, PISSN:2321:7308
4. Patel K, Pitrada J, Raval AK (2017) Effect of traditional and accelerated curing method on compressive strength of concrete incorporating with industrial waste. IJCRT 5(4):2336–2339. ISSN: 2320-2882
5. Das B, Gattu M (2018) Study on performance of quarry dust as fine aggregate in concrete. In: International conference advances in construction materials and structures (ACMS-2018) IIT Roorkee, Uttarakhand, India

6. Jyothi RN, Rao K (2019) Effect of accelerated curing on compressive strength of high strength concrete with fly ash. Int J Recent Technol Eng (IJRTE) 7(6c2):193–198. ISSN:2277–3878
7. Shivaprasad KN, Das BB (2018) Effect of duration of heat curing on the artificially produced fly ash aggregates. IOP Conf Ser: Mater Sci Eng 431(9):092013

## *List of Codes*

1. IS:9013-1978 Indian Standard Method of making, curing and determining compressive strength of accelerated-cured concrete test specimens
2. IS:10262-2019 Guidelines for Concrete Mix Design
3. IS 383-1970 Indian Standard Specification Coarse and Fine Aggregates from Natural Sources for Concrete
4. IS: 516-1959 Indian Standard Methods of Tests for Strength of Concrete
5. IS 2386(PART IV)-1963 Indian Standard Methods for Aggregates for Concrete

# Using Recycled Aggregate from Demolished Concrete to Produce Lightweight Concrete

### Abd Alrahman Ghali, Bahaa Eddin Ghrewati, and Moteb Marei

**Abstract** The need for urbanization continues all over the world, greatly increasing the need for concrete. On the other hand, concrete debris from building demolitions causes a major environmental obstacle to storage, and it is a big challenge to use it appropriately. The research aims to study the possibility of using an aggregate of recycled concrete from demolished buildings to produce no-fines lightweight concrete. After studying the properties of recycled aggregate and comparing it with natural aggregate, it was found that the specific gravity of the recycled aggregate was lower, but its wear and water absorption were greater than natural aggregate. We conducted laboratory experiments on five different mixtures of no-fines lightweight concrete from natural aggregate using different sizes of aggregate. In addition to conducting laboratory experiments on three different mixtures of no-fines lightweight concrete from recycled aggregate, using different sizes of aggregate, we found that the density of no-fines lightweight concrete which was produced by demolished concrete is 5% less than that of lightweight concrete produced by the natural aggregate, and the compression resistance is close to each other. The best sample of lightweight concrete free of soft materials (no-fines) from demolished concrete had a compression resistance of 130 kg/cm$^2$ and the density was 1704 kg/m$^3$ where W/C = 0.37 and aggregate size was between 12.5 and 19 mm.

**Keywords** Lightweight concrete · Lightweight aggregate · Recycled aggregate

---

A. A. Ghali (✉) · B. E. Ghrewati
Department of Civil Engineering, Siksha 'O' Anusandhan University, Bhubaneswar, India
e-mail: abboudghali@gmail.com

M. Marei
Department of Civil Engineering, Aleppo University, Aleppo, Syria

© The Author(s), under exclusive license to Springer Nature Singapore Pte Ltd. 2023
M. S. Ranadive et al. (eds.), *Recent Trends in Construction Technology and Management*, Lecture Notes in Civil Engineering 260,
https://doi.org/10.1007/978-981-19-2145-2_9

## 1 Introduction

Cities expand and urban development leads to facing debris ridding matter, every building that is designed has a certain period of service years. After that, we have to demolish and rebuild again, leading to the accumulation of large quantities of concrete debris and facing a new challenge to the community and environment. Therefore, scientists think about a specific methodology for disposing of this waste and converting it to valid materials, using it again in a different way to save the environment [1] (Fig. 1).

This process involves collecting demolition and construction waste, treating it and reusing it again. And it is returned to the life cycle and be a usable material for the same or other purposes, instead of being the cause of many problems on various levels. Lightweight concrete is known to have a density of less than 1800 kg/m$^3$ compared to normal concrete It has a density of 2240–2480 kg/m$^3$, and it has gathered the economic and practical benefits available in lightweight concrete in recent years, having an important place in the construction of facilities and the demand for manufacturing them continues to increase [2].

**Fig. 1** Demolished concrete structures

## 2 Research Objective

The research was conducted to study the possibility of using the aggregate of demolished concrete structures in the production of lightweight concrete with a density of not more than 1800 kg/m$^3$, by making a comparison between natural and recycled aggregate and determining the appropriate method for producing lightweight concrete, and the study of appropriate diameters for recycled aggregate and using additives to improve lightweight concrete specifications.

## 3 Literature Review

Cracking the concrete to produce coarse aggregate in order to produce new concrete is a universally popular feature for more environmentally friendly concrete, which in turn reduces the consumption of natural resources, as well as the requirements for waste dump. In order to study this possibility, it was necessary to get acquainted with some international experiences in this field and to identify the properties of materials resulting from the crushing of demolished structures.

It is important to recognize the difference between recycled aggregate and natural aggregate; in this way, the differences may be taken into account when using recycled aggregate in concrete.

For example, Tamadur [3] made a comparison between the natural aggregate and recycled aggregate from the automatic grinding of concrete cubes, and it was found that the largest nominal size for natural and recycled aggregates is 44 mm and it reached the following results (Table 1).

Another research for Athanas [4] used two types of coarse aggregate, namely natural aggregate (solid granite fracture) and recycled aggregate (crushing debris).

Table 2 shows the characteristics of the two types.

**Table 1** Comparison between natural and recycled aggregate according to Tamadur's study

| Specifications | Natural aggregate | Recycled aggregate |
|---|---|---|
| Saturated surface dry | 2.67 | 2.51 |
| Absorption | 1.17% | 5.15% |
| Bulk density, kg/m$^3$ | 1335 | 1270 |
| Wear rate | 19% | 32.4% |

**Table 2** Comparison between natural and recycled aggregates according to Athanas' study

| Aggregate type | Size (mm) | Density (g/cm$^3$) | Water absorption (%) | Los Angeles factor % |
|---|---|---|---|---|
| Natural | 5–15 | 2.7 | 1.46 | 16.2 |
| Natural | 15–25 | 2.7 | 1.46 | 16 |
| Recycled | 8–25 | 2.61 | 6.39 | 26 |

Table 3 Comparison between natural and recycled aggregates according to Akbari's study

| Specifications | Natural aggregate | Recycled aggregate |
|---|---|---|
| Bulk density (kg/m$^3$) | 1470 | 1451 |
| Specific gravity | 2.87 | 2.68 |
| Absorption | 1.7% | 7% |

Also, Akbari et al. [5] used two types of coarse aggregate, namely natural aggregate (solid granite fracture with a maximum grout of 20 mm) and recycled aggregate (breaking the cubes at the materials lab, where the maximum size was 20 mm and the minimum size was 4.75 mm.) Table 3 shows the characteristics of the two types.

## 4 Lightweight Concrete

Lightweight concrete has an expansion factor that increases the size of the mixture, leading to decrease in the dead load. Using lightweight concrete is increasing widely in countries such as the USA, the United Kingdom and Sweden. Lightweight concrete maintains large blanks and has a density between 300 and 1840 kg/m$^3$, which is 8–23% lighter than traditional concrete [6–11].

Lightweight concrete can be classified according to the obtaining method into three types:

- Concrete free of soft materials,
- Concrete of lightweight aggregate, and
- Foam concrete (Fig. 2).

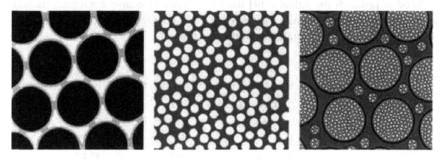

Fig. 2 Kinds of lightweight concrete

## 5 Experimental Study

Lightweight concrete free of soft materials was produced using Portland cement, water and aggregate. Two types of aggregate were used in this paper; natural aggregate and aggregate from destroyed facilities.

The natural aggregate was the crushed type of dolomitic origin (Fig. 3; Table 4). The recycled aggregate which has used was of two different types.

- **The first type**: the aggregate produced by crushing previous samples in the construction laboratory (Table 5).
- **The second type**: the aggregate produced by demolished concrete structures, and sorted into three types—soft, medium, and coarse.

Particle-size distribution was performed for the three types as shown in Fig. 4 (Table 6; Figs. 5 and 6).

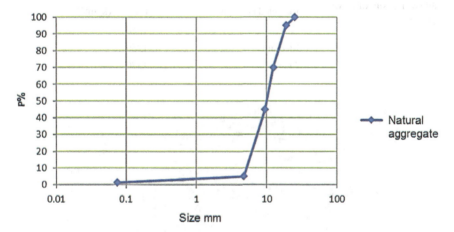

**Fig. 3** Particle-size distribution for natural aggregate

**Table 4** Physical and mechanical properties of natural aggregate

| Specific gravity | Density (kg/m$^3$) | Water absorption | Wear ratio |
|---|---|---|---|
| 2.825 | 1410 | 1.3% | 18.7% |

**Table 5** Properties of recycled aggregate from crushed samples

| Specific gravity | Density (kg/m$^3$) | Water absorption | Wear ratio |
|---|---|---|---|
| 2.6 | 1290 | 7.2% | 37% |

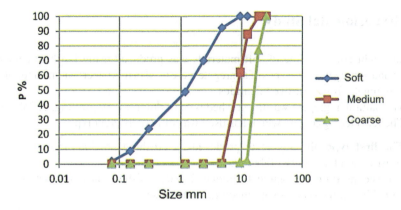

**Fig. 4** Particle-size distribution for recycled aggregate

**Table 6** Density of recycled aggregate

| Recycled aggregate | Density (kg/m$^3$) |
|---|---|
| Soft | 1340 |
| Medium | 1280 |
| Coarse | 1370 |

**Fig. 5** Mixing machine

## 5.1 Experiments of Lightweight Concrete Produced by Natural Aggregate

- **First mixture**: the aggregate size was between 8 and 25 mm and the cement to aggregate ratio was 1/6. W/C = 0.45 and cement calibre was 340 kg/m$^3$.

**Fig. 6** Breaking samples of the first mixture

Compression resistance was up to 147.6 kg/cm$^2$ after 28 days and the density was 1963 kg/m$^3$, and this density was deemed high for lightweight concrete free of soft materials.

- **Second mixture**: the aggregate size was between 12.5 and 25 mm and the same proportions of aggregates, cement and water as in the first mixture was used.

  Compression resistance was up to 108.5 kg/cm$^2$ after 28 days and the density was 1789 kg/m$^3$, and this density is acceptable for lightweight concrete free of soft materials.

- **Third mixture**: the aggregate size was between 12.5 and 19 mm and the same proportions of aggregates, cement and water as in the first mixture was used.

  Compression resistance was up to 108.6 kg/cm$^2$ after 28 days and the density was 1771 kg/m$^3$, and this density is acceptable for lightweight concrete free of soft materials.

- **Fourth mixture**: the aggregate size was between 12.5 and 19 mm and the same proportions of aggregates and cement as in the first mixture was used. But we added a high active plasticizer of 1.5% to the cement weight and added 35% of water to the cement weight with the same operability W/C = 0.35.

  Compression resistance was up to 188.2 kg/cm$^2$ after 28 days and the density was 1901 kg/m$^3$, and this density is acceptable for lightweight concrete free of soft materials.

**Table 7** Experimental results of lightweight concrete produced by natural aggregate

| Mixture | Size (mm) | W/C | Density (kg/m$^3$) | Compression resistance (kg/cm$^2$) |
|---------|-----------|------|--------------------|-----------------------------------|
| First   | 8–25      | 0.45 | 1963               | 147.6                             |
| Second  | 12.5–25   | 0.45 | 1789               | 108.5                             |
| Third   | 12.5–19   | 0.45 | 1771               | 108.6                             |
| Fourth  | 12.5–19   | 0.35 | 1901               | 188.2                             |
| Fifth   | 19–25     | 0.35 | 1833               | 103.2                             |

- **Fifth mixture**: the aggregate size was between 19 and 25 mm and the same proportions of plasticizer, aggregates, cement and water as in the fourth mixture was used.

  Compression resistance was up to 103.2 kg/cm$^2$ after 28 days and the density was 1833 kg/m$^3$. Both density and compression resistance decreased compared to the fourth mixture because of the big size of aggregate (Table 7).

  The results show that increasing the size range of aggregate leads to an increase in both compression resistance and density.

## 5.2 Experiments of Lightweight Concrete Produced by Recycled Aggregate

The recycled aggregate which has used was of two different types.

### 5.2.1 The First Type

The aggregate is produced by crushing previous samples in the construction laboratory. The aggregate size was between 9.5 and 25 mm and the cement to aggregate ratio was 1/6 and W/C = 0.47.

Compression resistance was up to 98.5 kg/cm$^2$ after 28 days and the density was 1778 kg/m$^3$ (Fig. 7).

### 5.2.2 The Second Type

The aggregate is produced by demolished concrete structures.

- **First mixture**: the aggregate size was between (12.5 and 19 mm) and the same proportions of aggregates, cement and water as in the first-type experiment was used.

**Fig. 7** Broken samples in construction labs

Compression resistance was up to 95 kg/cm$^2$ after 28 days and the density was 1686 kg/m$^3$. The density decrease was caused the decrease of recycled aggregate density.

- **Second mixture**: the aggregate size was between 9.5 and 12.5 mm) and the same proportions of aggregates, cement and water as in the first mixture was used.

  Compression resistance was up to 96 kg/cm$^2$ after 28 days and the density was 1699 kg/m$^3$.

- **Third mixture**: the aggregate size was between 12.5 and 19 mm and the same proportions of aggregates and cement as in the first mixture was used. But we added a high active plasticizer of 1.5% to the cement weight and added 37% of water to the cement weight W/C = 0.37.

Compression resistance was up to 130 kg/cm$^2$ after 28 days and the density was 1704 kg/m$^3$, and this density is acceptable for lightweight concrete free of soft materials (Fig. 8).

For recycled aggregate without adding plasticizer, the best density was 1686 kg/m$^3$ and the compression resistance was 95 kg/cm$^2$, where the size was between 12.5 and 19 mm, W/C = 0.47 and the cement to aggregate ratio was 1/6. However, after adding the plasticizer with W/C = 0.37, the density was 1704 kg/m$^3$ and the compression resistance was 130 kg/cm$^2$ (third mixture.) The plasticizer led to an increase in both density and compression resistance (Table 8; Fig. 9).

# 6 Conclusions

- The aggregate produced by demolished concrete structures has a lower density but has a greater rate of wear and absorption than natural aggregate.
- The compression resistance for lightweight concrete free of soft materials from natural aggregate was 108.6 kg/cm$^2$ and the density was 1771 kg/m$^3$, where W/C = 0.45, the size was between 12.5 and 19 mm, the cement to aggregate ratio was 1/6 and cement calibre was 340 kg/m$^3$. After adding the plasticizer, the compression

 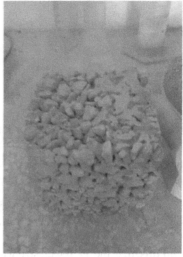

**Fig. 8** Breaking samples of the third mixture

**Table 8** Experimental results of lightweight concrete produced by recycled aggregate

| Aggregate type | Mixture | Size (mm) | W/C | Density (kg/m$^3$) | Compression resistance (kg/cm$^2$) |
|---|---|---|---|---|---|
| Laboratory samples | | 9.5–25 | 0.47 | 1778 | 98.5 |
| Demolished concrete | First | 12.5–19 | 0.47 | 1686 | 95 |
| | Second | 9.5–12.5 | 0.47 | 1699 | 96 |
| | Third | 12.5–19 | 0.37 | 1704 | 130 |

resistance increased to 188.2 kg/cm$^2$ and the density increased to 1901 kg/m$^3$, where W/C = 0.35 and the size between was 12.5 and 19 mm.
- The compression resistance for lightweight concrete free of soft materials from demolished concrete was 95 kg/cm$^2$ and the density was 1686 kg/m$^3$, where W/C = 0.47, the size was between 12.5 and 19 mm, the cement to aggregate ratio was 1/6 and cement calibre was 340 kg/m$^3$. After adding the plasticizer, the compression resistance increased to 130 kg/cm$^2$ and the density increased 1704 kg/m$^3$ where W/C = 0.37 and the size was between 12.5 and 19 mm. It was the best sample in this research because it had low density and good compression resistance.
- The compression resistance for lightweight concrete free of soft materials from demolished concrete was close to which is from the natural aggregate.
- The density for lightweight concrete free of soft materials from demolished concrete was less of 5% than which is from natural aggregate, because the recycled aggregate has less specific gravity.
- The plasticizer increased the compression resistance effectively.

Using Recycled Aggregate from Demolished Concrete ...

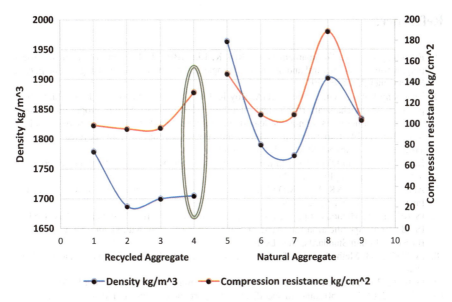

Fig. 9 Comparing results between recycled and natural aggregate

- We can use the lightweight concrete produced by the recycled aggregate in parking and pavement, and also we can use it in wooded pavement and drainage layers because it has good permeability (Fig. 10).

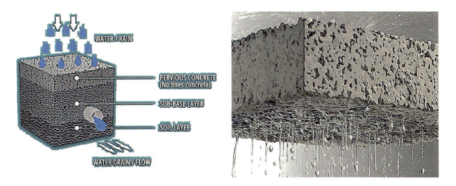

Fig. 10 Uses of lightweight concrete in pavement and drainage layers

# References

1. Rajamony Laila L, Gurupatham BGA, Roy K, Lim JBP (2021) Effect of super absorbent polymer on microstructural and mechanical properties of concrete blends using granite pulver. Struct Concr 22:E898–E915
2. Rajamony Laila L, Gurupatham BGA, Roy K, Lim JB (2021) Influence of super absorbent polymer on mechanical, rheological, durability, and microstructural properties of self-compacting concrete using non-biodegradable granite pulver. Struct Concr 22:E1093–E1116
3. Muqbil T (2014) Experimental study on the concrete produced from recycled aggregate. Tishreen Univ J Res Sci Stud 36(4):2 (in Arabic)
4. Konin A, Kouadio DM (2011) Influence of cement content on recycled aggregates concrete properties. Mod Appl Sci 5(1):23
5. Akbari YV, Arora NK, Vakil MD (2011) Effect on recycled aggregate on concrete properties. Int J Earth Sci Eng 4(6):924–928
6. IS 4031:1988. Methods of physical test for cement. Bureau of Indian Standards, New Delhi
7. IS 383:1970. Specification for coarse and fine aggregates from natural sourced for concrete. Bureau of Indian Standards, New Delhi
8. IS 2386:1963. Methods of test for aggregates for concrete. Bureau of Indian Standards, New Delhi
9. IS 10262:2009. Guidelines for concrete mix design. Bureau of Indian Standards, New Delhi
10. ASTM, C. (2017). 150 standard specification for portland cement, ASTM International
11. Snehal K, Das BB (2019) Mechanical and permeability properties of hybrid fibre reinforced porous concrete. Indian Concr J 93(1):54–59

# Recent Trends in Construction Materials

# A Study in Design, Analysis and Prediction of Behaviour of a Footbridge Manufactured Using Laminate Composites—Static Load Testing and Analysis of a Glass Fibre Laminate Composite Truss Footbridge

**Col Amit R. Goray and C. H. Vinaykumar**

**Abstract** Structural steel has been used extensively in civil engineering since time immemorial. However, despite being a material of choice, it suffers from being susceptible to the vagaries of nature in the form of corrosion. The high unit weight of steel has resulted in researchers trying and succeeding in developing materials that are stronger and lighter than steel. The present attempt is to study and utilise a very well-known material used in aerospace for civil engineering purposes. GFRP is known for its high specific strength and specific modulus, and these properties are studied and exploited in the present case to develop a footbridge. A bridge span of 4 m is manufactured using Glass Fibre-Reinforced Plastics. This footbridge has been designed to carry pedestrian traffic while at the same time being lightweight and modular. The bridge being lightweight and robust can be used in mountainous regions and jungle trails. The modular nature enables the bridge to be easily dismantled and each module is carried out by one man. The bridge was tested for static loading condition and the results are validated using FEM software. The close match between the results of numerical simulations matches the values obtained through in situ experimental testing. This study has proven that it is possible to design structural members for civil engineering by deriving the strength of a GFRP laminate using Equivalent Single-Layer (ESL) method. This study opens up the use of composite laminates for other related applications in civil engineering besides bridging, thus making it an exciting material for civil engineers in the future.

**Keywords** Footbridge · GFRP · Laminate · Deflection

---

C. A. R. Goray (✉) · C. H. Vinaykumar
Department of Civil Engineering, CME Research Center (Savitribai Phule Pune University), Dapodi, Pune 411031, India
e-mail: amitgoray@gmail.com

# 1 Introduction

Glass Fibre-Reinforced Plastics (GFRP) has been very commonly used in the aerospace industry since the early 1970s by NASA. They are lightweight and strong with a high specific modulus and strength. The manufacture of advanced composites using glass fibres involves embedding the fibres in a resin matrix. The glass fibre provides the strength while the epoxy resin provides a limited amount of flexibility as desired. Due to the method of manufacture of the composite and also the fact that glass 'fibres' display a property to resist only unidirectional tensile force, the method of construction/manufacture of a composite laminate attains great importance. Each application utilising an advanced composite material will require a unique process of manufacture, construction and layout of the lamina. This definitely leads to a comparatively higher cost for fabrication and also the fact that advanced composites are non-isotropic materials, becomes a major drawback. The advantage of advanced composites, however, lies in the fact that they provide a high value of specific modulus and also specific strength.

$$\text{Specific modulus} = E/\rho$$
$$\text{Specific Strength} = \sigma_{\text{ult}}/\rho$$

where
$E$—Modulus of elasticity,
$\rho$—Density of material,
$\sigma_{\text{ult}}$—Ultimate Tensile strength.

In the present instance, a need was felt to exploit the properties of advanced composites to design, construct and study the behaviour and utility of bridges using GFRP composites. These could be utilised for providing quick connectivity during emergencies in rural, mountainous regions and also for military use.

The reports and studies related to human-induced vibrations, namely JRC report on EUR 23984 EN [1], HIVOSS Studies [2], Sétra Norms [3] and also various papers have been referred to in order to analyse this new material. Pedestrian loading has been considered to be a periodic function in the form of a Fourier series. The static loading has been taken to reach a maximum of 4.071 kN/m$^2$ of the loading area.

# 2 Description of the Bridge

This footbridge is a simply supported truss structure. It is modular and can be dismantled into parts for easy carriage. Each bay can be broken up into three components to enable ease of carriage. A view of the bridge laid out prior to testing is shown in Fig. 1.

**Fig. 1** 04m bridge laid out for testing at CoE, Pune. The location of the deflection sensors can also be seen

The GFRP bridge has been designed and developed by utilising Glass Fibre-Reinforced Plastic pultruded sections and also purposely designed laminates. The aim of the design was to produce a bridge with the abilities as follows:

- Light in weight: Ease of handling and carriage of a single by two individuals only.
- Modular Design: Each bay should have be capable of being used independently while also being able to be stacked, unloaded and joined with other bays. The basic unit was hence designed as a single bay acting as a module.
- Ease of transport: The bay dimensions should enable ease of loading on the roof of an SUV and also interlocking.
- Simple to assemble at site.
- Ease of assembly: Minimal number of parts for assembling and launching the bridge.
- Easy to launch and de-launch.
- Easily transportable to desired places using helicopters.
- Can be reused after de-launch.
- No repair is needed during the lifecycle of bridge.
- If needed, parts and components can be replaced in situ.

    Excellent dynamic response properties.

## 3 Materials

In the manufacture of this bridge, GFRP was used in two forms, pultruded section of the GFRP was used to form the truss and a purposely designed laminate for the joints.

The footbridge has been constructed using pultruded sections and joints manufactured through a unique process. The bridge was tested at the Government College of Engineering, Pune. The prototype bridge was subjected to a series of load tests and the static loading was as per norms listed out by the RDSO, Lucknow [4].

Each pipe is constructed through pultrusion. The density of the pultrusion is **21.57 kN/m³**. The properties of the pultruded pipe are shown in Table 1.

The joints are constructed using various layers of fabric constructed using E-glass with varying orientations (+45/−45, +90/−90) with a thickness of 5 mm. The resin used is flame-retardant vinyl ester. Further treatment has been carried out to cater for damage due to UV radiation. The mechanical property of the joints was calculated by the method given in RDSO [4] and presented in Table 1.

The joint has been constructed using E-glass fabrics. E-glass fabric of Unidirectional (U), Biaxial (B) and Triaxial (T) orientation was used. CSM was also incorporated within the joint. To prevent damage due to pedestrian movement and environmental effects (UV radiation), gelcoat, PU coat were given. An anti-skid layer was provided to prevent any accidents. Quality control of the components was ensured and joints with voids were rejected. The inspection was carried out visually and also by measuring and comparing the weight of each component manufactured.

The composite was analysed and an ABD matrix was derived. The properties of this laminate were calculated and substituted into a software model for the individual components of the truss bridge.

The results of FEA analysis using this software model were thereafter compared with actual testing for the various components in order to confirm the properties and behaviour of each individual component of the Truss bridge.

**Table 1** Properties of GFRP pipes and laminate joints

| Title | Properties of GFRP | |
|---|---|---|
| | Pultruded pipe | Laminate joints |
| E1 (MPa) | 28,044.50 | 14,110.40 |
| E2 (MPa) | 13,611.20 | 10,923.80 |
| G12 (MPa) | 8524.42 | 5860.00 |
| G23 (MPa) | 3520.81 | 2466.13 |
| $\mu 12$ | 0.48 | 0.42 |
| Density (kN/m³) | 21.57 | 20.10 |

## 4 Load Tests

Load testing and deflection for pultruded pipe were carried out with a 3-point loading in laboratory and on the software FEA model. A photograph of the experimental 3-point setup is in Fig. 2. A comparative table of load versus deflections is given in Table 2. The comparative results between the experimental values and those obtained by FEA modelling are shown in Fig. 3.

As observed, the software model for loading matches the experimental results for the deflection of the pultruded pipe. These results were used to verify the software model. Thereafter, an experimental test for a fully rigged footbridge was carried out. This was for a 4 m span bridge loaded as per the norms of the Indian railways. The

**Fig. 2** 3-point load testing of pultruded pipe

**Table 2** Comparison of loading on pultruded pipe experimental versus FEA

| S. No. | Load (N) | Deflection (mm) | | Remarks |
|---|---|---|---|---|
| | | Actual | FEA | |
| 1 | 0 | 0 | 0 | |
| 2 | 135 | 6.875 | 10.042 | |
| 3 | 270 | 13.125 | 13.549 | |
| 4 | 405 | 20.000 | 23.592 | |
| 5 | 540 | 26.563 | 27.099 | |
| 6 | 675 | 40.625 | 40.648 | |
| 7 | 810 | 47.375 | 50.690 | |
| 8 | 945 | 54.313 | 54.198 | |
| 9 | 1080 | 61.563 | 64.240 | |
| 10 | 1215 | 68.750 | 67.747 | |
| 11 | 1349.46 | 76.250 | 77.789 | |

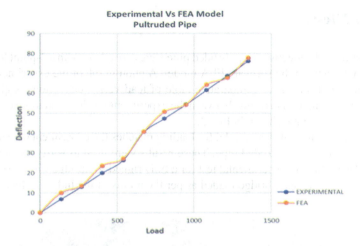

**Fig. 3** Graph experimental versus FEA results (pultruded pipe)

**Table 3** Comparative deflection values of the 04 m footbridge experimental versus FEA model

| S. No. | Result | Deflection (mm) | | | Remarks |
|---|---|---|---|---|---|
| | | Location 1 | Location 2 | Location 3 | |
| 1 | Experimental | 2.25 | 1.48 | 2.3 | 6.92 N load |
| 2 | FEA | 2.012 | 1.854 | 2.103 | 6.92 N load |

value of deflection after 24 h was also obtained to check for any residual deflection if any.

The bridge was tested for static loads [5] as well as pedestrian loads [6]. The deflection values obtained by experiment for the static loading and using the FEA model are mentioned in Table 3.

The deflection values are noted for the static loading. The loading on the bridge was also carried out by people jumping on the bridge simultaneously to simulate dynamic loading. However, the present paper only looks at static loading. The details of deflection due to static loading on the 4 m bridge are shown in Table 3 and Fig. 3.

## 5 Software Simulation and Analysis

While a number of methods exist to model the behaviour of composite laminates, only the equivalent single layer was used to calculate the properties of the material. Additionally, certain factors of safety were also incorporated to further refine the results of the analysis. This has been done in order to make the process of analysis and design simple and easier to understand for practising engineers. For constructing any civil structure, the next stage would be to set up ideal factors of safety.

This is because the aim is to establish norms for establishing thumb rules based on actual testing and failure.

The computer model of the bridge has been created using AutoCAD, thereafter meshing is carried out using Netgen. The model was then imported into Mecway FEA. The FEA analysis of the geometry of the bridge was analysed considering the components of the bridge as a Shell lamina. The properties of the various layers were input into the FEA model for the joints and the pipes based on validated theoretical values had been tested and the results were validated.

The mesh used for the analysis was a 2D mesh of an approximate size of edge as 25 mm, 182,808 nodes and 150,035 elements.

The supports were assigned by setting boundary conditions on elements at the bottom of the bridge where actual support was placed as simply supported.

After the mesh and the model were imported into the Mecway software, properties and boundary conditions were assigned, and the results of the static deflection as obtained in the FEA analysis are shown in Fig. 4 and closely matched the actual values obtained through the experiment. A comparison of the values of deflection for static loading obtained experimentally versus FEA is shown in Table 3.

There is a very close co-relation between the experimental and observed results. Thus, the same model can now be extrapolated to larger spans and exploited to arrive at a solution.

## 6 Results and Discussions

The aim of the study was to create an accurate FEA model for the static and dynamic testing of a bridge manufactured using GFRP. The model has been found to satisfactorily predict the behaviour of the GFRP footbridge in both static and dynamic loadings. The results of the static loading on the footbridge are being shared and have also formed the basis for accurately modelling the dynamic response and carrying out a modal analysis of the footbridge.

The in situ results obtained have been plotted against the results obtained from the FEM modelling of the bridge. Vertical displacement was measured and averaged out over the two sides of the bridge. The results of numerical simulation and actual test are close and indicate that the bridge was correctly modelled in the FEA software.

The FEA model is also accurate enough after incorporating the various factors of safety to predict the behaviour for loading over a 24-h cycle.

Deflection as predicted by the model varies within limits of 10–25% over the full span. This needs further refinement. But it can be seen to enable derivation of a **'Safety Factor'** for a similar design in the future. The results are encouraging and have been found to fairly predict the behaviour of a footbridge constructed using composites. The same can be considered while preparing a design table at a later date. However, considering the overall values of deflection as ranging between 2 and 4 mm, they are well within the permissible limits of bridge codes.

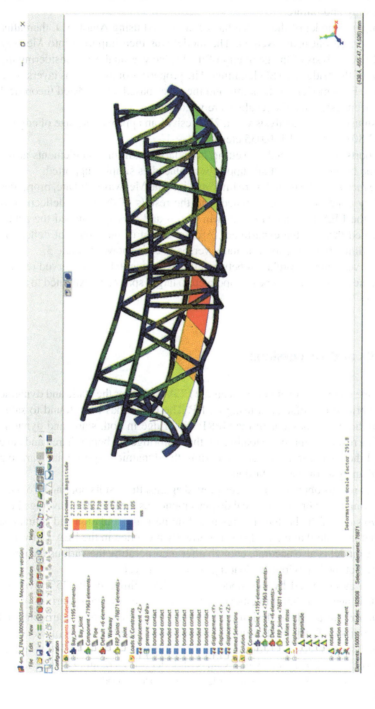

**Fig. 4** Results of FEA analysis of the 04 m footbridge

## 7 Conclusions

The study shows that Equivalent Single-layer (ESL) modelling can be used safely in estimating the strength and properties of the GFRP material. The findings of the study can be summarised as follows:

- The ESL model has an in-built factor of safety and values of deflection obtained theoretically are slightly larger than those obtained in the experimental study. This implies that the ESL model can be used along with derived factors of safety to enable the design of structures utilising GFRP.
- The ESL model provided a very accurate prediction of the deflection between experimental and theoretical values for the bridge. The variation in measured deflection was between 10 and 25%.
- The deflection pattern due to the presence of stiffer joints was well reflected in the FEA model. The software model has mimicked the actual behaviour observed during in situ testing.
- Long-term effects of creep need to be further studied in greater detail.
- The success in being able to predict the material behaviour of GFRP indicates the possibility of developing design tables for the footbridge.

## References

1. Design of lightweight footbridges for human induced vibrations—JRC first edition, May 2009 (EUR 23984 EN)
2. Murray TM, Allen DE, Ungar EE (1997) Steel design guide series 11: floor vibrations due to human activity. American Institute of Steel Construction
3. Sétra, Footbridges, Assessment of vibrational behaviour of footbridges under pedestrian
4. Loading TG (2006) Service d'Etudes Techniques des Routes et Autoroutes. France, Paris
5. Mechanics of composite materials: Autar K. Kaw. ISBN-13: 978-0849313431/ISBN-10: 0849313430
6. Rules specifying the loads for design of super-structure and sub-structure of bridges and for assessment of the strength of existing bridges issued by the Research Designs and Standards Organisation, Lucknow
7. IRC:6, para 209 for Pedestrian loads
8. Indian Standard Code for Steel. IS: 800-2000

# State-of-the-Art of Grouting in Semi-flexible Pavement: Materials and Design

**Hemanth Kumar Doma and A. U. Ravi Shankar**

**Abstract** Semi-flexible Pavement (SFP) is a composite pavement that consists of an open-graded friction course (OGFC) or porous asphalt mixture (PAM) having an air void content of 20–35%, grouted with cement paste/mortar with a fluidity of 10–16 s. The OGFC or PAM provides flexibility, skid resistance, and the grouting provides rigidity, capacity to carry heavy traffic without rutting, together to achieve a joint-free, rut-resistant pavement. The interconnected voids in the asphalt mixture filled with grout will be the secondary skeleton and help in load transfer, being the stone-on-stone contact with the primary skeleton. The review of supplementary cementitious materials (SCM), the formulation of the materials to meet the grouting design requirements, and the parameters to measure the efficiency are necessary to provide a more durable and fatigue-resistant pavement. The first part of the study discusses the mechanical properties of the materials, the design and preparation of the grouting. The grouting parameters and the contribution of grouting to SFP's performance are discussed in the later part. The review indicated that the marginal aggregates can also be used, and with the use of SCM, durability, and strength can be increased. Concerning the benefits of grouting in SFP, scope exists for further research to design and understand the grouting better, which helps SFP perform better.

**Keywords** Grouting · Semi-Flexible pavement · Supplementary cementitious materials · Mechanical properties

## 1 Introduction

Flexible pavement provides good riding quality, high skidding resistance, easy maintenance, but with rapidly increasing traffic, the occurrence of rutting and cracking has become obvious [14]. The rigid pavement has high fatigue life but needs more time to allow the traffic and poor driving comfort due to the joints [4]. There is a

---

H. K. Doma (✉) · A. U. Ravi Shankar
Department of Civil Engineering, National Institute of Technology Karnataka, Mangaluru, India
e-mail: dhemanthrgukt@gmail.com

© The Author(s), under exclusive license to Springer Nature Singapore Pte Ltd. 2023
M. S. Ranadive et al. (eds.), *Recent Trends in Construction Technology and Management*, Lecture Notes in Civil Engineering 260,
https://doi.org/10.1007/978-981-19-2145-2_12

need for a new type of pavement, such as SFP, to overcome the drawbacks and to enhance the performance of both flexible and rigid pavement [13]. The SFP or Grouted Macadam is a pavement that consists of the OGFC or PAM having an air void content of 20–35%, grouted with cement paste/mortar [15]. The OGFC or PAM provides flexibility, skid resistance, and the grouting provides rigidity, the capacity to carry heavy traffic without rutting, together to achieve a joint free, rut-resistant pavement [13, 17]. The OGFA or PAM possesses a large porous structure, ensuring the stone-on-stone contact to bear the traffic load and to provide the voids for grout to easily permeate.

## 2 Grouting

### 2.1 Materials

The grouting of SFP should have good flowability to fill the voids and be required to provide good compressive and flexural strength to sustain the loads and to resist the drying shrinkage. In other words, it should have high workability, permeating ability, and strength. Researchers investigated the grouts with various types of cement with different proportions of SCM such as fly ash (FA), silica fume (SF), and various dosages of superplasticizers (SP), which are used to achieve this goal. The two types of grouting materials were used such as the cement paste (CP) and cement mortar (CM), mainly consisted of cement, mineral powder, coal ash, and water [20]. Flexible admixtures such as neutral ethylene–vinyl acetate (EVA), SBR latex, asphalt emulsions were used to improve the flexibility of the grout [6]. The penetrants such as napthalene-based superplasticizer (NS), air-entraining superplasticizer (AS), and polycarboxylene-based superplasticizer (PS) were used to improve the flowability of cement paste (CP). The high-performance cement paste (HPCP) was prepared with the expansion admixture of UEA (UEA), air-entraining agent of ZY-99 triterpenoid saponins (AEA), and polycarboxylate-based TH-928 superplasticizer (PSP) [16]. The chemical admixtures were necessary to produce the high-performance cement slurry [6].

The sand finer than the 600 $\mu$m IS sieve size should be used in the preparation of grout (IRC). The fine sand passing through a 45 $\mu$m IS sieve can also be used so that the grout fills the voids in the asphalt mixture easily. The sand passing through a 300 $\mu$m sieve was used to prepare grout [8]. The sand content decreases, in general, the flowability of grout. However, the flowability was affected most with mineral powder (MP), water/cement (W/C) ratio, and the content of fly ash (FA), in comparison with the sand content [20]. The grout flowability can be increased with the addition of fly ash (FA) and silica fumes (SF) [10]. The FA should conform to IS:3812 with minimum of 65% passing the 45 $\mu$m IS sieve. These SCMs produce additional hydration products through pozzolanic reaction, leading to better strength and lower shrinkage than ordinary Portland cement (OPC) [8]. The drying shrinkage and the fluidity were improved with the increase in FA content, however, the flexural and compressive strengths were decreased [20]. The SF reduces the segregation and bleeding of grouting material and the shrinkage can be reduced with the use of aluminite powder [19]. The MP content improved the fluidity, the drying shrinkage, and reduced the strength of CP. The MP eliminates the bleeding during the preparation of the grout [20]. To obtain the asphalt's flexibility and cement's high strength, the cement asphalt emulsion paste (CAEP) prepared with asphalt emulsion, cement, finer sand, and chemical admixtures was used widely and has the great potential to be grout material in SFP.

## 2.2 Mechanical Properties

### 2.2.1 Fluidity

To determine the fluidity of grout mixtures, the flow cone test confirming to ASTM C939 can be used. The efflux time required for the grout volume of 1725 ml is noted as the fluidity. The Leeds flow cone test: The cone has a funnel of 200 mm upper diameter with an orifice of 18 mm that narrows down to 15 mm diameter over a length of 100 mm. The time required for the slurry volume of 1725 ml to flow through the cone was used to assess the fluidity. The fluidity of 9–11 s was recommended to penetrate a porous asphalt specimen of a depth of 100 mm [8]. Koting et al. [11] used 11–16 s flow time for 1 lt of grout as measured by the Malaysian mortar flow cone test to evaluate the different SPs. Fang et al. [6] limited the fluidity of the initial grouts, while formulating, to 15 s maximum based on testing the slurry formulations with the maximum allowable amount of admixtures and high W/C, but, accepted the fluidity range of 9–13 s. To evaluate the effect of the dosages and type of penetrant on the fluidity, three penetrants, i.e., NS, PS, and AS, at 1, 3, 5, 7, and 10% by mass of cement to the cement slurry, at the constant 0.6 W/C ratio, were used [6]. Figure 1 depicts the fluidity of cement slurry with NS was relatively stable, though couldn't meet the requirement. However, the dosages of PS had an insignificant effect on

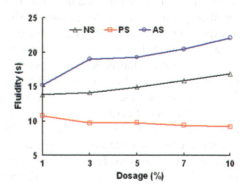

**Fig. 1** The effects of different dosages of penetrant on the fluidity of cement slurry [6]

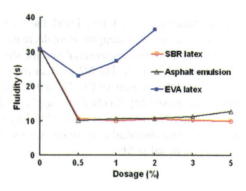

Fig. 2 The effects of flexible admixture on the fluidity of selected cement slurry [6]

the fluidity and the grout with AS had high fluidity. The dosage of 1% penetrant for both PS and NS was considered as optimal dosage and concluded that the PS had a significant effect on the fluidity than the AS and NS penetrants. Figure 2 depicted the effect of flexible admixtures at optimal W/C of 0.6 and PS of 1% and the fluidity was beyond the required value, hence, not recommended to prepare the grouts. The asphalt emulsion and SBR latex had met the fluidity requirement and had a similar effect on fluidity and concluded that the SBR latex could be used as a flexible admixture to prepare the fresh cement slurry.

Zarei et al. [19] added the CAEP in 20%, 40%, and 60% proportions of CAEP/C produced the grouts with good fluidity of 10–20 s while the cement paste exhibited the high fluidity due to the lower W/C ratio. Three additives such as PSP, UEA, and AEA were used and found that the PSP has a significant influence on fluidity followed by W/C ratio, AEA content, and UEA [16].

### 2.2.2 Compressive Strength

The prism samples, prepared using the rectangular steel molds with dimensions of 40 × 40 × 160 mm, were cast to determine the compressive strength. Hassan et al. [8] replaced the OPC with 5% of SF and achieved the compressive strength of 56 MPa on 1-day curing and increased to 115 MPa on 28 days. The FA/SF (used together) grout resulted in lower compressive strength of 28 MPa on 1-day curing, as expected because the FA has an adverse effect at early ages, but achieved higher strength upon 28 days of curing. The SBR affected the early strength development similar to FA but achieved lesser compressive strength than FA grout. The addition of either SBR or sand or both together reduced compressive strength significantly. Fang et al. [6] concluded that the compressive strength of hardened cement paste for 3, 7, and 28 days of curing, at the W/C of 0.6, decreased by 72%, 45%, and 8.7%, respectively, with the increase in dosage of SBR latex from 1%, 2%, and 3%. The flexural strength of hardened cement paste for 3 and 28 days of curing decreased by 4.5% and 24%, respectively, but for 7 days of curing, flexural strength was increased by 54%. The compressive strength of AE included grout at 7 and 28 days of curing were decreased in the range of 50–53% and 34–38%, respectively, in comparison with CP grout, with the increase in AE content [19]. There was no significant decrease observed in compressive strength with an increase in CAEP content, indicating that the dosage of CAEP had no significant influence on the strength. The increment rate of compressive strength from 7 to 28 days was about 38% for CAEP grout while the cement paste had very less increment. Pei et al. [16] concluded that the SP content and W/C ratio had the greatest influence on strength of the HPCP.

### 2.2.3 Flexural Strength

The specimens, prepared using the steel rectangular prism molds with the dimension of 40 × 40 × 160 mm, were cast to determine the flexural strength. Zarei et al. [19] observed that the flexural strength at 28 days was increased around 5–15% in comparison with CP grout, with the addition of CAEP. However, 7-day flexural strength decreased by about 40%. It was mentioned that the ratio of the flexural strength and the compressive strength of grouting paste increased with CAEP content. The ratio was the indication of the toughness of the grout material. The flexural strength was significantly affected by the W/C ratio and PSP content [16].

### 2.2.4 Bond Strength

The pull-off test was conducted on a Positest AT-A device to determine the bond strength between the asphalt-coated aggregate and the grout [2]. A homogenous uniform thin asphalt film of maximum 0.2 mm thickness was poured onto the preheated aggregate plate and a square glassy plastic mould having dimensions of 2 mm thickness, diameter of 21 mm and was positioned on the preheated aggregate plate. The bond between the grout material and asphalt was formed by pouring the grout into the circular holes. Eventually, after curing time, the hardened grouting was pasted with the metal pull-off stubs and then tested for the failure load to determine the bond strength or pull-off strength. The pull-off test was used to determine the bond strength between the asphalt and CAEP and observed that the bond strength increased with the increasing CAEP content, while the bond strength of CP was very less [16]. It is concluded that the interface between the hardened cement paste and asphalt coated aggregate was most critical to failure and can be reduced with the addition of CAEP.

### 2.2.5 Drying Shrinkage

The SFP specimens were immersed for 24 h in water for measuring the drying shrinkage. The length of the SFP specimen after wiping and further on 14 days of storage at room temperature was measured. The percentage reduction of the length will give the drying shrinkage. The drying shrinkage will be significantly influenced by the type of expansion admixture [16].

## 2.3 Composition and Formulation

### 2.3.1 Composition

The long molecules in the superplasticizer enclose the cement particles to make them repel each other by providing a negative charge, thereby attaining the higher flowability with a lesser W/C ratio. The one sulfonated naphthalene formaldehyde-based SP and two polycarboxylic ether polymer-based SPs were used to determine the effect of superplasticizer dosage and type on the flowability and the strength of the grout [11]. The dosage of the first type of polycarboxylic ether polymer-based SP and W/C ratio were varied from 0.5 to 2.0% by weight of grout and 0.24–0.50, respectively, to attain a fluidity of 11–16 s and observed that the higher dosages of SP increased the flowability irrespective of W/C ratio. The sulfonated naphthalene formaldehyde-based SP produced the grout mixtures of 136 s flow time and also observed as expected that the compressive strength was decreased with the increase in W/C. The W/C ratio was the important factor affecting the mechanical properties of the grout. The higher W/C causes the cracks due to the temperature shrinkage and the decrease in strength. Zachariah et al. [18] investigated the effect of the W/C ratio (0.5, 0.6, and 0.7) on the performance of the SFP and concluded that the residual porosity was decreased with the increase in W/C, which in turn increases the performance of SFP. However, the grout with a W/C ratio of 0.7 showed lesser compressive strength than grout with 0.5 irrespective of grout volume. Therefore, the W/C ratio should be selected based on fluidity without compromising the strength. Fang et al. [6] determined the optimum W/C of cement slurry as 0.6.

Pei et al. [16] determined the optimal W/C ratio for the grout having the additives, PSP, UEA, and AEA, based on the required range of fluidity, drying shrinkage, flexural and compressive strengths. When the W/C ratio varied from 0.51 to 0.54, the fluidity decreased from 25 to 16 s. A decrease of 5% and 5.7% was observed in the 7-day flexural and compressive strengths, respectively, with an increase in the W/C ratio to 0.57 from 0.51. Similarly, a decrease of 9% and 10.7% was observed in the 28 days' flexural and compressive strengths, respectively, with an increase in W/C ratio to 0.57 from 0.51. The drying shrinkage increased about 33% first with the increase of the W/C ratio from 0.51 to 0.54, then, reduced with a further increase in the W/C ratio. Based on the above mentioned analysis, the optimal range of W/C ratio was recommended as 0.55–0.57. The PSP content of 1% improved the flexural strength at 7 days and 28 days by about 12% with the corresponding strength of CP, but a significant improvement in the compressive strength and the drying shrinkage was not observed. The recommended amount of PSP was 1%. Up to the expansion admixture content of 10%, the admixture had no significant influence on the fluidity, flexural, and compressive strengths; however, the drying shrinkage was reduced by around 18% significantly. Therefore, the expansion admixture of 10% was recommended. The AEA content had a significant effect on the fluidity and increased the 7-day flexural strength with the increase in AEA content but had no influence on the compressive strength. However, 0.008% content was recommended due to the workability provided during the mixing process.

The grouted specimens should be cured for sufficient time to attain strength. The SFP samples were concealed with plastic sheets and then cured for 7 days at 20 ± 2 °C, relative humidity of 95%, and then cured at the same temperature with a relative humidity of 65 ± 5% for 28 days curing period [16]. After demolding, the SFP specimens were cured at room temperature indoors [16]. Hassan et al. [8] cured the specimens at 25 °C in a temperature-controlled chamber with 65% relative humidity until testing. Gong et al. [7] cured the SFP samples for 14 days curing period at 20 ± 2 °C and relative humidity of 95%.

### 2.3.2 Formulation

The composition of the grout materials affects the performance of SFP and the effect of each material should be carefully understood. The composition was done mostly, in general, on trial and error, which is a hectic laboratory work [6, 9]. The formulation of grout should be done in a scientific way to understand the effect of composition, ultimately the required grout can be achieved. Zhang et al. [20] analyzed the effects of grout composition with the preparation of two grouts, CP and CM, and determined the flowability, shrinkage, flexural, and compressive strengths. The influence factor levels of CP and CM were determined based on the literature. Tables 1 and 2 present the factor level of CP and CM. The approximate orthogonal arrays (12, $3^4\ 4^1$) and orthogonal Table L9 ($3^4$) were adopted for CP and CM respectively. The grout should have fluidity ranging from 10 to 14 s, a compressive strength of 10 to 30 MPa, and a flexural strength of a minimum of 3.0 MPa.

To achieve the fluidity of 10–14 s, the W/C ratio of 0.63, FA content of 20%, and MP of 20% were the best proportions for CP grout based on the analysis. The

**Table 1** Factor level of CP

| Factor level | W/C ratio | FA content (%) | MP content (%) |
|---|---|---|---|
| 1 | 0.48 | 0 | 0 |
| 2 | 0.53 | 10 | 10 |
| 3 | 0.58 | 20 | 20 |
| 4 | 0.63 | – | – |

**Table 2** Factor level of CM

| Factor level | W/C ratio | FA content (%) | MP content (%) | Sand content (%) |
|---|---|---|---|---|
| 1 | 0.55 | 0 | 0 | 10 |
| 2 | 0.60 | 10 | 10 | 15 |
| 3 | 0.65 | 20 | 20 | 20 |

fluidity was affected by the W/C ratio most and followed by FA content and MP content. Similarly, the W/C ratio of 0.65, FA content of 20%, MP content of 20%, and sand content of 10% were proportions for CM grout and the factors affecting the fluidity were of the same order as CP's factors except the sand content being the last. The W/C ratio was the most influential factor and the MP was the least factor of all. To achieve the flexural strength of more than 3 MPa at 7 days curing period and the compressive strength of 10–30 MPa at 7 days curing period for the CP, the W/C ratio of 0.48 should be used without FA content and MP content. It was concluded that the FA content had a great impact on the flexural strength while the W/C ratio had the most effect on the compressive strength. For 7 days flexural and compressive strengths of CM, the best proportion was W/C of 0.55, the sand content of 20%, with no FA content and MP content, and the FA and MP contents being the most influencing factors, respectively. To achieve the 28-day compressive strength of CM, the W/C of 0.55 and the sand content of 15% with no FA, MP contents were the best mix proportion and the MP was the great influence. The mixture prepared with a W/C ratio of 0.53, the FA of 20%, and the 0% MP content was the best mixture to reduce the drying shrinkage of CP. The W/C ratio of 0.60, the FA content of 10%, and the sand content of 20% were the best proportion to reduce the drying shrinkage of CM. The W/C ratio had a great influence on the drying shrinkage of the CP and the CM. To achieve the targeted values of fluidity, drying shrinkage, compressive, and flexural strengths of CP or CM, considering any one of the properties, the optimal proportions and combinations of W/C ratio, the content of MP, the content of FA, and the sand content were determined. The fluidity of grout was the main criteria and followed by the lesser drying shrinkage and higher strength, respectively, to select the optimal composition of grout. To determine the optimal W/C ratio, the plots of fluidity, ductility, shrinkage, and strengths of cement pastes (CP) for 7, 28 days against the W/C ratio were drawn and the optimal ranges of W/C ratio to meet the relevant standard requirement were determined. The recommended range of W/C ratio for CP is about 0.56 to 0.58. Similarly, the graphs of FA content against the ductility, shrinkage, ductility, and strengths of CP were drawn and the optimal FA content was determined as 10%. In the same way, the optimal mineral powder content of 10% was determined. In a similar way, recommended the composition of CM as a W/C ratio of 0.61–0.63, sand content of 15%, and 10% FA content as grouting material. It was concluded that the CP was more suitable as grouting materials than the CM. The CP grout was more stable than the CM when W/C was varied.

Pei et al. [16] used the three additives such as PSP, UAE, and AEA, to prepare the HPCP and the contents of the additive were set as shown in Table 3 based

**Table 3** Factor level of HPCP

| Factor level | W/C ratio | PCP content (%) | UEA content (%) | AEA content (%) |
|---|---|---|---|---|
| 1 | 0.57 | 0 | 0 | 0 |
| 2 | 0.54 | 0.5 | 10 | 0.008 |
| 3 | 0.51 | 1 | 12 | 0.012 |

**Table 4** The recommended range/ratio and selected ratios of HPCP

| Factor level Recommended value range | Sample Name | W/C ratio: 0.55–0.57 | PCP content: 0.5–1.0% | UEA content: 10% | AEA content: 0.008% |
|---|---|---|---|---|---|
| The recommended ratios of HPCP | HP1<br>HP2<br>HP3 | 0.56<br>0.56<br>0.56 | 0.5%<br>0.5%<br>1.0% | 10%<br>10%<br>10% | 0.008%<br>0.0%<br>0.008% |

on recommended dosages and relevant research. For each factor-level analysis, the orthogonal table L9 ($3^4$) was adopted. The three HPCP grouts, HP1, HP2, and HP3, were prepared with the obtained optimal ratios/ranges through an orthogonal test to evaluate the performance through the fluidity, drying shrinkage, and strength. The composition of HP1, HP2, and HP3 grouts were given in Table 4. HP2 and HP3 grouts performed well and were recommended as grouting materials while HP1 was omitted as flowability decreased significantly with time.

## 3 Evaluation and Contribution of Grouting

### 3.1 *Evaluation of Grouting*

Hou et al. [9] investigated the engineering properties of grouted macadam composite materials and recommended the fluidity of 9–11 s based on trial and error to have the void volume of 30% filled with the grout. Pei et al. [16] and Zhang et al. [20] evaluated the various grouts based on the fluidity, the drying shrinkage, and the strength to choose the best suitable grout. Zachariah et al. [18] prepared the porous asphalt mixtures (PAM) with 18%, 21%, and 24% air voids and observed the increment of 4–10% in Marshall stability of grouted mixtures with the increase in percentage air void from 18 to 24% and [5] reported the same. However, higher initial void content of asphalt mixtures was not always resulted in higher grouting because of poor interconnectivity of the voids in the asphalt mixtures. Therefore, the initial air void content was not a good parameter to evaluate the grouting. The interconnectivity of voids can be assessed through the permeability of the asphalt mixture. The higher permeability indicates that the grout can easily fill the voids. Setyawan [17] observed that the increase in binder content reduced the porosity, and so did the permeability. Generally, the water permeability of porous asphalt mixtures was about $1.44 \times 10^{-3}$–$1.61 \times 10^{-3}$ m/s. The PAM gradations were used to increase the porosity and permeability of asphalt mixtures, which makes grouts fill the voids easily. Though the permeability was not dependent on the porosity itself, the pore size also had a significant impact on the grouting. Ding et al. [5] studied the asphalt mixtures with different pore structures of the same air void content and observed that the mixture with homogenous grade had a good pore size and distribution, which reflected in better performance of asphalt mixtures. It was concluded that the large air void with homogenous grade should be adopted for SFP. Pei et al. [16] evaluated grouting flowability based on the three parameters; the volume of grout/unit area, grouting depth, and residual porosity of grouting composites.

1. Volume of grout/unit area

   The maximum volume of the grout was measured per unit area by preparing the asphalt mixture rutting samples to achieve targeted porosity.

2. Depth of grout

   The grouting depth represents the fluidity of grout and can be observed in both field and laboratory conditions. The rutting slabs were prepared and then grouted with the maximum volume of grout determined using a volume of grout/unit area. After the curing period, the samples were cored and observed for the grouting depth of grouts.

3. Residual porosity of SFP [12, 16];

   The residual porosity of SFP was determined based on the volumetric properties.

4. Grouting saturation degree [12]

The grouting saturation degree represents the grouted material quantity in the asphalt mixtures and is determined using Eq. (1).

$$S_g = \frac{(w_2 - w_1)}{\rho \times V \times V_a} \times 100\% \quad (1)$$

where $Sg$ = Grouting saturation degree (%); $w_2$ = sample weight before grouting (g); $w_1$ = sample weight after grouting *(g)*; $\rho$ = cement mortar's density (g/cm$^3$); $V$ = specimen's volume (cm$^3$); $Va$ = porosity of asphalt mixture (%).

A generalized increasing trend of the grouting volume with an increase in air void content was observed indicating that the grouting parameters were representing the grouting conditions well (Figs. 3, 4 and 5). The grouting volume and the grouting depth showed the variation at 18% air void of the asphalt mixture. The variation was reduced and attributed to the effective void content increment, which allowed the grout to permeate easily. The grout intensity was determined using image analysis on the sectioned cylindrical sample [18].

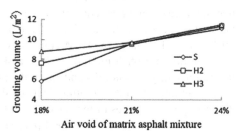

**Fig. 3** Variation of grouting volume

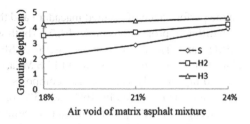

Fig. 4 Variation of the depth of grout

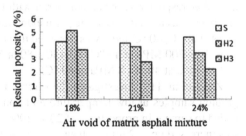

Fig. 5 Variation of volume of residual porosity

## 3.2 Contribution of Grouting

The aggregate skeleton in the asphalt mixture transfers the load through the contact points. The high internal friction among the aggregates increases with the number of contact points leads to the high rutting resistance. The internal friction contribution to rutting resistance is about 70%, which stated the significance of the aggregate skeleton of asphalt mixture in load transfer [1]. The internal friction decreases with the use of weaker aggregates and excessive bitumen content, leading to rutting [9]. Zachariah et al. [18] investigated the potential usage of the nonconventional aggregates in the CGBM, the crushed first-class brick aggregates (CFCBA), and the crushed over-burnt brick aggregates (COBBA). The grouted brick aggregate mixtures exhibited a higher tensile strength ratio (TSR) of more than 100%, i.e., 101.85%, reduced rut depth by about 35%, and increased the fatigue lives by about 144% in comparison with hot mix asphalt mixtures (HMA) samples. The rut reduction and fatigue life increment of grouted over burnt brick aggregate mixtures were 24.04% and 132.91%, respectively, while grouted first brick aggregate mixture exhibited the rut depth reduction of 3.4% and increment of 118.87% in fatigue lives. It was concluded that the marginal aggregates could be used and the over burnt brick aggregates could perform well. However, to avoid the negative influence on the hydration process in SFP samples, the investigation of brick aggregates' porous surface and the high water absorption was needed. Zachariah et al. [19] investigated the relationship between bitumen to cement grout mass ratio (B/CG) and the performance of grouted samples, i.e., rutting depth and fatigue lives and a good relationship with $R^2$ of 0.81 existed with fatigue life than the rutting depth. It was concluded that the strength and the amount of grout in CGBM affected the fatigue behavior while the quality of grout influenced the rutting behavior more. Hassan et al. [8] investigated the effect of cementitious grouts on the performance of grouted macadam and the five grouting mixes were prepared with different proportions of SF, FA/SF (both were used), Styrene-butadiene rubber (SBR), sand. The compressive strengths of SF and FA/SF grouted macadams were almost the same, i.e., 13.7 MPa and 13.8 MPa, even though the SF grouted macadam showed higher early strength. The compressive strengths of SF and FA/SF grouted macadam was increased by 16% and 13%, respectively, compared to the compressive strength of OPC-grouted macadam. Similarly, the 4–7% increment in indirect tensile stiffness modulus was observed and shrinkage strains were reduced by 22% and 18%, respectively, when compared with OPC-grouted macadam. The dynamic creep test was used to assess the permanent deformation resistance of grouted specimens at 40 °C, 60 °C and it was observed that the creep stiffness of grouted macadam were 170–200 MPa at 40 °C and 130–180 at MPa 60 °C where the asphalt mixture had 15 MPa at 40 °C, 8 MPa at 60 °C.

Zarei et al. [19] investigated the effect of CAEP on the cracking resistance of grouted samples at low temperature and compared SFP with CP. The CAEP was added to cement in proportions of 20%, 40%, and 60% by weight. The rutting depth was increased up to 2 mm with an increase in the CAEP content while SFP with CP was no rut material. Though the Marshall stability, retained Marshall stability, indirect tensile strength (ITS), and TSR were decreased with the increase in CAEP content, the flexural strength of SFP with CAEP 20% was increased to 6.43 MPa in comparison with SFP with CP. With increasing CAEP content, the flexural strength of SFP was reduced to 6.02 MPa. The pull-off strength of CAEP with asphalt mixture was increased to 42.55 kPa from 19.66 kPa of SFP with CP, which benefited the flexural strength of grouted samples. The 14% increment in flexural strain in comparison to SFP with CP was observed with the addition of 20% CAEP, but, the further increment in CAEP reduced the flexural strains by 20%. It was concluded that the grouting with a suitable modulus should be used to obtain the low-temperature resistant SFP. Cai et al. [3] studied the effect of the grouting and the fiber on the mechanical behavior of SFP and porous asphalt (PA) samples through the Acoustic Emission signal and dynamic modulus test. Figure 6. depicted that the SFP had a larger modulus than the PA mixture. The SFP and PA samples had higher storage modulus than the loss modulus demonstrating the lower cracking resistance. The storage modulus was decreased and the loss modulus was increased with the addition of fiber. Figure 7 depicted the failure strains of PA mixtures, with and without fibers, which were reduced by 34.7% and 44.7%, respectively. The 40% and 15% of the failure strains of SFP-30 mixture, with and without fibers, respectively, were reduced after grouting. The strength of SFP was not improved significantly with the addition of the fiber.

## 4 Conclusions

This study conducted a brief review on the performance and formulation of grouting materials and the following findings were observed.

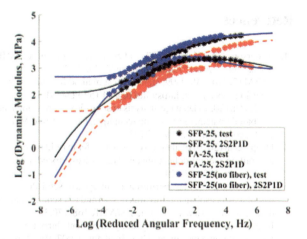

**Fig. 6** Typical master curves of SFP and PA at 15 °C

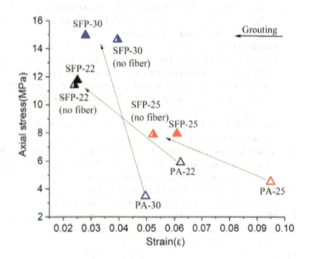

**Fig. 7** Peak stress and corresponding strain of SFP and PA

1. The nonconventional aggregates or the marginal aggregates can be used in SFP
2. The SCM and the additives improve the performance of grout when used at certain proportions/dosages.
3. The formulation of grouting materials should be used to obtain the required grout properties
4. The grouting parameters should be used to evaluate the grouting efficiency.

# References

1. Anderson DA, Kennedy TW (1993) Development of SHRP binder specification (with discussion). J Assoc Asphalt Paving Technol 62
2. ASTM (2014) D4541-09: standard test method for pull-off strength of coatings using portable adhesion. ASTM Int. https://doi.org/10.1520/D4541-17.2
3. Cai X et al (2020) Damage analysis of semi-flexible pavement material under axial compression test based on acoustic emission technique. Constr Build Mater 239:117773. https://doi.org/10.1016/j.conbuildmat.2019.117773
4. Deep P et al (2020) Evaluation of load transfer in rigid pavements by rolling wheel deflectometer and falling weight deflectometer. Transp Res Procedia 45(2019):376–83. https://doi.org/10.1016/j.trpro.2020.03.029
5. Ding Q et al (2011) The performance analysis of semi-flexible pavement by the volume parameter of matrix asphalt mixture mix design and test scheme. Adv Mater Res 168:351–356. https://doi.org/10.4028/www.scientific.net/AMR.168-170.351
6. Fang B et al (2016) Laboratory study on cement slurry formulation and its strength mechanism for semi-flexible pavement. J Test Eval 44(2). 907–913. https://doi.org/10.1520/JTE20150230
7. Gong M et al (2019) Evaluation on the cracking resistance of semi-flexible pavement mixture by laboratory research and field validation. Constr Build Mater 207:387–395. https://doi.org/10.1016/j.conbuildmat.2019.02.064
8. Hassan KE et al (2002) Effect of cementitious grouts on the properties of semi-flexible bituminous pavements. In: Proceedings of the fourth european symposium on performance of bituminous and hydraulic materials in pavement, pp 113–20
9. Hou S et al (2016) Investigation into engineering properties and strength mechanism of grouted macadam composite materials. Int J Pavement Eng 17(10):878–886. https://doi.org/10.1080/10298436.2015.1024467
10. IRC (2018) cement grouted bituminous mix surfacing for urban roads IRC SP 125 2019. E-Conversion—proposal for a cluster of excellence
11. Koting S et al (2007) Influence of superplasticizer type and doasage on the workability and strength of cementitious grout for semi-flexible pavement application. In: Proceedings of the eastern asia society for transportation studies, vol 6, pp 280–280
12. Luo S et al (2020) Open-Graded asphalt concrete grouted by latex modified cement mortar. Road Mater Pavement Des 21(1):61–77. https://doi.org/10.1080/14680629.2018.1479290
13. Mayer J, Mikael T (2001) Jointless pavements for heavy-duty airport application: the semi-flexible approach. In: Proceedings—international air transportation conference, pp 87–100. https://doi.org/10.1061/40579(271)7
14. Moghaddam TB et al (2011) A review on fatigue and rutting performance of asphalt mixes. Sci Res Essays 6(4): 670–82. Acad J
15. Oliveira JRM (2006) Grouted macadam-material characterisation for pavement design. no. May, The University of Nottingham
16. Pei J et al (2016) Design and performance validation of high-performance cement paste as a grouting material for semi-flexible pavement. Constr Build Mater 126:206–217. https://doi.org/10.1016/j.conbuildmat.2016.09.036
17. Setyawan A (2009) Design and properties of hot mixture porous asphalt for semi-flexible pavement applications. Media Teknik Sipil 5(2):41–46
18. Zachariah JP et al (2020) A study on the properties of cement grouted open-graded bituminous concrete with brick as aggregates. Constr Build Mater 256:119436. https://doi.org/10.1016/j.conbuildmat.2020.119436
19. Zarei S et al (2020) Experimental analysis of semi-flexible pavement by using an appropriate cement asphalt emulsion paste. Constr Build Mater 230:116994. https://doi.org/10.1016/j.conbuildmat.2019.116994
20. Zhang J et al (2016) Formulation and performance comparison of grouting materials for semi-flexible pavement. Constr Build Mater 115:582–592. https://doi.org/10.1016/j.conbuildmat.2016.04.062

# Utilization of Agro-Industrial Waste in Production of Sustainable Building Blocks

**S. S. Meshram, S. P. Raut, and M. V. Madurwar**

**Abstract** Increase in industrial waste has led to an environmental concern. Such types of waste are directly dumped on land, which can cause pollution. This waste should be used in the construction of building materials such as building blocks, which not only helps in reducing pollution but is also a cost-effective choice to design sustainable buildings. Brick is usually one of the essential construction materials, generally prepared from clay, shale, soft slate, calcium silicate, stones, etc. In view of utilizing agro-industrial waste, this study focused on the consumption of cupola slag and rice husk ash to develop sustainable building blocks. Cupola slag and rice husk ash were used in various proportions and cement is used as a binder with a constant ratio. The physicomechanical properties of developed sustainable building blocks were carried out and it meets the requirements of Indian standards. The results shows that the cupola slag–cement–rice husk ash mixture can be probably used in the manufacturing of economical bricks.

**Keywords** Rice Husk Ash · Cupola Slag · Sustainable building blocks · Physicomechanical properties

## 1 Introduction

Brick is a commonly used construction material across the globe. Because of its extraordinary properties such as low cost, high strength, and great durability, brick plays an important role in the construction industry [1]. Due to the continuously growing population, infrastructural needs are also increasing, which in turn increases the demand for construction materials. The application of various waste materials to develop building blocks reduces the exploitation of conventional construction

---

S. S. Meshram (✉) · S. P. Raut
Department of Civil Engineering, YCCE, Nagpur, India
e-mail: sangitameshram3@gmail.com

M. V. Madurwar
Department of Civil Engineering, VNIT, Nagpur, India

© The Author(s), under exclusive license to Springer Nature Singapore Pte Ltd. 2023
M. S. Ranadive et al. (eds.), *Recent Trends in Construction Technology and Management*, Lecture Notes in Civil Engineering 260,
https://doi.org/10.1007/978-981-19-2145-2_13

materials and provides a sustainable solution to the construction industry [2–9]. In India, about 960 million tons of solid wastes are currently produced annually as by-products during industrial, mining, and mining operations [10–15]. Nowadays, various industrial and agricultural wastes have been used to develop building blocks, which give sustainable solutions to the construction industry [16]. Cupola slag is a waste product manufactured by the cast iron industry, generated during the separation of the molten steel in cupola furnaces from the impurities [17]. In cupola furnaces, about 5–7% of the waste is produced during cast iron production. The industry is producing 50–3000 tons of C.I. Based on the size and specifications of the furnace [18], Cupola slag has been used as a partial replacement for cement with cupola slag in concrete [19]. Also, the behaviour of concrete was investigated using cupola slag as a partial replacement of both fine and coarse aggregate combinations in concrete [20]. Many researchers have utilized cupola slag as a replacement for cement, fine and coarse aggregates in concrete, pavement, and mortar. In addition, no research on cupola slag used in bricks is available. This can be utilized efficiently in building block manufacturing.

In this study, cupola slag as an industrial waste and a main raw material along with rice husk ash as an agricultural waste with a constant cement proportion were used. Cupola slag and rice husk ash were characterized physically and chemically to check their feasibility to be used as a raw material to develop sustainable building blocks. These developed bricks were subjected to various physicomechanical properties. This study is mainly focused on the efficient use of cupola slag to develop economical, sustainable, and high-performance building blocks which gives another solution for the management of industrial waste and pollution control.

## 2 Materials and Test Methods

### 2.1 Raw Materials and Preparation of the Brick Specimen

In this paper, cupola slag is used as industrial waste and rice husk ash as agricultural waste. Cement is used as a binding material. Though the cement percentage were kept constant while rice husk ash and cupola slag were used in different proportions to get a wide range of result. Rice husk ash is obtained when the rice husk is burned for a long time. Rice husk is a waste product obtained from rice farms. This husk can only be used for burning purposes and to generate heat. The ash obtained can only be used for ground filling purposes. Rice husk was collected from Bhandara, Maharashtra. The rice husk was dried under sunlight for 24 h. Rice husk burns slowly and hence needs time. Burning of rice husk requires 48–72 h. Out of all ash, only 30% of it was usable because ashes need to be sieved from a 300-μm sieve shown in Fig. 1. The sieved ash is used in the preparation of brick mix. Cupola slag is an industrial waste obtained from the steel industry, Kapilanshu Dhatu Udyog Pvt. Ltd., it is a pipe manufacturing factory situated in Kamthee, Nagpur. This can be only used

**Fig. 1** Rice husk ash sample

for landfilling purpose. Cupola slag is a thick material containing large iron filling in it. Slag needs to be sieved before using it in bricks. It was sieved from a 1.18 mm IS sieve as given in Fig. 2. Cement was of OPC grade 53 used as a binding material. It was used in minimum percentage and constant for all proportions.

The quantity required for raw materials was first calculated for different combinations of cupola slag, rice husk ash, and cement as per the dimension of the mold as 190 × 90 × 90 mm and densities of all the raw materials. As per the proportions given in Table 1, cupola slag was first dry-mixed with rice husk ash and cement. Then the water was incorporated into the dry mixture using hand mixing and resumed blending until it achieved homogeneous consistency. While casting brick, frog of dimension 140 × 60 × 10 mm was cast using the same dimensioned wooden piece. Bricks (Fig. 3) were prepared using appropriate molds. Then, the bricks were set aside for sun drying for 7 days under rain protected shed.

**Fig. 2** Cupola slag sample

**Table 1** Mix proportions of raw materials for bricks

| Mix | Cupola Slag (%) | Rice Husk Ash (%) | Cement (%) |
|---|---|---|---|
| M1 | 80 | 10 | 10 |
| M2 | 75 | 15 | 10 |
| M3 | 70 | 20 | 10 |
| M4 | 65 | 25 | 10 |
| M5 | 60 | 30 | 10 |

**Fig. 3** Brick specimen

## 2.2 Test Methodology

The chemical constituent of raw materials has been determined using X-ray fluorescence (XRF). The elemental composition of cupola slag and rice husk ash was investigated using an XRF spectrometer (Indian Bureau of Mines, Nagpur). The specific gravity of raw materials was determined as per IS 2386 (Part III):1963 [21]. The particle size distribution of the cupola was also determined. Densities of blocks are determined as per IS 2185 (Part 1): 2005 [22]. The compressive strength test for cupola slag brick specimens (Fig. 3) was carried out according to IS 3495 (Part 1): 1992 [23]. At a uniform rate of 14 N/mm$^2$ per minute, all developed bricks were subjected to compressive load until failure occurred. Water absorption for bricks was carried out based on the weight of bricks submerged in water for 24 h and oven-dried brick specimen.

# 3 Results and Discussion

## 3.1 Properties of Raw Materials

### 3.1.1 Chemical Composition

Table 2 shows the chemical composition of cupola slag and rice husk ash. Cupola slag consists of a significant amount of silica and alumina, which are suitable for

**Table 2** Chemical properties of the raw materials

| Chemical composition | Cupola slag (CS) | Rice husk ash (RHA) |
|---|---|---|
| Lime, CaO (%) | 15.91 | 1.31 |
| Silica, $SiO_2$ | 31.57 | 81.93 |
| Alumina, $AL_2O_3$ | 9.03 | 0.39 |
| Iron oxide, $Fe_2O_3$ | 18.58 | 0.35 |
| Magnesia, MgO | 1.52 | 0.44 |
| Sulphur trioxide, $SO_3$ | 0.88 | 0.59 |
| Sodium Oxide, $Na_2O$ | 0.42 | 0.09 |
| Potassium oxide, $K_2O$ | 0.64 | 0.57 |
| Titanium Oxide, $TiO_2$ | 1.58 | 0.05 |
| Pentium Oxide, $P_2O_5$ | 0.22 | 1.17 |
| Fluoride ion, F | 0.42 | 0.19 |
| Sodium Oxide, NaO | 0.42 | 0.09 |
| Chloride, CL | 0.06 | 0.57 |
| Strontium Oxide, SrO | 0.02 | – |
| Barium Oxide, BaO | 0.04 | – |
| Magnesium dioxide, $MnO_2$ | 0.51 | 0.22 |
| Chromium trioxide, $CrO_3$ | 0.04 | – |
| Nickel oxide, NiO | 0.01 | – |
| Cupric oxide, CuO | 0.02 | – |
| Rubidium Oxide, $Rb_2O$ | 0.02 | – |
| Loss of ignition, LOI | 18.42 | 8.86 |

**Table 3** Physical Properties of the cement, rice husk ash, and cupola slag

| S. No. | Tests performed | Cement | Test results | |
|---|---|---|---|---|
| | | | Rice husk ash | Cupola slag |
| 1 | Specific gravity | 3.15 | 2.19 | 1.36 |
| 2 | Density | 1440 kg/m$^3$ | 414 kg/m$^3$ | 1414.86 kg/m$^3$ |
| 3 | Standard consistency | 31% | – | – |
| 4 | Fineness | 6% | 15.03 | – |
| 5 | Water absorption | – | 9% | 11.5% |
| 6 | Initial setting time | 70 min | – | – |
| 7 | Final setting time | 135 min | – | – |

brick manufacturing. Slight amounts of calcium oxide, magnesium oxide, and iron oxides were also present in the cupola slag. A large amount of silica present in the cupola slag gives a better binding property to the brick specimen. Also, the quantity of silica present in rice husk ash is more. Calcium oxide present in cupola slag is higher as compared to rice husk ash which can contribute to pozzolanic activity. Cupola slag shows a higher loss of ignition as compared to rice husk ash which can increase the porosity of brick specimen. Rice husk ash showed fewer amounts of aluminum, calcium, and magnesium oxide.

### 3.1.2 Specific Gravity and Unit Weight

The specific gravities of CS and RHA were 1.36 and 2.19, respectively. CS has a comparatively lower specific gravity than RHA. The densities of CS and RHA were 1414.86 kg/m$^3$ and 414 kg/m$^3$, respectively, which would be able to assist in minimizing the weight of a single brick, resulting in a lighter and more economical structure. The physical properties of cement are given in Table 3.

### 3.1.3 Particle Size Distribution

Figure 4 shows the particle size distribution of cupola slag. It is observed that 95% of cupola slag particles were under the category of sand (Table 4). This indicates that the cupola slag particles can be used in making bricks. Only 5% of particles were comes under the category of silt. Table 4 shows there are zero particles under the category of gravel and clay.

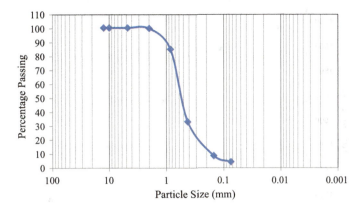

**Fig. 4** Particle size distribution for cupola slag

**Table 4** Cupola slag particle size distribution

| Percentage distribution | Size specifications (in mm) | Cupola slag |
|---|---|---|
| Clay | <0.002 | 0 |
| Silt | 0.002–0.075 | 5 |
| Sand | 0.075–2 | 95 |
| Gravel | >2 | 0 |

## 3.2 Mechanical Properties

### 3.2.1 Density

As per IS 2185 (Part 1):2005 [22], hollow (open and closed cavity) concrete blocks bearing units shall have a minimum block density of 1500 kg/m$^3$. These can be manufactured for minimum average compressive strengths of 3.5, 4.5, 5.5, 7.0, 8.5, 10.0, 12.5, and 15.0 N/mm', respectively, at 28 days. The increase in the percentage of CS in the mix resulted in a decrease in the density of bricks. The dry density decreased, that is, from 1689.9 to 1340.57 kg/m$^3$ by 20% as shown in Table 5 when the percentage of CS varied from 80 to 60% as per Fig. 5. This trend was observed in

**Table 5** Density for brick specimens

| Mix proportion | Density (kg/m$^3$) |
|---|---|
| M1 | 1689.9 |
| M2 | 1646.1 |
| M3 | 1540.57 |
| M4 | 1450.57 |
| M5 | 1340.57 |

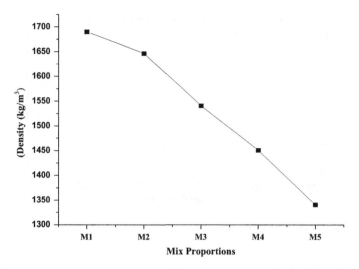

**Fig. 5** Density for brick specimens

density because of the lesser density of rice husk ash and cupola slag content present in the brick mix. This was useful in producing lightweight bricks. Therefore lighter bricks were more cost-efficient, more economical, and easy for transport.

### 3.2.2 Water Absorption

The quality and strength of bricks are significantly impacted by the water absorption. From Fig. 6, it can be observed that after incorporating cupola slag in the brick mix, the water absorption decreases from 12 to 6% as the percentage of cupola slag decreases as per Table 6. For all combinations water absorption is less than 15% as recommended by the IS 1077:1992 [24]. This can be used for higher brick classes and values less than 20% could be used for brick classes up to 12.5. The addition of CS can, therefore, be helpful in generating masonry bricks that are more durable, economical, and eco-friendly.

### 3.2.3 Compressive Strength

Compressive strength is an important parameter as it gives an idea about the strength and quality of the brick. The results for all the tested bricks are shown in Fig. 7. Table 7 indicates that the compressive strength decreases from M1 to M5 mixes, which is 4.4–3.8 N/mm$^2$ with a reduction of 12.5% (Fig. 8). Maximum strength was obtained for M1 (80 CS: 10 RHA: 10 cement) proportion mix. This is due to the chemical composition of raw materials. A lesser amount of CaO present in the chemical composition leads to the higher compressive strength of brick mixes integrating cupola slag.

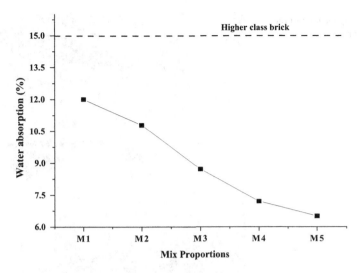

Fig. 6 Effect of CS on water absorption

Table 6 Water absorption for brick specimens

| Mix proportion | Water absorption (%) |
|---|---|
| M1 | 12 |
| M2 | 10.789 |
| M3 | 8.713 |
| M4 | 7.2 |
| M5 | 6.5 |

The minimum compressive strength requirements according to the Indian standard code 1077:1992 [24] are 3.5 N/mm$^2$. All brick specimens composed of cupola slag attained a compressive strength as per the IS recommendations. Also, it meets the requirements of I Class brick 4 N/mm$^2$. Hence, the developed blocks can be used for eco-friendly construction.

## 4 Conclusion

Dumping on land and disposal of cupola slag causes land-related contamination issues that directly impact the ecosystem. Therefore, its productive use to a certain degree in the manufacture of bricks provides an alternative treatment for such problems. Cupola slag, a by-product of the cast iron manufacturing industry was used in the production of brick along with rice husk ash as agro-waste with a constant cement ratio. The presence of silica and alumina in the cupola slag shows its potential to be used in the production of sustainable building blocks. From the 5 different

**Fig. 7** Compression test on brick specimen

**Table 7** Compressive strength of brick specimens incorporating CS

| Mix proportion | Compressive strength (N/mm$^2$) |
|---|---|
| M1 | 4.4 |
| M2 | 4.2 |
| M3 | 4.1 |
| M4 | 3.9 |
| M5 | 3.8 |

mixes with varying CS percentages such as 80–60%, M1 mix with 80% CS, 10% RHA, 10% cement was found to have a higher compressive strength of 4.4 N/mm$^2$. Minimum water absorption and density were obtained for brick mix M1 (60 CS: 30 RHA: 10 cement) as 6.5% and 1340.57 kg/m$^3$, respectively. Water absorption value for all the mixes was found to be less than 20%, hence satisfying the requirements as per standards. Therefore, cupola slag can be utilized in the preparation of sustainable building blocks. Cupola slag is a waste product obtained from the iron industry that can be potentially used in making bricks. Also, it reduces its disposal and landfilling problem which ultimately helps in minimizing land pollution. Based on experimental findings, it can be recommended that CS can be used in the manufacture of large-scale building blocks with the possibility of to improve the physicomechanical properties of bricks, contributing to economical, environmentally safe, and sustainable building construction.

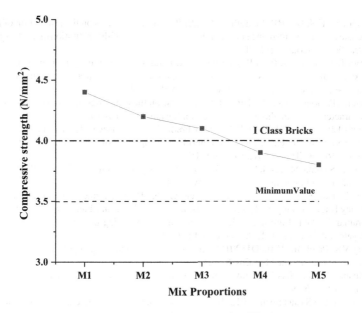

**Fig. 8** Compressive strength of brick specimens incorporating CS

## References

1. Zhang Z, Wong YC, Arulrajah A, Horpibulsuk S (2018) Review of studies on bricks using alternative materials and approaches. Constr Build Mater 188:1101–1118
2. Raut SP, Ralegaonkar RV, Mandavgane SA (2011) Development of sustainable construction material using industrial and agricultural solid waste: a review of waste-create bricks. Constr Build Mater 25:4037–4042
3. Shivaprasad KN, Das BB (2018) Determination of optimized geopolymerization factors on the properties of pelletized fly ash aggregates. Constr Build Mater 163:428–437
4. Goudar SK, Shivaprasad KN, Das BB (2019) Mechanical properties of fiber-reinforced concrete using coal-bottom ash as replacement of fine aggregate. Sustain Constr Build Mater Lect Notes Civil Eng 25
5. Sharath BP, Shivaprasad KN, Athikkal MM, Das BB (2018) Some studies on sustainable utilization of iron ore tailing (IOT) as fine aggregates in fly ash based geopolymer mortar. In: IOP conference: series, material science an engineering, vol 431, p 092010
6. Prasanna KM et al (2021) Fast setting steel fibre geopolymer mortar cured under ambient temperature. In: Recent developments in sustainable infrastructure. Springer, Singapore, 769–787
7. Prasanna KM, Tamboli S, Das BB (2021) Characterization of mechanical and microstructural properties of FA and GGBS-Based geopolymer mortar cured in ambient condition. In: Recent developments in sustainable infrastructure. Springer, Singapore, pp 751–768
8. Goudar SK, Das BB, Arya SB (2019) Microstructural study of steel-concrete interface and its influence on bond strength of reinforced concrete. Adv Civil Eng Mater 8(1):171–189
9. Goudar SK et al (2020) Influence of sample preparation techniques on microstructure and nano-mechanical properties of steel-concrete interface. Constr Build Mater 256:119242
10. Shivaprasad KN, Das BB (2017) Influence of alkali binder dosage on the efficiency of pelletization of aggregates from iron ore tailing and fly ash. Int J Eng Res Mechan Civil Eng 2(3):388–392

11. Shivaprasad KN, Das BB, Sharath SP (2020) Pelletisation factors on the production of fly-ash aggregates and its performance in concrete. In: Proceedings of the institution of civil engineers, construction materials, pp 1–20
12. Sharath BP, Das BB (2021) Production of artificial aggregates using industrial by-products admixed with mine tailings—a sustainable solution. In: Recent trends in civil engineering. Lecture Notes in Civil Engineering, vol 105, pp 383–397
13. Sumukh EP, Goudar SK, Das BB (2021) A review on the properties of steel-concrete interface and characterization methods. Smart Technol Sustain Dev 167–203
14. Sumukh EP, Goudar SK, Das BB (2021) Predicting the service life of reinforced concrete by incorporating the experimentally determined properties of steel–concrete interface and corrosion. Recent Trends Civil Eng 399–417
15. Pappu A, Saxena M, Asolekar (2007) Solid wastes generation in India and their recycling potential in building materials. J Build Environ 42:2311–232
16. Bories C, Borredon M, Vedrenne E, Vilarem G (2014) Development of eco-friendly porous fired clay bricks using pore-forming agents: a review. J Environ Manage 143:186–196
17. Balaraman R, Anne Ligoria S (2015) Utilization of cupola slag in concrete as fine and coarse aggregate. Int J Civil Eng Technol (IJCIET) 6(8):06–14
18. Mistry VK, Patel BR, Varia DJ (2016) Suitability of concrete using cupola slag as replacement of coarse aggregate. Int J Sci Eng Res 7(2)
19. Afolayan JO, Alabi SA (2013) Investigation on the potentials of cupola furnace slag in concrete. Int J Integr Eng 5:59–62
20. Balaraman R, NS Elangovan (2018) Behaviour of cupola slag in concrete as partial replacement with a combination of fine and coarse aggregates
21. IS 2386 (Part III): 1963—Methods of tests for aggregate of concrete
22. IS 2185(Part 1): 2005—Concrete Masonry Units- Specification
23. IS 3495 (Part 1): 1992—Methods of tests of Burnt Clay Building bricks
24. IS 1077: 1992—Common Burnt Clay Building Bricks—Specification

# Effect of Exposure Condition, Free Water–cement Ratio on Quantities, Rheological and Mechanical Properties of Concrete

**Mahesh Navnath Patil and Shailendra Kumar Damodar Dubey**

**Abstract** This research paper explores the effects of environmental exposure conditions, water/cement ratio and cement content on the rheological and mechanical properties of concrete. Even though the grade of concrete is identical, environmental exposure conditions with respect to water–cement ratio significantly controls the quantities of ingredients of concrete. Properties of fresh concrete such as slump value and compaction factor are changing with respect to cement content. It is possible to determine relations between them. Experimentation had been carried out under different water–cement ratios (0.3, 0.36, 0.4, 0.45 and 0.5) with and without the use of chemical admixture to establish graphical relationships. Mechanical properties are determined with respect to age and the effect of admixture. Suggested fresh concrete relations are quite suitable for concretes with and without utilization of chemical admixtures. Workability of concrete depends upon the dose of chemical admixtures such as superplasticizer and time-period after mixing of all ingredients of concrete. Sincere effort had been made to establish the relationship among slump loss, time-period after mixing and dosage of ingredients for different water–cement ratios through graphical representation. Relationship among water–cement ratio, quantity of cement per cubic meter, slump in mm and compaction factor are determined. The quantity of ingredients with respect to various environmental exposure conditions with and without chemical admixture is presented graphically. It was observed that slump values versus cement content and compaction factor show linear variation. Experiments show desired compressive strength couldn't be achieved without superplasticizer in case of higher grade concrete.

**Keywords** Environmental exposure condition · Water/cement ratio · Rheological and mechanical properties of concrete · Slump · Compacting factor

---

M. N. Patil (✉)
Research Scholar, Kavyitri Bahinabai Chaudhari North Maharashtra University, Jalgaon, India
e-mail: m.patil123@gmail.com

S. K. D. Dubey
Professor and Head of the Civil Engineering Department SSVPS BSD COE, Dhule, India
e-mail: dubey.dhule@gmail.com

© The Author(s), under exclusive license to Springer Nature Singapore Pte Ltd. 2023
M. S. Ranadive et al. (eds.), *Recent Trends in Construction Technology and Management*, Lecture Notes in Civil Engineering 260,
https://doi.org/10.1007/978-981-19-2145-2_14

# 1 Introduction

Workability defines the properties and behaviour of freshly mixed concrete. Consistency plays a vital role and defines the criteria in the concrete mix design. It affects the behaviour of concrete starting from placing to the hardened stage concrete.

The properties of concrete workability, consistency and finishing ability of concrete are difficult to measure, as they do not have any quantitative measurement. The slump cone test and compact factor test give a general idea of workability of concrete. The slump cone test is most widely used on site. The compaction factor test defines the compacting ability and finishing ability of fresh concrete [1]. Consistency and cohesiveness measure the finishing and compacting ability of concrete. Consistency represents the flowing ability of freshly mixed concrete. It is a measure of the wetness of the fresh concrete. The requirement of workability depends on the type of construction, method of placement, moulding shape of formwork and structural design [2].

Setting of concrete defines the transition between a fluid and a solid state of concrete. Period of setting starts when concrete loses its plasticity and is complete when concrete gain enough strength to withstand with self-weight and superimposed load [3]. The process of gaining strength is a continuous process later leading to a hardening stage [4]. Rheological properties of concrete change with respect to the initial and final setting time of concrete. Workability decreases with respect to time continuously. Energy consumption increases continuously in subsequent stages.

This study focuses on the effect of environmental exposure conditions; water–cement ratio on the quantity and properties of fresh concrete, which is still a topic of research. Mild, moderate, severe, very severe, extreme exposure are five environmental exposure conditions given in the guidelines of IS 456:2000. The water–cement ratios considered in this study are 0.3, 0.36, 0.4, 0.45, 0.5, 0.55 and 0.6. The slump test and compaction factor test are the indicators of workability. The workability of concrete very much depends upon the dose of superplasticizer and the time-period after mixing of all ingredients of concrete [5]. In this study, polycarboxylic ether polymer-based FOSROCK plasticizer is used. Efforts had been taken to establish the relationship between slump loss, time period after mixing and dosage of ingredients for different water–cement ratios through graphical representation. Workability of concrete decreases with a decrease in compaction factor. Relation between water–cement ratios, quantity of cement per cubic meter, slump in mm and compaction factor established through graphical representation with various environmental exposure conditions with and without the addition of superplasticizer.

Compressive strength is a performance indicator of concrete. In this study, experimental work was carried out to notice the optimum water–cement ratio for M40 grade concrete through crushing strength. Extreme environmental exposure condition is considered for mix design. Five different water–cement ratios of 0.3, 0.36, 0.4, 0.45, and 0.5 are used. Experimentation is carried out with and without a superplasticizer. The variation of compressive strength and failure load with and without plasticizer is shown graphically [6, 7].

## 2 Material Properties

The cement used is the Portland Pozzolana Cement (PPC) manufactured by Ambuja Cement Ltd. (Unit: Maratha Cement Works) confirming requirements as per IS: 1489 (Part1):2015 [8]. PPC cement used is blended with 32% of fly ash. The specific gravity of cement and bulk density of cement was determined experimentally and it was 2.76 and 1493.3 kg/m$^3$, respectively. The fine aggregate used is natural sand which passes over a 4.75 mm IS sieve, confirming to zone-I as per IS 383: 2016. It is clean and free from impurities with specific gravity of 2.52 and bulk density of 1900 kg/m$^3$. Coarse aggregate used is a natural stone retaining on 4.75 mm IS sieve and passing from 20 mm IS sieve as per IS 383: 2016 with specific gravity 2.88 and bulk density 1710 kg/m$^3$ Laboratory pan mixer of capacity 40 L was used to make the mixture. Technical specification of cement, superplasticizer, standard deviation, water content per cubic metre of concrete, proportionate volume of coarse aggregates tabulated below are essential for mix design as per IS 10262:2019 [9]. Potable water free from impurities as per the Indian Standard Code of practice used for mixing and curing of concrete specimens [7]. Superplasticizer is used to increase workability without affecting the strength of the concrete and without the use of additional water. High-range water reducing admixture under the brand name Auramix 200 manufactured by Fosroc was used. The vibrating table is used for compacting fresh concrete mix for a time-period of 1 min (Tables 1, 2, 3, 4, 5 and 6).

Table 1 Standard deviation (IS 10262:2019)

| S. No | Grade | Assumed Standard Deviation |
|---|---|---|
| 1 | M10 | 3.5 |
| 2 | M15 | 3.5 |
| 3 | M20 | 4 |
| 4 | M25 | 4 |
| 5 | M30 | 5 |
| 6 | M35 | 5 |
| 7 | M40 | 5 |
| 8 | M45 | 5 |
| 9 | M50 | 5 |
| 10 | M55 | 5 |
| 11 | M60 | 5 |
| 12 | M65 | 6 |
| 13 | M70 | 6 |
| 14 | M75 | 6 |
| 15 | M80 | 6 |

**Table 2** Water content per cubic metre of concrete (IS 10262:2019)

| S. No | Size of aggregates | Water content |
|---|---|---|
| 1 | 10 | 208 |
| 2 | 20 | 186 |
| 3 | 40 | 165 |

**Table 3** Cement content per cubic metre of concrete (IS 10262:2019)

| Exposure | W/C ratio | | Cement Content |
|---|---|---|---|
| | PCC | RCC | R.C.C.(kg) |
| Extreme | 0.4 | 0.4 | 360 |
| Mild | 0.6 | 0.55 | 300 |
| Moderate | 0.6 | 0.5 | 300 |
| Severe | 0.5 | 0.45 | 320 |
| Very Severe | 0.45 | 0.45 | 340 |

## 3 Environmental Exposure Conditions

In this study, the quantities of ingredients of concrete were calculated according to the norms given in IS 10262:2019. Environmental exposure conditions, water–cement ratio, and use of chemical admixtures such as superplasticizer affect the quantity of ingredients in concrete. Five different environmental exposure conditions are tabulated in Table 15 with reference to IS 456:2000 (Table 7).

Seven different water–cement ratios 0.3, 0.36, 0.4, 0.45, 0.5, 0.55, 0.6 were tried. The quantity of material (cubic metre) in different environmental exposure conditions along with the effect of admixture is represented graphically as below (Figs. 1, 2, 3, 4, 5, 6, 7, 8, 9, 10, 11, 12, 13 and 14).

## 4 Workability of Concrete

Workability describes the ease with which a fresh concrete can be assorted, placed, consolidated, and finished with the least loss in the consistency of concrete. Workability is one of the crucial properties of concrete, which has a direct impact on the quality of concrete in terms of compressive strength, placement and finishing operation.

### 4.1 Factors Affecting Workability

Properties, characteristics, proportion of raw material, admixture used in concrete and environmental exposure made an impact on workability and other properties of

**Table 4** Technical specification of cement used (as provided by the manufacturer) [8, 10, 11]

| S. No. | Characteristic | Requirement as per IS: 1489 (Part-I):2015 | Test results |
|---|---|---|---|
| *1.0 Chemical requirements (Clause 6)* | | | |
| 1.1 | Insoluble residue (%) | | 28.90 |
| | a. Maximum | $X + \frac{4.0(100-X)}{100}$ | |
| | b. Minimum | 0.6 X (where X is the declared percentage of fly ash in the given PPC) | |
| 1.2 | Magnesia (%) | 6.0 (Max) | 1.19 |
| 1.3 | Total sulphur as $SO_3$% | 3.5 (Max) | 2.02 |
| 1.4 | Loss on ignition % | 5.0 (Max) | 1.47 |
| 1.5 | Chloride content | 0.1 (Max) | 0.010 |
| | | 0.05 (Max.) for prestressed Structure | |
| *2.0 Physical requirements (Clause 7)* | | | |
| 2.1 | Fineness ($m^2$/kg) | 300 Min | 367 |
| 2.2 | Soundness | | |
| 2.2.1 | By Le Chatelier method (mm) | 10 (Max) | 1 |
| 2.2.2 | By auto calve test method (%) | 0.8 (Max) | 0.02 |
| 2.3 | Setting time | | |
| 2.3.1 | Initial (Minutes) | 30 (Min) | 190 |
| 2.3.2 | Final (Minutes) | 600 (Max) | 260 |
| 2.4 | Compressive strength in Mpa | | |
| 2.4.1 | 72 + - 1 h (3 Days) | 16 (Min) | 29.8 |
| 2.4.2 | 168 + - 2 h (7 Days) | 22 (Min) | 40.0 |
| 2.4.3 | 672 + - 4 h (28 Days) | 33 (Min) | 60.0 |
| 2.5 | Drying Shrinkage (%) | 0.15 (Max) | 0.06 |

**Table 5** Technical specification of superplasticizer used

| Base | Appearance | Density | pH value | Chloride content | Air entrainment | Compatibility |
|---|---|---|---|---|---|---|
| Polycarboxylic ether polymer | Light brown liquid | 1.07 + -0.02 kg/lit | 6 | Nil | Nil | All types of Portland cement |

concrete. Some of the factors affecting the workability of concrete are enlisted and described below.

**Table 6** Proportionate volume of coarse aggregates (IS 10262:2019)

| S. No. | Size of aggregate | Volume of coarse aggregates/total volume for Different Zones of Fine Aggregates ||||
|---|---|---|---|---|---|
| | | Zone-I | Zone-II | Zone-III | Zone-IV |
| 1 | 10 | 0.48 | 0.50 | 0.52 | 0.54 |
| 2 | 20 | 0.60 | 0.62 | 0.64 | 0.66 |
| 3 | 40 | 0.69 | 0.71 | 0.72 | 0.73 |

**Table 7** Environmental exposure conditions given in IS456:2000 [12]

| S. No. | Environment | Exposure condition |
|---|---|---|
| 1 | Mild | Concrete surfaces are protected against weather or aggressive conditions, except for those situated in the coastal area |
| 2 | Moderate | Concrete surfaces sheltered from severe rain or freezing whilst wet. Concrete exposed to condensation and rain Concrete continuously under water. Concrete in contact or buried under non-aggressive soil/groundwater. Concrete surfaces sheltered from saturated salt air in the coastal area |
| 3 | Severe | Concrete surfaces exposed to severe rain, alternate wetting and drying or occasional freezing whilst wet or severe condensation. Concrete completely immersed in seawater. Concrete exposed to the coastal environment |
| 4 | Very severe | Concrete surfaces exposed to seawater spray, corrosive fumes or severe freezing conditions whilst wet. Concrete in contact with or buried under aggressive sub-soil/groundwater |
| 5 | Extreme | Surface of members in the tidal zone. Members in direct contact with liquid/solid aggressive chemicals |

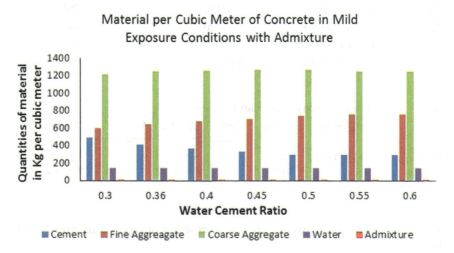

**Fig. 1** Quantities of material per cubic material in mild exposure condition with admixture

Effect of Exposure Condition, Free Water–cement Ratio ...

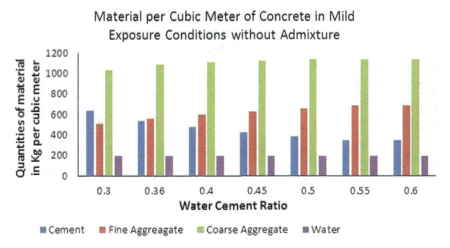

Fig. 2 Quantities of material per cubic material in mild exposure condition without admixture

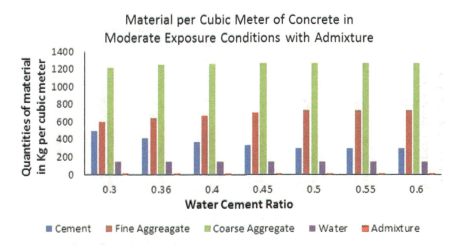

Fig. 3 Quantities of material per cubic material in moderate exposure condition with admixture

### 4.1.1 Water–Cement Ratio

Water–cement ratio has an impact on the compressive strength and durability of the concrete. A higher quantity of cement with a judicious amount of water during mix design reflects the desirable strength. Paste of cement and water forms coating over the surface of aggregate results in appropriate consolidated finishing. Insufficient quantity of water in concrete cannot fulfil the requirement of cement for the hydration process.

It may result in poor compressive strength. Lesser quantity of water also resists workability due to which, placement and finishing of concrete are affected. Vice

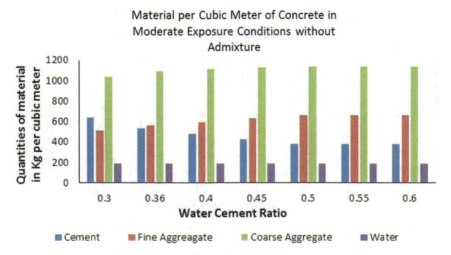

**Fig. 4** Quantities of material per cubic material in moderate exposure condition without admixture

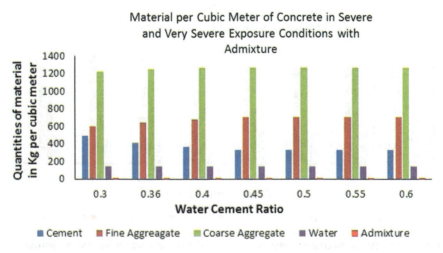

**Fig. 5** Quantities of material per cubic material in severe/very severe exposure with admixture

versa due to the excessive amount of water, workability increases but it may result in segregation and weaker concrete. The water–cement ratio should be selected as per the guidelines given in IS 456:2000 and IS 10262:2019 considering the desired strength and environmental exposure conditions.

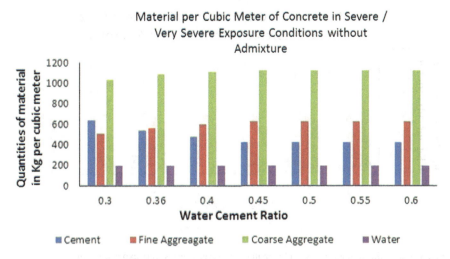

**Fig. 6** Quantities of material per cubic material in extreme exposure condition without admixture

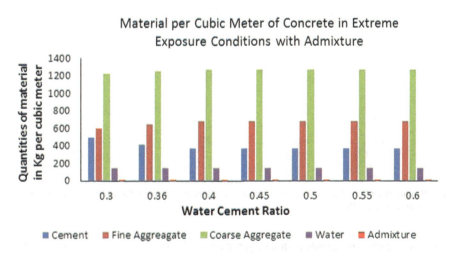

**Fig. 7** Quantities of material per cubic material in extreme exposure condition with admixture

### 4.1.2 Size and Shape of Aggregate

IS 383:2016 provides guidelines for coarse and fine aggregate of concrete. Larger size aggregates need more cement paste to cover the whole surface as compared to smaller size aggregates. According to IS 383:2016 and IS 456:2000, elongated, flaky and angular aggregates should be avoided due to larger surface area. Circular and rounded particles are also avoided. Though rounded aggregate has a smaller surface area, due to shape, aggregate interlocking cannot be attained. It invariably affects the

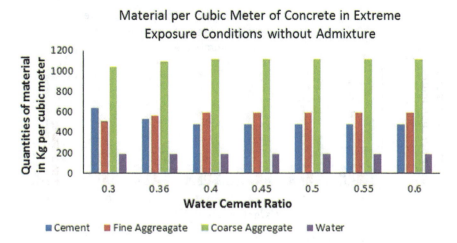

**Fig. 8** Quantities of material per cubic material in severe/very severe exposure without admixture

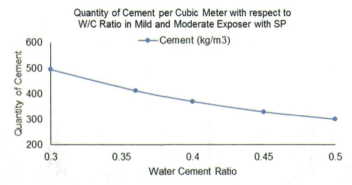

**Fig. 9** Quantification of cement per cubic meter for different w/c ratio at mild and moderate exposer condition with use of SP

strength parameters. Crushed stone aggregates with proper size proportion provide better bond as well as workable concrete [12, 13].

### 4.1.3 Admixtures

Admixtures like superplasticizers make concrete workable by reducing adhesion between cement and aggregate particles. Plasticizers make the concrete flowable without affecting the strength characteristics of concrete [14].

Effect of Exposure Condition, Free Water–cement Ratio ... 171

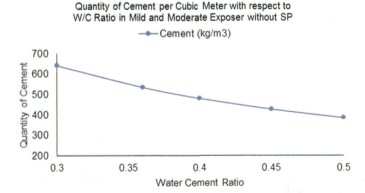

**Fig. 10** Quantification of cement per cubic metre for different water/cement ratios at mild and moderate exposure condition without use of SP

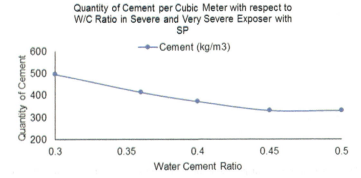

**Fig. 11** Quantification of cement per cubic metre for different water/cement ratios at severe and very severe exposure condition with SP

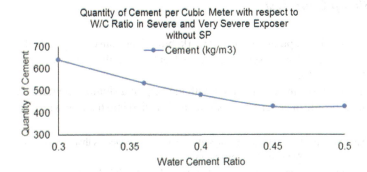

**Fig. 12** Quantification of cement per cubic metre for different water/cement ratios at severe and very severe exposure condition without SP

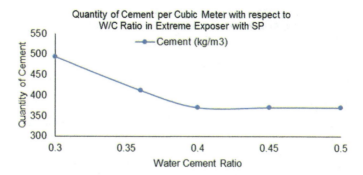

**Fig. 13** Quantification of cement per cubic metre for different water/cement ratios at extreme exposure condition with SP

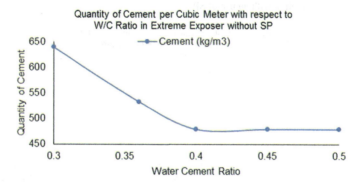

**Fig. 14** Quantification of cement per cubic metre for different weight/cement ratios at extreme exposure condition with SP

## 4.2 *Slump Cone Test*

Slump cone test is a popular measure of workability. The consistency of fresh concrete can be measured before it sets. It is carried out as per the guidelines given in IS 1199-1959 (Fig. 15).

Slump cone test can be performed with the help of a slump cone (Abrams cone). Based on the profile of concrete, the slump is classified into the following three types.

| 1 | Collapse Slump | Concrete gets collapsed completely. It indicates that the concrete is too wet and highly workable |
|---|---|---|
| 2 | Shear Slump | Concrete gets slipped off from the sides. It indicates a lack of cohesion in concrete, possibility of bleeding, and segregation of concrete. So not desirable for the strength and durability of concrete |
| 3 | True Slump | The whole mass of concrete slipped off evenly without any collapse or shear off. It is the desirable slump |

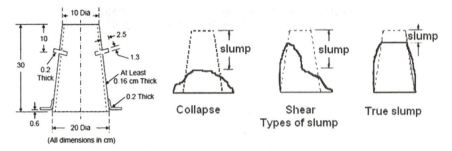

**Fig. 15** Dimensions of slump cone and types of slump (35)

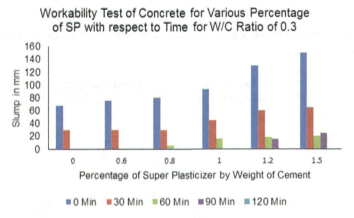

**Fig. 16** Workability test of concrete for various % of SP with respect to time for water/cement ratio of 0.3

### 4.2.1 Slump Loss Variation According to Time

Polycarboxylic ether polymer-based FOSROCK plasticizer was used in this study. Technical specifications are given in Table 5. The workability of concrete depends upon the dose of superplasticizer and time period after mixing of all ingredients of concrete. Sincere effort had been made to establish a relationship among slump loss, time period after mixing, and dosage of ingredients for different water–cement ratios through graphical representation (Figs. 16, 17, 18, 19 and 20).

## 4.3 Compaction Factor Test

It is a laboratory/field test used to check the workability of concrete. This test is preferred for medium to low workable concrete. It is carried out as per the guidelines of IS 1199-1959. It is more precise and accurate than the slump cone test. Compaction

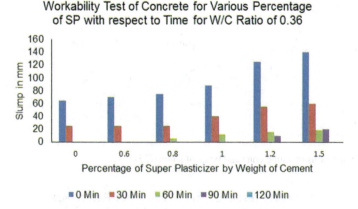

**Fig. 17** Workability test of concrete for various % of SP with respect to time for water/cement ratio of 0.36

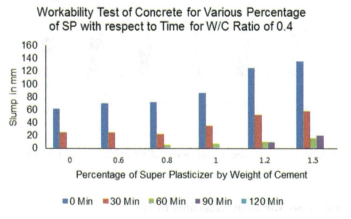

**Fig. 18** Workability test of concrete for various % of SP with respect to time for water/cement ratio of 0.4

factor is the ratio of the weight of semi-compacted concrete to fully compacted concrete. Workability of concrete decreases with a decrease in compaction factor. This study represents the relationship among water–cement ratio, quantity of cement per cubic meter, slump in mm and compaction factor through graphical representation with various environmental exposure conditions with and without the addition of superplasticizer (Figs. 21, 22, 23, 24, 25, 26, 27, 28, 29, 30, 31, 32 and 33).

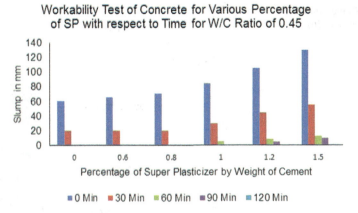

**Fig. 19** Workability test of concrete for various % of SP with respect to time for water/cement ratio of 0.45

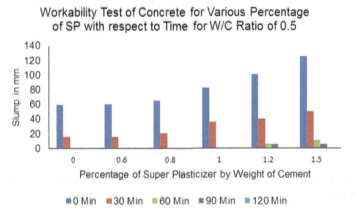

**Fig. 20** Workability test of concrete for various % of SP with respect to time for water/cement ratio of 0.5

## 5 Compressive Strength of Concrete

Compressive strength is a performance indicator of concrete. The test was performed as per the guidelines of IS 516:1959. Depending upon environmental exposure conditions, mechanical properties and durability requirements, concrete mix design are carried out to achieve the desired compressive strength. Compressive strength of concrete is the characteristic strength of a concrete cube of 15 cm size measured after 28 days.

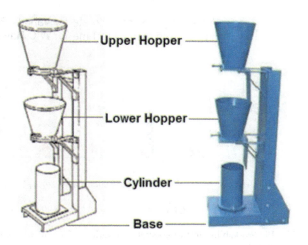

Fig. 21 Compaction factor test on concrete (36)

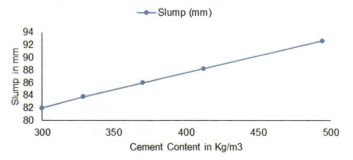

Fig. 22 Effect of cement content on slump in mild and moderate exposure condition with Super Plasticizer

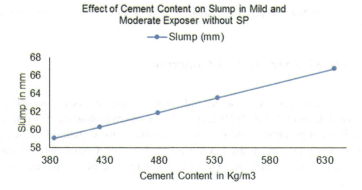

Fig. 23 Effect of cement content on slump in mild and moderate exposure condition without Super Plasticizer

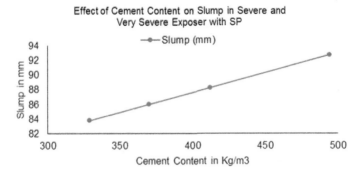

**Fig. 24** Effect of cement content on slump severe and very severe exposure condition with Super Plasticizer

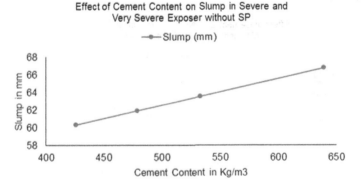

**Fig. 25** Variation of failure load with respect to age of concrete at different water/cement ratios without Super Plasticizer

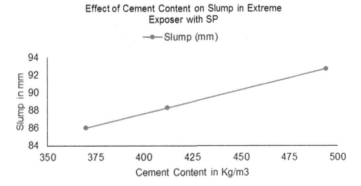

**Fig. 26** Effect of cement content on slump in extreme exposure condition with Super Plasticizer

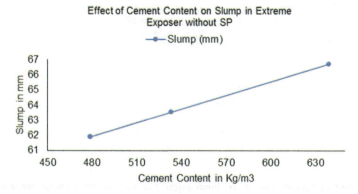

**Fig. 27** Effect of cement content on compaction factor in extreme exposure condition without Super Plasticizer

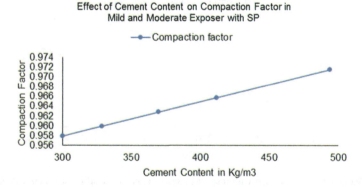

**Fig. 28** Effect of cement content on compaction factor in mild and moderate exposure condition with Super Plasticizer

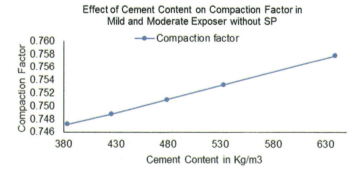

**Fig. 29** Effect of cement content on compaction factor in mild and moderate exposure condition without Super Plasticizer

Effect of Exposure Condition, Free Water–cement Ratio ... 179

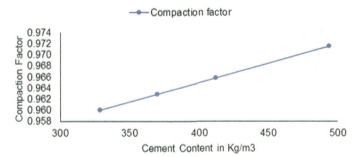

**Fig. 30** Effect of cement content on compaction factor in severe and very severe exposure condition with Super Plasticizer

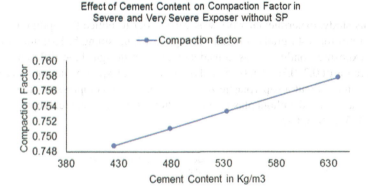

**Fig. 31** Effect of cement content on compaction factor in severe and very severe exposure condition without Super Plasticizer

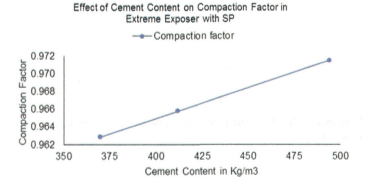

**Fig. 32** Effect of cement content on compaction factor in extreme exposure condition with Super Plasticizer

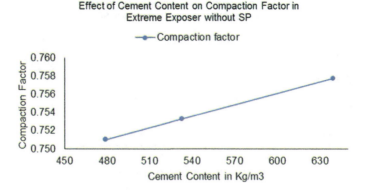

**Fig. 33** Effect of cement content on compaction factor in extreme exposure condition with Super Plasticizer

In this study, experimental work was carried out to notice the optimum water–cement ratio for M40 grade concrete through crushing strength. Extreme environmental exposure condition was considered for mix design. Five different water–cement ratios of 0.3, 0.36, 0.4, 0.45, and 0.5 were used. Experimentation was carried out with and without a superplasticizer. The variation of compressive strength and failure load with and without plasticizer was shown through graphical representation (Figs. 34, 35, 36 and 37).

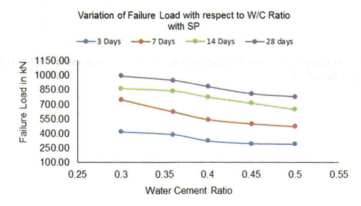

**Fig. 34** Variation of failure load with respect to age of concrete at different w/c ratio with use of Super Plasticizer

Effect of Exposure Condition, Free Water–cement Ratio … 181

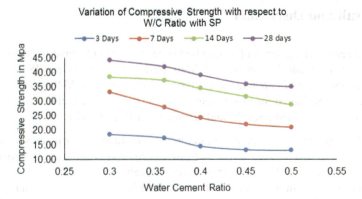

**Fig. 35** Variation of compressive strength with respect to age of concrete at different w/c ratio with Super Plasticizer

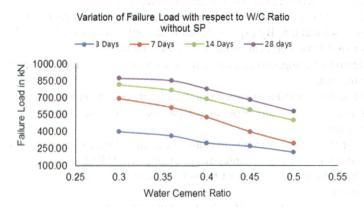

**Fig. 36** Effect of cement content on slump severe and very severe exposure condition without Super Plasticizer

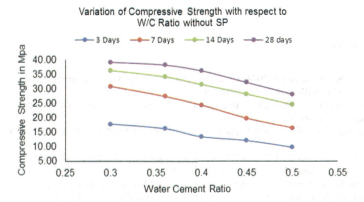

**Fig. 37** Variation of failure load with respect to age of concrete at different w/c ratio without using Super Plasticizer

## 6 Result and Discussion

In the present study, a series of testing had been carried out to investigate the rheological and mechanical properties of M40 grade concrete. As per IS 456:2000, the minimum grade of concrete for extreme exposure is M40. M40 grade standard concrete was considered for experimentation. The following conclusions are drawn based on the experimental results and guidelines of IS code [12].

1. Maximum free water–cement ratio for reinforced concrete structures with mild, moderate, severe, very severe and extreme environmental exposure condition are 0.55, 0.50, 0.45, 0.45 and 0.40, respectively.
2. Improvement in the workability of concrete is observed with the addition of superplasticizer. However, the cohesiveness of concrete is prejudiced with higher dosages of superplasticizer.
3. Superplasticizer can be used to reduce slump loss.
4. Slump cone test is performed on different water–cement ratios of 0.3, 0.36, 0.4, 0.45 and 0.5. By experimentation, the optimum dose of superplasticizer is found to be 1 percent by the weight of cement.
5. Slump and compaction factor increase with an increase in cement content per cubic metre.
6. A linear relationship between slump and cement content was obtained.
7. A linear relationship between compaction factor and cement content was obtained.
8. Desired compressive strength cannot be achieved without a superplasticizer.
9. Optimum values of compressive strength, i.e., M40 grade concrete can be achieved with water–cement ratios of 0.3 and 0.36 at extreme environmental exposure.
10. Water–cement ratio of 0.36 is the most appropriate. Though water–cement ratio of 0.3 gives desired compressive strength, workability of concrete is affected.

## References

1. Mehta PK, Monteiro PJ (2014) Concrete: structure, properties, and materials. McGraw-Hill, third edition. ISBN: 978-0-07-179787-0
2. Khayat KH (1999) Workability, testing, and performance of self-consolidating concrete. ACI Mater J 96(3):346–353
3. Pinto RC, Hover KC (1999) Application of maturity approach to setting times. ACI Mater J 96(6):686–691
4. Reinhardt HW, Grosse CU (2004) Continuous monitoring of setting and hardening of mortar and concrete. Constr Build Mater 18(3):145–154
5. Aruntas HY, Cemalgil S, Simşek O, Durmus G, Erdal M (2008) Effects of super plasticizer and curing conditions on properties of concrete with and without fiber. Mater Lett 62(19):3441–3443
6. IS:9013-1978, Indian Standard, Method of making, curing and determining compressive strength of accelerated-cured concrete test specimens

7. De la Verga I., Spragg RP, Di Bella C, Castro J, Bentz DP, Weiss J (2014) Fluid transport in high volume fly ash mixtures with and without internal curing. Cem Concr Compos 45:102–110
8. IS 1489 (Part-1): 1991, Portland-Pozzolana cement- specification
9. IS10262:2019, Concrete Mix Proporationing-Guidelines (Second revision).
10. IS 269:2013, Ordinary portland cement , 33 Grade-specification
11. IS 1489 (Part-1):2015, Portland pozolona cement fly ash based
12. IS 456:2000, Plain and Reinforced Concrete- Code of Practice.
13. IS 383:2016, Coarse and fine aggregates for concrete-specification (Third revision)
14. IS 9103: 1999, Concrete admixtures spcification

# Development of Sustainable Brick Using Textile Effluent Treatment Plant Sludge

**Uday Singh Patil, S. P. Raut, and Mangesh V. Madurwar**

**Abstract** A huge quantity of textile effluent treatment plant (TETP) sludge is generated from 21,076 units in India. Significant environmental impacts occur due to the landfilling of TETP sludge, such as land and water pollution. Thus, effective management of this sludge is important, which otherwise adds to the ever-escalating cost of disposal. With the rapid industrialization and urbanization, there is also an increased demand for construction materials to fulfill the shortage of housing in India (i.e., 18.78 million in urban and 43.9 million in rural). Brick is one of those significant construction materials, whose production is found to increase by 30% from 2000 to 2020 which consequently leads to an increase in carbon footprints. The present study, therefore, focuses on the effective utilization of TETP sludge to develop cost-effective, environmentally friendly bricks, which serve as an alternate solution for solid waste management, conservation of natural resources, and earning carbon credits. Sludge incorporated bricks were prepared with varying compositions of cement (6–24%), sludge (50–70%), and quarry dust (25%). Bricks are tested as per the Bureau of Indian Standards (BIS). The TETP sludge is characterized using X-ray fluorescence (XRF) and the properties of bricks were evaluated by conducting various tests such as compressive strength, water absorption, and density. Obtained results were also compared with commercially available fly ash bricks and clay bricks. The maximum strength of 4.2 $N/mm^2$ was observed for the combination of 24% cement, 51% sludge, and 25% quarry dust, which exceeds the value of 4 $N/mm^2$ for grade D, load-bearing units of IS: 2185(part1)—1979 and found to be greater than 3.5 $N/mm^2$ that meets the criteria of IS: 1077–1979 for bricks in load-bearing units. When textile sludge is used in the range of 50–57%, the water absorption value of bricks was found to be less than 20%, thus meeting the requirement of BIS. The resultant unit weight of the brick is also found to be lesser than the conventional bricks. Thus, it can be said that TETP sludge has the potential to develop sustainable

---

U. S. Patil (✉) · S. P. Raut
Department of Civil Engineering, Yeshwantrao Chavan College of Engineering, Nagpur, India
e-mail: patil.udaysingh4@gmail.com

M. V. Madurwar
Department of Civil Engineering, Visvesvaraya National Institute of Technology, Nagpur, India

bricks that meet the requirements of BIS, similar to the other sustainable materials, viz., concrete, mortar, etc., developed using other industrial wastes.

**Keywords** Sustainable brick · Material characterization · Physicomechanical properties · Sustainable construction material · Textile effluent treatment plant sludge

# 1 Introduction

In India, due to rapid growth, a huge amount of industrial waste is produced annually and gets accumulated, which is difficult to handle without proper management [1–10].

A substantial portion of the land is utilized for dumping these industrial wastes, thus causing soil and water pollution [11]. Therefore, effective management of this sludge is important which otherwise adds to the ever-escalating cost of disposal. Rapid urbanization has also increased the demand for building materials in urban and rural areas thereby exponentially rising the demand for new, economically viable building materials. As per the report on development alternatives [12] and technical group on an estimation of the urban /rural housing shortage [13], there will be a large demand for building materials to minimize housing shortages, i.e., 18.78 million housing shortages in urban areas with an approximate rural housing shortage of 43.9 million. It is seen that 90% of this scarcity applies to society's economically weaker and lower income classes, which is a cause of concern. The high construction costs cannot be afforded by an economically weaker segment, thus there is a need to increase the production of sustainable materials to satisfy these requirements. Consequently, an increase in output contributes to an increase in $CO_2$ emissions. It is observed that in India, about 22% of total $CO_2$ emissions are produced annually by the construction industry, of which 80% are mainly from industrial processes, involving the manufacture of steel, cement, bricks, and lime. One of the world's leading textile industries is the Indian textile industry. Approximately 21,076 textile units are distributed in India [14]. During the processing of textiles, a significant amount of water is used by the textile industry. As a result, an enormous amount of wastewater is produced, which needs to be properly handled before it is disposed of safely. In the effluent treatment plant units, this industrial wastewater is treated and a large amount of sludge is produced during the treatment process. This sludge is then disposed of for landfilling, resulting in pollution of the environment. The present study, therefore, focuses on the effective utilization of TETP sludge to develop cost-effective, environmentally friendly bricks that serve as an alternate solution for solid waste management, conservation of natural resources, and earning carbon credits.

## 2 Methodology for the Development of Sustainable Construction Material

### 2.1 Collection of Raw Materials

#### 2.1.1 Collection and Preparation of Sludge Sample

Dewatered and open-air dried textile sludge sample was collected from the Morarjee textile industry located in MIDC Butibori, Nagpur (Figs. 1 and 2). The sludge obtained from the industry was in semi-solid form, therefore it was sun-dried until

**Fig. 1** Image of dry textile effluent sludge

**Fig. 2** TETP sludge before pulverization

the sludge is completely dried. The dried samples were then ground and pulverized to convert them into fine powder. This material was then sieved through 300 μ for the proper bonding between cement and sludge. The percentage of various size particles in the textile effluent sludge sample is determined using sieve analysis. The coefficient of curvature and uniformity coefficient is also calculated.

### 2.1.2 Collection of Cement and Quarry Dust

Similarly, other raw materials such as cement and quarry dust are collected. The cement used is the Ordinary Portland Cement (OPC) of Grade 53 confirming to BIS, IS:12269 [15]. The consistency limit of cement and cement with different percentages of sludge is determined using standard Vicat apparatus IS: 4031-4 [16]. Initial and final setting time tests were also performed using the Vicats apparatus. A density bottle test is performed to measure the specific gravity of the cement and sludge IS: 4031-11 [16]. Quarry dust is a by-product, released from the cutting and crushing process of stone and was collected from Gupta Industries located at MIDC Buti Bori, Nagpur.

## 2.2 Chemical Characterization of Textile Effluent Treatment Plant Sludge

The sludge is characterized (Tables 2 and 3) for various physicochemical parameters using an X-ray fluorescence test (Fig. 3). The USEPA and toxicity characteristic leaching procedure (TCLP) (Table 4) tests were performed on developed bricks. The leaching test was carried out at ANACON Laboratories Pvt. Ltd., Butibori MIDC, to check the feasibility of textile sludge for the development of sustainable bricks.

## 2.3 Manufacture of Brick Specimens

Sludge integrated bricks with different cement and sludge compositions (6–30 % wt.) were prepared (Fig. 4) with a constant proportion of quarry dust (Table 1) in a mould of 230 × 150 × 1000 mm. The required quantities of raw materials were calculated for various combinations of cement, sludge with quarry dust, and then hand-mixed with a required quantity of water until a homogenous mixture is obtained. The material is then poured into the mould in three layers and immediately after the casting, each layer is tamped 25 times to expel the entrapped air present in it. (Fig. 5) After forming, the bricks were sun-dried for about 15 days. The compressive strength of bricks is evaluated as per IS 3495- 1, Cl.4.1.4. Three [17] samples were subjected to a compressive strength test in the Universal Testing Machine after 28 days of curing.

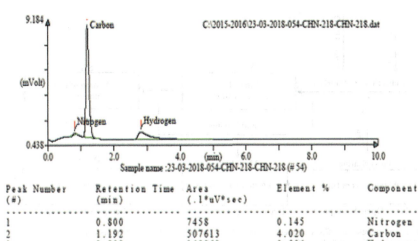

Fig. 3 Chemical characterization of TETP Sludge

Fig. 4 Formation of brick specimens

Table 1 Specimen ID with proportions

| S. No. | Proportions | ID |
|---|---|---|
| 1 | 6% cement+69% sludge+25% quarry dust | 6C94S25Q |
| 2 | 9% cement+66% sludge+25% quarry dust | 9C66S25Q |
| 3 | 12% cement+63% sludge+25% quarry dust | 12C63S25Q |
| 4 | 15% cement + 60% sludge+25% quarry dust | 15C60S25Q |
| 5 | 18% cement+57% sludge+25% quarry dust | 18C57S25Q |
| 6 | 21% cement + 54% sludge + 25% quarry dust | 21C54S25Q |
| 7 | 24% cement+51% sludge+25% quarry dust | 24C51S25Q |

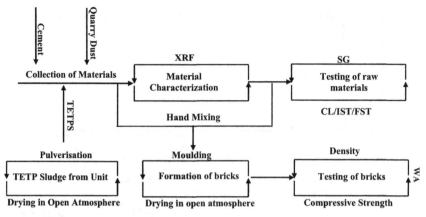

Fig. 5 Flowchart for development of sustainable bricks [19]

The average result of these specimens is taken as the brick's compressive strength. For the comparison of strength, different samples of commercially available bricks such as clay brick and fly ash brick have been taken. To obtain water absorption values, three samples of bricks were selected. The bricks are weighed in dry condition and immersed in water for 24 h. After 24 h, the brick is properly wiped and the weight is taken in wet condition. For the calculation of the percentage of water absorption, IS 3495- 2, Cl.4.1.4 [18] is referred.

## 3 Results and Discussion

### 3.1 Chemical Characterization of Textile Effluent Treatment Plant Sludge

From Table 2, the specific gravity of sludge obtained is 2.4. Total volatile solids of about 31.85% are found, which will result in an increase in ash content during incineration, hence it is not recommended as a technique for sludge disposal. During the treatment process of textile wastewater, the addition of excess lime makes the concentration of calcium oxide as 108.22 mg/l, which is considered one of the chief components of the sludge and affects the pH content. The pH of sludge is found to be 9.13, which shows its alkaline nature. The presence of various oxides (Table 3) shows its potential to develop a sustainable construction material. The present study is also focused on the concentration of heavy metals such as Cu, Ni, Cd, Pb, Zn, Co, and Cr in TETP sludge, which is commonly found in the textile effluent due to the usage of dyes and other chemicals. The presence of metals in TETP sludge is of concern

**Table 2** Characterization of textile effluent treatment plant sludge waste

| S. No. | Property | Values |
|---|---|---|
| 1 | Water content (%) | 28.72 |
| 2 | Specific gravity | 2.4 |
| 3 | pH | 9.13 |
| 4 | Average particle size | – |
| 5 | Cadmium(mg/kg) | 3.96 |
| 6 | Copper (mg/kg) | 57.48 |
| 7 | Total chromium(mg/kg) | 2.98 |
| 8 | Zinc (mg/kg) | 91.6 |
| 9 | Nickel (mg/kg) | 0.68 |
| 10 | Lead (mg/kg) | 12.1 |
| 11 | Ferrous (mg/kg) | 180.5 |
| 12 | Sulphates (mg/l) | 116 |
| 13 | Sulphides (mg/l) | BDLa |
| 14 | Calcium (mg/l) | 108.22 |
| 15 | Magnesium (mg/l) | 154.30 |
| 16 | Chlorides (mg/l) | 5445 |
| 17 | Total hardness as $CaCO_3$ (mg/l) | 905 |
| 18 | Total volatile solids | 31.85% |

**Table 3** Chemical composition of textile effluent treatment plant sludge waste (% By Mass)

| S. No | Composition | Sludge (%) |
|---|---|---|
| 1 | $SiO_2$ | 14.85 |
| 2 | $Al_2O_3$ | 2.87 |
| 3 | $Fe_2O_3$ | – |
| 4 | CaO | 21.04 |
| 5 | MgO | 9.53 |
| 6 | $K_2O$ | – |
| 7 | $Na_2O$ | – |
| 8 | $SO_3$ | – |
| 9 | $SO_4$ | 1.55 |
| 10 | $TiO_4$ | 1.12 |
| 11 | LOI | – |

because of its toxicity to aquatic and mammalian species. The possible sources of metals are incoming fibre, water, dyes, and chemical impurities. Some dyes include metals as an integral part of the dye molecule. The concentrations of heavy metals are compared with the CPCB guideline (Table 4) and found that the concentrations of all heavy metals, including chromium species, are within the regulatory limits

**Table 4** TCLP test result on TETP sludge

| S. No. | Composition | Test method | Limits as per CPCB guideline | Test result (mg/l) |
|---|---|---|---|---|
| 1 | Arsenic | USEPA test method | Max. 5 | absent |
| 2 | Barium | | Max.100 | 0.13 |
| 3 | Cadmium | | Max.1 | absent |
| 4 | Chromium | | Max.5 | absent |
| 5 | Lead | | Max.5 | absent |
| 6 | Manganese | | Max.10 | 0.07 |
| 7 | Mercury | | Max.0.2 | absent |
| 8 | Selenium | | Max.1 | absent |
| 9 | Silver | | Max.5 | absent |
| 10 | Ammonia | TCLP | Max.50 | 10.15 |
| 11 | Cyanide | | Max.20 | 0.02 |
| 12 | Nitrate | | Max.1000 | 8.51 |

indicating that the sludge is non-hazardous. Hence, this sludge can be explored for the possibility of reuse and recycling using some suitable technology rather than disposing of it in a landfill. It is also observed that the results of the characterization of TETP sludge vary from industry to industry and depend on the type of chemical used during the processing of textiles.

## 3.2 *Physical Tests on Raw Materials*

The physical properties of the cement and quarry dust are shown in Table 5. The specific gravity of cement obtained is 3.15 and that of quarry dust is 2.64. Results

**Table 5** Physical test results of cement and quarry dust

| S. No. | Tests conducted | Test results Cement | Quarry dust |
|---|---|---|---|
| 1 | Specific gravity | 3.15 | 2.64 |
| 2 | Density | 1440 kg/m$^3$ | 1650 kg/m$^3$ |
| 3 | Water absorption | – | 10.6 |
| 4 | Standard consistency | 30% | – |
| 5 | Initial setting time | 80 min | Not less than 30 min |
| 6 | Final Setting time | 125 min | Not more than 600 min |

**Fig. 6** Sieve analysis of TETP sludge

of grain size analysis of dried textile sludge are shown in Figs. 7 and 8. The particle size distribution of the sludge sample (Fig. 6) shows that the maximum percentage of sludge particles was retained on a 0.075 mm sieve. The curve shows that the sample of textile sludge consists of materials of all sizes. Effective sizes of the textile sludge before pulverization were obtained as D10 = 0.019 mm, D30 = 0.14 mm and D60 = 0.4 mm. Whereas, after pulverization, it is D10 = 0.014 mm, D30 = 0.13 mm and D60 = 0.3 mm. The values of uniformity coefficient Cu and coefficient of curvature Cc were 21.045 and 2.57, respectively. Whereas, after pulverization, the values of Cu and Cc obtained are 21.42 and 4.02, respectively.

## 3.3 Physico-Mechanical Tests on Brick

### 3.3.1 Density of Sustainable Bricks

The clay bricks normally have a bulk density of 1.8–2.0 g/cm$^3$. It is found that the dry density of the TETP sludge is lower, so the resulting unit weight of the material would get reduced when used as building materials. In the current research, similar findings have been found. From Fig. 9, it can be seen that with the increase in the cement and corresponding decrease in the textile sludge, the density of bricks increases.

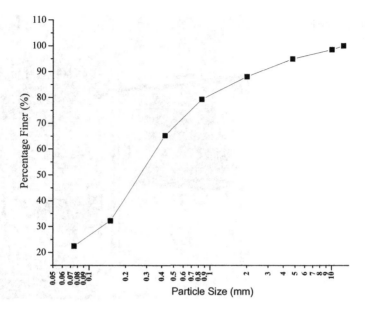

**Fig. 7** Grain size analysis before pulverizing

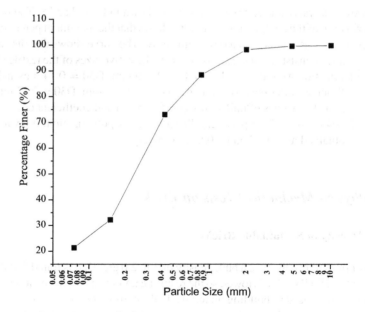

**Fig. 8** Grain size analysis after pulverizing

Development of Sustainable Brick Using Textile Effluent ...

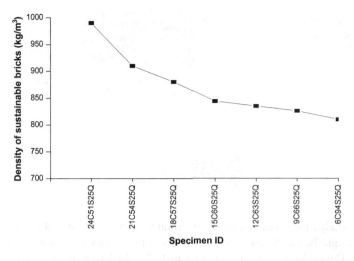

**Fig. 9** Density of sustainable bricks

### 3.3.2 Compressive Strength of Sustainable Bricks

Figures 10 and 11 demonstrate the results of the average compressive strength of bricks. It is observed from Fig. 10 that, when TETP sludge is used as a partial substitute for cement, the compressive strength of the bricks decreases as the percentage

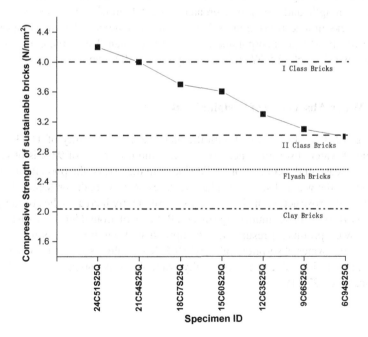

**Fig. 10** Compressive strength of sustainable bricks

**Fig. 11** Compressive strength test on third sample of the brick

of textile sludge increases. There is a maximum decrease of 29%, this reduction in strength might be due to fineness in particles of textile sludge compared to cement particles. The sludge is hydroscopic in nature [20], which increases the volume of the sample, so the demand for the water increases to preserve workability during mixing, thereby affecting the water–cement ratio. This may cause a reduction in the strength of bricks. The existence of chloride (Table 2) in the TETP sludge often reduces the quality of construction materials [21]. The obtained strength was found to meet the specifications of non-structural materials or components [22], as stated in various standards. The maximum strength of 4.2 N/mm$^2$ was observed for the combination of 24% cement, 51% sludge, and 25% quarry dust, which is greater than 4 N/mm$^2$ for grade D, load-bearing units of IS: 2185-1 [23], hence meets the criteria of required minimum strength, and found to be greater than 3.5 N/mm$^2$ as specified in IS: 1077 [24] for bricks in load-bearing units. The results also indicate that the compressive strength for the sludge incorporated brick is higher than that of the commercially available fly ash brick (2.5 N/mm$^2$) and clay brick (2.02 N/mm$^2$).

### 3.3.3 Water Absorption of Sustainable Bricks

From Fig. 12, it is observed that with the increase in the quantity of TETP sludge the value of water absorption increases. A maximum increase of 93% is observed. With an increase in the content of TETP sludge in the bricks, the porosity increases consequently the water absorption value increases [22]. Also, the presence of higher voids in the microstructure of sludge bricks subsequently increases the consumption of water [14]. With an optimum usage of TETP sludge of around 50%, the sustainable bricks showed promising results, i.e., compressive strength of 4.2 N/mm$^2$ with a corresponding water absorption value of 16%. This value is found to be less than the water absorption value of commercially available fly ash bricks, i.e., 18.43% and clay bricks, i.e., 27.11%.

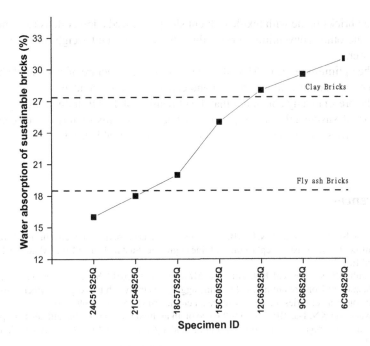

**Fig. 12** Water absorption value of the brick

## 4 Conclusions

The disposal of textile effluent treatment plant sludge has a detrimental effect on the environment. Thus, its effective utilization in the production of the bricks is carried out, which proves to be a good alternative solution to such problems. Thus following conclusions have been drawn based on the detailed experimental investigation:

1. The presence of calcium and magnesium in sludge suggests its potential as a partial substitute for the development of sustainable building materials.
2. The heavy metals are found to be within regulatory limits and can therefore be used directly as a raw material without producing any harmful effects.
3. Developed bricks showed the maximum strength of 4.2 N/mm$^2$ for the combination of 24% cement, 51% sludge, and 25% quarry dust, which exceed the value of 4 N/mm$^2$ for grade D, load-bearing units of IS: 2185(part1)–1979 [23] and found to be greater than 3.5 N/mm$^2$ that meets the criteria of IS: 1077–1979 [24] for bricks in load-bearing units.
4. The textile sludge incorporated bricks were found to have strength greater than conventional fly ash and clay bricks.
5. When textile sludge is used in the range of 50–57%, the water absorption value of bricks was found to be less than 20%, thus meeting the requirement according to standards.

6. The bricks made with textile effluent sludge showed a lesser density compared to the other conventional bricks thus the resultant unit weight of the brick is lesser.
7. The optimum range of 51% of sludge for the manufacture of sustainable bricks in combination with quarry dust and cement is recommended.
8. The present study concludes that the utilization of TETP sludge for the production of sustainable bricks provides promising results of physico-mechanical properties when compared with commercially available bricks.

# References

1. Goudar SK, Shivaprasad KN, Das BB (2019) Mechanical properties of fiber-reinforced concrete using coal-bottom ash as replacement of fine aggregate. Sustain Constr Build Mater Lect Notes Civil Eng 25
2. Sharath BP, Shivaprasad KN, Athikkal MM, Das BB (2018) Some studies on sustainable utilization of iron ore tailing (IOT) as fine aggregates in fly ash based geopolymer mortar. In: IOP conference: series, material science an engineering, vol 431, p 092010
3. Shivaprasad KN, Das BB (2017) Influence of alkali binder dosage on the efficiency of pelletization of aggregates from iron ore tailing and fly ash. Int J Eng Res Mechan Civil Eng 2(3):388–392
4. Shivaprasad KN, Das BB, Sharath PS (2020) Pelletisation factors on the production of fly-ash aggregates and its performance in concrete. Proc Inst Civil Eng Constr Mater 1–20
5. Sharath BP, Das BB (2021) Production of artificial aggregates using industrial by-products admixed with mine tailings—a sustainable solution, recent trends in civil engineering. Lect Notes Civil Eng 105:383–397
6. Snehal K, Das BB, Kumar S (2020) Influence of integration of phase change materials on hydration and microstructure properties of nanosilica admixed cementitious mortar. J Mater Civil Eng 32(6):04020108
7. Goudar SK, Das BB, Arya SB (2019) Microstructural study of steel-concrete interface and its influence on bond strength of reinforced concrete. Adv Civil Eng Mater 8(1):171–189
8. Goudar SK et al (2020) Influence of sample preparation techniques on microstructure and nano-mechanical properties of steel-concrete interface. Constr Build Mater 256:119242
9. Sumukh EP, Sharan KG, Das BB (2021) Predicting the service life of reinforced concrete by incorporating the experimentally determined properties of steel–concrete interface and corrosion. In: Recent trends in civil engineering. Springer, Singapore, pp 399–417
10. Sumukh EP, Sharan KG, Das BB (2021) A review on the properties of steel-concrete interface and characterization methods. In: Smart technologies for sustainable development. Springer, Singapore, pp 167–203
11. Balasubramanian J et al (2006) Reuse of textile effluent treatment plant sludge in building materials. Waste Manage 26(1):22–28. https://doi.org/10.1016/j.wasman.2005.01.011
12. Ministry of Rural Development (2011) Working group on rural housing for the 12th five-year plan. New Delhi, India
13. Ministry of Housing and Urban Poverty Alleviation. "Report of the technical group (TG-12) on urban housing shortage".2011, NBO, India
14. Shathika Sulthana Begum B et al (2013) Utilization of textile effluent wastewater treatment plant sludge as brick material. J Mater Cycles Waste Manage 15(4) 564–570. https://doi.org/10.1007/s10163-013-0139-4
15. BIS (Bureau of Indian Standard) IS: 12269:1987. Specification for 53 Grade Ordinary Portland Cement. BIS, New Delhi, India

16. BIS (Bureau of Indian Standard) IS: 4031(part 4):1980. Methods of Physical Tests for Hydraulic Cement. BIS, New Delhi, India
17. BIS (Bureau of Indian Standard) IS 3495–1992 (Part 1). Methods of Tests of Burnt Clay Building Bricks: Determination of Compressive Strength. BIS, New Delhi, India
18. BIS (Bureau of Indian Standard) IS 3495–1992 (Part 2). Methods of Tests of Burnt Clay Building Bricks: Determination of Water Absorption. BIS, New Delhi, India
19. Patil U, Raut SP, Ralegaonkar RV, Madurwar MV (2021) Sustainable building materials using textile effluent treatment plant sludge: a review. Green Mater 1–15. https://doi.org/10.1680/jgrma.21.00027
20. Kaur H (2019) Utilization of textile mill sludge waste in concrete—an experimental study. Int J Pure Appl Biosci 7(5):179–85. https://doi.org/10.18782/2320-7051.7615
21. Raghunathan T, Gopalsamy P, Elangovan R (2010) Study on strength of concrete with ETP sludge from dyeing industry. Int J Civ Struct Eng 1(3):379–389
22. Rahman, Md. Mostafizur, et al (2017) Textile effluent treatment plant sludge: characterization and utilization in building materials. Arab J Sci Eng 42(4):1435–42. https://doi.org/10.1007/s13369-016-2298-9
23. BIS (Bureau of Indian Standard) (1979) IS: 2185(1):1979. Specification for concrete masonry units - hollow and solid concrete blocks. BIS, New Delhi, India
24. BIS (Bureau of Indian Standard) IS: 1077–1979. Common Burnt Clay Building Bricks-Specification. BIS, New Delhi, India

# Utilization of Pozzolanic Material and Waste Glass Powder in Concrete

**Lomesh S. Mahajan and Sariputt R. Bhagat**

**Abstract** The material of glass has been used in various forms for versatile applications but it has a low life span compared to other materials. After the utilization of glass products, it has been either used as landfills or stored with stack piled. The landfills with broken glass products could not be the right choice, as it is a nonbiodegradable substance. Due to the strong need for alternative solution to landfills, the glass has been used in the concrete industry on a trial basis. For the concrete industry, mainly three places of waste glass is tried to suit, i.e. replacement to coarse aggregate, fine aggregate, and cement. However, the replacement of coarse aggregate has found lower results in compressive strength perspective. The similar phase of pozzolanic materials has seen in construction applications. The present work introduces the use of glass powder as a replacement for cement to assess the pozzolanic activity of fine glass powder in concrete. The experimentally evaluates its performance with other pozzolanic materials like silica fume and fly ash. The compressive strength study is conducted by considering specimens 15% and 30% replacement of cement by silica fume, fly ash, and glass powder. In addition, the particle size effect is evaluated using glass powder of size 150–100 μm. After tests, the results concludes that waste glass power has pozzolanic behavior. It reacts with lime at the early stage of hydration forming extra C-S-H gel and easily forming denser cement matrix and increases the durability property of concrete. When compared with fly ash mix concrete, fine glass powder concrete found slightly more strength.

**Keywords** Pozzolanic material · Cement · Fly ash · Silica fume · Glass

---

L. S. Mahajan (✉)
Research Scholar, Dr. Babasaheb Ambedkar Technological University, Lonere 402103, India
e-mail: loms786@gmail.com

S. R. Bhagat
Head, Department of Civil Engineering, Dr. Babasaheb Ambedkar Technological University, Lonere 402103, India
e-mail: srbhagat@dbatu.ac.in

© The Author(s), under exclusive license to Springer Nature Singapore Pte Ltd. 2023
M. S. Ranadive et al. (eds.), *Recent Trends in Construction Technology and Management*, Lecture Notes in Civil Engineering 260,
https://doi.org/10.1007/978-981-19-2145-2_16

# 1 Introduction

In India, about 960 million tons/year quantity of solid waste is thrown into the environment and out of this, about 21 million shares are of glass waste materials. Nowadays, these unutilized wasted glass materials are suitably addressed in the construction materials. Concrete is the indispensable complex material in the construction arena. The concrete industry depends on raw materials such as cement, sand, and aggregate. The mixture of concrete is generally prepared with Portland or pozzolana cement, sand, aggregate, and water content. All the civil engineering works are designed with concrete element to withstand the harsh environment. The use of pozzolanic substances is to minimize the reliability of cement and similarly, the waste material of glass is creditable to minimize the aggregate share in concrete. The global warning scenario becomes harmful due to huge cement production.

In the general case, glass does not create a nuisance in the environment because it is a non-polluted substance, but it causes harm to human beings. Pollution is created due to the dumping of glass on the land. The waste can be minimize the environmental hazards by using concrete as binding material. Glass material is categorized into nonbiodegradable substances. Thus, to sustain the environmental needs, we have to alter technology developed. Concrete could be the good place for accomplish such waste piece of glasses. The term "glass" contains several chemical uniformities including soda–lime silicate, alkali-silicate, and borosilicate glass. In few places, the powder has been used as concrete for cement and aggregate due to pozzolan property. The alkali content of cement is improved by adding the waste glass substance. Waste glass is also helpful in the ceramic and brick manufacturing industry. It benefits are decrease in the raw material requirement and protects landfill areas. The glass provides a way to recycle with recycling ratio near to 100% and can be safely used in concrete without scaring quality. de Moura et al. [7] stated that the fine glass powder improves the previous concrete's property. Waste glass powder associated with the type and size ratio provided a positive effect on concrete durability [11]. The long-term durability enhancement was seen due to pozzolanic reactivity. Mineral additives mass and types of additives are also equally responsible for the strength of concrete.

Enough work is carried out to explore the pozzolanic benefits in the concrete for enhancing their properties ([12–16]). Ternary blends of fly ash, silica fume, and ordinary Portland cement offers good properties over binary cementitious blends [10]. Hyeongi [5] were investigated silica-based industrial additives for cement substituent. Partial replacement upto 20% of cement weight was used for assessing mechanical property and found that glass powered mix concrete exhibited more compressive strength. The amorphous nature of glass compounds possesses reduction in porosity. Ali [1] found pozzolanic characteristics, but irrelevant effect on setting time and expansion test of blended material. The range for incorporation of glass powder was suggested up to 20% for getting an improvement in the strength of concrete. The glass causes the difficulties of damaging alkali-silica reaction. If the glass is superbly ground, powder achieves the pozzolanic belongings of cement which is essential for hydration. The cement is replaced by 5–20% of glass by dropping the cost of the binding material. Fly ash and bottom ash are useful for mortar strength [9]. The combination of cement, fly ash, and silica fume is also effective in the mortar for opting for beneficial workability and strength. The slag and fly ash benefits in concrete have been studied by Bijen [2], and alkali silica reaction was found beneficial in their investigation. Lam et al. [8] investigated the fracture behavioral effect of fly ash and silica fume and reported that enhancement in strength possible is due to silica fume–fly ash concrete.

This study focused on the pozzolanic use of waste glass powdered for a partial amount of cement in concrete. In addition, the effect of fly ash and silica use is evaluated. Various strategies have been adopted to improve the strength of fly ash concrete. Different approaches have been used to improve the early age strength of fly ash concrete. These consist of the addition of silica fume, high fineness fly ash use, and addition of hydrated lime [3, 4, 6].

## 2 Experimental Program

The mix of concrete M20 grade was chosen for the experimental program. OPC cement: 53 grade, Fly ash: F grade, and Silica fume: ASTM-C (1240) materials were used for the experimentation as cementations materials. The engineering properties of materials are presented in Table 1 and related consistency is presented in Table 2.

Table 1  Chemical properties of binder mixes

| Mineral additives | $SiO_2$ | $Al_2O_3$ | $Fe_2O_3$ | MgO | Cao | $K_2O$ | $Na_2O$ | $SO_3$ |
|---|---|---|---|---|---|---|---|---|
| Glass | 70.88 | 2.18 | 0.52 | 1.41 | 10.80 | 0.26 | 12.99 | 0.1 |
| Cement | 20.62 | 5.22 | 3.11 | 0.79 | 65.69 | 0.59 | 0.15 | 3.2 |
| Silica Fume | 96.21 | 0.28 | 0.48 | 0.45 | 0.35 | 0.49 | 0.28 | 0.18 |
| Fly ash | 33.44 | 19.55 | 5.28 | 6.54 | 26.27 | 0.1 | 1.44 | 2.39 |

Table 2  Normal consistency of binder mixes

| Mix designation | Description | Cement (g) | Silica fume (gm) | Fly ash (gm) | Glass powder (gm) | Consistency in (%) |
|---|---|---|---|---|---|---|
| C | Cement | 300 | 0 | 0 | 0 | 31.6 |
| C-S | C with 15% SF | 255 | 45 | 0 | 0 | 38.67 |
| C-F | C with 15% F | 255 | 0 | 45 | 0 | 39.3 |
| C-G | C with 15% G | 255 | 0 | 0 | 45 | 37.3 |

*C* Cement; *SF* Silica fume; *F* Fly ash; *G* Glass powder

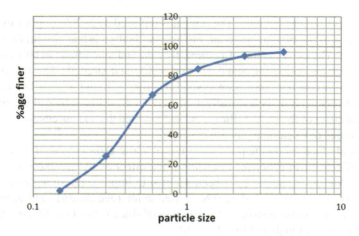

**Fig. 1** Particle size analysis of locally available river sand

Locally available Tapi river sand is used as fine aggregate in the present work. The details of particle size are shown in Fig. 1. Fifteen percent and 30% cement replacement concrete specimens are cast and tested in the compressive testing machine after 28 days and 56 days curing period. The particle size (150–100)μ effect of glass powder is investigated. Capillary absorption test is also performed to check the alkali aggregate reaction. IS: 4031 Part4 (1988) is used for normal consistency testing. The compressive strength of various mix condition cubes is reported in Table 3.

A compression examination typically provides an overall interpretation of the property of concrete because strength linked directly to hydrated cement paste. The strength enhancement in the concrete is significantly tested by the compression test. As the results obtained, compressive strength for 28 days seen medium range as compared to the combination of C-S and C-F. For the 30% partial replacement specimens, the C-S combination was found in the same range, while C-F was found to have 24.19% reduction. C-G combinations were found to have much lesser values

**Table 3** Compressive Strength Results

| Silica fume content (%) | Designation of samples | Compressive strength N/mm$^2$ (28 Days) | Compressive strength N/mm$^2$ (56 Days) |
|---|---|---|---|
| 30 | C(1) | 25.03 | 30.21 |
| 30 | C-S(1) | 25.48 | 29.59 |
| 30 | C-F(1) | 20.97 | 22.49 |
| 30 | C-G(1) | 17.89 | 22.68 |
| 15 | C(2) | 26.09 | 31.35 |
| 15 | C-S(2) | 27.87 | 29.98 |
| 15 | C-F(2) | 20.55 | 23.83 |
| 15 | C-G(2) | 21.01 | 24.53 |

for higher replacement, i.e., for 30%. This depletion goes 25.18% lesser at 15% replacement, while 28.52% reduced results obtained at 30% SF content. All obtained results of M20 designed concrete suitably addressed utility for a combination of 15% SF with cement and glass.

## 3 Conclusion

- Cement can be substituted by waste glass powder possessing pozzolanic behavior.
- Glass powdered concrete results are more committable with fly ash concrete due to fineness property.
- Silica fume is the better pozzolanic material and provides higher strength due to their smaller grain shape.
- The glass powdered, silica fume, and fly ash can easily replace cement in concrete up to 15% without failing original results.
- All obtained results of M20 designed concrete suitably addressed utility for a combination of 15% SF with cement and glass. The mechanical strength found in silica fume, fly ash, glass are adaptable materials as cement replacement by 15%. If considering the observed reading of 56 days curing period, the adaptive range of pozzalanic materials modified up to 30%.

## References

1. Aliabdo AA, Abd Elmoaty M, Abd Elmoaty, Aboshama AY (2016) Utilization of waste glass powder in the production of cement and concrete. Constr Build Mater 124:866–877
2. Bijen J (1996) Benefits of slag and fly ash. Constr Build Mater 10(5):309–314
3. Barbhuiya SA, Gbagbo JK, Russell MI, Basheer PAM (2009) Properties of fly ash concrete moidfied with hydrated lime and silica fume. Int J Constr Build Mater 23:3233–3239
4. Chindaprasirt P, Chai J, Sinsri T (2005) Effect of fly ash fineness on compressive strength and pore size of blended cement paste. Cem Concr Res 27:425–428
5. Lee H, Hanif A, Usman M, Sim J, Oh H (2018) Performance evaluation of concrete incorporating glass powder and glass sludge wastes as supplementary cementing material. J Cleaner Prod 170:683–693
6. Jeyakumar M, Salman AM (2011) Experimental study on sustainable concrete with the mixture of low calcium fly ash and lime as partial replacement of cement. Adv Mater Res 250–253
7. de Moura JMBM, Pinheiro IG, Aguado A, Rohden AB (2021) Sustainable pervious concrete containing glass powder waste: performance and modelling. J Cleaner Prod 316:128213
8. Lam L, Wong YL, Poon CS (1998) Effects of fly ash and silica fume on compressive and fracture behavior of concrete. J Cem Concrete Res 28:271–283
9. Mahajan LS, Bhagat SR (2020) The contribution of bottom ash toward filler effect with respect to mortar. Lect Notes Civil Eng 72:145–154
10. Thomas MDA, Shehata MH (1999) Use of ternary cementitious systems containing silica fume and fly ash in concrete. Cem Concrete Res 29:487–495
11. Dong W, Li W, Tao Z (2021) A comprehensive review on performance of cementitious and geopolymeric concretes with recycled waste glass as powder, sand or cullet. Resour Conserv Recycl 172:105664

12. Sahoo S, Das BB (2019) Mineralogical study of concretes prepared using carbonated flyash as part replacement of cement. Sustain Constr Build Mater519–529. Springer, Singapore
13. Shetti AP, Das BB (2015) Acid, alkali and chloride resistance of early age cured silica fume concrete. Adv Struct Eng 1849–1862. Springer, New Delhi
14. Sahoo S, Das BB, Mustakim S (2017) Acid, alkali, and chloride resistance of concrete composed of low-carbonated fly ash. J Mater Civil Eng 29(3):04016242
15. Sahoo S et al (2015) Acid, alkali and chloride resistance of high volume fly ash concrete. Indian J Sci Technol 8(19):1
16. Das BB, Pandey SP (2011) Influence of fineness of fly ash on the carbonation and electrical conductivity of concrete. J Mater Civ Eng 23(9):1365–1368

# Recent Trends in Construction Technology and Management

# Integrating BIM with ERP Systems Towards an Integrated Multi-user Interactive Database: Reverse-BIM Approach

M. Arsalan Khan

**Abstract** In the context of rapid technological advancements, the Architectural, Engineering and Construction sectors (AEC) are preparing themselves towards the enhanced level of co-ordination. This requires integration at every phase of the construction process, including design, processes, engineering services, fabrication, construction and maintenance phases—all together to help implement Level-3 Building Information Modeling (BIM). In this regard, the current understanding of BIM models is largely restricted to the graphical representation of the services, such as architectural model, structural model, plumbing model, etc. However, such decisions are confined to their respective departments before a 'blueprint' is released; for example, a design office may require changes to be incorporated into an architectural drawing or vice-versa. Such conflict of interests or process repetitions, particularly during planning, design and implementation stages, are identified as the main sources of significant delays in the overall project. For an integrated project delivery to overcome the fragmentation or loss of information, this paper proposes an architecture and demonstrates the feasibility of the implementation of Enterprise Resource Planning (ERP) system as a method to integrate the BIM processes among different departments that are using a heterogeneous environment of software packages. It is identified that the concept of ERP can be utilized not only to generate a read-only information database, but also rather reversibility in processes can be achieved by developing an interactive (read and write) database format to reach a consensus-based decision between seemingly scattered departments. A wide range of tasks is identified to be integrated during the planning, design, construction and operation phases. This will eliminate unnecessary time delays, for example, by minimizing or eliminating rotation of 'blueprints', enhancing the transfer of knowledge and overcoming the information-fragmentation. With a worked-out example, it is established that the available products on BIM-ERP integration can be expanded to directly

M. A. Khan (✉)
Department of Civil Engineering, Z.H. College of Engineering and Technology, Aligarh Muslim University, Aligarh, Uttar Pradesh 202001, India
e-mail: mohd.arsalan.khan@hotmail.co.uk

School of Architecture, Building and Civil Engineering, Loughborough University, Loughborough LE11 3TU, UK

integrate structural or other heterogeneous software, but this requires realignment of processes.

**Keywords** Building information modelling · Enterprise resource planning · Collaborative working · SMEs · Systems integration

# 1 Introduction

Utilising the strength of computers in recent years, after the introduction of CAD software as a major breakthrough to generate graphical representation, researchers are introducing multiple dimensions to a drafting tool to call it Levels of BIM, moving from BIM Level 1 (models, objects, collaboration) to Level 3 (transactable, interoperable data), which is seen as an object-intelligent architectural CAD tool [1, 2]. The work of [3–5] identifies key literature and examples to contribute towards the current significance of the BIM model explaining consistent efforts that are being made to incorporate up to seven dimensions. That is, a 3D model (BIM Level 1) is evolved through the incorporation of virtual construction and space-conflict identification method(s), time and cost estimation, sustainability and maintenance aspect in BIM Level 2 that can provide point solutions. Mark Bew and Mervyn Richards model [6] suggests that BIM Level 3 maturity requires building lifecycle management through Integrated Web Services (BIM Hub development); now the progress is resulting in the launch of standard BS EN ISO 19650 [7].

With the progress of research work into BIM technologies, Hosseini et al. [8] asserted that the BIM and collaboration on BIM-assisted construction projects are subsets of ICT (Information and Communication Technology). In order to foster collaboration, Sackey et al. [9] maintained that the members of BIM-assisted construction projects are to become constant leaders. For example, Merschbrock [10] attempted to enhance collaboration through advanced collaboration technologies. Although, Dassault Systems [11] are seemingly the first to suggest BIM Level 3 integration; however, the initiation and adoption of such methodologies require additional expertise, there are no official leaders in the building industry to take a decisive first step; meanwhile, the architects and engineers find themselves caught within the nitty–gritty of the procurement logic on a project-by-project basis that does not support long-term development and sustainable drives. Thus, an ERP platform needs to be project and asset centric, and support functions specific to the industry. Meanwhile, such solutions should also be flexible as per their deployment needs so as to cater to the needs of users to achieve the desired solution at their own pace.

For example, the communication gaps and delays particularly at designer's, owner's and contractor's end in case of small building projects [12]. Although within the AEC market, these require extensive data-driven processes, however, the adoption of integration systems (ERP/EAM/project management) may still be widely feasible within micro, small and medium-sized enterprises (SMEs).

Putting concisely the focus of this paper, it is proposed that ERP integration techniques may be extended to handle the core tasks of AEC sectors. This can be integrated with BIM; for this, a re-engineering process is also proposed. It is suggested that this may be a cost-effective approach, particularly for SMEs, whose structure is relatively easier to revamp and implement ERP-BIM integration. Through a worked-out example, it is demonstrated that this would allow to-and-fro coordination between previously isolated departments before reaching a final decision which would now be binding from the initial stage of the project, achieved through the involvement of all the parties from the beginning of the project life cycle.

## 1.1 ERP System and Integration

ERP referred to as 'Enterprise-wide Information Systems' [13] or 'Enterprise Systems' [14, 15] is defined as 'configurable information systems packages' that integrate information and information-based processes within and across functional areas in an organization [16]. ERP systems are also defined as 'computer-based systems designed to process an organization's transactions and facilitate integrated and real-time planning, production and customer response' [17]. According to Davenport [15], 'ERP comprises of a commercial software package that promises the seamless integration of all the information flowing through the company– financial, accounting, human resources, supply chain and customer information'.

Integrated ERP systems have been widely assisting industries, such as manufacturing, to manage demand and supply-chain processes through recording of orders, procurements, controlling production and inventory, maintaining financial records and managing distribution as demonstrated through a block diagram by Zeng [18] shown in Fig. 1. Enterprise application integration (EAI) can be used to link applications that are at the disposal of the end-user with the main database of the system end; example package: SAP ERP. Gartner Group explains that EAI is the 'unrestricted

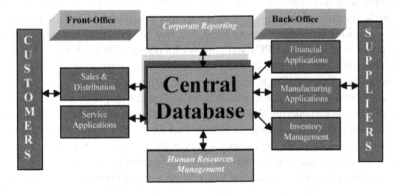

**Fig. 1** ERP system concept. *Source* [18]

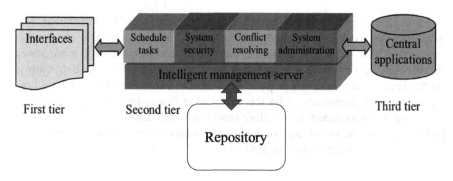

**Fig. 2** A 3-tier architecture to construct-ERP. *Source* Edited after [20]

sharing of data and business processes among any connected application or data sources in the enterprise'. In the words of the research firm Gartner, EAI is the 'unrestricted sharing of data and business processes among any connected application or data sources in the enterprise' [19].

Projects in the AEC sector require wide expertise due to which often a number of clients are attached to a project using different digital platforms. As presented in Fig. 2, Shi and Halpin [20] presented a 3-tier architecture for ERP to be implemented in construction firms.

Whereas, using case studies, Ahmed et al. [21] established that 'A standard, best methodology for implementation does not exist. It is up to each company to approach ERP in the way that best fits its business needs'. With this in effect, this paper proposes that integration of BIM platforms within AEC sectors is also feasible that could be linked with estimating scopes of work, timescales, risks, quality, cost control and managing subcontracts. This means that the implementation of an ERP system as a truly integrated platform needs to address the following:

- Engineering, procurement and construction companies (EPCs) should be able to meet the needs of operator/owner typically engaged in creating digital assets through 3D CAD tools. The associated data with the CAD model can be modified with the progress of the project. This model serves as the basis for BIM processes.
- Specifications are preferably extracted from the BIM environment for tendering purposes to prospectus EPCs and contractors responsible for building an asset. Bidding out of the project is down in parts as per the scope of work.
- Companies often use BoQs as the basis for estimation of quantities which are based on various historical approaches centred on the scopes of work. A move or trend towards considering the BIM model for estimation is changing traditional estimation approaches. Using the BIM model for the purpose is proving efficient as many factors go into developing a successful bid, these include: subcontract work, plant and equipment rental, material and labour resources.
- Managing subcontract is essential to the construction industry, the need of which is required when the main contractor starts sub-bidding out the parts of the

project to subcontractors. In relation to the discrete manufacturer, this is similar to component sourcing.
- As the project progresses, emphasis is given to coordination that is done through available tools, costs that need to be checked, and managing project milestones and timelines. This requires the need to track the old data and update with the new one that can be cycled through and into the BIM environment.
- On the completion of construction and commissioning of the asset, it is handed over to the operator/owner. Enterprise asset management (EAM) system's role is to facilitate in this process and ensure smoothness of the data transfer. Through 'data drops', in contrary to final document 'handover', this is helping to build asset data repositories and thus leveraging BIM processes.

To avoid data loss due to the use of different software tools within the architecture sector, Oh et al. [22] developed a BIM checker and BIM modeller to help integrate many people with a common BIM server. Whereas, the objective of our proposal or hypothesis extends to associate itself to remind or to signify the fact that the main aim of BIM is not to be confined within any specific AEC office(s) to create an information/decision-making database that is disconnected, it is rather to interlink the flow of data within all sectors of AEC to the satisfaction of customers. Subsequently, with the use of new digital information systems within AEC, BIM can be a key approach towards adopting a truly integrated database, one such concept is proposed in this article.

## 2 Method

As mentioned by AIA [23], process re-engineering may be required to eliminate the trade-off among cost, time and quality to suit the main purpose of BIM tools to maximize lean construction by reducing wastage and maximizing value at the project delivery level. Therefore, for an integrated project delivery (IPD), a heterogeneous environment created by the use of software tool systems requires an establishment of the common database in a format easily apprehensible by all parties. Shen et al.[24] have identified that to enable data integration in heterogeneous environments, a common neutral model acts as the most feasible solution. This technique roots out errors and inconsistencies during the data recreation process and enables the enrichment and re-use of data throughout the building lifecycle. Recently, BIM has demonstrated large potential to become a platform for design and construction integration [25]; however, it is argued by Howard and Bjork [26] maintaining that BIM solutions are too complex in nature and their application has a limited scope initially. In addition, the observations suggest that although their application has evolved from a focussed toolset, however, it is evidently restricted to the idea of data models organizing the data of certain domains of interest for clients. To address this, the author suggests that in making use of the available resources or common software, there has to be an approach for systems (and AEC sectors) integration that can be adopted depending upon a case-to-case basis such that maximum tasks of integration are understood (and achieved) at the beginning of any project.

As reported by Sun et al. [27], over 30% of completed projects suffer from bad quality and around 30% of construction is simply reworked, while the client is largely dissatisfied due to overspending and delays of over 50% of construction projects. Therefore, to target this gap and to overcome the one-way project execution approach of [28], a procedure in our study suggests both-way communication to encourage swift reversibility in decision-making. That is, for example, an architectural drawing or/and an engineering drawing would only be finalized through the mutual consent of all the offices/sectors involved that now can have a facility to instantaneously review the drawings (a traditional BIM approach) as well as to simultaneously record the desired changes onto the drawings (a proposed approach) before finalizing a blueprint (reaching decision) by either office. The author assumes that this approach is likely to resolve areas of conflict of interest between different sectors at the early stages of the construction lifecycle. Figure 3 identifies some possible changes (process re-engineering) between a proposed model and the model of Anumba et al. [28] used by primary BIM users (decision-makers) in areas of control and decision-making.

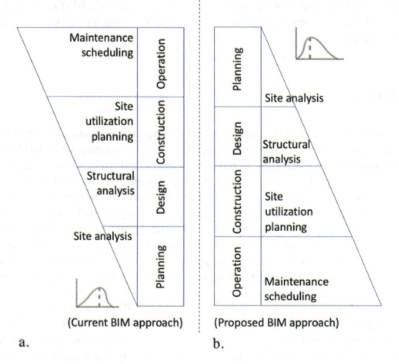

**Fig. 3** Pyramid of BIM Uses throughout a Building Lifecycle: **a** Current approach (inverted pyramid), **b** proposed approach (reverse-BIM, balanced pyramid), noting changes in MacLeamy's curve

Reshuffling the order of processes, Fig. 3b looks like a pyramid of co-ordination from the side and would look more like a pizza chart if seen from the top. The planning and site analysis stage lies at the ground level to indicate that maximum work needs to be done at this stage (indicates the foundation work at the base), it is then followed by Design and Construction stages on the next higher level, and the Operation stage forms the top to indicate the reduced amount of hassle (in management and control) left in co-ordination all the ongoing activities during a life span of a project. Such activities include 3D coordination, Design reviews, Site work. This could easily be compared when the ongoing practice of BIM activities, as indicated by Anumba et al. [28], are put into this form as shown in Fig. 3a, showing an inverted pyramid (hence, less balanced), where planning and site analysis are given least emphasis at the start of the project (foundation); thereby, shifting the stress of workload from Planning to Operation, leading to wastage and proving costly once a need of reconstruction arises that could otherwise have been resolved beforehand using process re-engineering of Fig. 3b through enhanced (two way) co-operation. Noteworthy is also the fact that, for a generic project consisting of Pre-design, Schematic design, Design development, Construction documentation, Procurement, Construction Administration, and Operation, the cost–time graph would be reversed with the choice of BIM practice; an example of how this works is demonstrated by Patrick MacLeamy [29] through Patrick MacLeamy curve. The applicability of MacLeamy's curve to support automation in industries is also demonstrated by Delavar [30] through BIM models. And, with the adoption of our proposed model (of Fig. 3b), an architectural drawing or/and an engineering drawing would now be finalized through the mutual consent of all the office bearers involved that can now have a facility to review the drawings (a conventional BIM approach), to record the desired changes onto the drawings (a current approach) and to simultaneously over-write changes using a heterogeneous system of software (a proposed approach) before finalizing a blueprint (reaching decision) by either office from early stages of the construction lifecycle. This allows the peak of the MacLeamy's curve to shift further towards the early stages of the project. However, the development or selection of a suitable system integration methodology is needed to implement the suggested BIM approach. Addressing this, Fig. 4 proposes an outlook on ERP-BIM integration plan towards systems integration. Different software in AEC offices are now connected through common database using ERP integration.

## 3 Working Example and Technical Details

To establish an understanding of the hypothesis for developing an ERP system linking the two key offices of architecture and designing, the author of this study has picked two legacy software packages (being widely adopted), that is, AutoCAD (2006) by

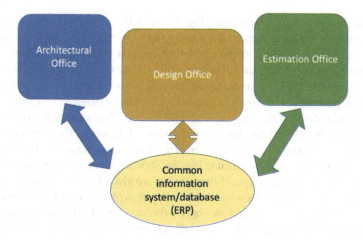

**Fig. 4** Construct on the integration of ERP—BIM technologies

architects and STAAD Pro (v8i) by designers; that are integrated here for collaboration with the help of the most widely used open-source agent system development tool JADE (Java Agent Development Framework (version 7u11-windows-i586); compiled through Eclipse open-source compiler software. Following the outlook shown in Figs. 4 and 5, this required developing three adaptor codes linking: AutoCAD, STAAD Pro and file for cost estimation and commissioning purposes into a common data pool format so that the output information is not just read-only but is also writable on input file formats adopted by both the architecture and design software hereby selected. In this aspect, the flowchart of Fig. 5 identifies areas of data flow and interlinking of tasks from the start of the project through to completion.

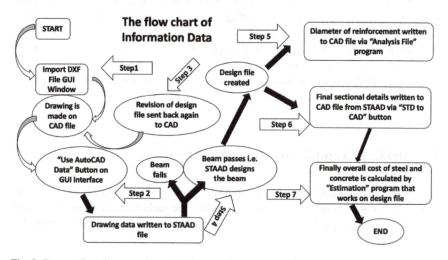

**Fig. 5** Process flow diagram of new BIM approach using example

For example, once a design is not satisfactorily reached, it can be sent back to the architectural office for reconsideration before a final agreement is made. Thereby, in a multidisciplinary context, BIM-ERP integration will further be able to bridge the information loss in real time by allowing all the stakeholders to pro-actively make changes to, add to and refer back to all the information and share concerns required to make informed decision(s) during the lifecycle of the project. Therefore, the conflicts can be well addressed beforehand through the integration of ERP system into BIM technologies.

In order to integrate architectural and design stages, the system is divided into six classes; out of these, three are main classes (Beam, AnalysisFile and EstimationGUI) and objects of the other three (File class, File2 class and File3 class) are called from among one of the main class. Java code programs for these classes are written. The class inherits the JFrame class of Java swing and makes use of objects of JPanel, JTextField, JLabel and JButton, etc., to represent a desirable GUI window. On pushing 'OK' button (working example follows later below), the object of 'File' class is called which reads coordinates of length and width in both STAAS and CAD files along with other necessary information. The object of 'File2' class is called on clicking 'Use AUTOCAD DATA' button, this writes the coordinates of length and width extracted from AutoCAD file onto Staad file format along with necessary information. While pushing 'STAAD to CAD' button made available on the GUI interface, the object of 'File3' class is called upon to acquire the coordinates from Staad file and be re-written over AutoCAD file format. Therefore, here the subclasses of Beam class are meant for mainly architectural drawing (without design elements). 'EstimationGUI', a main class, also extends the JFrame class to create a GUI interface to show the required information necessary for cost estimation after the final drawing/blueprint for site execution (after design analysis) is agreed upon by architecture and design offices.

A trial beam Sect. (450 mm × 100 mm × 100 mm having 4 numbers 8 mm diameter bars to start with) under 10 kN load (4-point bending problem) for lab experiment is taken (refer to Fig. 6) to be tested on our common information system. To facilitate the understanding of the procedure, Fig. 6 is linked with the steps mentioned previously in Fig. 5. First of all the plan and cross-section is drawn on AutoCAD software. Once an agreement on design is reached, the sectional details are passed onto the STAAD software, which then designs a final section (now changed to 750 mm × 150 mm × 150 mm, 10 mm diameter bars) ready for blueprint. Once the final drawing is submitted by the design office, the estimate of the entities, namely concrete and steel, will be calculated automatically by the estimation interface. The same methodology is applied to produce tests on beams [31–33].

## 4 Conclusion

An integrated and scalable solution was needed to deliver functionalities for any project, the lifecycle issues lacked in-depth understanding that needed to be addressed

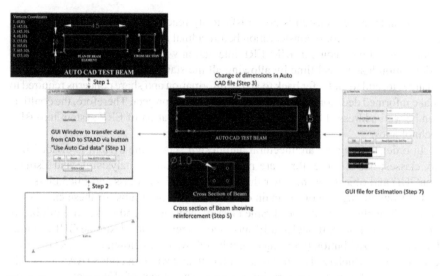

**Fig. 6** Reverse-BIM-ERP verification through RC beam design

during modelling and planning, and the ability to assess the impact and uncertainties of risks and failures was not mature due to a lack of evaluation knowledge and availability of limited tools. After years of R&D, the willingness of AEC sectors and policymakers to embrace software systems to promote the concepts of integration and collaboration using data-driven platforms such as the adoption of Building Information Modelling (BIM) tools to better manage the entire asset lifecycle (towards BIM Level 3), an exceeding need is felt to integrate enterprise-class systems to further the adoption of 3D design and tools of product life-cycle management. To this end, this work presented a process re-engineering methodology (reverse BIM approach) to integrate BIM with ERP system to allow the creation of interdisciplinary consortia with to-and-fro coordination within AEC sectors for SMEs. The idea is demonstrated with the help of an example of designing a reinforced concrete (RC) beam.

Additionally, the drawbacks and lessons learned can be expanded to develop the presented concept at practical scale:

- The study is conducted using only two software packages used in the AEC industry namely AutoCAD and STAAD Pro.
- Although, the methodology proposed herewith considered the practicality of the problem by adopting two popular software (AutoCAD and STAAD Pro) for lab testing; however, no practical exercise of sizable scale is conducted to verify the approach mainly due to two reasons: (1) lack of funds, (2) leaves scope for further study.
- The beam element we considered in developing the common information system is not an actual element that is being used in the construction industry.

- Also, there can be a slight difference between the design data provided by STAAD for the beam and between manual calculations; however, this falls under software-specific limitations. This would affect the connected steps, such as the estimation of quantities of concrete and steel.

Therefore, when it comes to a system of heterogenous software, data pooling and integration can be a tedious task; however, ERP integration across AEC offices is a way forward and can be achieved within SMEs.

**Acknowledgements** Part of the software work was analyzed by Mr. S. Zameer under the guidance of the author of this paper M. Arsalan Khan [PhD, MSc Engg, ACI (Faculty member, USA), IStructE (Graduate, UK), MIES (Chartered with Institution of Engineers Singapore), IACM (Spain), IET (UK)]. The author is extremely thankful for the invaluable support from his late mother.

# References

1. Murphy M, McGovern E, Pavia S (2013) Historic Building Information Modelling – Adding intelligence to laser and image based surveys of European classical architecture. ISPRS J Photogrammetry Remote Sens 76:89–102
2. Thomsen C, Darrington J, Dunne D (2010) Managing integrated project delivery. Virginia, USA
3. Altun M, Akcamete A (2019) A method for facilitating 4D modeling by automating task information generation and mapping. In: Advances in informatics and computing in civil and construction engineering. Springer International Publishing, Cham, pp 479–486
4. Antwi-Afari MF, Li H, Pärn EA, Edwards DJ (2018) Critical success factors for implementing building information modelling (BIM): a longitudinal review. Autom Constr 91:100–110
5. Koo B, Fischer M (2000) Feasibility study of 4D CAD in commercial construction. J Constr Eng Manag 126(4):251–260
6. BIM Industry Working Group (2011) A report for the government construction client group—building information modelling (BIM) working party strategy paper. Communications (March):107
7. Cdbb (n.d.) GSL 1 Intro—Centre for digital built britain
8. Hosseini MR, Chileshe N, Zuo J, Baroudi B (2015) Adopting global virtual engineering teams in AEC projects. Constr Innovation 15(2):151–179
9. Sackey E, Tuuli M, Dainty A (2015) Sociotechnical systems approach to BIM implementation in a multidisciplinary construction context. J Manag Eng 31(1):A4014005
10. Merschbrock C (2012) Unorchestrated symphony: The case of inter-organizational collaboration in digital construction design. Electron J Inf Technol Constr Royal Inst Technol 17:333–350
11. Marvy R (2014) What Is BIM Level 3? 3D perspectives. https://blogs.3ds.com/perspectives/what-is-bim-level-3/. 24 Feb 2019
12. Shalwani A, Lines BC (2020) An empirical analysis of issue management in small building construction projects. Int J Constr Educ Res
13. Milford M, Stewart G (2000) Are ERP implemenations qualitatively different from other large systems implementations? In: AMCIS 2000 proceedings
14. Chen IJ (2001) Planning for ERP systems: analysis and future trend. Bus Process Manage J 7(5):374–386
15. Davenport TH (1998) Putting the enterprise into the enterprise system. Harvard Bus Rev 76(4):121–131. Harvard University Graduate School of Business Administration, Boulder, CO, USA

16. Kumar K, Van Hillegersberg J (2000) ERP experiences and evolution. Commun ACM 43(4):22–26
17. O'Leary DE (2000) Enterprise resource planning systems : systems, life cycle, electronic commerce, and risk. Cambridge University Press
18. Zeng Y (2010) PPT—risk management for enterprise resource planning system implementations in project-based firms. Digital Repository at the University of Maryland, University of Maryland, College Park, MD 20742-7011
19. Gartner (2001) Enterprise applications- adoption of e-business and document technologies: 2000–2001 Europe
20. Shi JJ, Halpin DW (2003) Enterprise resource planning for construction business management. J Constr Eng Manag 129(2):214–221
21. Ahmed SM, Ahmad I, Azhar S, Mallikarjuna S (2003) Implementation of enterprise resource planning (ERP) systems in the construction industry. American Society of Civil Engineers, Reston, VA, Construction Research Congress, pp 1–8
22. Oh M, Lee J, Hong SW, Jeong Y (2015) Integrated system for BIM-based collaborative design. Autom Constr 58:196–206
23. AIA (2007) Integrated project delivery: a guide California Council National. The American Institute of Architects
24. Shen W, Hao Q, Mak H, Neelamkavil J, Xie H, Dickinson J, Thomas R, Pardasani A, Xue H (2010) Systems integration and collaboration in architecture, engineering, construction, and facilities management: a review. Adv Eng Inf 24(2):196–207
25. Young WN, Jones SA, Bernstein HM (2007) SmartMarket report on building information modeling: transforming design and construction to achieve greater industry productivity. McGraw-Hill Construction
26. Howard R, Björk B-C (2008) Building information modelling—experts' views on standardisation and industry deployment. Adv Eng Inf 22(2):271–280
27. Sun M, Sexton M, Senaratne S, Chung P, El-Hamalawi A, Motawa DI, Lin Yeoh M, Plc B, Atkins W (2004) Managing changes in construction projects
28. Anumba C, Dubler C, Goodman S, Kasprzak C, Kreider R, Messner J, Saluja C, Zikic N (2010) BIM project execution planning guide—version 2.0. Computer Integrated Construction Research Program, The Computer Integrated Construction Research Group, Pennsylvania
29. CURT (2004) Collaboration, integrated information and the project lifecycle in building design. Constr Oper WP-1202
30. Delavar M (2017) BIM assisted design process automation for pre-engineered buildings (PEB). Electronic Thesis and Dissertation Repository, The University of Western Ontario
31. Khan MA (2021) Towards key research gaps in design recommendations on flexurally plated RC beams susceptible to premature failures. J Bridge Eng
32. Khan MA (2022) Bond parameters for peeling and debonding in thin plated RC beams subjected to mixed mode loading–Framework. Adv Struct Eng 25(3): 662–682. https://doi.org/10.1177/13694332211065184
33. Khan MA (2021). Towards progressive debonding in composite RC beams subjected to thermomechanical bending with boundary constraints–a new analytical solution. Compos Struct 274114334. https://doi.org/10.1016/j.compstruct.2021.114334

# Application of Game Theory to Manage Project Risks Resulting from Weather Conditions

**Abd Alrahman Ghali and Vaishali M. Patankar**

**Abstract** Construction projects are distinguished from other projects, and in that, they are greatly affected by emergencies such as inclement weather conditions, drilling and backfilling surprises and supply delays. These incidents are known as risks, which usually have negative consequences and constitute an obstacle to the completion of the project as they may cause delays and increase in cost, efficiency loss and quality deterioration. In this research, we have gathered information on the most important risks that could hinder the completion of the project, in general, through previous studies and conducted a questionnaire survey among the engineers to collect the largest number of risks faced by the construction projects in Bhubaneswar, India. After collecting the information, it is analysed and modelled according to the game theory for probabilistic decision-making which helps in choosing the best strategy to address risks and reduce their negative impact on the engineering project. The focus of the research will be on the risks arising from weather conditions in Bhubaneswar, India as a case study.

**Keywords** Game Theory · Risks management · Risk evaluation · Weather risks

## 1 Introduction

Risk management is the way to identify all future challenges that may negatively or positively affect the progress of the project. Negative risks that may stop or delay the project for various reasons are the most dangerous and must be identified. The possibility of occurrence of these risks and the degree of its impact on the project will determine the mechanism of dealing with it when it occurs, either by accepting it, or mitigating its effects. As for the positive risks, it is an assisting factor for the project manager. For example, the decrease in the prices of some materials, as measured by the market performance for a period of time or during the period similar to previous

---

A. A. Ghali (✉) · V. M. Patankar
Department of Civil Engineering, Siksha 'O' Anusandhan (Deemed to be University), Bhubaneswar, India
e-mail: abboudghali@gmail.com

© The Author(s), under exclusive license to Springer Nature Singapore Pte Ltd. 2023
M. S. Ranadive et al. (eds.), *Recent Trends in Construction Technology and Management*, Lecture Notes in Civil Engineering 260,
https://doi.org/10.1007/978-981-19-2145-2_18

years may help the project manager to save some sums and expenses and use them elsewhere in the project. At the beginning of any project, any risks that may affect the project must be identified, arranged in terms of impact on the project from the top to the bottom, and plans are drawn up to avoid these risks, especially to those of the highest rank because they affect the time and cost of the project. The engineer must study the risks, study their impact on the project and provide solutions and proposals that help address these challenges and reduce their impact on the project as much as possible. This is known as a risk management plan [1]. 'Most project risks are common concern to project participants; the industry has shifted from risk transfer to risk reduction' [2]. 'Currently the Government of India has proposed a risk rating system that will help the developers to develop projects at a faster pace by taking quick decisions. Each rating agency will have its own methodology to rate projects' [3]. Jayasudha and Vidivelli [3] provided a study 'that should assist management in identifying activities where there is a risk of Financial, Time and Construction aspects and hence provide a basis for management to take objective decisions on the reduction of risk to an agreed level' [3].

There are many practical methods that help in analysing the risks that could affect the engineering project, and they also help in choosing the most appropriate strategy that ensures the success of the project and its completion with the least possible losses. We will present in this paper one of these methods, which is game theory. This research aims to review the most important steps involved in risk management, and then model these steps according to the methodology of game theory, which helps in choosing the best strategy to address risks and reduce their negative impact as much as possible on the engineering project. The focus of the research will be on the risks resulting from weather conditions and weather.

## 2 Research Methodology

- Information gathering: through a reference study of the most important risks that may hinder the completion of the project according to the drawn plan, as well as strengthening the information obtained by conducting a questionnaire among the engineers, in order to collect the largest number of proposals and solutions that can be used to address risks and reduce their negative impact on the project, information gathering is done.
- Modelling: after collecting the information, it is analysed and modelled according to game theory methodology in probabilistic decision-making.

## 3 Literature Review

After we know that risk is one of the most vital problems that affect projects and that all vital projects are exposed to risks, it is possible, through a review of the literature, to determine the types of these risks more accurately. Some of them are simple and fade and do not affect the progress of projects. Some of them develop and turn into a problem with dimensions, and a plan must be drawn up to fix it quickly. A small percentage may damage the entire project. The aim of studying the risks is to ensure that the third type can completely disappear when well planned by the risk management. Through previous studies, in general, we can classify the risks facing construction projects into these categories:

1. *Physical risks*: represented by workers who are not technically qualified, or an accident due to lack of safety measures or the supply of materials that are not valid, or do not comply with specifications
2. *Environmental risks*: they occur as a result of very bad and harsh weather conditions, or as a result of the difficulty in complying with the laws and prevailing environmental legislation, or as a result of environmental disasters (flood or earthquake) or unexpected pollution, or as a result of the difficulty in accessing the site.
3. *Design Risks*: Errors in design, inconsistencies between bill of quantities and specifications, or as a result of architectural design mismatch with structural.
4. *Logistical risk*: a result of high competition occurs during bidding, inaccurate project scheduling, insufficient labour availability, poor communication with the contractor and using new devices for the first time without training
5. *Financial risk*: it occurs as a result of late payment of payments according to the contract, and lack of cash flow control
6. *Legal risks*: legal disputes during the construction phase between the project parties, difficulty in obtaining licenses and starting work and lack of clarity in work legislation.
7. *Risks related directly to implementation*: difference in actual and contractual quantities, changes in design, reducing work quality versus time commitment, delays and technical problems with contractors, the difference between implementation and specifications as a result of misunderstanding the plans and specifications and the lack of documentation of change orders resulting during work.
8. *Political risks*: insecurity during project implementation, unexpected wars, political pressures and changes in prevailing laws.
9. *Administrative risks*: change in management methods, problems in managing resources, lack of information, poor communication between parties and planning incomprehensible due to project complexity [4–17].

## 4 The questionnaire

The questionnaire was conducted among a group of civil engineers and architects in public and private sector companies, and the response was made by 33 engineers. In the first phase, the most important risks that may be exposed to engineering projects in Bhubaneswar were identified. Table 1 shows these dangers in order of importance.

**Table 1** Risks ranked according to the importance

| Risks | Sorted by importance |
|---|---|
| Harsh weather conditions | 1 |
| Accident due to lack of safety measures | 2 |
| Supply of non-conforming materials | 3 |
| Technically unqualified labour | 4 |
| Land pollution due to work | 5 |
| Difficulty accessing the site | 6 |
| The mismatch between quantities, plans and specifications | 7 |
| The fluctuation of productivity rates for workers and equipment | 8 |
| Design errors | 9 |
| Inaccuracy in calculating quantities | 10 |
| Assigning the design to an inept office | 11 |
| Design mismatch (structural, architectural) | 12 |
| Difficulty and cost of adhering to environmental legislation | 13 |
| Environmental disasters | 14 |

Knowing that the engineer's answer was weighted according to the number of years he worked in the construction field.

In the second phase of the questionnaire, the focus was on the risk (harsh weather conditions). Where the cases shown in Table 2 related to weather conditions affecting the project were proposed. And depending on the results recorded in weather stations for the year 2020, which we obtained from the University of (Siksha 'O' Anusandhan), and by asking the engineers to determine the number of days in each month that pertains to each weather condition, and by adopting the weighted average, we obtained the results as shown in Table 3.

In addition, the questionnaire showed the duration of stoppage for each of the main project operations, depending on the weather situation, Table 4. Where the following classification was adopted: total stop, large-stop, partial stop and non-stop.

**Table 2** Weather conditions

| The description | Status | code |
|---|---|---|
| Strong winds, hurricanes, rain and thunderstorms | Bad weather | $A_1$ |
| Very hot weather | Moderately bad weather | $A_2$ |
| Stable or semi-stable weather | Good weather | $A_3$ |

**Table 3** Weather conditions by month

|  | Status |  |  |  |  |  |
|---|---|---|---|---|---|---|
|  | $A_1$ |  | $A_2$ |  | $A_3$ |  |
|  | days | $P(A_i)$ % | days | $P(A_i)$ % | days | $P(A_i)$ % |
| January | 6 | 0.194 | 2 | 0.064 | 23 | 0.742 |
| February | 6 | 0.207 | – | – | 23 | 0.793 |
| March | 6 | 0.193 | 3 | 0.096 | 22 | 0.709 |
| April | 6 | 0.2 | 11 | 0.366 | 13 | 0.433 |
| May | 7 | 0.225 | 4 | 0.129 | 20 | 0.645 |
| June | 14 | 0.466 | 1 | 0.033 | 15 | 0.5 |
| July | 14 | 0.451 | – | – | 17 | 0.584 |
| August | 19 | 0.612 | – | – | 12 | 0.387 |
| September | 15 | 0.5 | – | – | 15 | 0.5 |
| October | 15 | 0.5 | – | – | 16 | 0.516 |
| November | 3 | 0.1 | – | – | 27 | 0.9 |
| December | 10 | 0.323 | – | – | 21 | 0.677 |

**Table 4** The duration of the project work stoppage depending on the weather condition

| The mission | Downtime depending on weather conditions |  |  |
|---|---|---|---|
|  | Bad weather ($A_1$) | Moderately bad weather ($A_2$) | Good weather ($A_3$) |
| Soil work | Total stop | Total stop | Non-stop |
| Structural work | Total stop | Large-stop | Non-stop |
| Outer brick | Total stop | Non-stop | Non-stop |
| Inner brick | Partial stop | Non-stop | Non-stop |
| Outer finishing | Total stop | Non-stop | Non-stop |
| Inner finishing | Partial stop | Non-stop | Non-stop |

# 5 Responding to Risk Using Game Theory

### I. Principles of Game Theory

Game theory is concerned with studying the issues of duelin and competition between two or more parties, and these two parties have different goals, and each party acts consciously, thoughtfully and opposes the actions of the other party in order to achieve its goal optimally. The set of options available to the player to act and confront the other party is called a set of strategies, and it changes from one step to another. If the number of these strategies is finished, the game is finished, otherwise, it is called an endless game. The game is called zero-sum if the sum of what one side wins is equal to what the other loses in the considered step. If the strategy makes the largest possible profit (or the least possible loss) when the game is repeated as many times as possible, it is called a perfect strategy. The player can choose a single strategy and he can also choose several strategies together, which is then called a composite strategy, in this case, each individual strategy constituting the composite strategy is associated with a probability frequency distribution.

## II. Payoff Matrix

Assuming we have a double-ended game:

(A) has the following strategies: $A_1, A_2 .... A_m$
(B) has the following strategies: $B_1, B_2 .... B_n$.

If player A chooses strategy $A_i$ and player B chooses strategy $B_j$, then the outcome of the game for A is a specific number $a_{ij}$. The $a_{ij}$ values make up the payoff matrix—Table 5. The minimum price for the game is calculated according to the principle of (MaxMin). A chooses the largest possible profit and B what makes A's profit the lowest i.e., $\alpha_i = $ Max (Min $a_{ij}$). The maximum price of the game is calculated according to the principle of (MinMax). B chooses the smallest winnings for player A, and player A matches him by choosing the largest that player B has set for him. i.e., $\beta_i = $ Min (Max $a_{ij}$). The game is stable if the lower price of the game equals its higher price and the corresponding strategy for both is the best appropriate reaction to the game being studied ($V = \alpha_i = \beta_j$). if the lower and higher prices are not equal, then we resort to determining the probability distribution for each proposed work strategy, and the final solution will determine the percentage that should be used with each proposed strategy (Table 5).

Game theory is usually used within the focus of responding to risk, specifically in planning to avoid and reduce risk. We will study the application of game theory to hazards arising from weather and weather conditions, as the fluctuations of weather and weather conditions (such as snowfall, heavy rain, frost, strong winds, extreme heat) usually delay the project and increase its cost, and it may also cause a decrease in quality, especially during specific times of the year. Here, nature is the opponent for project management in planning the risk response.

Strategies (alternatives) available for player 1 (nature) can be proposed for weather conditions affecting the project as shown in Table 2. Strategies available to the second

**Table 5** Payoff Matrix (a risk aversion matrix)

| A\B | $B_1$ | $B_2$ | .... | $B_n$ | $\alpha_i$ |
|---|---|---|---|---|---|
| $A_1$ | $a_{11}$ | $a_{12}$ | ... | $a_{1n}$ | $\alpha_1$ |
| $A_2$ | $a_{21}$ | $a_{22}$ | ... | $a_{2n}$ | $\alpha_2$ |
| .... | ... | .... | .... | .... | ...... |
| $A_m$ | $a_{m1}$ | $a_{m2}$ | .... | $a_{mn}$ | $\alpha_m$ |
| $\beta_j$ | $\beta_1$ | $\beta_2$ | .... | $\beta_n$ | |

player (project manager) represent the probable month to start the project as shown in Table 6.

### III. Probability of occurrence ($P_i$)

Using the questionnaire results shown in Table 3, the probability of occurrence of each of the three previous cases ($A_1, A_2, A_3$) was calculated and it was coded $P(A_i)$ as it appears as a percentage.

### IV. Impact probability ($F_j$)

The project usually consists of a set of tasks and activities, some of which are affected by very bad weather only, and some of them are affected by the weather, even if it is moderately bad, and some of them are not affected by the weather, whatever its condition, in other words, it is not necessary for the mission ($A_1$) to be affected by the weather even if it is bad. Total stoppage means that the effect will be 100%, and if the stoppage is partial, it will be matched by a partial percentage of the damage according to the studied mission. Based on the results of the questionnaire shown in Table 4, the probability of an impact was determined for each mission, as shown in Table 7, as a percentage.

### V. Project damage (game price $a_{ij}$)

The elements of the $a_{ij}$ payoff matrix in Table 5 represent the percentage of delay in the project according to the studied weather condition, i.e., equal to the total amount of damage to the tasks located on the critical path (Project Damage = Total Damage to the missions on the Critical Path). The damage size for mission $i$ in weather condition $A_j$ = probability of occurrence of weather condition $A_j$, which is $P(A_i)$ * probability of impact $F_{jk}$. [18, 19].

## 6 Practical Application

Figure 1 shows a brief bar chart for the critical path of an apartment building project in addition to the duration of each task per day according to the months of the year and considering 26 working days in every month.

Payoff matrix: the game price for each $A_i$ state according to strategy $j$ is $a_{ij} = \sum P_{ik} * F_{kj}$

$$A_{11} = (15*1 + 11*1)*0.194 + 26*1*0.207$$
$$+ 26*1*0.193 + 26*1*0.2 + 26*1*0.225$$
$$+ 26*1*0.466 + (20*1 + 6*1)*0.451$$
$$+ (24*1 + 2*0.2)*0.612 + (18*0.2 + 8*1)*0.5$$
$$+ (10*1 + 16*0.2)*0.5 = 77.6688$$

**Table 6** Strategies of the project manager

| January | February | March | April | May | June | July | August | September | October | November | December |
|---|---|---|---|---|---|---|---|---|---|---|---|
| $B_1$ | $B_2$ | $B_3$ | $B_4$ | $B_5$ | $B_6$ | $B_7$ | $B_8$ | $B_9$ | $B_{10}$ | $B_{11}$ | $B_{12}$ |

**Table 7** Impact potential

| $F_{jk}$* | Downtime depending on weather conditions | | |
|---|---|---|---|
| The mission | Bad weather ($A_1$) | Moderately bad weather ($A_2$) | Good weather ($A_3$) |
| Soil work | total stop 100% | total stop 100% | non-stop 0% |
| Structural work | total stop 100% | large-stop 70% | non-stop 0% |
| Outer brick | total stop 100% | non-stop 0% | non-stop 0% |
| Inner brick | partial stop 20% | non-stop 0% | non-stop 0% |
| Outer finishing | total stop 100% | non-stop 0% | non-stop 0% |
| Inner finishing | partial stop 20% | non-stop 0% | non-stop 0% |

*$F_{jk}$ The probability of affecting mission $J$ when $K$ occurs

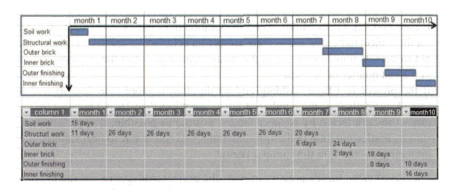

**Fig. 1** Gantt chart for the considered project

By applying the principle of equalizing the lowest and highest price of the game, we note that the least delay in the project corresponds to that the starting month is October (Table 8).

## 7 Conclusions

Using the previous methodology and by adopting two main variables which are the expected total project duration and the method of its spread, the following rules were concluded:

**Table 8** The payoff matrix for each of the three cases ($A_1$, $A_2$ and $A_3$) depending on the proposed start-up month over an entire year

|  | A1 | A2 | A3 | $\beta$ |
|---|---|---|---|---|
| January | 77.6688 | 12.8096 | 0 | 77.6688 |
| February | 80.524 | 11.3568 | 0 | 80.524 |
| March | 86.4456 | 11.7888 | 0 | 86.4456 |
| April | 85.5516 | 11.2566 | 0 | 85.5516 |
| May | 87.068 | 3.5289 | 0 | 87.068 |
| June | 86.4344 | 0.7491 | 0 | 86.4344 |
| July | 79.6096 | 0.896 | 0 | 79.6096 |
| August | 73.3352 | 1.1648 | 0 | 73.3352 |
| September | 66.0832 | 2.5088 | 0 | 66.0832 |
| October | 61.4908 | 8.036 | 0 | 61.4908 |
| November | 62.1724 | 11.3792 | 0 | 62.1724 |
| December | 71.7116 | 12.383 | 0 | 71.7116 |
| $\alpha$ | **61.4908** | 0.7491 | 0 | **61.4908** |

- If the project is a floor building and its duration does not exceed 1 year, in order for the project to be affected by weather factors as little as possible, it is preferable not to start its implementation in the winter season and start the implementation in the last quarter of the year.
- If the project is a floor building and its duration is more than 2 years and is spread vertically (i.e., the number of floors is not less than 3), and in order for the project to be affected by weather factors as little as possible, it is preferable to start its implementation in the spring months, and if it is spread horizontally, it is preferable to start it in the summer period.

The use of game theory greatly supports the project manager in making the best decision regarding the preferred project start time. Reducing the risk of weather factors during the project implementation period was studied based on game theory, but the rest of the risks over the project period, such as financing risks, soil surprises risks and also risks of disruption of renewable and non-renewable resources have not been studied, and this could be the focus of future research in this field.

# References

1. Project Management Institute (2017) A guide to the project management body of knowledge. Sixth edn
2. Tang W, Qiang M, Duffield CF, Young DM, Lu Y (2007) Risk management in the Chinese construction industry. J Constr Eng Manag 133(12):944–956
3. Jayasudha K, Vidivelli B (2016) Analysis of major risks in construction projects. ARPN J Eng Appl Sci 11(11):6943–6950

4. Miller MB (2018) Quantitative financial risk management. Wiley
5. Hopkin P (2018) Fundamentals of risk management: understanding, evaluating and implementing effective risk management. Kogan Page Publishers
6. Damnjanovic I, Reinschmidt K (2020) Data analytics for engineering and construction project risk management. Springer, Cham, Switzerland
7. Baloi D, Price AD (2003) Modelling global risk factors affecting construction cost performance. Int J Project Manage 21(4):261–269
8. Belassi W, Tukel OI (1996) A new framework for determining critical success/failure factors in projects. Int J Project Manage 14(3):141–151
9. Bunni NG (2003) Risk and insurance in construction. Routledge
10. Dikmen I, Birgonul MT, Han S (2007) Using fuzzy risk assessment to rate cost overrun risk in international construction projects. Int J Project Manage 25(5):494–505
11. Iyer KC, Jha KN (2005) Factors affecting cost performance: evidence from Indian construction projects. Int J Project Manage 23(4):283–295
12. Saqib M, Farooqui RU, Lodi SH (2008) Assessment of critical success factors for construction projects in Pakistan. In: First international conference on construction in developing countries, pp 392–404
13. Tah JH, Carr V (2000) A proposal for construction project risk assessment using fuzzy logic. Constr Manag Econ 18(4):491–500
14. Wang J, Yuan H (2011) Factors affecting contractors' risk attitudes in construction projects: Case study from China. Int J Project Manage 29(2):209–219
15. Zavadskas EK, Turskis Z, Tamošaitiene J (2010) Risk assessment of construction projects. J Civ Eng Manag 16(1):33–46
16. Zayed T, Amer M, Pan J (2008) Assessing risk and uncertainty inherent in Chinese highway projects using AHP. Int J Project Manage 26(4):408–419
17. Zeng J, An M, Smith NJ (2007) Application of a fuzzy based decision making methodology to construction project risk assessment. Int J Project Manage 25(6):589–600
18. Cox LA Jr (2009) Game theory and risk analysis. Risk Anal Int J 29(8):1062–1068
19. Neogy SK, Bapat RB, Dubey D (ede) (2018) Mathematical programming and game theory. Springer Singapore

# Environmental Impact Analysis of Building Material Using Building Information Modelling and Life Cycle Assessment Tool

**Kunal S. Bonde and Gayatri S. Vyas**

**Abstract** The construction industry consumes a large number of resources and energy, producing a negative impact on the environment. Nowadays, the world is facing problems of climate change, ozone layer depletion, temperature rises, etc. with an increase in greenhouse gas emissions. In India to get an environment clearance certificate for any project, environmental impact assessment is most important, which is measured by means of life cycle impact assessment. The aim of this research work is to measure the impact of various building materials by evaluating energy consumption at different stages throughout the life span of the building. This work follows a life cycle assessment framework based on the ISO 14040 and ISO 14044 guidelines within the available databases using Autodesk Revit (building information modelling tool) and One-Click LCA (life cycle assessment tool) integration. The main objectives of the research work are: (1) to compare concrete frame structure and steel frame structure; (2) to compare conventional and modern building materials, on the basis of their environmental impact. To achieve the objectives of the work, a real-life case study of the residential apartment building is selected and analysis of energy consumption is done. The results from the energy consumption calculation indicate that steel frame with modern building materials produces a less negative impact on the environment as compared to concrete frame and conventional building material. This research work is limited to the environmental impact of building frame type and material selection during the early design phase of the building. It encourages the application of steel frames as well as modern building materials to protect the natural environment and climate change.

**Keywords** Building information modelling · Concrete frame · Environmental impact assessment · Life cycle assessment · Steel frame

---

K. S. Bonde (✉)
College of Engineering, Pune, India
e-mail: bondeks19.civil@coep.ac.in

G. S. Vyas
Department of Civil Engineering, College of Engineering, Pune, India

© The Author(s), under exclusive license to Springer Nature Singapore Pte Ltd. 2023
M. S. Ranadive et al. (eds.), *Recent Trends in Construction Technology and Management*, Lecture Notes in Civil Engineering 260,
https://doi.org/10.1007/978-981-19-2145-2_19

# 1 Introduction

In recent years, the usage of energy in buildings has been on the rise, due to the population increase, economic growth, and extremely luxurious life. The construction industry consumes a large number of natural resources and hence contributes to increasing greenhouse gas emissions and climate change. Due to this, the world is witnessing the depletion of natural resources as well as non-renewable energy resources. According to Tusher et al. [13], globally 27% of overall energy consumption and 17% of $CO_2$ production were due to the housing sector. Jin et al. [2] stated that, globally, 20–40% of total energy usage is due to the building sector. Sandberg et al. [12] suggested that new sustainability requirements have been progressed by the environmental impact from the built environment regarding the building's performance throughout life cycle cost and energy. According to the World Green Council, buildings are using 40% of energy, 32% of natural resources, and are responsible for 36% of greenhouse gas emissions. India ranks 3rd in electricity consumption as of 2018 with 1181 kwh/ capita/ year use. In India, 80% of electricity is produced from fossil fuel. This burning of fossil fuels produces harmful emissions that affect the environment. Through these activities, GHG emissions such as $CO_2$, $CH_4$, $N_2O$, etc. develop at the higher layer and result in global warming.

## 1.1 Building Information Modelling

Building information modelling (BIM) tools help architects, designers, and construction engineers to create 3D, 4D (time schedule), 5D (cost estimation), 6D (sustainable design) model for a better understanding of the building construction process. As per the National BIM Standard Project Committee in the United States of America (2017), BIM is defined as "it is a digital representation of physical and functional characteristics of a facility. A BIM is a shared knowledge resource for information about a facility forming a reliable basis for decisions during its life-cycle; defined as existing from earliest conception to demolition".

3D model of the building is developed with great precision and optimum level of detailing using Autodesk Revit. It is the platform where architecture, MEP (mechanical, electrical, plumbing), and structural engineers work together. It is a BIM tool that is capable of making detailed plans using the material library available in its database. The building model is developed in Revit and transferred to life cycle assessment (LCA) software for energy analysis. In this study, an apartment building model is developed and LCA is done throughout the construction stages. The different stages of construction are shown in Fig. 1.

**Fig. 1** Building construction phase [9]

## 1.2 Life Cycle Assessment

Life cycle assessment is used to measure the environmental impact of various types of projects. It helps in the decision-making process prior to the construction stage. There are various software for life cycle assessment. In this research work, One-Click LCA software, which is a plugin to Autodesk Revit is used for energy analysis. Life cycle assessment methodologies follow ISO 14040 (principle and formwork) and ISO 14044 (defining goal and scope). It considers the life cycle right from the extraction of raw material to the demolition stage for environmental impact calculation. The life cycle stages and energy involved in a particular stage are shown in Table 1. Pre-building phase consists of activities right from the extraction of raw material to transportation at the site and the energy involved in this phase is embodied energy.

## 2 Literature Review

Nowadays, LCA is one of the most popular tools for supporting environmental decision-making for products, services, and technologies [2]. LCA aims to determine the environmental impact of the product right from raw material extraction to the demolition stage [3]. LCA research methodology is divided into four phases according to ISO 14040 and ISO 14041 [4]: (1) Aim and scope definition: specifying the goal, scope, and limit of LCA research [1]; (2) Inventory analysis (life cycle inventory): energy and material data is collected and established the inventory

**Table 1** LCA phase, activities, and energy involve

| Life cycle phase | Activities | Energy involves |
|---|---|---|
| Pre-building phase | Extraction of raw materials, manufacturing, packaging, and transporting to the site | Embodied energy |
| Building phase | Construction, installation, operation, and maintenance | Operational energy |
| Post-building phase | Demolition and recycling | Demolition energy |

input or output [5]; (3) Environmental impact assessment (life cycle impact assessment): calculation of the flow to be assessed for an overall impact category [1]; (4) Interpretation: quantification and evaluation of the life cycle inventory and life cycle impact assessment study findings, and originating conclusions and suggestions [9]. The aim of this research work is to perform a life cycle assessment of building material from extraction to demolition stage. This can be done by integrating BIM and LCA for analysis of building frame and material.

## 2.1 LCA in the Building Sector

LCA has been applied to the rating of sustainable green buildings, which encouraged building sustainability assessment in practice [7]. Various stages of construction produce different negative impacts on the environment. From the study, it was observed that during the manufacturing and operating stages of building life cycle, the negative impact on the environment is more as compared to other stages [8]. The use of LCA in the construction sector is multifaceted and includes the ranking of sustainable green buildings, energy assessment, refurbishment of buildings, costs, social assessment, and carbon emissions [10]. With the ultimate goal of formulating strategies to reduce the use of primary energy in buildings, LCAs have been applied to the energy evaluation of buildings. The environmental impact of a building is dependent on the type of building frame. Steel building frames produce less negative impact on the environment as compared to concrete building frames [9]. Building materials can contribute significantly to achieving sustainable development goals. Use of modern building materials and removing typical building materials help us to achieve a sustainable design in the build environment. Modern building materials in building construction help to achieve 13 sustainable development goals and 25 targets of sustainable development goals [11].

## 2.2 BIM and LCA Integration

BIM is used to model energy-efficient building. Autodesk Revit has an extensive material library, which is helpful to find the most efficient building material. LCA helps to calculate the energy consumed in different stages of life cycle of a building right from raw material extraction to disposal and end of life of a product [3]. Najjar et al. [8] in their work integrated BIM and LCA for calculating the environmental impact of various building materials. To validate the results of the study, they take one real-life case study of an office building. Results show that during the manufacturing process and operation stage, impact of negative energy is more. Lu et al. [6] in their work integrated BIM and LCA for calculating carbon emissions. A case study of a hospital building is considered for validation. Results of this work show that RCC in the construction stage and HVAC in the operational stage contribute a large amount of $CO_2$. Life cycle assessment helps in decision-making at the early design phase of the building by performing operational energy calculation and checking the impact of building material [9]. Najjar et al. [9] calculate operational energy throughout the life cycle of a building. To achieve this objective, one case study of three different profile buildings (low, normal, and high) was taken. Analysis results show that exterior wall and windows are most accountable for energy.

## 3 Research Methodology

The flowchart of the research work is shown in Fig. 2. It consists of six stages. Goal, scope, and system boundary is defined in the preliminary stage. After defining boundary, building design takes place in Revit, and quantity takeoff takes place for inventory analysis. This quantity takeoff file is then imported into One-Click LCA software for energy analysis. The results of energy analysis are compared in the interpretation stage.

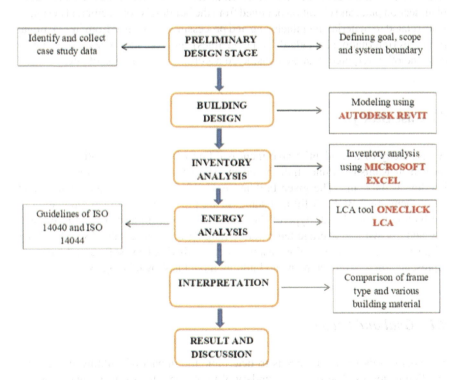

**Fig. 2** Flowchart of research work

## 3.1 One-Click LCA

One-Click LCA operates on the basis of a web-based interface that can be integrated into compatible open standard applications and as a Revit software plug-in tool. This instrument follows the international criteria for the study of the LCA. For EN 15978, EN 15804, ISO 21931-1, ISO 21929-1, and ISO 14040, the One-Click LCA is independently accepted. In addition, an international database of Environmental Product Declarations has been developed to target the European market. It also enables the performance of building materials to be chosen, modified, and simulated. It has been used by researchers, who have researched LCA and BIM integration using various methods, such as the One-Click LCA and Tally.

## 3.2 Data Collection

Real-life case study data is collected which includes the bill of materials, each floor plan, section plan, and elevation obtained from the building's contractor and designer. Data related to material manufacturing and processing is obtained from international resources (international standards and guidelines) and from local resources. Climate data and other required data are considered according to the region of site.

## 4 Case Study

A case study of a residential apartment building named 'Gagan Nulife', which is located in Kamshet, Pune, India is taken. The gross floor area of the apartment building is 4400 m$^2$. The given building is G+10. There are 4 flats per floor and all of them are 2 BHK. A lift and staircase are provided for vertical circulation in the building. The climate type of the Pune region is semi-arid bordering tropical wet and dry with an average temperature ranging between 20 and 28°. Modelling of the apartment building is done using Autodesk Revit a BIM software. A pictorial representation of a typical plan and building model is shown in Fig. 3.

### 4.1 Goal and Scope

The goal of this research work is to determine the impact of building frames and building materials. The system boundary of LCA consists of stages right from raw material extraction, manufacturing, construction, operation, and end of life. Attention should be given to various impact categories and other data that need to be inputted. Scope of this work is to analyse buildings at the early design stage with

Environmental Impact Analysis of Building Material Using ... 239

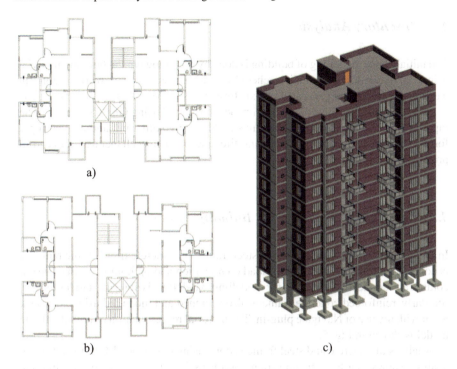

**Fig. 3** **a** Plan of odd floor **b** Plan of even floor **c** Revit model

different materials to find out environment sustainable alternatives. It helps to get an environmental clearance certificate. The layout of the system boundary is shown in Fig. 4. The layout of the system boundary is defined from raw material extraction to the disposal stage.

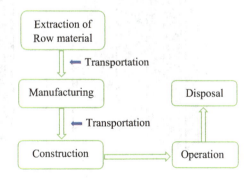

**Fig. 4** Layout of system boundary

## 4.2 Inventory Analysis

The initial phase modelling of building is done by assigning typical building material using Autodesk Revit. Quantity scheduling of various items of work is done in Revit and the report is generated in 'txt' format. Using Microsoft Excel, 'txt' file is converted into 'xlsx'. Environmental impact is assessed using One-Click LCA. The input given to One-click LCA software should be in gbxml, xml, and manual input format. Energy use, water use, building life, and other miscellaneous data need to provide to the software.

## 4.3 Calculation Result of the Building Frame

In this research work, comparison of steel frame and concrete frame of the building is done. Steel frame is analysed for load combination and optimum design of beam, column and connections are used for modelling. Modelling is done in Autodesk Revit. Similarly, reinforced concrete frame is also designed and modelled using Autodesk Revit with the use of Navigate plug-in. The pictorial representation of both the frame model is shown in Fig. 5.

Analysis of concrete and steel frame is done using One-Click LCA. Result of the analysis shows that the concrete frame produces 3.23 kg $CO_2e/m^2$/year while the steel frame produces 2.55 kg $CO_2e/m^2$/year. A comparison of both the frame is done using Microsoft Excel which is shown in Fig. 6.

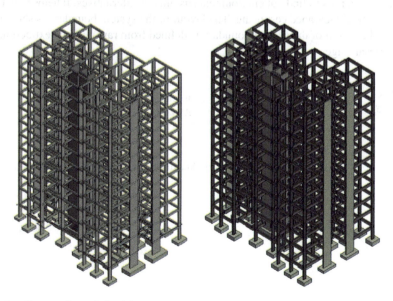

**Fig. 5** **a** Concrete frame **b** Steel frame

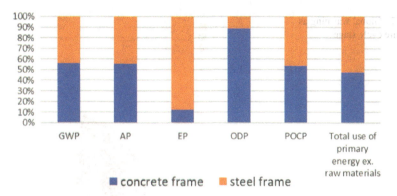

**Fig. 6** Environmental impact comparison of concrete and steel frame

As per the above result data, it is observed that the impact of the steel frame is less than a concrete frame and the social cost of carbon is also minimum. But steel frame produces more impact in the category of eutrophication potential and total use of primary energy.

## 4.4 Calculation Result of Building Material

The whole building model with steel frame and typical building material is analysed and the result of the analysis shows that 45.46 kg $CO_2e/m^2$/year is produced in the environment. The overall electricity use is 80,000 kwh/year. From all the devices, the electricity consumption by other types of machinery and HVAC is more. Fuel in the form of LPG consumption is at the rate of 7497.6 kg/year and in the form of diesel is 4559 l/year. Water consumption rate is 20,000 $m^3$/year. Of all the life cycle stages, operational stage produces more impact on the environment. Global warming potential as per life cycle stage is shown in Fig. 7.

Of all typical building materials, clay bricks produce more impact on the environment. The most contributing material towards global warming is shown in Fig. 8.

To reduce the negative environmental impact of building materials, various material alternatives are analysed to find out the best sustainable option. Alternatives of wall, window, door, slab, insulation, and partition material are analysed for their embodied energy, and results are shown in the following figures.

**Fig. 7** Global warming as per life cycle stage

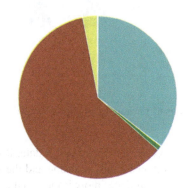

| No. | Resource | Cradle to gate impacts (A1–A3) | Of cradle to gate (A1–A3) |
|---|---|---|---|
| 1 | Clay bricks | 2 576 tons CO₂e | 73.3 % |
| 2 | Structural steel profiles, generic | 449 tons CO₂e | 12.8 % |
| 3 | Ready-mix concrete | 282 tons CO₂e | 8.0 % |
| 4 | Slide-open aluminium frame window | 81 tons CO₂e | 2.3 % |
| 5 | Ready-mix concrete | 71 tons CO₂e | 2.0 % |
| 6 | Ready-mix concrete, normal strength, generic | 43 tons CO₂e | 1.2 % |
| 7 | External wood door | 13 tons CO₂e | 0.4 % |

**Fig. 8** Most contributing materials towards global warming

## 5 Interpretation

Interpretation is the last stage of the LCA methodology. After energy analysis of various building materials and after doing their comparison, it is observed that sand-lime brick (13.1%), wood door (3.3%), double glazing window with wooden frame (13.4%), composite steel deck floor (6.6%), gypsum board partitioning system (3.3%), and cellulose insulation (0.7%) are the sustainable materials that produce less negative impact on the environment.

## 6 Results and Discussion

The 3D model of an apartment building is developed in Revit. Steel frame and concrete frame are analysed in One-Click LCA and the result shows that steel frame produces 2.55 kg $CO_2e/m^2$/year and concrete frame produces 3.23 kg $CO_2e/m^2$/year. The comparison of both the frame is shown in Fig. 6. Similarly, different material alternatives are analysed for their suitability for the environment and the result of the analysis is shown in Figs. 9, 10, 11, 12, 13 and 14. Sand lime brick, wood door, double glazing window with wooden frame, composite steel deck floor, gypsum board partitioning system, and cellulose insulation are the sustainable alternatives to the typical building material.

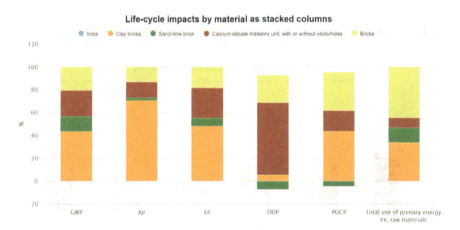

**Fig. 9** Comparison of alternative brick materials

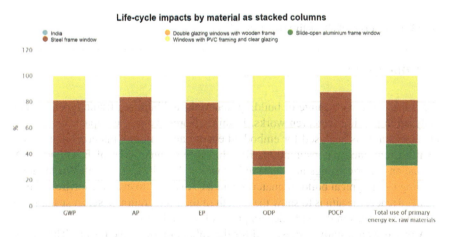

**Fig. 10** Comparison of alternative window materials

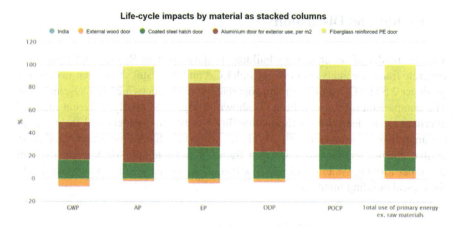

**Fig. 11** Comparison of alternative door materials

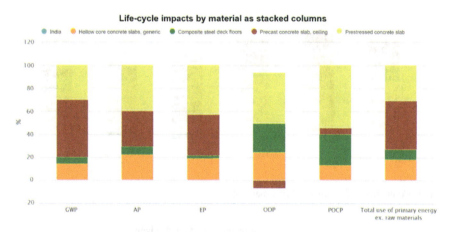

**Fig. 12** Comparison of alternative slab materials

## 7 Conclusion

The impact of a wide range of building materials and building frame types are not analysed in previous research works. In this research work, the impact of concrete and steel frame is analysed for embodied energy and building material is analysed for life cycle energy consumption. To achieve this, integration of BIM and LCA is done and shortcomings are observed properly. Autodesk Revit helps to model a building using typical building material and LCA is used to evaluate environmental impact. Hence, the aim is to show the benefit of their integration to select sustainable building materials in the early design phase of the building.

A real-life case study is presented for the validation of all the results. This study highlighted the impact of building frames and building materials. One-Click LCA is

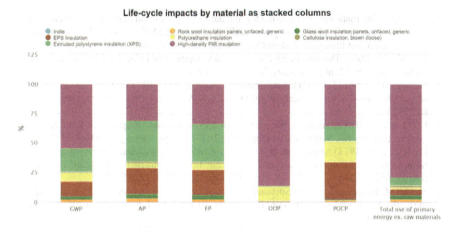

**Fig. 13** Comparison of alternative insulation materials

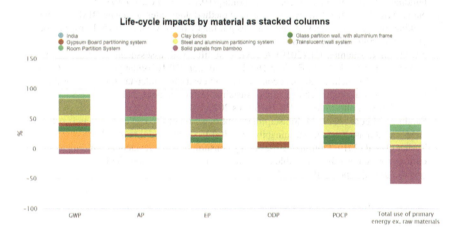

**Fig. 14** Comparison of alternative partition material system

used to calculate impact which is a plug-in to Autodesk Revit. The problem regarding One-Click LCA software is that there is only a limited material database of India. This work is limited to the selection of building frame as well as building material at an early design phase of building. Future work should focus on the design aspect of building, the use of renewable energy resources and green building certification.

# References

1. Gardezi SSS, Shafiq N (2019) Operational carbon footprint prediction model for conventional tropical housing: a malaysian prospective. Int J Environ Sci Technol 16(12):7817–7826. https://doi.org/10.1007/s13762-019-02371-x

2. ISO (2006) International Organization for Standardization. Environ Manage Life Cycle Assess Principles Framework 14044, Ruoyu J et al (2019) Integrating BIM with building performance analysis in project life-cycle. Autom Constr. https://doi.org/10.1016/j.autcon.2019.102861
3. Klöpffer W (2006) The role of SETAC in the development of LCA. Int J Life Cycle Assess 11(SPEC,1):116–22. https://doi.org/10.1065/lca2006.04.019
4. Kluppel HJ (1998) ISO 14041: environmental management—life cycle assessment goal and scope definition—inventory analysis. Int J Life Cycle Assess 3(6):301. https://doi.org/10.1007/BF02979337
5. Lee J et al (2018) A study on the analysis of $CO_2$ emissions of apartment housing in the construction process. Sustainability (Switz) 10(2):1–16. https://doi.org/10.3390/su10020365
6. Lu K et al (2019) Development of a carbon emissions analysis framework using building information modelling and life cycle assessment for the construction of hospital projects. Sustainability (Switz) 11(22):1–18. https://doi.org/10.3390/su11226274
7. Mateus R, Bragança L (2011) Sustainability assessment and rating of buildings: developing the methodology SBToolPT-H. Build Environ 46(10):1962–1971. https://doi.org/10.1016/j.buildenv.2011.04.023
8. Najjar M, Karoline F, Palumbo M et al (2017) Integration of BIM and LCA: evaluating the environmental impacts of building materials at an early stage of designing a typical office building. J Build Eng 14:115–26. https://doi.org/10.1016/j.jobe.2017.10.005
9. Najjar MK et al (2019) Life cycle assessment methodology integrated with BIM as a decision-making tool at early-stages of building design. Int J Constr Manage 1–15. https://doi.org/10.1080/15623599.2019.1637098
10. Nwodo MN, Anumba CJ (2019) A review of life cycle assessment of buildings using a systematic approach. Build Environ 162. https://doi.org/10.1016/j.buildenv.2019.106290
11. Omer MAB, Noguchi T (2020) A conceptual framework for understanding the contribution of building materials in the achievement of sustainable development goals (SDGs). Sustain Cities Soc 52:101869. https://doi.org/10.1016/j.scs.2019.101869
12. Sandberg M et al (2019) Multidisciplinary optimization of life-cycle energy and cost using a BIM-Based master model. Sustainability (Switz) 11:2. https://doi.org/10.3390/su11010286
13. Tushar Q et al (2019) Optimizing the energy consumption in a residential building at different climate zones: towards sustainable decision making. J Cleaner Prod 233:634–649. https://doi.org/10.1016/j.jclepro.2019.06.093

# Enhancing the Building's Energy Performance through Building Information Modelling—A Review

Dhruvi Shah, Helly Kathiriya, Hima Suthar, Prakhar Pandya, and Jaykumar Soni

**Abstract** Building Information Modelling (BIM) is a technique that facilitates the Architect, Engineering, Construction, and Operation (AECO) practitioners to model the building with detailed information in the early design phase. BIM is increasingly in demand as it provides a reliable platform for deciding the life cycle of a building at any time. We can say that BIM is the process of building a building twice. In the planning phase, the building is created in the virtual environment. All the procurement management phases, i.e., planning, building, scheduling, costing, operating and managing are simulated. The operation phase of any building has different aspects such as locating building components, facilitating real-time data access, space management, energy management, demolition and renovation. This paper focuses on the energy management segment. With ongoing increases in energy costs, rising climatic consequences and updating building guidelines worldwide, AECO practitioners progressively have to consider the building's energy performance while planning. The buildings' energy utilization pattern depends upon various factors like daylight, HVAC system, location, orientation, resident's behaviour, building material and weather. Current energy analysis systems are satisfactory up to limited levels; hence, there is a need for more efficient energy simulation systems. BIM-integrated energy simulation techniques can play a crucial role in efficiently analysing the building's performance and enhancing it. These methods are termed as Building Energy Modelling (BEM). Different BEM tools work based on input data and considerations, significantly affecting a particular tool's capacity. In this paper, different techniques are discussed accordingly. The energy modelling process is carried out by transferring the building data to a BEM tool in an acceptable file format. This research provides deep insights into the advantages and disadvantages of various

---

H. Kathiriya · H. Suthar · P. Pandya · J. Soni (✉)
Civil Engineering Department, L.J. Institute of Engineering and Technology, LJ University, Ahmedabad, Gujarat, India
e-mail: jay.soni_ljiet@ljinstitutes.edu.in

D. Shah
Civil Engineering Department, Pandit Deendayal Energy University, Gandhinagar, India
e-mail: dhruvi.smtcl21@pdpu.ac.in

© The Author(s), under exclusive license to Springer Nature Singapore Pte Ltd. 2023
M. S. Ranadive et al. (eds.), *Recent Trends in Construction Technology and Management*, Lecture Notes in Civil Engineering 260,
https://doi.org/10.1007/978-981-19-2145-2_20

methods that contribute to future research and development of advanced methods and software programs.

**Keywords** Building information modelling (BIM) · Building energy modelling · Energy simulation · Energy performance

## 1 Introduction

The awareness that construction applications now face globalism, sustainability and atmospherical concerns other than that evolving legalization and skill requirements for the information era has triggered the technology known as Building Information Modelling (BIM) [12]. Building Information Modelling (BIM) is the technique of building a building twice. It is a process in which all the construction phases are simulated in a virtual environment. BIM is called the n-d process as it has many more dimensions along with 3D. 4 and 5 Ds stand for scheduling and cost evaluation of the project, respectively. 6th D is for facility management and 7th is for security and surveillance. It allows engineers to predict the building's performance in the early design stage [4]. BIM is quickly being adopted by the Architectural, Engineering, and Construction (AEC) industry as such techniques can boost productivity throughout the life cycle of the building. For example, the owner, design team and general contractor can use BIM to predict the requirements of the project, analyse the project scopes and intensify the coordination with a contractor and supplier, respectively. It can be a helping hand for facility managers by predicting the performance of the facility during the operation phase [5]. It is nowadays being used progressively as an arising technique to design, construct and operate the building in many countries [20]. BIM can be proven as a qualitative technique as it allows visualization, analysis, better documentation and easy modification in the design. Also, It allows different stack holders of the buildings to get valuable data from the model to fulfil efficient decision-making and economic project [4]. In short, BIM can be a handy tool for every building phase. In this article, the operation stage is taken under consideration as it contributes to about 60% of the total cost throughout the wheel of life of the building [18]. The operation phase of a building passes through various stages such as locating building components, visualization, marketing, space management and energy management. This article focuses on the energy management aspect.

Technological evolution affects social psychology, leading to variations in consumed energy [19]. The energy consumption pattern of a building is highly responsible for daylight and location [11], HVAC system, orientation, resident's behaviour, building material and weather. The overuse and misuse of energy resources are responsible for severe ecological and economic problems, including global warming [8]. With ongoing multiplying energy rates and the boosting concern about weather change, the idea of green building is a significant aspect of a sustainable energy future for the world. The significance of energy consumption evaluation has increased, but traditional statistics calculations are still done for estimation [16].

BIM-integrated energy simulation techniques can play a crucial role in efficiently analysing the building's performance and enhancing it. These methods are termed as Building Energy Modelling (BEM). A building energy model is a BIM model that performs energy consumption simulation. To prepare BEM models, various energy simulating tools have been developed and used. The building is first modelled using a modelling tool, this model is exported to the BEM tools in the accepted file format, and various design alternatives considering the energy utilization of the buildings are observed. The best alternative is finalized for the other construction process. This paper focuses on the file format for the data exchange between BIM and BEM. The focused formats are IFC (Industry Foundation Classes) and gbXML (Green Building Extensible Markup Language). These tools have been on the market since the 1980s. They are now improved to model and simulate more complicated and detailed systems. The energy modelling tool of a building can traditionally simulate the original structure and building materials along with complicated HVAC networks and innovative energy reservation techniques such as advanced window therapy for sun shading and renewable energy systems. But the interoperability between the BEM tool and 3D modelling tools is still challenging. The challenges are also discussed later in this article. The following segments of the article are literature review and methodology followed by discussion and conclusion.

## 2 Literature Review

It is believed that the planning stakeholders can be significantly benefited from BEM if BIM is implemented at the preliminary planning stage. The design alternatives in the context of energy utilization and thermal ease can be explored at this stage. The comparison between various available options can provide precise data regarding the energy consumption of the building. Dong et al. [7] said that BEM software programs could be segregated majorly into two parts. The one developed by the US Department of Energy and the other is private. For instance, eQUEST, Autodesk Green Building Studios(GBS), DesignBuilder and EnergyPlus are developed by the US Department of Energy. Whereas IES Virtual Environmental and Trace 700 are developed by private firms. Various BEM tools work according to different assumptions and details provided, influencing the performance and quality of results. For example, EnergyPlus is more flexible than BLAST and DOE-2 [6].

The BIM-based BEM can be considered a trending research topic. Some of the methods that have been developed are now used in practice. The main element of BIM-integrated BEM is data transfer. The building data such as geometry, thermal load, space heating, shading, daylight, space and thermal properties of materials are needed to be accurately transferred from BIM tool to BEM tool. There are four available primary methods for data transfer, i.e., Methods based on (1) Industry Foundation Classes (IFC), (2) Green Building Extensible Markup Language (gbXML), (3) Modelica Building Energy Modelling methods and (4) others. This article mainly focuses on the development and future research needed for the first two methods.

For the last decade, researchers and practitioners have been developing various methods to exchange the building data from BIM to the IFC-compatible BEM tools. A tool was developed to extract the data concerning the geometry of the facility from IFC to transfer it in ASCII format, preserving the data alteration rules of EnergyPlus [3]. Giannakis et al. [9] developed a semi-automatic technique using the IFC BIM model based on an algorithm to generate geometric data for EnergyPLus. An algorithm named 'Common Boundry Intersection Projection' (CBIP) was developed to produce the boundary of the geometry of the building from the IFC file. The results showed that this method provides accurate results [17]. For the transfer of material properties, an object-based technique was developed [13]. Kim et al. [14] developed an IFC-based converter that converts the building data within ArchiCAD and Revit. Laine et al. [16] developed an interface for EnegyPlus for the transformation of IFC-HVAC data to IDF and vice versa. These tools are efficient for data alteration but it is found that building geometry needs modification before simulation.

Abanda and Byers [1] examined the effect of orientation on the energy consumption process of building. The researchers used Revit and GBS for energy simulation. The results were relatively quick without regeneration of building geometry. The effect of window size, location and orientation on the energy performance of the building was examined in a study using GBS and Revit via the gbXML database [15]. To help the designer to estimate the daylight and energy usage, [2] innovated a system in Dynamo with gbXML format and a cloud-based program. Ham and Golparvar-Fard [10] developed a method for the ease of modifying thermal properties through 3D thermography using gbXML supported BIM tool. The methods compatible with the gbXML format are comparatively quick and do not require a large proportion of building geometry change. Also, to compare the results of energy consumption between various design alternatives, it is not required to get back to the BIM tool.

## 3 Methodology

To evaluate the building's energy behaviour through building Information modelling, the first step is to generate a BIM model of the facility to be constructed. The BIM model is not only a 3D model of a facility, but it also comprises material details, the building's location, type of building, the orientation of the building and all the aspects that affect the actual facility. After modelling, the building data are transferred to the BEM tool via accepted file format. BEM tool analyses the building's energy performance, gives results and compares the results of various design alternatives. The most possible and energy-efficient design is selected for the execution. IFC and gbXML are two various data extensions for data transfer.

IFC was initially proposed by the International Association for Interoperability (IAI). It is the only 3D object-oriented model used by BIM for information exchange and allows the exchange of details among collaborative groups. Autodesk Green Building Studios developed the gbXML based on the extensible markup language.

The XML represents a rich, generic, purposeful and validatable data format for storing and transmitting text and other details on and off the web service.

## 4 Discussion

Building Information Modelling is a technique evolving the decision-making segment of the construction sector. It can help enhance the collaboration between various stakeholders of a facility and provide a reliable base for decision-making during any building stage throughout its wheel of life. The operation phase of the building accounts for a large share of cost among all the phases so it is needed to enhance the performance of the building through various aspects. The overuse of energy has been increasing with the increase in demand and the time spent in the house. The construction sector is counted for considerable global energy utilization. Nowadays, Green Building has become a trending topic among researchers and practitioners due to the awareness and rising cost of energy sources. Building Energy Modelling Integrated with Building Information Modelling can perform a crucial part in analysing and enhancing the building energy performance. In this article, the methods of BIM-integrated BEM are focused. The exchange of information between BIM and BEM tools is an essential consideration as it is responsible for the accuracy of the results followed by decision-making. Both the medium, IFC and gbXML, are accurate up to a limited level. gbXML format is accurate for the simple building shapes. At the same time, IFC can deal with complex designs also. IFC requires comparatively more time for the analysis, and the gbXML format supports a rapid process. The gbXML file format is presently initiated to facilitate information exchange between BIM-supporting software programs and energy analysis tools. Compared to the IFC format which focuses on acquiring a complete and non-specific path to manage the information exchanging problems of a whole construction project, the gbXML format acquires particularity in the energy simulation discipline [7]. The literature review concluded that most of the gbXML-based techniques allow the successful exchange of geometrical, material and thermal data. It can be said that gbXML-based techniques have improved execution and touch a degree of semi-automated BIM-integrated BEM methods, where several building informations is altered from BIM, and the other necessary inputs need to be recorded by users in the user interface.

## 5 Conclusion

This article focuses on the methods of integrating BIM and BEM. Compared to manual methods used to date, BEM can provide accurate results for efficient decision-making concerning the building's energy performance. There are more than 400 tools for BIM and energy analysis. The BIM model data needs to be efficiently transferred to the BEM tool to achieve accurate energy performance prediction of the building. The data that needs to be transferred is geometry, material, space type, thermal zone, space load and HVAC system. This paper discusses two methods (IFC-based and gbXML-based) for the information exchange. Many researchers have tried developing various tools in compliance with IFC or gbXML format, but in the end, it is needed to check and modify several BEM data manually for reliable energy modelling. The main reason is the lack of proper interoperability between BIM and BEM tools. A BIM model generally consists of high-level data, which is much more complicated for BEM to interpret. Future research should also focus on the design engineer's aspect of utilizing Building Information Modelling integrated with Building Energy Modelling and reliable transfer of BIM data to the BEM tool. This research gives an insight into the significance of BIM-integrated BEM. It is recommended that BEM should be an integral part of the design phase for economic and ecologic decision-making.

## References

1. Abanda FH, Byers L (2016) An investigation of the impact of building orientation on energy consumption in a domestic building using emerging BIM (building information modelling). Energy 97:517–527. https://doi.org/10.1016/j.energy.2015.12.135
2. Asl MR, Bergin M, Menter A, Yan W (2014) BIM-Based parametric building energy performance multiobjective optimization. In: 32nd ECAADe conference, vol 224, p 10. http://autodeskresearch.com/pdf/bimparametric.pdf
3. Bazjanac V (2009) Implementation of semi-automated energy performance simulation: building geometry. In: Managing IT in construction/managing construction for tomorrow, pp 613–20. https://doi.org/10.1201/9781482266665-84
4. Borrmann A, König M, Koch C, Beetz J (2018) Building information modeling: Why? What? How? Build Inf Model Technol Found Ind Pract 1–24. https://doi.org/10.1007/978-3-319-92862-3_1
5. Bryde D, Broquetas M, Volm JM (2013) The project benefits of building information modelling (BIM). Int J Project Manage 31(7):971–980. https://doi.org/10.1016/j.ijproman.2012.12.001
6. Crawley DB, Lawrie LK, Winkelmann FC, Buhl WF, Joe Huang Y, Pedersen CO, Strand RK et al (2001) EnergyPlus: creating a new-generation building energy simulation program. Energy Build 33(4):319–331. https://doi.org/10.1016/S0378-7788(00)00114-6
7. Dong B, Lam KP, Huang YC, Dobbs GM (2007) A comparative study of the IFC and GbXML informational infrastructures for data exchange in computational design support environments. In: IBPSA 2007—international building performance simulation association 2007, pp 1530–37. https://www.researchgate.net/publication/285494452_A_comparative_study_of_the_IFC_and_gbXML_informational_infrastructures_for_data_exchange_in_computational_design_support_environments
8. Enshassi A, Ayash A, Mohamed S (2018) Factors driving contractors to implement energy management strategies in construction projects. J Financ Manag Prop Constr 23(3):295–311. https://doi.org/10.1108/JFMPC-09-2017-0035
9. Giannakis GI, Lilis GN, Garcia MA, Kontes GD, Valmaseda C, Rovas DV (2015) A methodology to automatically generate geometry inputs for energy performance simulation from IFC BIM models. In: 14th International conference of IBPSA—building simulation 2015, BS 2015, conference proceedings, pp 504–11

10. Ham Y, Golparvar-Fard M (2015) Mapping actual thermal properties to building elements in GbXML-Based BIM for reliable building energy performance modeling. Autom Constr 49:214–224. https://doi.org/10.1016/j.autcon.2014.07.009
11. Hviid CA, Nielsen TR, Svendsen S (2008) Simple tool to evaluate the impact of daylight on building energy consumption. Sol Energy 82(9):787–798. https://doi.org/10.1016/j.solener.2008.03.001
12. Jaradat S (2014) Educating the next generation of architects for interdisciplinary BIM environments. Charrette 1(1):127–136
13. Kim H, Shen Z, Kim I, Kim K, Stumpf A, Jungho Y (2016) BIM IFC information mapping to building energy analysis (BEA) model with manually extended material information. Autom Constr 68:183–193. https://doi.org/10.1016/j.autcon.2016.04.002
14. Kim I, Kim J, Seo J (2012) Development of an IFC-Based IDF converter for supporting energy performance assessment in the early design phase. J Asian Archit Build Eng 11(2):313–320. https://doi.org/10.3130/jaabe.11.313
15. Kim S, Zadeh PA, Staub-French S, Froese T, Cavka BT (2016) Assessment of the impact of window size, position and orientation on building energy load using BIM. Procedia Eng 145:1424–1431. https://doi.org/10.1016/j.proeng.2016.04.179
16. Laine T, Karola A (2007) Benefits of building information models in energy analysis. Energy 8. http://www.irbnet.de/daten/iconda/CIB8170.pdf
17. Lilis GN, Giannakis GI, Rovas DV (2017) Automatic generation of second-level space boundary topology from IFC geometry inputs. Autom Constr 76:108–124. https://doi.org/10.1016/j.autcon.2016.08.044
18. Liu R, Issa RRA (2016) Survey: common knowledge in BIM for facility maintenance. J Perform Constructed Facil 30(3). https://doi.org/10.1061/(ASCE)CF.1943-5509.0000778
19. Røpke I, Christensen TH (2012) Energy impacts of ICT—insights from an everyday life perspective. Telematics Inform 29(4):348–361. https://doi.org/10.1016/j.tele.2012.02.001
20. Wong AKD, Wong FKW, Nadeem A (2013) Comparative roles of major stakeholders for the implementation of BIM in various countries. J Chem Inf Model 53(9):1689–99

# Analysis of Clashes and Their Impact on Construction Project Using Building Information Modelling

**Samkit V. Gandhi and Namdeo A. Hedaoo**

**Abstract** Building Information Modelling (BIM) is an intelligent digital database containing all required construction features that offer stakeholders, and AEC professionals the knowledge they can use across the project life cycle to plan, design, manage and make successful decisions. The cooperation between the teams by coordinating 3D designs through BIM is necessary for the final design to be free of clash. The literature indicates that the need for BIM has increased and that some countries have also mandated its use in their construction projects. This paper aims to investigate the root cause of clashes occurring in construction work and their quantification using building information modelling. The research method implemented in this study is the preparation of 3D architectural, structural and MEP (Mechanical, Electrical and Plumbing) Revit model and the analysis of clashes using BIM software for buildings in India. The result from the BIM software showed the different types of clashes that affect the construction mainly, hard clashes which occurs due to the intersection of the component from different disciplines, soft clashes due to lack of buffer space around equipment's and workflow, or 4D clashes focused on the conflict between workflow and scheduling of equipment. This study concludes with the detailed quantification of clashes, their causes and impact of occurrence on building construction projects.

**Keywords** Building information modelling · Clash · Detection · Quantification · Scheduling

## 1 Introduction

In the design stage, 70% of the cost of building construction is committed; therefore, it is necessary to incorporate multidisciplinary expertise throughout the building

S. V. Gandhi (✉)
Construction Management, College of Engineering Pune, Pune, India
e-mail: gandhisv19.civil@coep.ac.in

N. A. Hedaoo
College of Engineering Pune, Pune, India
e-mail: nah.civil@coep.ac.in

© The Author(s), under exclusive license to Springer Nature Singapore Pte Ltd. 2023
M. S. Ranadive et al. (eds.), *Recent Trends in Construction Technology and Management*, Lecture Notes in Civil Engineering 260,
https://doi.org/10.1007/978-981-19-2145-2_21

construction process. The fact that there are so many diverse people working in the Architectural Engineering and Construction (AEC) industry is one of the problems according to a fragmented system of management in various sections of the project (structure, facilities, design, delivery, etc.). Each of these components must be based on a common concept and follow the same requirements as the others; they are a core component of the understanding that helps the project to be created together. If knowledge is not reliable and detailed enough, it will influence the next steps. Therefore, it is very common for existing buildings to lack suitable records, alter orders and a built-drawing. Because of its ability to cater to all stakeholders involved in the project, Building Information Modelling (BIM) is an important element in the integrated management of the project. BIM has the capacity to provide all stakeholders involved in the construction process with access to project-related information. BIM can assist owners, designers and builders to develop and coordinate the design of building systems and the planning of construction work, their manufacturing and construction processes and their operating and maintenance processes, as well as the decommissioning of their installations.

**Building Information Modelling (BIM)** is a sophisticated 3D model-based approach that offers architects, developers and construction practitioners the expertise and resources they need to help plan, design, create and maintain buildings and infrastructure. "The use of a shared digital representation of a built object (including buildings, bridges, roads, process plants, etc.) to facilitate design, construction and operation processes to form a reliable basis for decisions". ISO 19650 (2019). In all stages of project implementation, BIM can be used. It also progresses through multiple so-called Levels of Development (LOD) as a project evolves through different delivery stages. The use of BIM as a way of promoting cooperation and enhancing implementation performance and project quality has been promoted by several governments in recent years (Fig. 1).

**Clash Detection**—In the film industry, the initial concept of detecting spatial conflicts with objects originated. Developers were immediately faced with an entirely new dilemma when programing the first video games. When modelling their trails, they realized the inevitability of the collision of solid elements. Therefore, when two entities were touching each other, they had to define what to do. This effectively led to the creation of a new program code that would describe the general actions of each object involved in the dispute. The fundamental idea was to see how two objects might collide in three-dimensional space. In the case of a collision with another entity being detected by the system, the solution was to add certain object attributes and compile behaviour rules. Civil engineers began to struggle with those concerns by using BIM as a result of the accelerated growth of information technology. Clash identification for the purpose of smooth collaboration consists of the process of examining and identifying disputes within various BIM models. Collision detection or teamwork are other expressions for collision detection. This method has an effect on significant design choices and, when used correctly, will help ensure a smooth building implementation. At various stages of the project, collisions will occur.

Analysis of Clashes and Their Impact on Construction Project ...

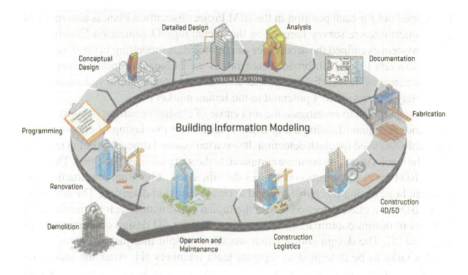

**Fig. 1** BIM workflow. *Courtesy* Hagerman and Company

They are better resolved at the outset of the process, when it is most cost-effective. Some techniques were followed by the Built Environment to establish automatic clash identification and conflict element indicator. As time went by, this BIM clash detection system developed greatly. Today, science is actively designing tools that are more modern. These tools tend to assist in operations such as model processing and general concept making.

## 2 Literature Review

It is important to go for literature reviews from various journals, books, newsletters and magazines to find the attributes. By finding attributes, we will be able to know the efficiency, utilization of new technologies in construction for various purposes. BIM is a new trend which is changing the face of the construction industry. To move with the new trend, it is necessary to study and understands the impact of this adoption of new technologies in the construction sector. Impact on the construction sector, new policies to adopt, the main workforce to be recruited, cost effects of these types of various doubts will be cleared from the literature survey. Also, gaps find out in past research will be researched in this research.

BIM-based design collaboration was discussed in a recent publication with the involvement of designers, general contractors and engineers [1]. It is important to first set the team's ground rules, and confirm that each trade can use the LOD, then the workflow of design coordination can be processed. A construction management team with proven positions specific to BIM, as well as the requisite expertise that should be mapped out for each position in the BIM Project Execution Plan, is also required [2]. A questionnaire survey focused on the Technology-Organization-Environment (TOE) system examined the acceptance of building data modelling in the developing Indian markets [3], which broadly categorizes the influencing factors along these three dimensions. It was discovered that architectural companies in India had yet to completely leverage BIM's potential in the Indian market [4].

This paper aims to investigate the root cause of clashes occurring in construction work and their quantification using BIM. Traditionally overlaying of drawings on a light table was used for clash detection. It was a tedious and time-consuming process. Also, the results were less accurate compared to the software-based method. Thus, the use of BIM for clash detection enhances the efficiency of building construction and generate huge saving through early and accurate collision detection. The emerging use of BIM as a clash detection tool in the Indian construction sector is remarkable. In terms of design coordination, a 10-step method for 3D design coordination was proposed [5]. The design coordination workflow is split into just six steps, which enables tasks to be delegated to separate team members [1]. After the individual discipline-specific models have been developed, each BIM author must perform internal correspondence before sending their individual models to the BIM manager. After that, a federated model (which includes individual discipline-specific models) will be developed. It is also possible to define clashes as true positives, false negatives or false positives. Real positives are conflicts that have been identified as clashes that are in fact clashes, while false negatives are conflicts that have not been identified as clashes It is also possible to define clashes as true positives, false negatives or false positives [6]. True positives are the conflicts identified as clashes as actually they are, false negatives are those which have not been registered are clashed, while false positives are clashes identified wrongly [7]. This designation aims to raise awareness of the accuracy of clash identification, either to minimize potential risks or to improve speed. The presence of positive relationships between clash identification and architecture, buildability and building efficiency is then confirmed [8].

The research carried out by Leite and Akinci [9] has compared types of clashes identified in their manual coordination process and through automatic clash detection using a BIM in Navisworks. Results from this study show a general high recall and low precision for the automatic clash detection process, and low recall and high precision for the manual approach. As a result, the root causes of clashes were investigated using the PAS 1192-2 design process requirements [10]. The isolated practice was shown to be the primary cause of high incidences of clashes related to 3D BIM Mechanical, Electrical and Plumbing (MEP) structures. They recommend that the common data environment should be unsegregated and capable of facilitating interdisciplinary and remote cooperation in an effective manner.

## 3 Types of Clashes

### 3.1 Hard Clashes

Hard clash is a type of conflict that involves only geometric issues. A collision in which the geometry of one part may or may not overlap that of another, but the distance between them is less than the tolerance. Hard conflicts, owing to both the specific sequence of components and their null dynamics, are perhaps the most readily recognizable of all forms of conflicts (Autodesk).

According to Seo et al. [11], these entities are assumed to be static and, thus, their permanent installation and sufficient interaction with the rest of the system can be predicted. Hard conflicts are, for example, piping system collisions with the walls or other structural components without identifying the opening.

### 3.2 Soft Clashes

Soft clash occurs when the element does not receive the required spatial or geometrical tolerance, or when its buffer zone is breached. In this case, in the control process, space, connection, installation and manipulation requirements are considered. "A clash in which the selection geometry of one component may or may not directly interfere with that of the selection of another component, but is less than the set tolerance gap" [12]. For example, for secure and simple maintenance access, an air conditioning unit can require buffer space from a beam. If left unattended, soft clashes can lead to maintenance issues as well as safety concerns.

### 3.3 Workflow or 4D Clashes

Contractor scheduling conflicts, delivery of equipment and services and conflicts over the general workflow timetable are all examples of workflow or 4D clashes. Unlike hard and soft clashes, 4D clashes result from scheduling clashes of interdisciplinary activities that eventually reduce the efficiency of the entire construction industry. Engineers may face challenges that can have a significant impact on ongoing work caused by different timetables or coordination plans. Since one dispute has a cascading effect on many fields and can bring work to a standstill, contractors cannot afford 4D disputes; it is necessary to prefer BIM. The control framework investigates these problems for 4D conflicts [13]. Basically, 4D collisions happen mostly during construction and have little to do with the structure's design. An example of a 4D clash is insufficient distance between the crane and the structure during construction.

## 4 Research Methodology

From the literature survey, the need of using BIM in the design and coordination process was known and its benefits were identified. In order to illustrate the proposed clash detection work, the process is applied to the commercial office building in the central Maharashtra region. The building was designed as a multipurpose building for various office spaces. The data is collected in the form of 2D CAD drawings (structural, architectural, MEP) and other construction and interior details. The data available was analysed and separate 3D BIM models were developed from the 2D CAD drawings for all the 3 disciplines as architectural, structural and MEP. The ISO 19650 guideline, which was based on UK PAS 1192, specifies the criteria for achieving BIM Level 2 by developing a structure for collaborative working and knowledge requirements in BIM.

Considering LOD 300, the 3D architectural model was prepared using Revit 2020. As a base guide, the 2D plan was imported into the Revit. Using the reference, a plan is created taking into account the origin of the Revit to which the levels with the appropriate height and wall thickness have further been applied. The Revit families have allocated different building materials and architectural installations.

The 3D structural model is prepared using the same Revit 2020 software with the aid of the 3D architectural model and RCC plan available. The grid line for the creation of the model is mainly fixed. The location of columns and beams with the coordinates was then set with regard to the architectural model and the structural plan available. Then, as the slab height was determined by floor height, the roof was covered with galvanized sheets of iron. The connection between the beams, columns and all other structural members was provided in the Revit for RCC frame connections using the plug-in advance steel connection. The model prepared was been reviewed by the structural expert for maximum accuracy.

The preparation of the MEP model was the most critical and challenging modelling task. As it encompasses a wide range of components from various contractors, which is the primary reason for clashes. For ease of planning, separate 3D models were prepared for each system along with HVAC from the 2D drawings and information available for mechanical, electrical and plumbing systems. With the expert recommendations, other miscellaneous components such as cable trays, ducts, hangers were also added. In one federated model, these all-separate 3D models were integrated and all systems as a single model were linked together.

The clash detection analysis of these models was done using Navisworks Manage 2020 software. In this, the different models were merged clash detective test was carried out between two at a time as structural versus MEP, structural versus architectural, architectural versus MEP. The clash test between different components of the same model (i.e., piping with electrical) or different models from the same level also can be done. The clashes can be sorted based on hard, hard conservative, soft (clearance) and duplicate clash in the clash detective test. The elaboration of the clash test is shown in the data analysis part. The tolerance for each test can be set as per the requirement, it controls the severity of the clashes reported and filter out

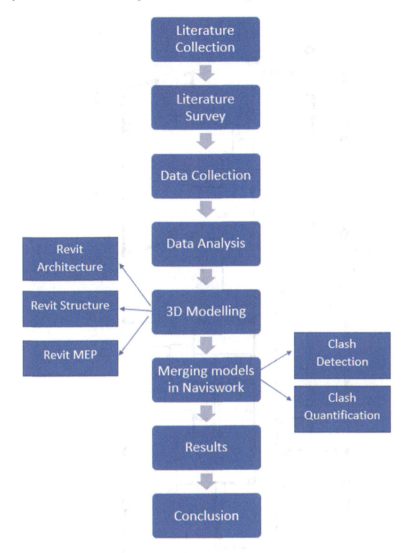

**Fig. 2** Research framework

negligible clashes. Thus, after the clash test, the report of each test can be exported in excel format with the clash coordinates for ease of understanding (Fig. 2).

## 5 Data Collection

To study the effective use of BIM, a real-life case study of the multi-purpose buildings was selected. Details of the case study are as follows:

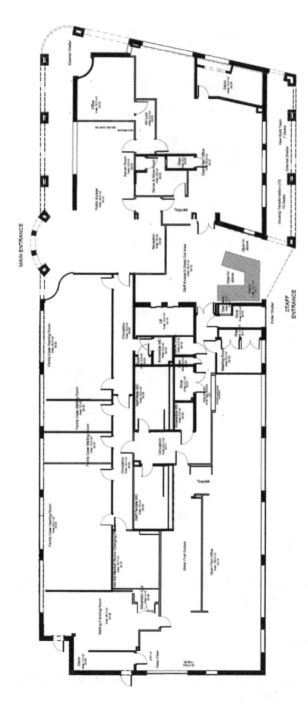

**Fig. 3** 2D AutoCAD plan of office building

Location—Ahmednagar, Maharashtra India.
No. of Floors—G + 3.
Area of each Floor—1080 m$^2$.
Total Floor Area—4320 m$^2$.
Building type—Commercial office building.
Details Available—Floor Plan of all floors in 2D (Fig. 3).

## 6 Data Analysis

The collected data was categorized according to requirements and data analysis was carried out at various stages. Using Revit 2020 software, the 3D BIM-based Structural, Architectural and MEP modelling was primarily carried out. After the model was prepared, Navisworks Manage 2020 was used to conduct clash detection and further clash quantification based on various clashes and models (Figs. 4, 5 and 6).

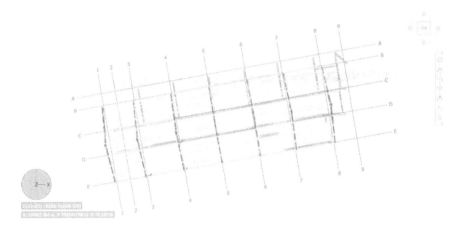

**Fig. 4** BIM based 3-dimensional model of structural component

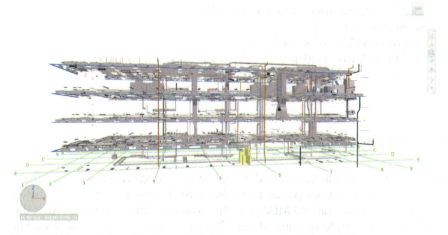

**Fig. 5** BIM based 3-dimensional model of MEP components including HVAC system

**Fig. 6** BIM-based three-dimensional model of architectural component

## 7 Result and Discussion

See Figs. 7, 8, 9, 10, 11, 12 and 13.

Analysis of Clashes and Their Impact on Construction Project ... 265

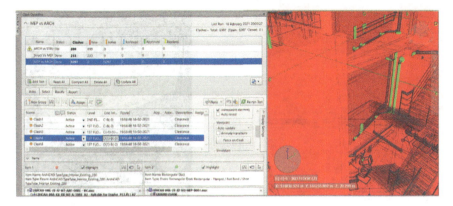

**Fig. 7** Clash detection analysis between architectural and MEP models using Navisworks

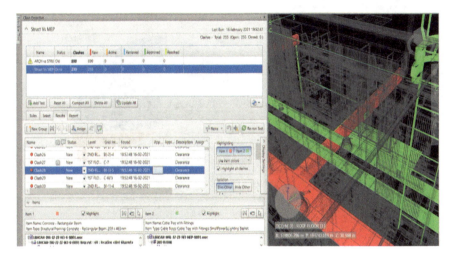

**Fig. 8** Clash detection analysis between structural and MEP model using Navisworks

## 7.1 Observation from Results

From the data analysis and the results obtained from the data, we can observe that after the total 9 clash detection test between different models of architectural, structural and MEP using Navisworks out of total hard clashes maximum of 72% (2575) were caused between Architectural versus MEP models, 22% (793) between structural versus architectural models and only 6% (228) in structural versus MEP models. It can be seen that the maximum collision is between architectural and MEP components geometry and the lowest is between the structural and MEP components. The clash detective test for the soft/clearance clashes shows relatively same results showing 67% (2363) between architectural versus MEP, 26% (899) between structural versus

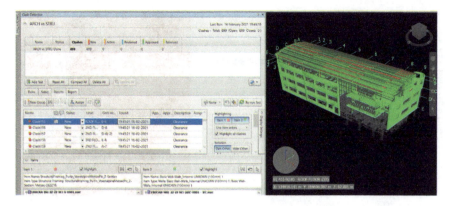

**Fig. 9** Clash detection analysis between architectural and structural models using Navisworks

**Fig. 10** Soft clash report

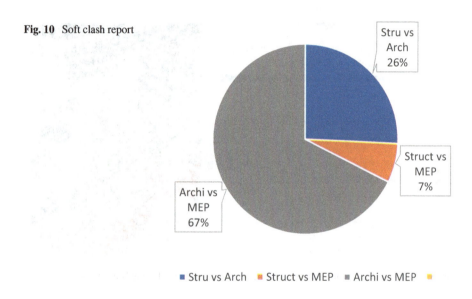

architectural, and only 7% (233) structural versus MEP. It shows that the geometry of the following clashes doesn't directly intersect but comes between the tolerance limit. We can further notice the occurrence of duplicate clashes having twin geometry and at the same location is also more for the architectural versus MEP test but comparatively negligible in numbers.

Thus, a comparative analysis of results shows that architectural and MEP components and their collision are the main reason for the occurrence of clashes and need to be focused primarily on designing and construction work.

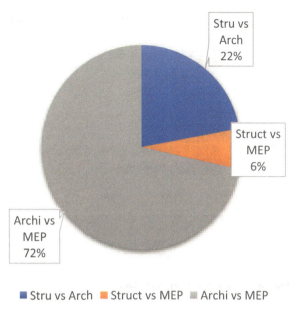

**Fig. 11** Hard clash report

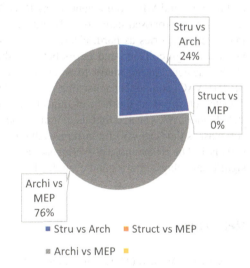

**Fig. 12** Duplicate clashes report

## 8 Conclusion

After the review of the clash detection with the application of technology-based on BIM, the study shows that the incorporation of 3D models is important for better visualization in construction planning; therefore, pointless errors can be avoided to reduce the risk of cost overruns in the construction process. The clash detection terminology was explained with its types. 3D modelling was done for structural,

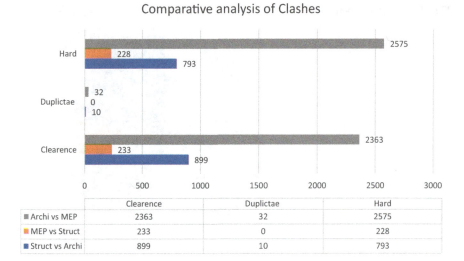

**Fig. 13** Comparative analysis of clashes

architectural and MEP components using Revit2020 software. A Simple clash detection test was proposed using BIM-based Navisworks software. Clashes are sorted into three categories as hard, clearance and duplicate. Clash detective test identified the design collision and errors between the 3 models, with overall 3596 hard clashes detected in this study. Most clashes are detected between architectural and MEP (HVAC) component consisting of 2575 clashes caused. The main cause for the detected clashes was the lack of multidisciplinary coordination between different stakeholders. We can see that the count of the hard clash is nearly the same as the soft clashes thus can say circumstances that led to the hard clash may also cause the soft clash. The clashes pose possible risks, according to cited sources. High spending, significant delays, injuries and even deaths may be caused by them.

# References

1. Arayici Y, Aouad G (2011) Building information modelling (BIM) for construction lifecycle management. In: Construction and building: design, materials, and techniques, March, pp 99–117
2. Abanda FH, Vidalakis C, Oti AH, Tah JHM (2015) A critical analysis of building information modelling systems used in construction projects. Adv Eng Softw 90:183–201. https://doi.org/10.1016/j.advengsoft.2015.08.009
3. Ahuja R, Sawhney A, Jain M, Arif M, Rakshit S (2020) Factors influencing BIM adoption in emerging markets—the case of India. Int J Constr Manag 20(1):65–76. https://doi.org/10.1080/15623599.2018.1462445
4. Sawhney A, Singhal P (2014) Drivers and barriers to the use of building information modelling in India. Int J 3-D Inf Model 2(3):46–63. https://doi.org/10.4018/ij3dim.2013070104

5. Staub-French S, Khanzode A (2007) 3D and 4D modeling for design and construction coordination: issues and lessons learned. ITcon 12(Sept 2006):381–407. http://www.itcon.org/2007/26
6. Tommelein ID, Gholami S (2012) Root causes of clashes in building information models. In: IGLC 2012—20th conference of the international group for lean construction, 1(510)
7. Tixier AJP, Hallowell MR, Rajagopalan B, Bowman D (2017) Construction safety clash detection: identifying safety incompatibilities among fundamental attributes using data mining. Autom Constr 74:39–54. https://doi.org/10.1016/j.autcon.2016.11.001
8. Hartmann T (2010) Detecting design conflicts using building information models : a comparative lab experiment. In: Proceedings of the CIB W78 2010: 27th international conference—Cairo, Egypt, 16–18 Nov
9. Leite F, Akinci B (2009) Identification of data items needed for automatic clash detection in MEP design coordination: construction research congress 2009, pp. 416–425
10. Akponeware AO, Adamu ZA (2017) Clash detection or clash avoidance? An investigation into coordination problems in 3D BIM. Buildings 7(3):1–28. https://doi.org/10.3390/buildings7030075
11. Seo J-H, Lee B-R, Kim J-H, Kim J-J (2012) Collaborative process to facilitate BIM-based clash detection tasks for enhancing constructability. J Korea Inst Build Constr 12(3):299–314. https://doi.org/10.5345/jkibc.2012.12.3.299
12. Reddy P (n.d.) BIM for building owners and developers: making a business case for using BIM on projects
13. Matejka P, Sabart D (2018) Categorization of clashes and their impacts on construction projects. In: Engineering for rural development, 17, pp 827–835. https://doi.org/10.22616/ERDev2018.17.N102

# Predicting the Performance of Highway Project Using Gray Numbers

**Supriya Jha, Manas Bhoi, and Uma Chaduvula**

**Abstract** Earned Value Analysis is a methodology used to monitor project performance in terms of time, scope and cost and also to deal with uncertain situations that come within. Uncertainty is a part of construction project and sometimes these situations can cause a great loss in the project's success. Recently, to deal with uncertain situations a different approach has been developed to predict the project performance in a non-deterministic way, i.e., using gray interval numbers. A framework using gray interval numbers has been developed to predict the project performance and hence this study aims at using the framework to predict the performance of a real-life highway project of total duration of approximately 2 years. The analysis involves the verbal directed data from the site by the experts which were denoted as gray interval numbers. The results indicate that the project is under budget as the CPI is 1.06 and ahead of schedule as the SPI is 1.2. The results also show the worst case scenario that the project may exceed the budget as CPI is 0.83 and may run behind the schedule as SPI is 0.69. The outcomes of the study are in the form of range which provides the overall profile of the project and also helps the project team members to not always be accurate or deterministic with the outcomes. Since the construction sector was majorly hit by an uncertain event, i.e., COVID-19, this study can be very helpful in determining the performance after facing such a huge gap.

**Keywords** Project management · Earned value analysis · Gray numbers · Performance control

## 1 Introduction

Project management is an integral part of the construction sector and it involves making a project to be a success. Earned value analysis (EVA) is an important key to the project management which in turns predicts the overall performance of the project. Many methods of EVA have been developed like earned schedule, earned schedule

S. Jha (✉) · M. Bhoi · U. Chaduvula
Pandit Deendayal Petroleum University, Gandhinagar, India
e-mail: supriyajha534@gmail.com

© The Author(s), under exclusive license to Springer Nature Singapore Pte Ltd. 2023
M. S. Ranadive et al. (eds.), *Recent Trends in Construction Technology and Management*, Lecture Notes in Civil Engineering 260,
https://doi.org/10.1007/978-981-19-2145-2_22

min–max, earned duration management, etc., which are all deterministic methods. Timur et al. [1] discussed about the conventional Index-based (IB) EVM method, how it restricts the accuracy of forecasting Cost Estimate at Completion (CEAC) of a project by presuming that it only depends on the past available data and having insufficient amount of data at an early stage of the project, they suggested a new approach for CEAC which is a blend of index-regression technique to foresight the final cost of the impending projects. This strategy depends on ES-based IB equation altered by combination of Gompertz Growth Model (GGM) via non-linear regression analysis which gives more exact and detailed CEAC. Batselier et al. [2] compared three primary deterministic methods of EVM and their reciprocated blend on an actual project. The two methods i.e., ESM-1 and ESM-SPI(t) (where the performance index is selected as 1 and SPI respectively) were used as the addition in EVM and both of them combine rework and activity sensitivity in EVM time foresight, they both are performance-aimed techniques. Rupak [3] suggested a gray concept technique that will assist in acknowledging and creating project interrelationships and also to keep up with the uncertainty. This concept is believed to be an effective technique to work out the R&D project portfolio preferences.

Li et al. [4] aims at developing a different non-deterministic approach, i.e., gray-based approach for the selection of the ideal supplier. The outcomes show that gray theory works great with complex and uncertain situations and this proposed method is quite efficient and effective. Li et al. [5] discussed about the most common method for complex decision making process, i.e., TOPSIS (technique for order performance by similarity to ideal solution). They have proposed to extend this method by including Minkowski distance function to enhance the limitations of the TOPSIS method followed by the addition of gray numbers to deal with the uncertain situations involved and lastly to amalgamate it with the expert opinion decision process. Oztaysi [6] focussed on IT field to select the best communicating platform for both the internal and external affairs. They have proposed an amalgamation of Analytical Hierarchy Process (AHP) with Gray-TOPSIS model which enables the evaluation of the weightage of each factor and deals with the ranking including uncertainties. Li et al. [7] aims at developing a framework for assessing safety performance of the organizations that may involve all the risk-based factors. They aimed at broadening the conventional analytical hierarchy process to gray numbers systems to deal with the uncertain factors in the assessment. Amirian and Sahraeian [8] aims at approaching a different technique for project selection and scheduling. They have converted a pure integer model into gray model as gray numbers have distinctive type of relations with the variables other than the deterministic variables.

Durga et al. [9] compared the manual and computer-aided technique and showed that software could be easily managed and leads to time-saving and quicker decision making, also showed the track progress of the ongoing project. Paulo et al. [10] blended schedule performance and schedule adherence as an index to enhance the duration foresight precision where schedule adherence is an extension to EDM concept that leads to addition of c-factor with respect to schedule adherence which is not cost-related. Ibrahim et al. [11] discussed about the how the risk variables influence the EVM technique for evaluating the performance of infrastructure in the

Australian context, aims at recognizing the potential risk variables which influence the performance at the implementation stage, i.e., design and delivery stage. Perez et al. [12] suggested two new metrics and their related foresight techniques, i.e., $ES_{min}$ and $ES_{max}$, these metrics highlight the preference of the activities and distribution of resources, these metrics can be separately used from EVM structure as they are not dependent on any primary metrics of EVM technique. Nadafi et al. [13] suggested a new approach to decide the project completion time and cost using verbal data. Nadafi et al. [13] introduced interval gray numbers for time and cost foresight of the project where the verbal data is assigned to gray numbers and the calculation is carried out accordingly.

Since project control is a major part of the construction project Earned Value Analysis (EVA) is very much necessary to monitor and track the project at different stages to predict the early warnings and how to cope up with them. Recently, gray numbers have been introduced in EVM technique to predict project performance where gray numbers are known for uncertain situations and unknown information. For uncertain situations, risk management for all the construction projects has been carried out but some uncertainties still remain which are unpredictable. For now-situation, due to COVID-19, construction projects were shut down for a long time and a great loss has been occurred in terms of time and cost. Hence, to have a bigger picture and insufficient amount information in hand, prediction of the projects' health condition is quite difficult. In such situation, gray numbers are of great use where partial information is known and from that much information foresight of the project condition in comprehensive view can be obtained.

Nadafi et al. [13] suggested a new approach to decide the project completion time and cost using verbal data and introduced interval gray numbers using a numerical example. This study contributes to improving the efficiency of the methodology by applying it to a real-life highway project and predicting the performance at different stages to avoid delays and shortage of resources. Since gray numbers are known for unpredictability and insufficient amount of data, this approach is quite useful for now-situation in the construction industry. Also, for more accurate results and indicating the effectiveness of gray numbers, intermediate values have also been considered in this study to show that the outcome will not cross the range, which will better assist the project members in making accurate decisions for management.

## 2 Research Methodology

This paper aims at predicting the performance of the project with a different approach which involves non-deterministic input, i.e., verbal directed data. The methodology involves gray interval numbers which will assist in predicting the performance of the project. Gray interval number is defined as a range of data that can be used in situations where data is incomplete or insufficient. Gray interval number can be defined by ⊗ symbol and it provides the worst and best case scenarios of the project. Gray numbers provide a wholesome profile of the project progress with minimal data and as a result, help the project members to be prepared for the worst and best case scenarios of the project. For the analysis, a real-life highway project with a total cost of approximately Rs. 680 (millions) and total duration of approximately 2 years has been taken into account. The progress of the activities was directed by the project members from the site verbally which will be denoted as gray interval numbers accordingly as shown in Table 1.

The verbal data for the progress of the activities on the site has been collected through project members and assigned to the gray numbers accordingly as shown in Table 2 [13].

The gray interval numbers are then multiplied with the cost of each activity and earned value of the activities was evaluated. The summation of the earned value of each activity gives the overall earned value of the project. Earned value indicates the actual work performed in the project at the particular tracking period. Similarly, the verbal data for the planned progress of the activities has been collected and gray interval numbers have been assigned accordingly as shown in Table 3.

The gray interval numbers are then multiplied by the cost of each activity and planned value of activities were evaluated. The summation of the planned value of each activity gives the overall planned value of the project at that particular tracking period. Planned value indicates the work scheduled to be completed at that particular period. After evaluation of the planned value and earned value, cost performance index (CPI) which indicates the cost regulation of the project has been evaluated. It determines how much we have earned for the work completed in comparison with the actual cost spent. It also indicates whether the project is under budget/ over budget or on budget. CPI can be calculated as shown below:

$$\otimes CPI = \otimes EV/AC \tag{1}$$

where AC = Actual cost spent in the overall project.

Similarly, schedule performance index (SPI) indicates the time regulation of the project has been evaluated. It determines how much work has been completed as compared to the planned schedule in terms of time. It also determines whether the project is behind schedule/ahead of schedule or on schedule. SPI can be calculated as shown below:

**Table 1** Gray numbers and corresponding verbal terms

| Gray numbers | Verbal term |
|---|---|
| [0.0–0.1] | Very low |
| [0.1–0.3] | Low |
| [0.3–0.4] | Less than Medium |
| [0.4–0.5] | Medium |
| [0.5–0.6] | More than medium |
| [0.6–0.9] | High |
| [0.9–1.0] | Very high |

**Table 2** Earned value estimation by gray numbers

| Activity | | Percentage of real progress as verbal | Cost (in millions) | Gray interval number | ⊗EV = Cost × P |
|---|---|---|---|---|---|
| *Construction in Kesinga to Bhavanipatna* | | | | | |
| 1.1 | Culverts | Very high | 15 | [0.9–1.0] | [13.5–15] |
| 1.2 | Earthwork and Subgrade preparation | Very high | 28 | [0.9–1.0] | [25.2–28] |
| 1.3 | GSB | Very high | 29 | [0.9–1.0] | [26.1–29] |
| 1.4 | WMM | Very high | 29 | [0.9–1.0] | [26.1–29] |
| 1.5 | DBM | Very high | 60 | [0.9–1.0] | [54–60] |
| 1.6 | BC | Not started | 0 | [0–0] | [0–0] |
| 1.7 | Road side drain | Not started | 0 | [0–0] | [0–0] |
| 1.8 | Road signages and boards | Not started | 0 | [0–0] | [0–0] |
| *Construction of Amath to kesinga* | | | | | |
| 2.1 | Culverts | Low | 134 | [0.1–0.3] | [13.4–40.2] |
| 2.2 | Minor bridges | Not started | 0 | [0–0] | [0–0] |
| 2.3 | Embankment and subgrade preparation | Less than medium | 22 | [0.3–0.4] | [6.6–8.8] |
| 2.4 | GSB | Less than medium | 22 | [0.3–0.4] | [6.6–8.8] |
| 2.5 | WMM | Less than medium | 22 | [0.3–0.4] | [6.6–8.8] |
| 2.6 | DBM | Not started | 0 | [0–0] | [0–0] |
| 2.7 | BC | Not started | 0 | [0–0] | [0–0] |
| 2.8 | Road side drain | Not started | 0 | [0–0] | [0–0] |
| 2.9 | Construction of toll plaza and admin building | Not started | 0 | [0–0] | [0–0] |
| 2.10 | Road signages and boards | Not started | 0 | [0–0] | [0–0] |
| 2.11 | Protection work | Not started | 0 | [0–0] | [0–0] |
| 2.12 | Crash barriers | Not started | 0 | [0–0] | [0–0] |
| 2.13 | Miscellenous work | Not started | 0 | [0–0] | [0–0] |
| *Acceptance* | | | | | |
| 3.1 | Testing | Not started | 0 | [0–0] | [0–0] |
| 3.2 | Handover | Not started | 0 | [0–0] | [0–0] |
| 3.3 | Finish | Not started | 0 | [0–0] | [0–0] |

(continued)

**Table 2** (continued)

| Activity | | Percentage of real progress as verbal | Cost (in millions) | Gray interval number | $\otimes EV = \text{Cost} \times P$ |
|---|---|---|---|---|---|
| | | | 406 | | [178.1–227.6] |

$$\otimes SPI = \otimes EV / \otimes PV \qquad (2)$$

Thus, by calculating CPI and SPI, we can predict the project performance using gray numbers in terms of cost and time which will assist the project members to be prepared for the worst and best case scenarios and act accordingly to avoid any uncertainties at any given point of time.

## 3 Analysis and Results

In this paper, we have aimed at approaching a different method for earned value analysis for predicting the project performance and avoid the uncertainties at various stages. For the analysis, a real-life highway project with total cost of approximately Rs. 680 (millions) and total duration of approximately 2 years has been taken into account and earned value analysis has been performed using gray interval numbers. Gray interval numbers were used to provide an overall satisfactory profile of the project when there is insufficient amount of data or there is shortage of time. Gray numbers assist in providing worst and best case scenarios for the project so that the project members will be prepared for both the scenarios and avoid any delays or resource losses in the project. As mentioned above, verbal data has been collected for the project progress from the site which is graphically shown in Fig. 1. Figure 1 shows the worst case scenarios of the percentage progress of the activities as per gray numbers based on verbal directed data for the ninth month of the project as tracking period.

From Fig. 1, it has been observed that progress of activity 6th seems to be planned to get completed by 10% for that particular tracking period but it has been not started yet and similarly for the progress of activity 12th seems to be planned to get completed by 60% but it has been completed till 30% for that particular tracking period. Thus, this progress shows the worst case scenarios of the project which may affect the completion time and cost of the project.

Figure 2 shows the best case scenarios of the percentage progress of the activities as per gray numbers based on verbal directed data for the ninth month of the project as tracking period. From this, it has been observed that the progress of activity 10th seems to be on schedule and similarly progress of activity 14th seems to be completed by 20% but it has been completed till 40% at that particular tracking period which indicates better performance of the project.

Calculation of CPI and SPI of the project using gray interval numbers:

**Table 3** Planned value estimation by gray numbers

| Activity | | Percentage of planned progress as verbal | Cost (in millions) | Gray interval number | ⊗PV = Cost × P |
|---|---|---|---|---|---|
| *Construction in Kesinga to Bhavanipatna* | | | | | |
| 1.1 | Culverts | Very high | 15 | [0.9–1.0] | [13.5–15] |
| 1.2 | Earthwork and subgrade preparation | Very high | 28 | [0.9–1.0] | [25.2–28] |
| 1.3 | GSB | Very high | 29 | [0.9–1.0] | [26.1–29] |
| 1.4 | WMM | Very high | 29 | [0.9–1.0] | [26.1–29] |
| 1.5 | DBM | Very high | 60 | [0.9–1.0] | [54–60] |
| 1.6 | BC | Low | 65 | [0.1–0.3] | [6.5–19.5] |
| 1.7 | Road side drain | Very low | 14 | [0–0.1] | [0–1.4] |
| 1.8 | Road signages and boards | Not started | 0 | [0–0] | [0–0] |
| *Construction of Amath to kesinga* | | | | | |
| 2.1 | Culverts | Low | 134 | [0.1–0.3] | [13.4–40.2] |
| 2.2 | Minor bridges | Not started | 0 | [0–0] | [0–0] |
| 2.3 | Embankment and subgrade preparation | High | 22 | [0.6–0.9] | [13.2–19.8] |
| 2.4 | GSB | Medium | 22 | [0.4–0.5] | [8.8–11] |
| 2.5 | WMM | Low | 22 | [0.1–0.3] | [2.2–6.6] |
| 2.6 | DBM | Not started | 0 | [0–0] | [0–0] |
| 2.7 | BC | Not started | 0 | [0–0] | [0–0] |
| 2.8 | Road side drain | Not started | 0 | [0–0] | [0–0] |
| 2.9 | Construction of toll plaza and admin building | Not started | 0 | [0–0] | [0–0] |
| 2.10 | Road signages and boards | Not started | 0 | [0–0] | [0–0] |
| 2.11 | Protection work | Not started | 0 | [0–0] | [0–0] |
| 2.12 | Crash barriers | Not started | 0 | [0–0] | [0–0] |
| 2.13 | Miscellenous work | Not started | 0 | [0–0] | [0–0] |
| *Acceptance* | | | | | |
| 3.1 | Testing | Not started | 0 | [0–0] | [0–0] |
| 3.2 | Handover | Not started | 0 | [0–0] | [0–0] |
| 3.3 | Finish | Not started | 0 | [0–0] | [0–0] |
| | | | 406 | | [189–259.5] |

**Fig. 1** Real versus planned progress of activities (worst case scenario)

**Fig. 2** Real versus planned progress of activities (best case scenario)

Taking into consideration the calculation of Tables 2 and 3 we have:

$$\otimes EV = [178.1 - 227.6]$$

$$\otimes PV = [189 - 259.5]$$

Figure 3 shows worst case scenario for Earned value versus Planned value of the activities for that particular tracking period and it has been observed that for activity 6th no value has been earned yet as per schedule some value was planned to be spent, hence such case shows the worst case scenario and assists the project members to be prepared with contingency plan.

Similarly, Fig. 4 shows the best case scenario for Earned value versus Planned value of the activities for that particular tracking period.

From EV and PV, we can further calculate the CPI and SPI and Estimate at completion (EAC) for predicting project performance and completion time and cost for the ninth month of the project as tracking period. Hence,

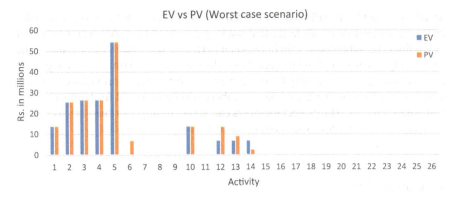

**Fig. 3** EV versus PV of the activities (worst case scenario)

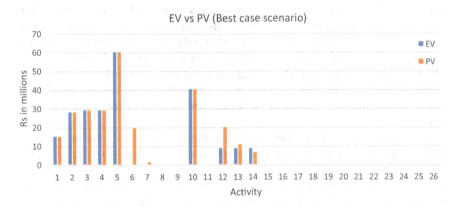

**Fig. 4** EV versus PV of the activities (best case scenario)

$$\otimes \text{CPI} = \otimes \text{EV}/\text{AC}$$
$$= ([178.1 - 227.6])/214.7 = [0.83 - 1.06] \quad (3)$$

$$\otimes \text{SPI} = \otimes \text{EV}/\otimes \text{PV}$$
$$= [178.1 - 227.6]/[189 - 259.5] = [0.69 - 1.20] \quad (4)$$

$$\otimes \text{EAC} = \text{BC}/\otimes \text{CPI}$$
$$= [415.1 - 530.4] \quad (5)$$

Thus, from the results of CPI and SPI, it has been observed that project is going under budget for that particular tracking period as CPI shows 1.06 on higher side which indicates under budget as CPI > 1. Similarly, project is going ahead of schedule for that particular tracking period as SPI shows 1.20 on higher side which indicates ahead of schedule as SPI > 1. So overall it can be said that the project is under budget and ahead of schedule for that tracking period. Also, EAC seems to be out of the limit, i.e., 406 doesn't lie in between [415.1–530.4] which clearly shows the under budget scenario of the project. Thus, the outcomes determine that the project may worsen but not exceeding 0.69 and may perform better; it provides the overall satisfactory result so that the project members will be prepared for both worst and best-case scenarios. The project members can act accordingly to keep the project on track and avoid any serious uncertain situation without further loss.

## 4 Conclusion

This study aims at predicting the project performance of an infrastructure project using interval gray numbers. Interval gray numbers are used when determinant value, i.e., true value is not known, where data is insufficient. They assist in providing an overall view of the project profile and help in better decision-making process. Interval gray numbers are based on a non-deterministic approach other than the other methods, which are deterministic where determinate value, i.e., true value is known. Since lack of data availability is obvious in recent times; interval gray numbers are an effective way to predict the performance of the project and assist in better decision-making process.

Previously, using gray numbers to predict project performance has been carried considering a numerical example. Hence, this study covers the gap by applying the methodology on a real-life project by extending it to validate the theory to a better level by considering intermediate values to precisely provide the outcomes. A real-life commercial project has been considered and analysis has been done based on a non-deterministic approach, which indicated the project to be over budget and ahead of schedule as CPI = [0.83–1.06] and SPI = [0.69–1.20].

- Interval gray numbers are an effective way to predict the project performance and having a better decision-making process.
- Interval gray numbers are based on a non-deterministic approach unlike other methods which are deterministic in nature, i.e., true value or determinate value is known.
- Interval gray numbers provide the outcomes in a range which indicates the worst and best case situations of the project assisting the project members to be prepared for both situations and take better decision to keep the project on track.
- Considering intermediate values makes the interval gray numbers theory more efficient to understand the situation and precisely monitor and track the project performance to avoid any delays or shortages and losses of resources.

## References

1. Narbaev T, Marco AD (2013) Earned value-based performance monitoring of facility construction projects. J Facil Manage 11(1):69–80
2. Batselier J, Vanhoucke M (2015) Evaluation of deterministic state-of-art forecasting approaches for project duration based on earned value management. Inter J proj Manage 4(3)
3. Bhattacharyya R (2015) A grey theory based multiple attribute approach for R&D project portfolio selection. Fuzzy Inf Eng 7(2):211–225
4. Li GD, Yamaguchi D, Nagai M (2007) A grey-based decision making approach to the supplier selection problem. Math Comp Modell 46:573–581
5. Li YH, Lee PC, Chang TP, Ting HI (2008) Multi-attribute decision making model under the condition of uncertain information. Autom Constr 17:792–797
6. Oztaysi B (2014) A decision model for information technology selection using AHP integrated TOPSIS-Grey: the case of content management systems. Knowl Based Syst 70:44–54
7. Li C, Chen K, Xiang X (2015) An integrated framework for effective safety management evaluation: application of an improved grey clustering measurement. Exp Syst Appl 42(13):5541–5553
8. Amirian H, Sahraeian R (2017) Solving a grey project selection scheduling using simulated shuffled frog leaping algorithm. Comp Indus Eng 107:141–149
9. Sruthi MD, Aravindan A (2019) Performance measurement of schedule and cost analysis by using earned value management for a residential building. Mater Today: Proc 33:524–532
10. Andrade PA, Martens A, Vanhoucke M (2019) Using real project schedule data to compare earned schedule and earned duration management project time forecasting capabilities. Autom in Constr 99:68–78
11. Ibrahim MN, Thorpe D, Mahmood MN (2019) Risk factors affecting the ability for earned value management to accurately assess the performance of infrastructure projects in Australia. Constr Inno 19(4):550–569
12. Pérez PB, Ablanedo ES, Melià DM, Cruz MCG, Bargues JLF, Pellicer E (2019) Earned schedule min-max: two new EVM metrics for monitoring and controlling projects. Autom Constr 103:279–290
13. Nadafi S, Moosavirad SH, Ariafar S (2019) Predicting the project time and costs using EVM based on gray numbers. Eng Constr Arch Manage 26(9):2107–2119

# COVID-19—Assessment of Economic and Schedule Delay Impact in Indian Construction Industry Using Regression Method

**Soniya D. Mahind and Dipali Patil**

**Abstract** Coronavirus (COVID-19) is a Global pandemic and has severe impacts on the economy and schedule delay of the various industries including the construction sector. The construction industry is the second largest industry after agriculture in India. The current pandemic caused all projects and onsite construction work to halt abruptly which raised the need for the researcher to analyze and be prepared for such crises in the future. *This research investigates* the net impact of the effect of COVID-19 on Indian construction projects by considering essential aspects and their net impact on the total cost and delay of the project. The risk factors *impacting the construction industry* are gathered from available resources and observation. The empirical study was done using a questionnaire survey with 62 responses collected from well-known construction companies from India. The significance of the delay and cost factors was investigated using the relative importance index method and regression modeling. From factor analysis, we identified that cost factors such as conditions of clients, delay in handing over the project, transportation cost, unforeseen circumstances, cost for site hygiene, and delay factors like lockdown period, labor shortage, delay in supply, late payments for completed work and material shortage are majorly affecting on economy and delay of the project. The regression models established will help to predict the impact of a factor on the cost and delay of the project. This data will help the client, project manager, and contractor to confront such disruption in the future.

**Keywords** COVID-19 · Economy · Regression analysis · Schedule delay

---

S. D. Mahind (✉)
Construction Management, College of Engineering Pune, Pune, India
e-mail: mahindsd19.civil@coep.ac.in

D. Patil
College of Engineering Pune, Pune, India
e-mail: dlp.civil@coep.ac.in

## 1 Introduction

Coronavirus (COVID-19) was recognized as a pandemic by the world health organization (WHO) on 11th March 2020. This pandemic has severe impacts on the economy and delays on the various industries including the construction industry. This global pandemic not only caused illness and fatalities but also destroyed the global economy. India is also attempting to respond to the post-COVID-19 reality challenges, which have come to define a new normal for our economy and society as a whole, as in most parts of the world. After agriculture, the construction industry in India is the second-largest employer and is therefore vital to the economic stability of the country. It accounts for about 8% of the nation's GDP with an industry size of INR 10.5 trillion and employs close to 57.5 million people [1]. The infrastructure and construction industries, which are mainly responsible for the development of the nation, are already facing the headwinds of the COVID-19 pandemic and it cannot be assumed to be removed from its harmful effects. Besides, the unorganized and fragmented nature of the construction sector is likely to worsen this effect. The construction industry is expected to undergo a simultaneous downturn in both supply and demand due to this pandemic. Since the sector is mainly affected by infrastructure projects, it is predicted that current levels of uncertainty, weak business and consumer sentiments, loss of revenue, and the diversion of government funds to COVID-19 management would be seriously affected. The productivity of the construction sector mainly depends on man, material, and money and is directly related to the cost and time of the project. These key parameters were again subcategorized into various factors which were affected to various degrees due to COVID-19.

The study aims to analyze the impact of COVID-19 on construction projects by considering essential aspects and their net impact on the total cost and duration of the project and provide remedies or suggestions effectively to confront such disruption in the future. There are few factors that are majorly affecting the cost and the total duration of construction projects on the different construction phases. In this research, we are going to collect and assess those factors to understand their impacts on the Indian construction Industry.

## 2 Literature Review

There are ongoing challenges due to COVID-19 in the construction industries of Kabul, Afghanistan due to the COVID-19 pandemic. Research identifies the initiatives that should be taken by the government to overcome the situation. The mining sector is not immune to these impacts, and the crisis may have important short-, medium- and long-term repercussions for the industry. The productivity of the construction sector mainly depends on man, material, and money and is directly related to the cost and time of the project. COVID-19 is impacting construction and employment under various scenarios such as investment and economy. The survey was done to check the influence of this pandemic on projects in India considering materials, man, and machinery [2]. The COVID-19 pandemic has placed labor-intensive industries at risk, including a typical one in the construction industry. Results obtained from the study give strategies to reduce and prevent the spread of COVID-19 in the construction industry of China [3]. COVID-19 Pandemic has slowed down the progress of industry because of reverse migration of labors causing the shortage, stoppage in various construction activities, the effect on order book execution rate, and working capital problems for firms that are financially vulnerable [1]. Time and cost overruns in Nigeria have been recognized as the most significant factors responsible for the abandonment and failure of contractors. Delay is a condition in which the contractor and the owner of the project contribute jointly or severally to the failure to complete the project during the initial or the contract duration stipulated or decided [4]. Factor analysis is performed for the study and quantitatively analyzed the underlying structure of the causes of variance [5]. The Relative Importance Index (RII) method is used by Larsen et al. [6] to check the effect of factors schedule delay, cost overrun, and quality level. The research explores, firstly through a questionnaire survey of construction professionals, the effects of construction delay on the implementation of the project and, secondly, the effects of delay on the cost and time of completion by an empirical approach. Doloi [7] identified the many factors that contribute to construction delays, use statistical methods to determine the relationship between these variables, and use a regression model to estimate the influence of these factors on construction delays in the Indian construction industry.

There is a lack of research availability due to the ongoing COVID-19 pandemic. The factors which were considered for the previous study of cost and duration of the construction project are not concerning the COVID-19 situation. Few findings are available on the impact of COVID-19 on the construction industry in Indian context. There is no mathematical model established to indicate the relationship of schedule delays and economy with their respective attributes in such a pandemic situation.

## 3 Research Methodology

See Fig. 1.

## 4 Data Collection

### 4.1 Preparation of Questionnaire

An important phase in the research's success is the identification of important attributes for the study and the creation of a questionnaire. The questionnaire survey is adopted as it is the most popular way of data collection, especially for large sample

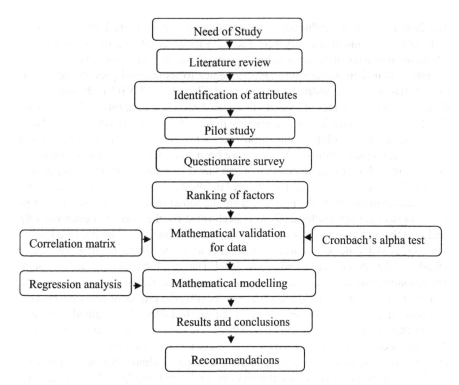

**Fig. 1** Flow chart of research methodology

size. Total 17 attributes of schedule delay were identified under 5 broad categories such as materials and equipment related, manpower related, project management related, government related, owner related and 14 attributes of economy were identified under 4 broad categories materials and equipment related, Labor related, owner related, contractor related. We have used a structured questionnaire for this survey for easy analysis. It is the most helpful in a situation in which the score of the respondent can be compared to a distribution of scores from some well-defined category. Factor analysis was used to give rank to the attributes and regression analysis was used to develop a predictive model based on the best-fit attributes. The pilot study is mainly used for the identification of factors and validation and is done prior to the execution of the main study. To conduct this pilot study a small set of respondents having a sample size of sixteen including owners, contractors, Professional experts and developers will share their views on factors. For checking the influence of factors on the Economy and Schedule of the project, the questionnaire is prepared with the help of the Likert scale. It is most useful in a situation wherein it is possible to compare the respondent's score with a distribution of scores from some well-defined group. On a Likert scale, the respondent is asked to respond to each of the statements in terms of several degrees, here we have used a 5-point scale.

**Fig. 2** Type of organization of respondents

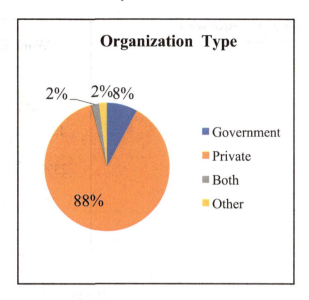

## 4.2 Respondent's Profile

The required quantitative data is obtained with the help of google form which is feasible and safe during this pandemic situation. Google form consists of three sections out of which 1st section is of general information of respondents including their name, type of organization, name of the organization, qualification, work experience. And remaining two sections asking to rate the impact of the factors affecting on Economy and Schedule Delay. Total 62 responses were considered from the project managers, construction managers, research scholars, and seniors engineers from well-known construction companies and institutes in India such as Larsen & Turbo, Shapoorji Pallonji & Co. Pvt. Ltd., B.G.Shirke Const. Tech Pvt (See Figs. 2, 3 and 4).

## 5 Data Analysis

### 5.1 Ranking of Attributes

According to Faridi and El-Sayegh [8] the mean and standard deviation of each attribute, are not suitable to judge overall rankings because they don't represent any relationship between them. Hence the analysis was done by using the Relative Importance Index (RII) for providing ranks to the factors affecting on economy and delay of the construction industry collected and verified from literature. This method is used by Larsen et al. [6] to check the effect of factors schedule delay, cost overrun,

**Fig. 3** Work experience of respondents

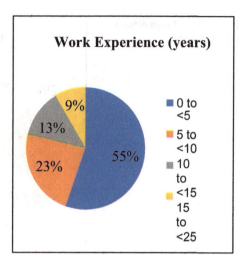

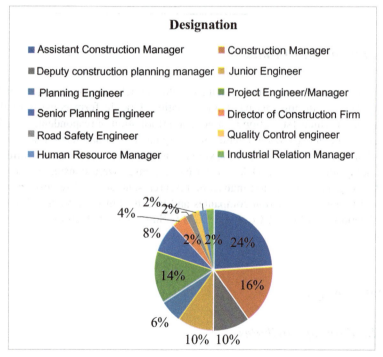

**Fig. 4** Designation of respondents

**Table 1** Thumb rule for Cronbach's alpha test

| Cα | Reliability |
|---|---|
| >0.9 | Excellent |
| 0.9 > Cα > 0.8 | Good |
| 0.8 > Cα > 0.7 | Acceptable |
| 0.7 > Cα > 0.6 | Questionable |
| 0.6 > Cα > 0.5 | Poor |
| 0.5 > Cα | Undesirable |

and quality level (Table 1).

$$RII = \sum_{i=1}^{5} W_i/(A \times N)$$

where $W_i$ = Total sum of each factor (1 to 5 scale), $A$ = highest possible weight used for study, $N$ = Total number of respondents for each variable.

Tables 2 and 3 show the ranking of the project attributes in descending order based on their relative importance calculated using the equation above. Along with the ranks, the RII scores are displayed in the column. The most important traits are represented by the greatest RII values, while the least important traits are represented by the lowest RII values.

**Table 2** Ranking of attributes affecting the economy of construction projects

| Code | Risk factors | RII | Rank |
|---|---|---|---|
| E12 | Economic conditions of clients (the main source of funding) | 0.812 | 1 |
| E11 | Delay in handing over the project (penalties) | 0.796 | 2 |
| E3 | Transportation cost (scarcity in the local availability) | 0.768 | 3 |
| E14 | Unforeseen circumstances (if labors get infected) | 0.764 | 4 |
| E6 | Cost for site hygiene (sanitization at a construction site) | 0.752 | 5 |
| E2 | Labor cost (extra wages) | 0.74 | 6 |
| E1 | Material cost (due to their shortage) | 0.736 | 7 |
| E13 | Extent of variations (variation in market) | 0.736 | 8 |
| E8 | Implementation of new policies (policies related to COVID-19) | 0.716 | 9 |
| E9 | Liquidity risk (cannot meet short-term debt obligations) | 0.716 | 10 |
| E10 | Interest rate during the project period (fewer investments) | 0.708 | 11 |
| E4 | Medical cost (personal protective equipment) | 0.692 | 12 |
| E7 | Hiring charges of plants and machinery (unavailability of operators and equipment) | 0.68 | 13 |
| E5 | On-site labor accommodation cost (for migrant workers) | 0.672 | 14 |

**Table 3** Ranking of attributes affecting the schedule delay of the construction project

| Code | Risk factors | RII | Rank |
|---|---|---|---|
| D7 | Lockdown period (the total duration of projects ultimately increases) | 0.868 | 1 |
| D2 | Labor shortage (causes pending of construction work) | 0.82 | 2 |
| D3 | Delay in supply (supply chain is disturbed) | 0.812 | 3 |
| D17 | Late payments for completed work (causes late start for next work) | 0.78 | 4 |
| D1 | Material shortage (shortage of material turns into the halting of the project) | 0.732 | 5 |
| D5 | Implementation of new safety guidelines (takes some time to implement causes delay) | 0.732 | 6 |
| D10 | Delay in obtaining permits from the municipality (work was on halt in the period) | 0.732 | 7 |
| D4 | Productivity at work (the efficiency of labor reduced due to fear) | 0.724 | 8 |
| D13 | Inadequate client's finance (lack of funds available) | 0.7 | 9 |
| D6 | Unavailability of plant and machinery (causes transportation from another place) | 0.696 | 10 |
| D9 | Slow mobilization of equipment (the free movement of traffic is restricted) | 0.692 | 11 |
| D15 | The complexity of the project (inexperienced contractors are not able to handle) | 0.676 | 12 |
| D16 | Building permits approval (delay for approval) | 0.668 | 13 |
| D12 | Suspension of work by owner (nonavailability of resources required to continue the project) | 0.66 | 14 |
| D14 | Contractor's planning skills (inexperienced contractors are not able to handle) | 0.66 | 15 |
| D11 | Poor communication and coordination with other parties (insufficient information and misunderstandings) | 0.648 | 16 |
| D8 | Poor site management and supervision (due to such unforeseen situation) | 0.644 | 17 |

## 5.2 Mathematical Validity of Factor Analysis

After the factors have been ranked, it is vital to double-check that factor analysis measured what was intended to be assessed, i.e. that the attributes in each factor describe the same measure within target dimensions collectively [7]. It is acknowledged that if attributes compose the specified factor, they should reasonably correlate with one another, but no need to be perfect. We can determine the level of correlation among distinct variables by using MS Excel to calculate Pearson correlation. Tables 4, 5 and 6 lists the Pearson correlation coefficients.

**Table 4** Cronbach's alpha test results

|  | Economy | Schedule delay |
|---|---|---|
| No. of risk factors | 14 | 17 |
| Sum of factor variances | 9.299621297 | 21.92381377 |
| Variance of the total score | 98.9664 | 181.3316 |
| Cronbach's $\alpha$ | 0.97572735 | 0.934038926 |

**Table 5** Results of multiple regression analysis for economy factor

|  | Coefficients | Standard error | $p$-value |  |  |
|---|---|---|---|---|---|
| Intercept | −0.891 | 0.221 | 0.0003 | $R$ Square | 0.97397 |
| E1 | 0.137 | 0.045 | 0.0042 | Adjusted $R$ Square | 0.963558 |
| E2 | 0.111 | 0.039 | 0.0068 | $F$ | 93.54396 |
| E5 | 0.178 | 0.046 | 0.0005 |  |  |
| E6 | 0.137 | 0.041 | 0.0019 |  |  |
| E7 | 0.101 | 0.044 | 0.0277 |  |  |
| E8 | 0.135 | 0.043 | 0.0033 |  |  |
| E9 | 0.088 | 0.033 | 0.0124 |  |  |
| E10 | 0.119 | 0.050 | 0.0231 |  |  |
| E11 | 0.146 | 0.044 | 0.0022 |  |  |
| E12 | 0.093 | 0.038 | 0.0215 |  |  |
| E14 | 0.157 | 0.045 | 0.0014 |  |  |

**Table 6** Results of multiple regression analysis for schedule delay factor

|  | Coefficients | Standard error | $p$-value |  |  |
|---|---|---|---|---|---|
| Intercept | −1.641 | 0.2578 | 1.6E−07 | $R$ Square | 0.95731 |
| D3 | 0.143 | 0.0631 | 2.9E−02 | Adjusted $R$ Square | 0.946365 |
| D17 | 0.205 | 0.0502 | 2.1E−04 | $F$ | 87.45812 |
| D1 | 0.224 | 0.0583 | 4.4E−04 |  |  |
| D5 | 0.197 | 0.0609 | 2.5E−03 |  |  |
| D10 | 0.179 | 0.0571 | 3.3E−03 |  |  |
| D4 | 0.187 | 0.0377 | 1.4E−05 |  |  |
| D13 | 0.153 | 0.0393 | 3.8E−04 |  |  |

### 5.2.1 Cronbach's Alpha Test

Cronbach's alpha test was done on the complete data as well as attributes in each component, as indicated in Table 1, for reliability analysis, which is essential to validate the build of the model over time (i.e. consistency of measured characteristics and scale). C$\alpha$ can have a value ranging from 0 to 1, with a higher value indicating stronger internal consistency and vice versa. Doloi [7] suggested that the following ranges can be used as a thumb rule in most situations:

$$\text{Coefficient of Cronbach}(\alpha) = (N/(N-1)) \times (1 - (V/A))$$

where $N$ = Number of factors, $V$ = Sum of the variance of all factors, $A$ = Variance of the total score.

### 5.2.2 Regression Analysis

As previously stated, the factors that affect schedule delays and economy are further investigated using linear multiple regression to establish a prediction model in the Indian construction scenario. While factor analysis shows the presence of clusters of strong correlation coefficients with observable underlying dimensions, these dimensions (or factors) have no predictive value on the phenomena being studied. Multiple regression analysis, on the other hand, fits the complete dataset without taking into account the correlation matrix's common variance ($R^2$ value) [9]. $R^2$ is a useful criterion for evaluating the intensity of the relationship between the predictive or dependent variable and the independent variables. Some of the important underlying assumptions in the predictor variables should be met before conducting the regression analysis. Before starting the regression analysis, the parametric test was performed by analyzing the homogeneity of variance (Levene's test) on the selected attributes [9] Since the results of Levene's test are relevant ($p < 0.05$), the null hypothesis that the variances between variables are zero is rejected, and the regression analysis is viable. The attributes derived from the factor analysis, as shown in the analysis were used as independent variables.

## 6 Results

### 6.1 Impact of Attributes Affecting Economy

Out of the factors considered, the top five factors affecting the economy of the construction project, E12 (Economic conditions of clients (the main source of funding)) which has the highest RII, i.e. 0.812. As indicated by the decrease in RII values, this was followed by E11 (Delay in handing over the project), E3 (Transportation Cost), E14 (Unforeseen circumstances), E13 (Extent of variations) (Table 2).

## 6.2 Impact of Attributes Affecting Schedule Delay

Of the factors that affect schedule delay, the highest-ranked factor is D7 (lockdown period) with an RII of 0.868. As indicated by the decrease in RII values, this was followed by D2 (Labor shortage (causes pending of construction work)), D3 (Delay in supply), D17 (Late payments for completed), D1 (Material Shortage) (Table 3).

## 6.3 Correlation Matrix of the Attributes

Table 4 indicates the person bivariate correlations between different attributes in all components are more than 0.4 in the majority of cases. We can ensure that factors produced by factor analysis contain properties that are related based on these findings [7] The correlation matrices for factors, economy and schedule delay are helpful to determine the relationship between the attributes of respective categories.

## 6.4 Cronbach's Alpha

From Table 4, the Cronbach's alpha coefficient for the data related to the Economy of the construction is 0.976, and for data related to Schedule Delay is 0.934, which indicates that the consistency of data for both of them is excellent. Hence the data collected for this research is highly reliable. This test is applied to validate the consistency and reliability of the collected data.

## 6.5 Regression Model

The effects on the overall Economy and Schedule delay of the project are the dependent variables, which were asked separately to each respondent focusing on a specific project. For categorical variables, these attributes are added to the regression model in steps. As a result, the regression model used to assess the aggregate effect of individual attributes on delay can be summarized as follows:

$$Y_i = \beta_0 + \beta_1 X_{1i} + \beta_2 X_{2i} + \beta_3 X_{3i} + \cdots + \beta_j X_{ji} + \varepsilon_i$$

where $Y_i$ = Dependent term(one of the factors, economy, and delay); $\beta_0$ = Constant and intercept at $Y$-axis; $\beta_i$, i.e. ($\beta_1$ to $\beta_j$) = Determined regression coefficients; $X_i$, i.e. ($X_1$ to $X_j$) = Values of independent variables or attributes; $\varepsilon_i$ = Error which is a random variable with mean 0; $i$ = index of the performance of attributes being predicted; and $k$ = number of predictor variables.

The adjusted $R^2$ values and the variation from $R^2$ values indicate how much the model generalizes the dependent variable's predictive power [7]. If the difference between $R^2$ and adjusted $R^2$ is less indicates the stronger model. Hence in an ideal case, the $R^2$ and adjusted $R^2$ should be the same values. The values in Tables 5 and 6 are acceptable for appropriate strengths based on the excellence of the model fit.

### 6.5.1 Economy Factor

The final regression model for the impact on Economy can be expressed as:

$$\begin{aligned}\text{Extra Cost}(\%) = &-0.891 + 0.137 * (\text{Material Cost}) + 0.111 * (\text{Labor Cost}) \\ &+ 0.178 * (\text{On site labor accomodation}) \\ &+ 0.137 * (\text{Cost for site hygiene}) \\ &+ 0.101 * (\text{Hiring charges of plants and machinery}) \\ &+ 0.135 * (\text{Implementation of new policies}) \\ &+ 0.088 * (\text{Liquidity risk}) \\ &+ 0.119 * (\text{Interest rate during project period}) \\ &+ 0.146 * (\text{Delay in handing over the project}) \\ &+ 0.093 * (\text{Economic conditions of clients}) \\ &+ 0.157 * (\text{Unforeseen circumstances})\end{aligned}$$

Where all the values of attributes causing extra cost are considered to be in percentage.

As the regression model in Table 5 demonstrates, on-site labor accommodation (E5) and unforeseen circumstances (E14) have maximum impact on the economy of the Indian construction projects in crisis due to COVID-19. Delay in handing over the project (E11) caused by unforeseen circumstances (E14). The Material cost (E1) and hiring charges of plants and machinery (E7) has been increased due to unstable market condition and shortage of locally available resources. The cost for site hygiene (E6) and Implementation of new policies (E8) is mainly because of the pandemic situation. Liquidity risk (E9), Interest rate during the project period (E10), Economic conditions of clients (E12) these costs are mainly related to the owners and affect the project economy as they are the source of funding. As labor is a key resource of the construction industry to continue the work which is on halt, an accommodation facility is provided. It seems that labor has more impact on the economy of the project, i.e. labor cost (E2).

### 6.5.2 Schedule Delay Factor

The final regression model for the impact on Schedule Delay of Indian construction Project can be expressed as:

$$\begin{aligned}
\text{Delay(months)} = &-1.641 + 0.143 * (\text{Delay in supply}) \\
&+ 0.205 * (\text{Late payments for completed work}) \\
&+ 0.224 * (\text{Material shortage}) \\
&+ 0.197 * (\text{Implementation of new safety guidelines}) \\
&+ 0.179 * (\text{Delay in obtaining permits from municipality}) \\
&+ 0.187 * (\text{Productivity at work}) \\
&+ 0.153 * (\text{Inadequate client's finance})
\end{aligned}$$

Where all the values of attributes contributing to the delay of the project are considered to be in months. The above regression model implies material shortage (D1), late payment for completed work (D17), and Implementation of new safety guidelines (D5) is majorly affecting parameters on schedule delay of construction projects in India during the COVID-19 pandemic. Productivity at work (D4) is the attribute having an impact on the construction industry irrespective of any situation as previous findings [9] also show a similar impact. In this pandemic, poor productivity is mainly due to the fear of getting infected among laborers. The Lockdown situation causes a delay in obtaining permits from the municipality (D10). Delay in supply (D3) causes a delay in the schedule of the project. Money is an important resource in any project hence inadequate clients' finance (D13) has an impact on delays in the construction industry.

## 7 Conclusion

The findings from this study showed the significance of factors on the economy and schedule delay of projects in the Indian Construction industry. This pandemic situation hampered all four M's of the construction industry, i.e. man, material, money, and machinery. The lockdown period is mainly responsible for the scheduled delay of construction work and this is proved by both quantitative and qualitative data. And the extension of it causes an increase in the total duration and cost of projects. Late payments of completed work cause the late start of the successive activity. Since the construction industry is one of the labor-intensive industries, turned into their shortage which caused the proposed construction work to be pending or moving at a slow rate. Hence at some sites, on-site labor accommodations were provided which required extra investment. The economic conditions of clients have a great impact on the economy of the project. Though the RERA of various states in India has extended the completion period of all registered projects, delay in handing over the project has majorly affected the economy as it causes penalties if unable to complete the project in time.

This study is validated by the available reports on the impact of COVID on Indian Construction. *Impact on Construction Sector in India Table of Contents* [2] indicates that in a pessimistic case increase in the total cost of the project 4.28% and the average value we have identified is 4.886%. According to the report, the underdevelopment projects were delayed by a minimum of 2–3 months and the findings from the study show the value of 3.465 months. The corresponding values are from June 2020 and have been increasing due to the extension of lockdown conditions.

Due to lockdown, the supply chain is disturbed and causing a delay in the construction industry. This data will help to construction manager, consultant, site engineer, and owner to get the idea of majorly affecting factors and prepare some mitigation strategies to confront such situation in the future.

## 8 Recommendations

1. Provide onsite accommodations facility for labors and motivate them, will help to reduce the reverse migration of labors and also to increase the productivity of their work.
2. Making small groups of workers and assigning work in alternate shifts to reduce the chances of infection and to implement new safety norms such as social distancing.
3. Establishment of the digital platform for office workers to reduce public gatherings and avoid direct contact.
4. Identifying alternate sources of resources so that the risk of their shortage can be minimized. The site should be flexible enough to store the resources to tackle possible supply chain interruptions.
5. Continuous evaluation of profitability in such situations will help to overcome financial crises with the help of mitigation strategies.
6. An alternative solution to minimize over-reliance on labor is the adoption of automation, mechanization and off-site production.
7. Design and enforcement of a risk management system for early detection of the infected worker to avoid further spread. This can be achieved with the help of daily health checkups.
8. For new projects to be sanctioned, impact factor of such situation and clause related to equal risk sharing should be considered in a contract.

# References

1. Building again—brick by brick COVID-19: construction sector in India: growth paused temporarily (2020), 3(April)
2. Impact on construction sector in India Table of contents (2020), May
3. Zheng L, Chen K, Ma L (2021) Knowledge, attitudes, and practices toward COVID-19 among construction industry practitioners in China. Front Public Health 8(January):1–9. https://doi.org/10.3389/fpubh.2020.599769
4. Aibinu AA, Odeyinka HA (2006) Construction delays and their causative factors in Nigeria. J Constr Eng Manag 132(7):667–677. https://doi.org/10.1061/(asce)0733-9364(2006)132:7(667)
5. Wambeke BW, Hsiang SM, Liu M (2011) Causes of variation in construction project task starting times and duration. J Constr Eng Manag 137(9):663–677. https://doi.org/10.1061/(asce)co.1943-7862.0000342
6. Larsen JK, Shen GQ, Lindhard SM, Brunoe TD (2016) Factors affecting schedule delay, cost overrun, and quality level in public construction projects. J Manag Eng 32(1):04015032. https://doi.org/10.1061/(asce)me.1943-5479.0000391
7. Doloi H (2009) Analysis of pre-qualification criteria in contractor selection and their impacts on project success. Constr Manag Econ 27(12):1245–1263. https://doi.org/10.1080/01446190903394541
8. Faridi AS, El-Sayegh SM (2006) Significant factors causing delay in the UAE construction industry. Constr Manage Econ 24(11):1167–1176.https://doi.org/10.1080/01446190600827033
9. Doloi H, Sawhney A, Iyer KC, Rentala S (2012) Analysing factors affecting delays in Indian construction projects. Int J Project Manage 30(4):479–489. https://doi.org/10.1016/j.ijproman.2011.10.004

# Comparison of Afghanistan's Construction and Engineering Contract with International Contracts of FIDIC RED BOOK (2017) and NEC4—ECC

**Mohammad Ajmal and C. Rajasekaran**

**Abstract** Construction and engineering contracts are the core documents used to set out contractual relations, rights and obligations of parties, payment systems, communication lines, and dispute avoidance/resolution procedures in construction and engineering projects. Thus the importance of these documents is obvious. In this study, the Standard Bidding Document (SBD) for the Procurement of Small Works (PSWs) of the Government of Afghanistan (which's the approximate cost is up to AFN 500 million) is chosen. It is to be compared with the Conditions of Contract for Construction—FIDIC RED BOOK—2nd Edition (2017) and New Engineering Contracts (NEC4) Engineering and Construction Contract (ECC) 4th Edition (2017). This study was conducted using different research papers, theses and books, which were relevant to the research, and by deep study of the above three mentioned types of documents of contracts. After that, the comparison of Afghanistan's SBD for the PSWs with the FIDIC RED BOOK (2017) and NEC4—ECC (2017) in different parts with each other was carried out. After this comparison, the achieved results show that Afghanistan's SBD for the PSWs has shortcomings, lack of some essential clauses and/or sub-clauses, insufficient information in some clauses which may lead to disputes, errors and/or mistakes in naming of some clauses, lack of flexibility, and unbalanced risk-sharing.

**Keywords** Construction and engineering contracts · Standard bidding document · FIDIC RED BOOK (2017) · NEC4—ECC · Afghanistan's construction and engineering contract

---

M. Ajmal (✉) · C. Rajasekaran
Department of Civil Engineering, National Institute of Technology Karnataka, Surathkal, Mangalore 575025, India
e-mail: mohammadajmalbah@gmail.com; mohammadajmal.192cm031@nitk.edu.in

C. Rajasekaran
e-mail: bcrajasekaran@nitk.edu.in

© The Author(s), under exclusive license to Springer Nature Singapore Pte Ltd. 2023
M. S. Ranadive et al. (eds.), *Recent Trends in Construction Technology and Management*, Lecture Notes in Civil Engineering 260,
https://doi.org/10.1007/978-981-19-2145-2_24

# 1 Introduction

## 1.1 General

A contract is an agreement between two or more parties that gives rise to rights and responsibilities which will be applied according to the law system applying to the contract. Often, the law system applying to a contract will be specified within the contract itself [12].

Contracts may differ widely in type, from a very simple verbal contract to a very complicated written contract. In many cases, parties are going to take a standard form of contracts as an initial point. These contracts will typically set out the work to be performed under the contract in some detail, the price to be paid or the basis on which to measure the amounts due, and commonly the rights and obligations of the parties entering into the contract.

As far as possible, a properly written and well-prepared construction contract should aim to make sure clarification and assurance of the specific responsibilities and rights of the parties', provide the parties with the methods to be followed in certain cases or to achieve a particular outcome, and allocate the risk of damage or loss happening to the project precisely and thoroughly so that each party recognizes their risks and consequences of those risks accurately [12].

Unbalanced conditions in construction and engineering contracts have been identified as one of the significant problems in project execution. Traditionally, the text of contract documents is drafted by the owner (client). It usually results in a manual that is sometimes drafted to favor one of the parties—owner—entering into a contract. An unbalanced contract harms all parties at all times, and therefore, this type of contract is not a workable contract [18]. Accordingly, the related entities to procurement and those preparing contract documents may take in mind to draft it carefully and don't make any bias to any party in the contract.

## 1.2 Scope of the Study

The scope of this study is to compare the following standard contracts documents with each other:

(1) SBD for the PSWs of the Government of Afghanistan (GoA) of unit price (admeasurement) contracts with BoQ (which's the approximate cost is up to AFN 500 million).
(2) FIDIC RED BOOK 2nd edition (2017), according to unit price contracts.
(3) NEC4—ECC (2017), main option B, priced contract with Bill of Quantities (BoQ).

In all the above contracts, the construction and engineering works are designed by the Owner (Employer/Client/Entity) without any small designs, i.e., design of temporary works, etc.

## 1.3 Research Problems

There are some questions that are to be answered by doing this research:

(1) What are the similarities and differences of SBD for the PSWs of the GoA with FIDIC RED BOOK (2017) and NEC4—ECC (2017)?
(2) What are the shortcomings of SBD for the PSWs of the GoA, according to FIDIC RED BOOK (2017) and NEC4—ECC?
(3) What improvements can be made to SBD for the PSWs of the GoA, according to FIDIC RED BOOK (2017) and NEC4—ECC?
(4) Which one of them is more useful, more flexible, and to complete the project with better performance?

## 1.4 Objectives of the Study

The objectives of this research which are to be achieved by this research, are as below:

(1) to help with international engineering and construction entities and companies who are interested in getting contracts in Afghanistan,
(2) help them to know about Afghanistan's engineering and construction contract and its differences and similarities with FIDIC RED BOOK (2017) and NEC4—ECC, and
(3) to help the procurement authorities of the GoA and other countries to improve the process and conditions of contracts for better performance, good management, dispute avoidance, dispute resolution, and successful completion of projects.

## 1.5 Research Methodology

The methodology which is used in this study is as below:

- We did this study by using different research papers, theses, books, academic presentations, and some reliable electronic sources, which were relevant to our research, and
- by deep studying of SBD for the PSWs of Afghanistan, FIDIC RED BOOK (2017) and NEC4—ECC.

- After the above two steps, we compared SBD for the PSWs of the GoA with the FIDIC RED BOOK (2017) and NEC4—ECC (2017) in different parts with each other.

## 2 Literature Review

### 2.1 General

A good and well prepared Standard Contract Document (SCD) or Standard Bidding Document (SBD) will lead to reduce tendering cost, reduce the cost of negotiating the contract [25], improve co-operation between the parties entering into a contract, and other stakeholders, sense of responsibility, reduction in delays as well as improvement in productivity of the contract parties, reduce the incidence of disputes between them [18], provide a more significant stimulus to good project management [15], and can help to avoid the cost over-runs in some projects [16]. In fact, some conditions included in SBDs can cause conflicts and disputes, and some contracts could cause more disputes than others [7].

Internationally, many types of construction and engineering contracts are in use today, from which the two most commonly used forms are FIDIC and NEC3 [2]. Fortunately, we chose the latest editions of both of these two standard form contracts: FIDIC RED BOOK 2nd edition (2017) and NEC4—ECC 4th edition (2017). The FIDIC forms are commonly used and are well recognized as providing balanced forms of contract [12], and NEC3 is also used internationally. It is becoming a serious competitor to FIDIC [23]. More than 30 countries worldwide including the UK are using NEC. Thus, this indicates its reputation [3].

A study by Besaiso et al. [3], which evaluates the efficiency of FIDIC and NEC in reduction of disputes in the Palestine construction projects, compared FIDIC with NEC in six aspects: clarification and easiness, risk allocation and managing, force majeure and prevention occasions, physical and weather conditions risks, variations, and project organization (engineer vs. project manager).

It suggests that both standard contracts have admirable and necessary features for all parties of a contract and can be effectively used everywhere. Although, both contracts have certain areas of concern and sometimes restrictions. It expresses that NEC has possibly many benefits over FIDIC, mainly in clarification, variations, risk managing, and objective measurement of weather and field conditions risks. Its results reveal that disputes may be minimized by contract, and NEC seems to be more capable of doing so than FIDIC.

The engineer (FIDIC)/project manager (NEC) position has advantages and restrictions in both contracts; although, in Palestine, it appears that the role of project manager under NEC is more practical than the engineer in FIDIC. The force majeure or prevention occasions are unconvincing in both forms of contracts. The authors further express that FIDIC has the advantages of precedence and familiarity, widespread reputation, and confirmation by many governments, international development banks and organizations [3].

A study by Rasslan and Nassar [22] compared the NEC and FIDIC standard forms of contracts and evaluated their suitability for Egypt's construction projects. It expresses that, FIDIC standard form of contract is used in a large number of the most important construction projects and in major infrastructure projects which are financed by the USAID (US Aid for International Development) and World Bank.

The authors express that variations, errors and omissions, and delays are the three most influential reasons for disputes arising during any construction project in Egypt.

Its results conclude that both FIDIC and NEC are well-designed, but NEC3 has more advantages over FIDIC. To manage a project's time and cost, the NEC3 provides a forward-looking proactive environment. It is a flexible, clear and easily understandable contract compared to FIDIC, which has many references to follow when reading any clause and contains lots of legal terminologies. The early warning system in NEC encourages identifying problems so that parties can establish an early resolution for upcoming issues [22].

Lord et al. [17] recommended three pillars of modern contract and examined specific provisions within the NEC3. These three pillars are: "fairness", "payment operating mechanism", and "functions and role of project participants". After examining, they found that the NEC is almost wholly compliant with these three pillars. Heaphy [13] comprehensively compared FIDIC and NEC3-ECC. He found that NEC3-ECC encourages collaborative behaviors, good project management and proactive risk management. Rooney and Allan [24] concluded that NEC3-ECC has performed better than other standard forms of contract in terms of time and cost predictability. He further discussed that NEC3-ECC has provided improved cooperation, flexibility, enhanced management practices and risk sharing in the UK highways projects [16].

## 2.2 Afghanistan Construction and Engineering Contracts

The National Procurement Authority (NPA) of the GoA has drafted and prepared two types of SBDs; the

- first one is SBD for the Procurement of Small Construction and Engineering Works (PSWs). Which's the approximate cost of procurement is up to AFN (500) millions [21].
- the second one is SBD for the Procurement of Large Construction and Engineering Works (PLWs). It is for procurements which's the approximate cost is more than AFN (500) millions.

It has been prepared to be used for the procurement of unit price or rate (admeasurement) type of works, through International Tender Process in projects, which the public funds finance.

These SBDs are not appropriate for lump-sum contracts without substantial changes to the method of payment and price adjustment, and the BoQ, Schedule of Activities, and so onward [20, 21].

In this study, we chose the SBD for the PSWs of the GoA (Unit Price Contract with BoQ). It is drafted and prepared by the NPA of the GoA according to International Multilateral Development Institutions' SBDs. The Procurement Entities of the GoA should use it for the PSWs financed by public funds.

## 2.3 FIDIC Contracts

The French acronym FIDIC corresponds to the Fédération Internationale des Ingénieurs-Conseils (in English; International Federation of Consulting Engineers) [4].

In 1957, for the first time, a contract was published by the FIDIC under the name of Conditions of Contract for Works of Civil Engineering Construction. It became popular because of its red cover as the RED BOOK. The latest edition of FIDIC RED BOOK was published in 2017.

There is a significant feature of FIDIC standard contracts that FIDIC does not generally allow its standard form contracts to be amended except a special license is negotiated. Generally in FIDIC, with the amending of Particular Conditions, the contract's standard form must be amended. With a logical arrangement of clauses, its standard contracts are consistent and coherent [25].

FIDIC RED BOOK is recommended for construction and engineering works designed by the owner (employer) or by his representative, the engineer. In this type, the contractor constructs the works according to a design provided by the owner. However, the works may have some part(s) that is(are) to be designed by the contractor [4, 9].

## 2.4 NEC4 Contracts

NEC is a wholly-owned subsidiary of the Institution of Civil Engineers (ICE), the NEC owner and developer.

The NEC contracts are the only suite of standard contracts designed to assist and encourage good management of the projects in which they are used. NEC standard contracts are used successfully around the world in both public and private sector projects; this trend seems set to continue at an increasing pace.

Each of the NEC standard forms of contract has the below characteristics:

- Its use facilitates and motivates good management of the relationships between the parties to the contract.
- It can be used for a wide variety of types of work and in any place.

- It is a simple and clear document, using a structure and language that is straightforward and easily understood.
- Its objective is far-sighted cooperation between all contributors to the project [10, 11, 15].

The existence of a number of optional clauses in NEC contract caused it to be adequately flexible in its terms to permit for any form of project delivery [14, 25].

## 3 Comparison of SBD for the PSWs of the GoA with FIDIC RED BOOK (2017) and NEC4—ECC

The above comparison is made in the below parts:

### 3.1 Structuring and Arrangement

There are some differences in structuring and arrangements of clauses and sub-clauses between these standard contracts. They are as follows.

#### 3.1.1 SBD for the PSWs of the GoA

SBD for the PSWs of the GoA divided its General Conditions into five parts. The first part is General Issues. It has named the other three parts, like the triple constraint triangle of project management—time, scope, and budget—the three parts are Time Control, Quality Control, and Costs Control. The fifth part is about the Termination or Completion of the Contract. It has a total of 62 clauses.

#### 3.1.2 FIDIC RED BOOK (2017)

FIDIC RED BOOK (2017) divided its General Conditions into 21 clauses which in the last edition of 1999 it was 20 clauses. One clause addition is the 20th Clause of the previous edition division into two separate clauses with some additions and subtractions. A good point is the Dispute Adjudication Board (DAB) changing in the previous edition to Dispute Avoidance and Adjudication Board (DAAB) in this edition. FIDIC has arranged its second to sixth clauses as the persons or groups involved in a project like the Employer, Contractor, Engineer, Subcontracting(Subcontractor), Staff and Labour. Employer or Contractor's Termination has separate clauses, but in NEC4—ECC, it is under one core clause, Termination. The structuring and arrangement are good, but if someone wants to know about any clause or sub-clause when he is reading one clause or sub-clause, it is referenced and linked with another clause or sub-clause; when he/she refers there, it is referenced to another one, which creates a little bit confusion in some parts. SBD for the PSWs of the GoA also has this referencing, but not as much as in FIDIC, and even in referencing to the other clauses, there are several mistakes, and the number of referenced clauses and/or sub-clauses are wrong. Referencing to other clauses and/or sub-clauses is significantly less in NEC4—ECC, and more of their clauses and sub-clauses are written independently and clearly.

### 3.1.3 NEC4—ECC

NEC4—ECC has nine Core Clauses which are the same for every contract of NEC4. These clauses are General, The Contractor's Main Responsibilities, Time, Quality Management, Payment, Compensation Events, Title, Liabilities and Insurance, and Termination. For the flexibility of this contract's usage, it has six Main Option Clauses for the six different types of contracts from the point of view of funding the project and measuring and pricing the works done. Here in this study, our case is about the priced contracts with the BoQ because the SBD for the PSWs of the GoA is generally for this type of contracts. The part of Resolving and Avoiding Disputes has three Options, W1, W2, and W3. NEC4—ECC has added the new clause of W3 and with some changes in W1 and W2 from the last edition of NEC3. It has 21 Secondary Option Clauses, and it is not necessary to use any of them. Any combination may be used other than those stated there in NEC4—ECC [1, 6, 8–11, 21, 22].

## 3.2 Terminology and Expressions

There are some differences and similarities of some expressions and terminology between these standard contracts; they are as follows.

### 3.2.1 SBD for the PSWs of the GoA

SBD for the PSWs of the GoA uses Entity[*] instead of the owner, employer, or client. Entity's representative for the project is the Project Manager. This SBD uses the word Bid and Bidding instead of Tender and Tendering in FIDIC RED BOOK and NEC4—ECC. In both this SBD and FIDIC RED BOOK (2017), the word Day means a calendar day.

### 3.2.2 FIDIC RED BOOK

FIDIC RED BOOK uses the word Employer (mostly), Owner (less) and sometimes Client. The Engineer represents the Employer (owner). FIDIC provides a central role for the Engineer in the certification process during the performance of the works. It means that the certification process is performed by the Engineer. The Engineer can appoint a representative of him named as Engineer's Representative or Assistant. FIDIC uses the word Consultant instead of Designer in NEC.

### 3.2.3 NEC4—ECC

NEC4—ECC uses the word Client, but in NEC3, it was Employer. The Project Manager represents the Client and sometimes Supervisor can be the representative of the Project Manager. NEC4—ECC uses the expression of Scope instead of Works Information in NEC3—ECC, which is called the Specification and Drawings in the other two documents [1, 5, 6, 9–11, 21, 25].

*Entity: it is defined in Law on Procurement of the GoA as, Entity are the ministries, general departments, independent government commissions, municipalities, enterprises, their secondary units, other budgetary units, state-owned companies, and joint ventures [19].

## 3.3 Errors or Mistakes about the Naming of some Clauses

There are some errors or mistakes in SBD for the PSWs of the GoA about the naming of some clauses, some of them are as follow.

### 3.3.1 Certification by the Project Manager [18th Clause]

The name of the clause expresses that it will be about all certification that the Project Manager does, but it is not the case. It is about the Contractor's responsibility for designing temporary structures, preparing its specification and drawings, and submitting it to the Project Manager for approval and certification. And if needed to be certified by any third party, the Contractor should certify it. If it were named Contractor's Responsibilities about Temporary Structures, it would be better and express the actual meaning of the clause.

### 3.3.2 Safety [19th Clause]

It is better to be changed to Health and Safety Responsibilities of the Contractor. This clause of SBD for PSWs is discussed below in sub-sub Sect. 3.4.3.

### 3.3.3 Testing [32nd Clause]

It is about the tests which are not mentioned in the technical specifications and are beyond the contract. It is ordered by the Project Manager to be performed by the Contractor. If defects are detected, the contractor will pay its cost; if there was no defect, its cost will be paid by the Entity and counted in Compensable Events [42nd Clause]. The name of this clause shall change to Testing beyond Contract or Testing out of Contract. The new name will be most suitable for the information in this clause.

### 3.3.4 Tax [43rd Clause]

It is not about Tax, but about Changes in Tax and its result on the contract price.

### 3.3.5 Completion [53rd Clause]

It is not about the Completion or date of Completion, but about the Certification of Completion of Construction Works.

### 3.3.6 Dispute Resolution [62nd Clause]

If the amendments in the sub-sub Sect. 3.4.4 are added to this clause, then its name shall be changed to Dispute Avoidance and Resolution, which will express its comprehensive meaning [21].

## 3.4 Lack of some Essential Clauses and/or Sub-clauses

There are a lack of some essential clauses and/or sub-clauses in SBD for the PSWs of the GoA, which are as follows.

### 3.4.1 Subcontracting [7th Clause]

It has some information about the subcontracting and some are referenced to the Law on Procurement and Procurement Procedure of the GoA. One essential sub-clause that is not included in both SBD for the PSWs of the GoA and NEC4—ECC, but is in FIDIC RED BOOK (2017) is the duration for the response of the Engineer about the acceptance or rejection of the Subcontractor proposed by the Contractor. This duration is 14 days from the submission of the proposal about the Subcontractor. Suppose the Engineer does not respond within this duration after getting this proposal by giving notice to the proposed Subcontractor. In that case, the Engineer shall be considered to have given his/her consent.

Notice to the Engineer should be given by the Contractor not less than 28 days before the commencement of each Subcontractor's work on the site, and not less than 28 days before the intended date of the commencement of work [9–11, 21].

### 3.4.2 Entity's Risks [10th Clause]

In this clause, besides the information included in it, it is missing some issues which are given in detail in NEC4—ECC. If it is added to this clause, it will complete the missing part of it. These are written as below:

- Damage to or loss of materials and plant provided by the Entity or others on the Entity's behalf to the Contractor; until the Contractor has received and accepted them.
- Damage to or loss of any equipment and the works, plant and materials retained on the site by the Entity after a termination, except damage or loss because of the activities of the Contractor on the site after the termination [10, 11, 21].

### 3.4.3 Safety [19th Clause]

It expresses that the Contractor is responsible for the safety of all activities on the site. It does not have sufficient information about safety. Also, it did not mention anything about the health issues of the personnel on the site. NEC4—ECC has also given less information in this case. FIDIC RED BOOK (2017) has given more detail in this part which, if added under this Clause of SBD for the PSWs of the GoA, it would make it better and clear. These are as follows:

The Contractor shall submit a health and safety manual to the Engineer for information, within 21 days of the start date and before starting any construction on the site; which has been mainly prepared for the works, the site, and other places (if any) where the Contractor intends to perform the works.

The Contractor shall:

- conform with all the applicable health and safety laws and regulations;
- conform with all the applicable health and safety responsibilities specified in the contract;
- take care of all persons' health and safety permitted to be on the site and other places where the works are being performed. Keep the site and other places where the works are being performed clear of unnecessary obstructions to avoid hazard to these persons;
- provide any temporary works (including fences, guards, footways, and roadways) that may be necessary, because of the execution of the works, for the use and safety of the owners and of public and occupants of adjacent property and land.

The name of this clause of SBD for the PSWs shall also be changed to express its clear meaning as written above in sub-sub Sect. 3.3.2 [9–11, 21].

### 3.4.4 Disputes Resolution [62nd Clause]

The second sub-clause of it is given as a separate clause—Disputes [24th Clause]—which should be removed, because it is already there in the 62nd Clause. The Dispute Resolution clause has some information, but not as much as to remove ambiguity and make it clear. This clause does not have any information for creating the Dispute Avoidance/Adjudication Board (DAAB), which can avoid or resolve the arisen disputes during the contract period. It also does not have any procedure to avoid the disputes or any stimulus for the cooperation between the Entity, Contractor, and Project Manager to lead to avoid the incidence of disputes between them. Both FIDIC RED BOOK (2017) and NEC4—ECC have good information and conditions; if added to the Dispute Resolution clause, it will improve [9–11, 21].

The name of this clause of SBD for the PSWs shall also be changed to express its comprehensive meaning as written above in sub-sub Sect. 3.3.6.

### 3.4.5 SBD for the PSWs of the GoA

The SBD for the PSWs of the GoA also lacks some of the essential clauses and/or sub-clauses as follows: staff and labor; delayed payment; suspension of works; contractor's entitlement to suspend work; lacks some essential information about termination by the Entity; lacks information about force majeure or exceptional events; lacks sufficient information about measurement and evaluation; claims; lacks adequate details on testing and inspection [9–11, 21].

If these clauses and/or sub-clauses are added/amended in the SBD for the PSWs of the GoA. In that case, it will lead to a better standard form of contract, clear and easily understandable, and avoid or minimize the arising disputes due to the ambiguity and insufficiency of information.

## 3.5 Flexibility

Flexibility is a good and admirable feature of a standard contract. As mentioned in the last sections, the SBD for the PSWs of the GoA is for the admeasurement or unit price (priced contract with BoQ) contracts, which is in the scope of our study too. It can also be used for the lump sum contracts but considering some mentioned terms, which are to be amended in the SBD for the PSWs [21]. On the contrary, the NEC4—ECC is more flexible in this case and can be used for different types of contracts like; priced contract with Activity Schedules (AS), priced contract with BoQ, target contract with AS, target contract with BoQ, cost-reimbursable contract, and management contract [10, 11]. If procurement authorities of the GoA make an effort, they can draft and prepare this SBD for the PSWs so that it can be more flexible and can be used for different types of contracts as NEC4—ECC is used.

## 3.6 Risk-Sharing

Balanced risk-sharing is also a desirable feature of a standard form of a contract. All modern contracts try to allocate the risk between the parties clearly. Good standard contracts allocate risk to the party who can best manage them [25]. The FIDIC RED BOOK (2017) and NEC4—ECC have balanced risk-sharing between the Employer and the Contractor. However, the SBD for PSWs' risk-sharing is approximately more in favor of the Entity, which increases the risk part of the Contractor. The SBD for the PSWs of the GoA shall be amended and prepared by the procurement authority of the GoA, to balance its risk-sharing between the Entity and the Contractor and to do not harm any party by unbalanced risk-sharing [9–11, 21].

## 4 Conclusion

The above study concluded that standard forms of contracts and their balanced conditions are of most importance.

In this study, the SBD for the PSWs of the GoA compared with the other two international standard forms of contracts FIDIC RED BOOK (2017) and NEC4—ECC. As a result of this research, the conclusion is written briefly as below:

- Balanced conditions are very essential for each contract, and unbalanced conditions led to arise disputes. FIDIC RED BOOK (2017) and NEC4—ECC are well prepared and balanced than the SBD for the PSWs of the GoA.
- The structuring and arrangement of Afghanistan's SBD for the PSWs is good but needs to be updated and amended to become better structured.
- There are some differences in terminology and expressions between these three standard forms of contracts.
- There are some errors and mistakes in the naming of some clauses in the SBD for the PSWs of the GoA, which the procurement authorities of the GoA should correct them for more clarity and easy understanding.
- There are a lack of some essential clauses and/or sub-clauses in the SBD for the PSWs of the GoA, which creates ambiguity and could cause disputes. The procurement authorities of the GoA may take in mind to add those clauses and/or sub-clauses mentioned in this research and other which are not mentioned here. It will cause to avoid or minimize the incidence of disputes and conflicts in the contract.

- NEC4—ECC is more flexible than FIDIC RED BOOK (2017) and SBD for the PSWs of the GoA. This flexibility is in case of usage of the standard form for different types of contracts from the aspect of its financing.
- The SBD for the PSWs of the GoA does not have balanced risk-sharing between the Entity and the Contractor, but FIDIC RED BOOK (2017) and NEC4—ECC are more balanced in this case.

## References

1. Barnes M et al (2013) NEC3 engineering and construction contract, 3rd edn. Institution of Civil Engineers (Thomas Telford Ltd), Glasgow
2. Berry PA (2013) NEC3 or FIDIC? Which is most effective in managing the most commonly identified areas of dispute within construction?. Heriot-Watt University, Edinburgh
3. Besaiso H, Wright D, Fenn P, Emsely M (2016) A comparison of the suitability of FIDIC and NEC conditions of contract in Palestine. In: The construction, building and real estate research conference of the royal institution of chartered surveyors
4. Booen PL (2000) The FIDIC contracts guide (conditions of contract for construction, conditions of contract for plant and design-build and conditions of contract for EPC/Turnkey projects), 1st edn. FIDIC (International Federation of Consulting Engineers), Geneva
5. CMS (2017) NEC4; a closer look at the changes in the ECC. CMS Cameron McKenna Nabarro Olswang LLP, London
6. Cousins P, Reed C, Nicholson T (2006) Guidance notes for the engineering and construction contracts (NEC3), 3rd edn. Institution of Civil Engineers (Thomas Telford Ltd), Glasgow
7. Fenn P, Lowe DJ, Speck C (1997) Conflict and dispute in construction. Constr Manage Econ 6(15):513–518
8. FIDIC (1999) Conditions of contract for construction (for building and engineering works designed by the employer), 1st edn. FIDIC (International Federation of Consulting Engineers), Geneva
9. FIDIC (2017) Conditions of contract construction—for building and engineering works designed by the employer, 2nd edn. FIDIC (International Federation of Consulting Engineers), Geneva
10. Garratt M et al (2017) NEC4 engineering and construction contract, 4th edn. Bell & Bain Limited, Glasgow
11. Garratt M et al (2017, June) NEC4 engineering and construction contract, 4th edn. Institution of Civil Engineers (Thomas Telford Limited), Glasgow
12. Godwin W (2013) International construction contracts, 1st edn. Wiley-Blackwell, West Sussex
13. Heaphy I (2013). NEC versus Fidic. In: Proceedings of the ICE—management, procurement and law, vol 166, pp 21–30
14. Hughes K (2019) Understanding the NEC4 ECC contract. Routledge (Taylor & Francis Group), London and New York
15. Kilburn K, Cornelius A (2014) Using NEC3-what you need to know
16. Lau CH, Mesthrige JW, Lam PTI, Javed AA (2019) The challenges of adopting new engineering contract: a Hong Kong study. Eng Constr Archit Manag 26(10):2389–2409
17. Lord WE, Liu A, Zhang S, Tuuli, MM (2010). A modern contract: developments in the UK and China. In: Proceedings of the ICE—management, procurement and law, vol 163, pp 151–159
18. MoSPI (2005, April) Contract document for domestic bidding (Ministry of Statistics and Programme Implementation). Chandu Press, New Delhi
19. NPA (2016, Sept) Law on procurement. Ministry of Justice, Kabul
20. NPA (2017, May) Standard bidding document for procurement of large works. National Procurement Authority (NPA), Kabul

21. NPA (2018) Standard bidding document for procurement of small works. National Procurement Authority (NPA), Kabul
22. Rasslan ND, Nassar AH (2017) Comparing suitability of NEC and FIDIC contracts in managing construction project in Egypt. Int J Eng Res Technol (IJERT) 6(6):531–535
23. Roe M (2011) Presentation to the Danish society for construction and consulting law. Pinsent Masons, London
24. Rooney P, Allan R (2013) A case study of changing procurement practices on delivery of highways projects. In: Smith SD, Ahiaga Dagbui DD (eds) Procs 29th annual ARCOM conference, 2–4 September 2013, Reading, UK. Association of Researchers in Construction Management, pp 779–788
25. Shnookal T, Charrett D (2010) Standard form contracting; the role for FIDIC contracts domestically and internationally. In: Australia, Society of construction law conference, pp 1–30

# Comparative Analysis of Various Walling Materials for Finding Sustainable Solutions Using Building Information Modeling

**Amey A. Bagul and Vasudha D. Katare**

**Abstract** In this modern age of the construction industry, the building sector has consumed around 35–50% of energy. Sustainable development has therefore become an essential factor to reduce the impact on the environment. The traditional brick used as a walling material has been a major contributor to the environmental adversity so finding an alternative has become the need of the hour. In this paper, the comparative analysis of various walling materials is done to determine the most suitable and sustainable one. The analysis is done using Building information modelling (BIM). The comparison is carried out by creating a 3D building model in Revit 2019. The sun-path analysis, heating and cooling load analysis, Wind analysis and shading/lighting analysis are performed. Technical, social, environmental, and economic aspects are also considered for comparative analysis. Analytic hierarchy process (AHP) is used as the decision-making process to determine the best walling material. The comparative analysis concludes that walling materials using waste are useful for achieving the goal of sustainable solution.

**Keywords** Sustainability · Walling materials · Building information modelling · Analytic hierarchy process · Revit

## 1 Introduction

In several countries, unsustainable growth in the construction industry is regarded as a major environmental hazard. [9] The construction industry in India alone accounts for 22% of overall greenhouse gas (GHG) emissions into the atmosphere. [17] To overcome this limitation, alternative construction materials are developed by using

---

A. A. Bagul (✉) · V. D. Katare
Department of Civil Engineering, COEP, Pune, Maharashtra 411005, India
e-mail: ameybagul@ymail.com

V. D. Katare
e-mail: vdk.civil@coep.ac.in

industrial and agro wastes are good options [6]. In building design and its construction, energy plays an important role. Many aspects like quantity and type of materials used, location and orientation of the building, parameters of spaces and zones are must be considered to estimate the overall energy need of a building [3, 13]. Materials play an important role in engineering design. Material selection is one of the most difficult problems in product design and production, and it is also crucial for the success and competitiveness of manufacturing organisations. Buildings use large quantities of raw materials that consume a significant amount of energy [20]. For sustainable building construction, the preference of building materials with a minimum amount of energy is of primary importance, as embodied energy in the manufacturing process contributes to a significant fraction of the overall energy usage of the building [19]. Clay bricks and fly ash bricks are significant sources of greenhouse gas emissions due to the use of coal and cement in its processing. For sustainable solutions, reusable and chip alternative options need to be studied [2, 4]. REVIT Architecture, based on parametric building modeling technology, is a purpose-built instrument for knowledge modeling [12]. As a result, it not only graphically portrays the design, but also collects and uses the data required to help sustainable design to deliver its optimum advantages. For analysis and documentation of various design choices, a model created in REVIT will be useful [7, 8, 10]. Building Information Modelling (BIM) is transforming the conventional way construction is delivered [1]. Therefore, the conversion of large energy-consuming structures into energy-efficient sustainable structures is important and it takes time to establish a system for sustainable building analysis using modern tools and technologies [16]. Sustainable construction meets social, environmental and economic needs in a balanced manner [5]. The present study evaluates the use of different wall bricks made up of industrial and agricultural waste to produce renewable and energy-efficient walling materials. The selected masonry units can be used as an alternative to the better thermal resistance and energy efficiency of traditional bricks. Technical, economic, environmental and social aspect were considered for comparison. The most sustainable brick along the selected is obtained by using analytic hierarchy process (AHP) method. The analytic hierarchy process (AHP) is one of the multi-criteria decision-making methods that can be used to help in the selection of the best strategic solution [18]. This study will assist in the identification of alternative, sustainable materials and technologies for walling materials that can minimise economic, environmental, and social impacts such as energy consumption and cost.

## 2 Methodology

### 2.1 Materials

For this study, five types of brick are selected with different material properties. Literatures are refried to find physical, thermal properties of the brick. Traditional

**Table 1** Technical and social properties of selected bricks

| Brick type | Technical properties | | | Social property |
|---|---|---|---|---|
| | Size (Inmm$^3$) | Compressive strength (MPa) | Density (Kg/m$^3$) | Thermal conductivity (W/m–K) |
| (Clay bricks) CB | 230 × 105 × 75 | 03.10 | 1900 | 0.811 |
| (Co-fired blended ash) CBA bricks | 230 × 100 × 90 | 14 | 1330 | 0.55 |
| (Paper and RHA waste bricks) RHAB | 230 × 105 × 80 | 15.00 | 588 | 0.35 |
| (Fly ash bricks) FB | 230 × 105 × 75 | 03.12 | 1550 | 1.05 |
| (Cotton and paper waste bricks) CWB | 230 × 105 × 80 | 23.64 | 598 | 0.25 |

bricks like clay brick and fly ash bricks are considered along with cotton and paper waste bricks (CWB) [14], co-fired blended ash (CBA) brick [11], paper and RHA waste bricks (RHAB) [15]. Rate and environmental factors are calculated by survey and analysis. The standards of bricks are considered as per Bureau of Indian Standard (BIS), 2002. Table 1 shows the technical and thermal properties based on social comfort factor of selected Bricks.

## 2.2 Building Model

The location of study model is located in Aurangabad, Maharashtra, India. Single story building having a built up area of 71 m$^2$ is used as a study model. 24-hour usage residential unit is considered for building model. Figure 1 shows the plan view of building model. The location and building definition are considered to achieve accurate heating and cooling information.

Using Revit 2019; The 3D model of house is created (Fig. 2). For each type of selected brick, the properties of the wall were changed in Revit library. Remaining units were kept the same to justify the comparative analysis. The input data is taken from the literature. The rooms and spaces are provided for analysis purpose. The space-generated study model is shown in (Fig. 3). Energy analysis model is prepared in Autodesk Revit 2019 for performing sustainability analysis. The Energy analysis model is exported in gbXML, which will be analyzed for sun-path analysis, Wind analysis, Lighting/shading analysis and for calculating heating and cooling loads.

Using the data from result, heating and cooling loads analysis, wind analysis, and, finally, calculating energy usage building facility are performed in Autodesk Green building Studio.

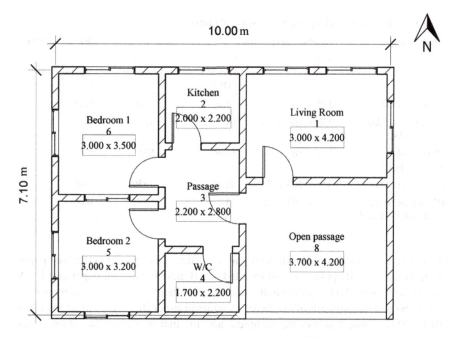

**Fig. 1** 2D view of floor plan of house

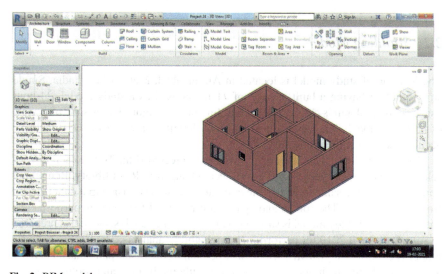

**Fig. 2** BIM model

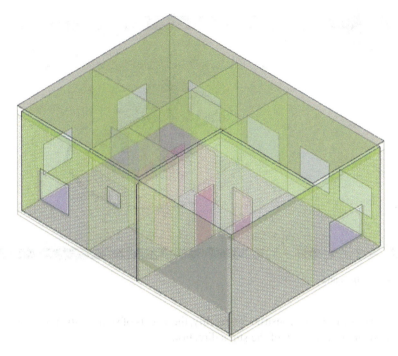

**Fig. 3** Space generated study model in REVIT architecture

## 2.3 Analysis

For the comparative analysis of walling material Technical, social, environmental, and economic aspects are considered. Analytic hierarchy process (AHP) is used as a decision-making process to determine the best walling material. the four criteria that is technical, social, environmental, and economic aspect are considered with five alternative bricks. The matrix is formed and by using Saaty's Ratio Scale; the values are considered. A judgmental matrix is used to calculate the priorities of the corresponding criteria.

## 3 Result

### 3.1 Sun Path Analysis

The outcome of the sun path study will be useful for evaluating the suitable position for the electricity generation solar PV panel system, solar water heaters and other solar equipment mounted on the building facility to optimise the usage of solar energy. Sun path analysis is carried out using Autodesk Insight. To achieve thermal

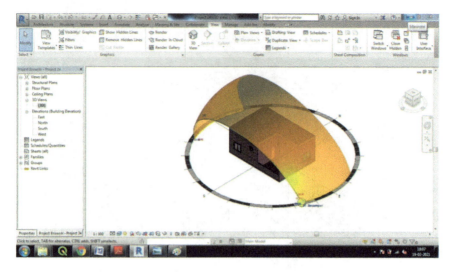

**Fig. 4** Sun path analysis

and visual comfort in a construction facility, the effects of solar analysis may be used. Figure 4 shows the study of the sun's direction.

### 3.2 Shading/Lighting Analysis

In any construction plant, for proper day lighting plan, the shading study is very helpful. It will minimise energy costs with little to no additional expenditure on systems. Also helps to decrease power costs, prevent eyestrain in the workplace and may be useful to avoid excessive absorption of summer heat. The location of openings such as doors, windows, etc., is determined using the shading analysis. Figure 5 shows the study of the shading/lighting analysis.

### 3.3 Wind Analysis

The most appropriate technique for inducing passive cooling is natural ventilation. Appropriate air circulation in a facility helps to maintain comfortable temperatures. In a building facility, proper air circulation and ventilation ensure that the contaminated air is replaced repeatedly by cleaner and fresher air. This analysis is based on the Annual wind roses study. Considering 16 cardinal directions, wind roses are made. On a radial scale, the wind frequency is represented with strength of 0, 2, 4, 6, 8, 10, 12, 14, 16 and 18% of the time as shown in Fig. 6.

Comparative Analysis of Various Walling Materials ... 321

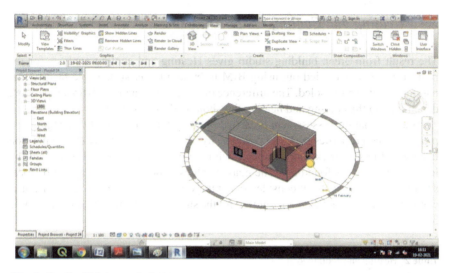

Fig. 5 Shading/lighting analysis

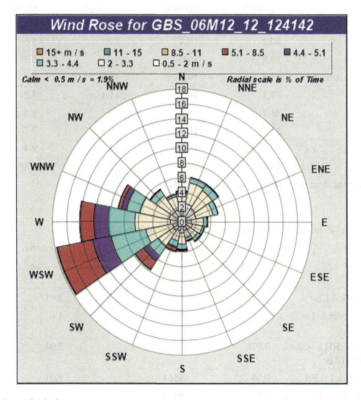

Fig. 6 Annual wind roses

## 3.4 Heating and Cooling Loads Analysis

The result from green building studio gives Heating and cooling analysis for the project which was carried out using BIM to build a better sustainable facility that naturally heated and cooled. The difference of the dry bulb and wet bulb temperature depending on the cooling and heating threshold is shown in Table 2.

The sustainability assessment is carried out only for brickwork. by changing and providing the properties such as brick density, thermal conductivity (k), dimension from wall library. Bricks are compared with the primary case model. peck cooling load for each case is found out. The market values of bricks are finding out by local survey. The average value is considered. For the value analysis, local brick distributer, contractor and civil industry are referred for the survey. The values are shown in Table 3.

**Table 2** Heating and cooling load

| Threshold (%) | Cooling | | Heating | |
|---|---|---|---|---|
| | Dry bulb (°C) | MCWB (°C) | Dry bulb (°C) | MCWB (°C) |
| 0.1 | 40.2 | 19.4 | 8.8 | 5.5 |
| 0.2 | 39.8 | 19.0 | 9.9 | 5.3 |
| 0.4 | 39.6 | 19.0 | 10.8 | 6.5 |
| 0.5 | 39.3 | 18.7 | 11.1 | 7.2 |
| 1 | 38.7 | 18.4 | 12.3 | 8.4 |
| 2 | 37.7 | 18.1 | 13.4 | 9.1 |
| 2.5 | 37.3 | 18.8 | 13.7 | 9.4 |
| 5 | 35.5 | 18.2 | 15 | 10.8 |

*Mean coincident wet bulb temperature (MCWB)

**Table 3** Environmental and economic value of selected bricks

| Brick type | Environmental properties | | Economic Properties |
|---|---|---|---|
| | EE (MJ/Brick) | Peck cooling load demand (W) | Cost/unit (INR) |
| (Clay bricks) CB | 4.19 | 11,243 | 7.61 |
| (Co-fired blended ash) CBA bricks | 1.71 | 9650 | 4.6 |
| (Paper and RHA waste bricks) RHAB | 0.078 | 7212 | 7.04 |
| (Fly ash bricks) FB | 0.09 | 8561 | 9.01 |
| (Cotton and paper waste bricks) CWB | 0.078 | 5418 | 5.68 |

**Table 4** Saaty's ratio scale

| Evaluation | Scale/Judgment |
|---|---|
| 1 | Equal importance of both elements |
| 3 | Moderate importance of one element over another |
| 5 | Strong importance of one element over another |
| 7 | Very strong importance of one element over another |
| 9 | Extreme importance of one element over another |
| 2, 4, 6, 8 | Compromises between the previous judgments |
| 1/2, 1/3, 1/4, 1/5, 1/6, 1/7, 1/8, 1/9 | If activity i has one of the above nonzero numbers assigned to it when compared with activity j, then j has the reciprocal value when compared with i |

*Source* Saaty (1977, 1980)

**Table 5** Ranking of brickwork

| Rank | Technical | Social | Environment | Economic |
|---|---|---|---|---|
| 1 | CWB | CWB | CWB | CBA |
| 2 | CBA | CBA | RHB | CWB |
| 3 | RHB | CB | FB | RHAB |
| 4 | FB | RHB | CBA | CB |
| 5 | CB | FB | CB | FB |

Using the obtained value the AHP method is used for analysis. AHP is a valuable method for dealing with tough choices, and it can help to set goals and make the best possible decision. AHP has the potential to use qualitative and quantitative criteria, the organised manner in which decisions are taken, which allows for good decision traceability, and the quality assurance offered by consistency indices are all benefits of this system. Five alternatives with four criteria are considered for AHP analysis. The Saaty's Ratio Scale is considered for comparative valuation as Shown in Table 4. The obtained result is shown in Table 5.

# 4 Conclusion

In The modern era of construction; sustainability becomes an essential part to balance energy and nature. The solid waste produced by agriculture and industrial waste can be a new solution for modern bricks. The Cotton and paper waste brick brick have the highest sustainability index. co-fired blended ash) CBA bricks and (Paper and RHA waste bricks) RHAB are also effective solutions on clay Brick. The Cotton and paper waste bricks have the highest sustainability index. which is 60% efficient in thermal conductivity, 26% cost efficient than Clay brick. (co-fired blended ash) CBA bricks and (Paper and RHA waste bricks) RHAB are also effective solution on clay Brick. The Building Information Modelling (BIM) is the advanced tool that helps to understand the Problems and risk in panning and pre-construction stage. It provides accurate results with statically and visually. The AHP method helps to effectively deal with the issue so that it is easier for the decision maker to take appropriate actions. The comparative analysis concludes that walling materials using waste are useful for achieving the goal of sustainable solutions.

## References

1. Abdelalim AM (2019) Integrating BIM-based simulation technique for sustainable building design. In: International congress and exhibition sustainable civil infrastructures: innovative infrastructure geotechnology. Springer, Cham
2. Adewale BA et al (2018) Dataset on cost comparative analysis of different walling materials in residential buildings in a developing economy. Data Brief 19:1918–1924
3. Ajayi SO et al (2015) Life cycle environmental performance of material specification: a BIM-enhanced comparative assessment. Int J Sustain Build Technol Urban Dev 6(1):14–24
4. Ashby MF et al (2004) Selection strategies for materials and processes. Mater Des 25(1):51–67
5. Azhar S, Brown J, Farooqui R (2009) BIM-based sustainability analysis: an evaluation of building performance analysis software. In: Proceedings of the 45th ASC annual conference, vol 1, No. 4
6. Ceranic B, Latham D, Dean A (2015) Sustainable design and building information modelling: case study of energy plus house, Hieron's Wood, Derbyshire UK. Energy Procedia 83:434–443
7. Charef R, Alaka H, Emmitt S (2018) Beyond the third dimension of BIM: a systematic review of literature and assessment of professional views. J Build Eng 19:242–257
8. Chong HY (2016) Comparative analysis on the adoption and use of BIM in road infrastructure projects. J Manage Eng 32(6):05016021
9. Dissanayake DMKW, Jayasinghe C, Jayasinghe MTR (2017) A comparative embodied energy analysis of a house with recycled expanded polystyrene (EPS) based foam concrete wall panels. Energy Build 135:85–94
10. Farghaly K et al (2018) Taxonomy for BIM and asset management semantic interoperability. J Manage Eng 34(4):04018012
11. Gavali HR, Ram S, Ralegaonkar RV (2018) Evaluation of energy efficient sustainable walling material. In: Urbanization challenges in emerging economies: resilience and sustainability of infrastructure. American Society of Civil Engineers, Reston, VA, pp 227–233
12. Jung I, Kim W (2014) Analysis of the possibility of required resources estimation for nuclear power plant decommissioning applying BIM
13. Kim H, Anderson K (2013) Energy modeling system using building information modeling open standards. J Comput Civ Eng 27(3):203–211
14. Rajput D, Bhagade SS, Raut SP, Ralegaonkar RV, Mandavgane SA (2012). Reuse of cotton and recycle paper mill waste as building material. Constr Build Mater 34:470–475
15. Raut SP, Sedmake R, Dhunde S, Ralegaonkar RV, Mandavgane SA (2012). Reuse of recycle paper mill waste in energy absorbing light weight bricks. Constr Build Mater 27(1):247–251
16. Sadeghifam AN et al (2016) Energy analysis of wall materials using building information modeling (BIM) of public buildings in the tropical climate countries. J Teknologi 78(10)
17. Shams S, Mahmud K, Al-Amin M (2011) A comparative analysis of building materials for sustainable construction with emphasis on $CO_2$ reduction. Int J Environ Sustain Dev 10(4):364–374
18. Saaty TL (2008) Decision making with the analytic hierarchy process. Int J Serv Sci 1:83–98

19. Thormark C (2006) The effect of material choice on the total energy need and recycling potential of a building. Build Environ 41(8):1019–1026
20. Udawattha C, Halwatura R (2017) Life cycle cost of different walling material used for affordable housing in tropics. Case Stud Constr Mater 7:15–29

# Studies on Energy Efficient Design of Buildings for Warm and Humid Climate Zones in India

**Santhosh M. Malkapur, Sudarshan D. Shetty, Kishor S. Kulkarni, and Arun Gaji**

**Abstract** In recent years, significant efforts have been made to improve energy efficiency and decrease energy consumption. The idea of energy efficiency in structures is related to the energy supply needed to achieve desirable environmental conditions that minimize energy consumption. The residential sector is liable for a significant piece of the energy consumption in the world. Most of this energy is used in cooling, heating and natural ventilation systems. In this work, the energy analysis of a residential building is carried out by varying building envelope parameters such as aspect ratio, orientation and window to wall ratio of the building in order to lessen the energy demand for the cooling and heating. The building is modeled in a software tool and various parameters are assigned to study the thermal efficiency in a warm and humid climate zone in India. The results indicate that a square building with an aspect ratio of 1:1 is more thermally efficient structure and North–South orientation of the building is better than East–West orientation. Also increasing window to wall ratio decreases the thermal efficiency of the building. The findings of this work would be helpful in design phase of an energy-efficient residential building.

**Keywords** Energy efficiency · Warm and humid climate · Orientation · Aspect ratio · Window to wall ratio

## 1 Introduction

Designing an energy-efficient structure is one of the best methods to reduce energy costs in buildings. To design energy-efficient buildings, structural components and building envelope parameters must be optimized. It is critical to recognize the basic factors that are directly related to heat transfer processes. The theoretical design phase

---

S. M. Malkapur (✉) · S. D. Shetty · A. Gaji
Department of Civil Engineering, Basaveshwar Engineering College, Bagalkot, Karnataka 585102, India
e-mail: smmcv@becbgk.edu

K. S. Kulkarni
AR & P Division, CSIR-CBRI, Roorkee, Uttarakhand 247667, India

© The Author(s), under exclusive license to Springer Nature Singapore Pte Ltd. 2023
M. S. Ranadive et al. (eds.), *Recent Trends in Construction Technology and Management*, Lecture Notes in Civil Engineering 260,
https://doi.org/10.1007/978-981-19-2145-2_26

of a structure is the best ideal opportunity to incorporate feasible systems. Energy-efficient design methods provide additional value that benefits the end-user. An organized plan dependent on energy-saving standards diminishes economic expenses all through the lifecycle of the structure because of its lower energy consumption [1, 2]. Since there are additionally less $CO_2$ discharges into the environment all through the structure's yearly energy utilization, this is advantageous to society also. Energy saving is a high need in developed nations. Consequently, energy efficient measures are as a rule progressively executed in all possible areas.

## 2 Literature Review

The energy saving is a high need in the entire world. The residential sector is a major contributor to the total energy consumption. A large amount of energy is utilized in the cooling and heating systems. Energy efficient structures can be constructed by studying and applying the measures required for achieving it. The measures include shape of the building, building orientation, building envelope system, cooling, and heating, etc., of residential structures [3]. It is possible to improve the energy efficiency of buildings without any additional cost. Buildings should be oriented properly to get the maximum passive solar energy. Morrissey et al. [4] concluded that the concept of passive solar energy should be incorporated in design stage to improve the energy efficiency of buildings. The effect of passive parameters, for example, building shape and building orientation on heating demand has been theoretically examined by Aksoy and Inallib [5] by selecting a cold region (Elazig region) of Turkey. Building orientation was varied from 0° to 90°. It is concluded that structures with a square shape are more advantageous with respect to energy efficiency. Ourghi et al. [6] examined the effect of the shape for office building on its yearly cooling and total energy use. A streamlined examination technique is developed dependent on point-by-point investigations using a few blends of building geometry, glazing type, glazing area, and climate. An immediate connection has been made between relative minimization and complete structure energy use just as the cooling energy prerequisite.

Eskin and Turkmen [7] studied the electricity use in commercial buildings in Turkey. The interactions between various conditions, control systems, and heating/cooling loads in office buildings in the four significant climatic zones in Turkey were studied. This study is used to examine energy conservation opportunities on annual cooling, heating and total building load at four major climatic zones of Turkey. Jazayeri and Aliabadi [8] investigated energy efficiency in the cold and semi-arid climate of Shiraz, Iran. They evaluated energy efficiency in two stages. In the first stage, larger windows with NS facade were considered and in second stage all facades of equal window-to-wall ratio and different building aspect ratios were considered. When different facades are varied, the optimal aspect ratio of a building can be different for different Window to wall ratio (WWR). Friess et al. [9] studied the energy demand of residential villas in Dubai. Minimum insulation levels for external wall and roof wall (U-value = 0.57 w/m$^2$-k) and reinforcement concrete frame was considered non-insulated. This work studied the effect of thermal impact on the structure's energy consumption by using a software model. Simulation results showed that with suitable outside wall protection methodologies alone, energy savings of up to 30% are obtained. Studies were also carried out on energy simulation tools and the strengths and weaknesses of each tool were studied [10]. Most of the studies have focused on a particular envelope component in a generic building. There is a lack of comparative study of the relative efficiency and impact of passive design strategies.

## 3 Methodology

### 3.1 Model and Parameters

In this study, a residential single-family residence house is considered and the effects of different building envelope parameters are studied on the energy efficiency of the building. A single-family house with one floor and basement is modeled for study. Building overview details are provided in Table 1. The house is located at Gadag, Karnataka, which falls under climate Zone 1B: warm and humid as per ECBC [11]. Table 2 provides information on the geography and climate of Gadag. The floor plan is shown in Fig. 1 and the building envelope parameters which have been used as the input parameters in this study are presented in Table 3.

A residential building with three aspect ratios 1:1, 1:1.5 and 1:2 same building area 111.30 m$^2$ is considered for analysis. Hall, bedroom, kitchen, stairway, and

Table 1 Building overview details

| Variable | Value |
|---|---|
| Total built-up area | 111.3025 m$^2$ |
| Building use | Single-family detached |
| Number of floors | Ground floor |
| Construction type wood-framed | Advance framing |
| Heating | Gas furnace + Electricity |
| Cooling | Air conditioner—ducted split system |
| Domestic hot water (DHW) | Electricity and Gas boiler with storage tank |
| Aspect ratio (building size) | 1:1 (10.55 m × 10.55 m)<br>1:1.5 (8.62 m × 12.93 m)<br>1:2 (7.45 m × 14.96 m) |
| Orientation | North–South, East–West |
| Window to wall ratio (WWR) | 30, 40, and 50% |

**Table 2** Details of geography and climate of Gadag

| Variable | Value |
|---|---|
| City/state | Karnataka, Gadag |
| Climate zone | 1B (warm and humid) |
| Latitude | N 15°53' |
| Longitude | E 76°02' |
| Elevation | 650.0 m |
| Heating design | 18 °C baseline |
| Cooling design | 12 °C baseline |
| Building orientation | 0° from the true north and 90° from east |
| Wind speed | 8.2 |

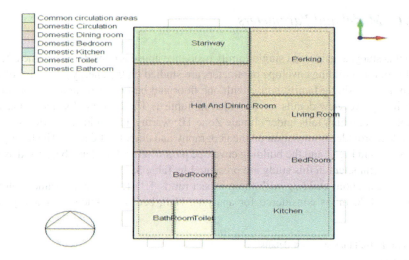

**Fig. 1** Building floor plan

parking, etc., suitable dimensions of length and breadth are set out as given in Table 4.

## 3.2 Internal Loads and Schedules

The residential building of single-family with an occupancy of five people is considered for the present study. Details of activity, maximum occupancy, equipment load and lighting loads in each of the rooms are provided in Table 5.

**Table 3** Building envelope parameters

| Model | Aspect ratio | Orientation North–South and East–West | Window to wall ratio % |
|---|---|---|---|
| Model 01 | 1:1 | | 30, 40, 50 |
| Model 02 | 1:1.5 | | 30, 40, 50 |
| Model 03 | 1:2 | | 30, 40, 50 |

**Table 4** Building geometry

| Room | Activity | | | | | | |
|---|---|---|---|---|---|---|---|
| Aspect ratio | | 1:1 | | 1:1.5 | | 1:2 | |
| Building size | | 10.55 m × 10.55 m | | 8.62 m × 12.93 m | | 7.45 m × 14.96 m | |
| Length in m | Breadth in m | $L$ (m) | $B$ (m) | $L$ (m) | $B$ (m) | $L$ (m) | $B$ (m) |
| Hall/dining room | Eating and Drinking Seated quiet | 5.7 | 2.56 | 4.67 | 4.45 | 3.84 | 5.92 |
| Master bedroom | Sleeping | 4.2 | 3 | 3.45 | 3.0 | 3.05 | 3.2 |
| Bedrooms | Sleeping | 4.24 | 2.7 | 3.35 | 3.0 | 3.0 | 3.6 |
| Kitchen | Cooking | 4.24 | 2.2 | 4.57 | 3.2 | 3.84 | 3.5 |
| Stairway | Light manual work | 5.7 | 2.0 | 4.67 | 2.2 | 3.84 | 2.22 |
| Parking and open | Light manual work | 4.24 | 2.24 | 3.35 | 3.15 | 3.0 | 4.07 |
| Living room | Entry | 4.24 | 2.2 | 3.35 | 2.6 | 3.0 | 2.8 |
| Bathrooms and toilet | Standing/Walking | 2.025 | 2.0 | 1.65 | 2.3 | 1.475 | 2.5 |

## 3.3 Building Envelope

The building envelope is the physical boundary between the outside and inside environments encasing a structure. It is comprised of a series of components and frameworks that shield the inside space from the impacts of the environment such as

**Table 5** Activities and schedules (general energy code or ECBC, 2017)

| Room | Activity | Maximum occupancy | Equipment load (W/m$^2$) | Lighting load (W/m$^2$) |
|---|---|---|---|---|
| Hall/dining room | Eating and drinking seated quiet | 5 | 3.06 | 5 |
| Master bedroom | Sleeping | 2 | 3.58 | 3 |
| Bedrooms | Sleeping | 1 per room | 3.58 | 3 |
| Kitchen | Cooking | 2 | 30.28 | 5 |
| Stairway | Light manual work | 5 | 2.16 | 5 |
| Parking | Light manual work | 5 | 1.57 | 5 |
| Living Room | Entrée | 1 per room | 1.57 | 2 |
| Bathrooms and toilet | Standing/Walking | 1 per room | 3.28 | 3 |

precipitation, wind, temperature, humidity and also ultraviolet radiation. The internal environment is comprised of the occupants, building materials, lighting, machinery and the HVAC system. Improving the structure envelope of houses is perhaps the most ideal approach to improve energy efficiency. This home is modeled using advanced framing techniques. Layer-by-layer details of the wall, roof, and floors are provided. Using these details, creating custom layers and, if necessary, materials using the Design-Builder software. Building envelope construction details are considered as per ECBC 2017 (Fig. 2).

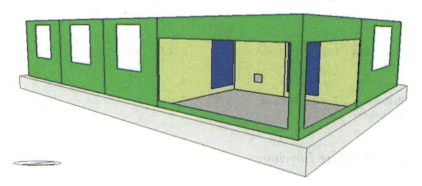

**Fig. 2** Building envelope selected for the present study

## 3.4 Mechanical Systems

The building is considered to be centrally heated and cooled. The heating is provided through a gas furnace and the cooling system is a central split system. Efficiency details of the HVAC and SHW systems are provided as per ECBC 2017 (Table 6). The building is modeled in a software tool (Design builder) and various parameters are assigned to study the thermal efficiency in a warm and humid climate zone in India. Three envelope parameters namely aspect ratio (1:1, 1:1.5 and 1:2), orientation (North–South and East–West) and window-wall ratio (30, 40 and 50%) are varied by using three typical models and their energy efficiencies are evaluated.

Following are the details of the three models considered in this study:

**Model 1**—Single-family residential building with outer dimensions of 10.55 m × 10.55 m is considered for Model 1. Aspect ratio is kept as 1:1 and both orientations North–South and East–West are considered. Window to wall ratio is varied as 30, 40 and 50% (Fig. 3).

**Model 2**—Single-family residential building with outer dimensions of 12.93 m × 8.62 m is considered for Model 2. Aspect ratio of 1:1.5 and orientations in both

**Table 6** Mechanical system details (general energy code or ECBC 2017)

| Variables | Values |
|---|---|
| *Heating load* | |
| System type | Furnace |
| Fuel type | Electricity + Natural gas |
| Heating system efficiency (AFUE) | 80% |
| Maximum supply air temperature (AT) | 35 °C |
| Maximum supply air humidity ratio | 0.0149 |
| Heating capacity system | 26 kW |
| *Cooling load* | |
| System type | Central air conditioning using a split system |
| Fuel type | Electricity |
| Cooling system EER | 12.00 |
| Cooling system SEER | 17.50 |
| Cooling system capacity | 14 kW |
| *Domestic hot water system (DHW)* | |
| System type | A storage hot water system (standalone) |
| System fuel | Natural gas + Electricity |
| Energy factor | 0.82 |
| Hot-water delivery temperature | 65 °C |
| Mains supply temperature | 10 °C |

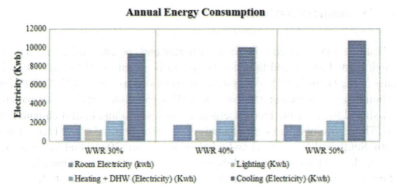

**Fig. 3** Energy consumption for model-1

North–South and East–West direction are considered. Window to wall ratio is varied as 30, 40 and 50%.

**Model 3**—Single-family residential building with outer dimensions of 14.96 m × 7.45 m is considered for Model 3. Aspect ratio is kept as 1:2 and both orientations North–South and East–West are considered. Window to wall ratio is varied as 30, 40, and 50%.

## 4 Results and Discussions

In this section, the results of the thermal analysis of the three models are presented and discussed. The thermal efficiencies are discussed in terms of electricity consumption for lighting, heating, DHW, cooling, annual energy consumption and temperature.

### 4.1 Thermal Energy Efficiency of Model-01

The results of the thermal energy efficiency for model-1 are presented in Table 7. It is seen that the room electricity requirement of 1722.28 kWh is constant for different window to wall ratios. When lighting electricity is considered, small variations are seen for different window to wall ratios. It is observed that for window to wall ratio of 30% lighting electricity is found to be 1197.14 kWh, 1179.70 kWh for 40% and 1169.16 kWh for 50%. Heating and DHW electricity remained constant for different window to wall ratios. The cooling electricity is found to vary for different window to wall ratios. It is seen that for window to wall ratio of 30% cooling requirement is found to be 9314.08 kWh and for 40% and 50%, the cooling requirements are found to be 10007.93 kWh and 10699.93 kWh respectively. There is a large variation in the annual energy consumption for different window to wall ratios. The annual energy consumption is found to be 14405.11 kWh, 15081.55 kWh and 15762.98 kWh for

Studies on Energy Efficient Design of Buildings ...

**Table 7** Energy consumption results model-1

| Description | | Values | | |
|---|---|---|---|---|
| Aspect ratio | | 1:1 | | |
| Orientation | | NS and EW | | |
| Window to wall ratio | | 30% | 40% | 50% |
| Room Electricity (kWh) | | 1772.28 | 1772.28 | 1772.28 |
| Lighting (kWh) | | 1197.14 | 1179.70 | 1169.16 |
| Heating + DHW (Electricity, kWh) | | 2171.61 | 2171.61 | 2171.61 |
| Cooling (Electricity, kWh) | | 9314.08 | 10,007.93 | 10,699.93 |
| Annual energy consumption | | 14,405.11 | 15,081.55 | 15,762.98 |
| Temperature in °C | Outside | 36.60 | 36.60 | 36.60 |
| | Inside | 33.10 | 33.51 | 34.00 |

window to wall ratio of 30%, 40% and 50%, respectively. In general, it can be seen that irrespective of the orientation, the energy consumption is found to increase with increasing window to wall ratios. This is true due to the fact that there will be increased thermal energy transfer between inside and outside environments with increased window to wall ratios and vice versa.

The energy consumption model is presented in Fig.4. The difference in the cooling electricity requirement is found to be lesser by 693.85 kWh and 693.85 kWh for a window to wall ratio of 30% and 40% respectively in comparison with 50%. It is observed that the annual energy electricity requirement is found to be reduced by 676.44 kWh and 681.43 kWh for window wall ratio of 30% and 40% as compared to 50%.

From Fig.4, it is observed that the inside temperature increases by increasing the window to wall ratio for a constant outside temperature of 36.6 °C. The inside temperatures observed are 33.1, 33.5 and 34 °C for window to wall area ratios of 30%, 40% and 50%, respectively.

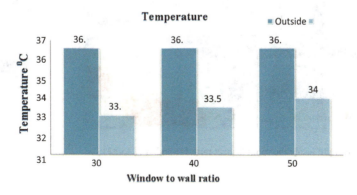

**Fig. 4** Temperature variations between inside and outside environments (model-01)

**Table 8** Annual energy consumption for WWR

| Window to wall ratio (%) | Annual energy consumption (kWh) | Increase in % |
|---|---|---|
| 30 | 14,405.11 | – |
| 40 | 15,081.55 | 4.7 |
| 50 | 15,762.98 | 4.5 |

Annual energy consumption for model-01 is found to increase for higher window wall ratios. The annual energy consumption is found to increase by 4.7 and 4.5% for window to wall ratios of 40% to 50% respectively as shown in Table 8. Also, daylight and sunlight dispersion map of model-01 is shown in Figs. 5, 6 and 7.

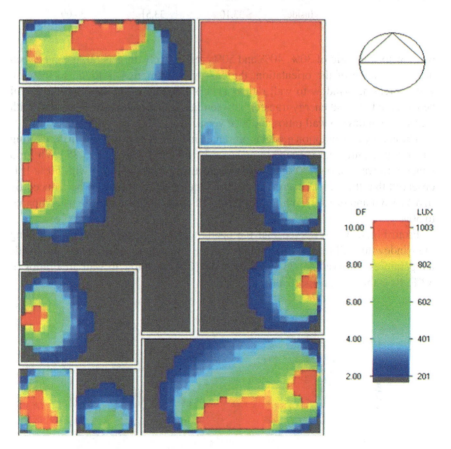

**Fig. 5** Daylight inside the room (aspect ratio 1:1 and WWR30%)

# Studies on Energy Efficient Design of Buildings ...

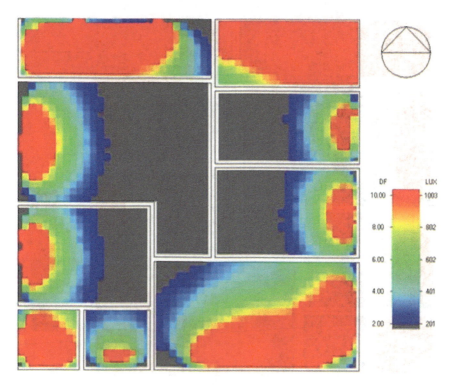

**Fig. 6** Daylight inside the room (aspect ratio 1:1 and WWR40%)

## 4.2 Thermal Energy Efficiency of Model-02

The results of the thermal energy efficiency for model-2 are presented in Table 9 and Fig. 8. It is seen again that the room electricity requirement of 1530.4 kWh is constant for different window to wall ratios. Heating and DHW electricity did not vary for different window to wall ratios. When lighting electricity is considered, small variations are seen for different window to wall ratios. More variations are seen for cooling electricity for different window to wall ratios. It is seen that for window to wall ratio of 30%, cooling requirement is found to be 9314.08 kWh and for 40% and 50%, the cooling requirement are found to be 10007.93 kWh and 10699.93 kWh respectively. There is a large variation in the annual energy consumption for different window to wall ratios which can be seen in Table 9. The cooling electricity is found to increase significantly for increased window to wall ratios. Also with respect to orientation, North–South orientation is found to be better than the East–West orientation in improving the overall thermal efficiency of the building.

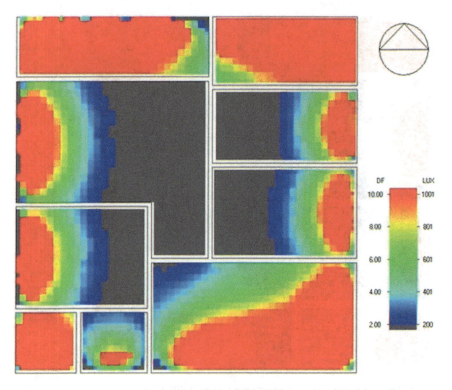

**Fig. 7** Daylight inside the room (aspect ratio 1:1 and WWR50%)

**Table 9** Energy consumption results of model-2

| Description | Values | | | | | |
|---|---|---|---|---|---|---|
| Aspect ratio | 1:1.5 | | | | | |
| Orientation | North–South | | | East–West | | |
| Window to wall ratio | 30% | 40% | 50% | 30% | 40% | 50% |
| Room electricity (kWh) | 1530.4 | 1530.4 | 1530.4 | 1530.4 | 1530.4 | 1530.4 |
| Lighting (kWh) | 1158.5 | 1145.8 | 1139.5 | 1161.8 | 1146.7 | 1139.0 |
| Heating + DHW (electricity, kWh) | 2386.6 | 2386.6 | 2386.6 | 2386.6 | 2386.6 | 2386.6 |
| Cooling (electricity, kWh) | 10,937.0 | 11,429.1 | 11,893.9 | 11,352.5 | 11,764.6 | 12,180.4 |
| Annual energy consumption (electricity, kWh) | 16,012.5 | 16,491.9 | 16,950.3 | 16,431.7 | 16,828.3 | 17,236.4 |
| Temperature (°C) Outside | 36.6 | 36.6 | 36.6 | 36.6 | 36.6 | 36.6 |
| Inside | 34.25 | 34.50 | 34.75 | 33.77 | 34.01 | 34.24 |

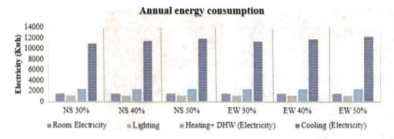

**Fig. 8** Energy consumption model-2

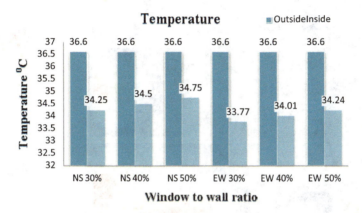

**Fig. 9** Temperature variations between inside and outside environments for model-2

From Figure 9 it is observed that inside temperature increases with increasing window to wall ratio for constant outside temperature of 36.6 °C. Also, the North–South orientation is better than East–West orientation with respect to annual energy consumption (Fig. 10). On the other hand, East–West orientation is slightly better in reducing the inside temperatures of the building. Also, daylight and sunlight dispersion map of model-02 is shown in Figs. 11, 12 and 13 (Table 10).

## 4.3 Thermal Energy Efficiency of Model-03

The results of the thermal energy efficiency for model-3 are presented in Table 11. It is seen again that the room electricity requirement and heating and DHW electricity do not vary for different window to wall ratios. When lighting electricity is considered, small variations are seen for different window to wall ratios. For cooling electricity, considerable variations are seen for different window to wall ratios. It is seen that with increasing window to wall ratios, the cooling requirement increases irrespective of the orientation. There is a large variation in the annual energy consumption for

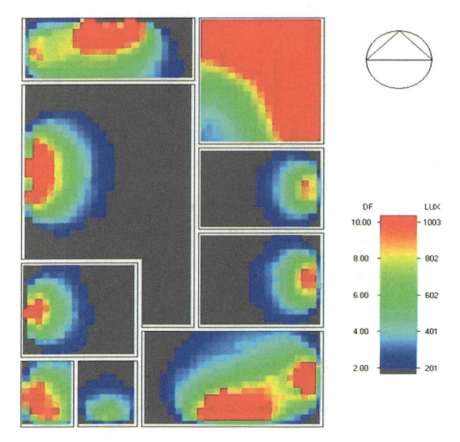

**Fig. 10** Daylight inside the room (aspect ratio 1:1.5 and WWR30%)

different window to wall ratios, which can be seen in Table 11 and Fig 13. Also with respect to orientation, North–South orientation is found to be better than the East–West orientation in improving the overall thermal efficiency of the building.

The cooling electricity values of building with North–South orientation are found to be 12576.8 kWh, 13367.2 kWh and 14411.0 kWh for the window to wall ratio of 30, 40 and 50%, respectively. Similarly, for East–West orientation the cooling electricity values are found to be 12883.7 kWh, 13483.0 kWh and 14416.1 kWh for the window to wall ratio of 30, 40 and 50%, respectively (Fig. 14).

From Fig. 15, it is observed that inside temperature increases with increasing window to wall ratio for both North–South and East–West orientations for a given outside temperature. For the outside temperature of 36.6 °C, the inside temperatures are found to be 34.5, 34.61, and 34.65 °C for North–South orientation and for window to wall ratio of 30, 40, and 50% respectively. Similarly, for East–West orientation the inside temperatures are found to be 33.75, 33.9, and 34.1 °C for window to wall ratio

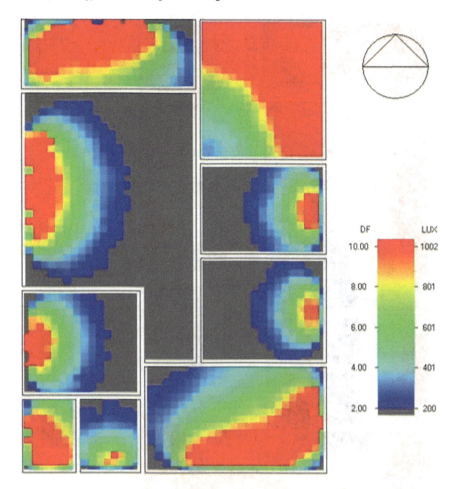

**Fig. 11** Daylight inside the room (aspect ratio 1:1.5 and WWR40%)

of 30, 40, and 50%, respectively. Daylight and sunlight dispersion map of model-03 is shown in Figs. 15, 16 and 17.

Also, an overall comparison of electricity usage, annual energy consumption and temperature for different Aspect Ratios are presented in Figs. 18, 19 and 20 (Table 12).

## 5 Conclusion

(1) In warm and humid climate conditions, the aspect ratio of 1:1, for a square building, energy consumption is 8–12% lesser in comparison with aspect ratio of 1:1.5.

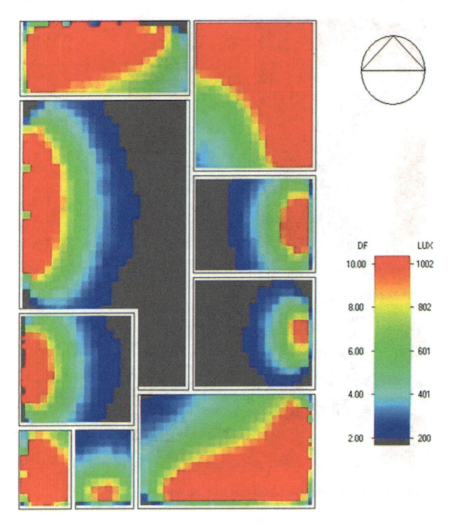

**Fig. 12** Daylight inside the room (aspect ratio 1:1.5 and WWR50%)

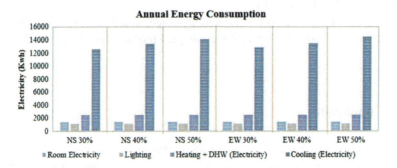

**Fig. 13** Energy consumption chart model-3

**Table 10** Annual energy consumption for NS and EW model-02

| Window to wall ratio (%) | Annual energy consumption (kWh) ||  Difference in % |
|---|---|---|---|
|  | NS | EW |  |
| 30 | 16,012.5 | 16,431.3 | 2.61 |
| 40 | 16,491.9 | 16,828.3 | 2.03 |
| 50 | 16,950.3 | 17,236.4 | 1.68 |

**Table 11** Energy consumption results model-3

| Description | | Values | | | | | |
|---|---|---|---|---|---|---|---|
| Aspect ratio | | 1:2 | | | | | |
| Orientation | | North–South | | | East–West | | |
| Window to wall ratio | | 30% | 40% | 50% | 30% | 40% | 50% |
| Room electricity (kWh) | | 1448.9 | 1448.9 | 1448.9 | 1448.9 | 1448.9 | 1448.9 |
| Lighting (kWh) | | 1131.9 | 1172.0 | 1167.83 | 1179.6 | 1170.9 | 1166.5 |
| Heating + DHW (electricity, kWh) | | 2516.8 | 2516.8 | 2516.8 | 2516.8 | 2516.8 | 2516.8 |
| Cooling (electricity, kWh) | | 12,576.8 | 13,367.2 | 14,111.0 | 12,883.7 | 13,483.0 | 14,416.1 |
| Annual energy consumption (electricity, kWh) | | 17,674.4 | 18,504.9 | 19,244.5 | 18,029.0 | 18,619.6 | 19,548.3 |
| Temperature (°C) | Outside | 36.6 | 36.6 | 36.6 | 36.6 | 36.6 | 36.6 |
| | Inside | 34.5 | 34.61 | 34.65 | 33.75 | 33.9 | 34.1 |

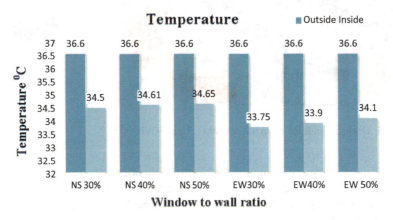

**Fig. 14** Temperature variations between inside and outside environments for model-3

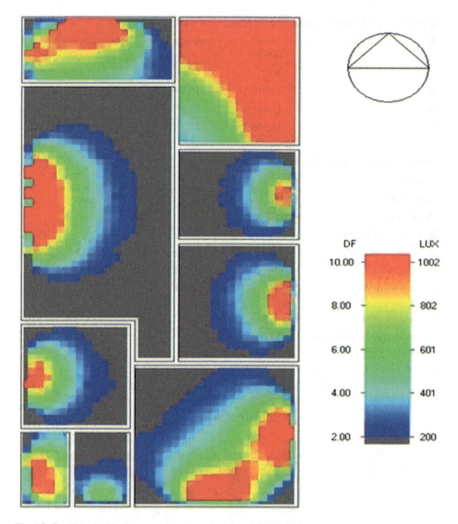

**Fig. 15** Daylight inside the room (aspect ratio 1:2 and WWR30%)

(2) Annual energy consumption for North–South orientation is found to be lesser as compared to the East–West orientation for window to wall ratio 30%, 40%, and 50% respectively. The trend remains same for increased aspect ratios of 1:1.5 and 1:2.
(3) A window to wall ratio increases overall energy consumption of the building. However, the energy required for lighting decreases.
(4) Every 10% increase in the window to wall ratio, the inside temperature increases by 0.25–0.5 °C.

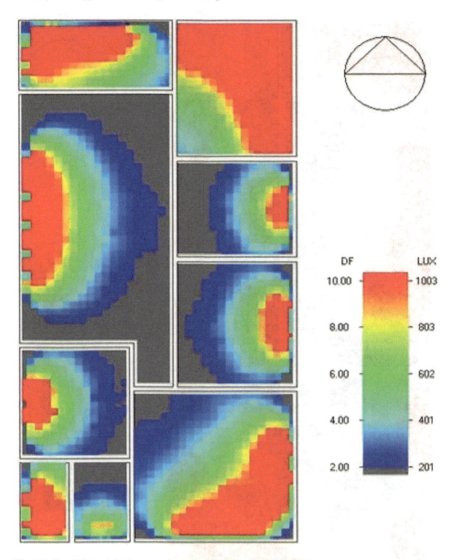

**Fig. 16** Daylight inside the room (aspect ratio 1:2 and WWR40%)

The study is conducted based on the analytical approach using a software tool. Hence, further study involving actual experimentation can be conducted to validate the obtained results.

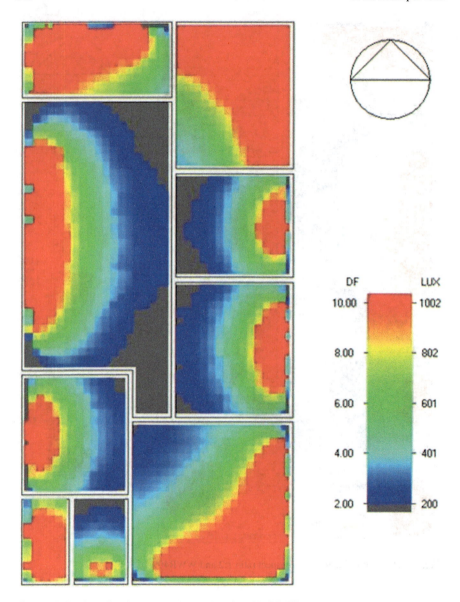

**Fig. 17** Daylight inside the room (aspect ratio 1:2 and WWR50%)

Studies on Energy Efficient Design of Buildings ... 347

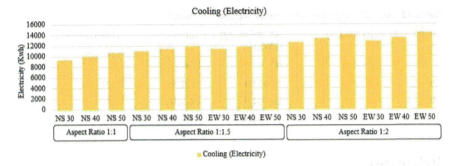

**Fig. 18** Cooling load for all the models

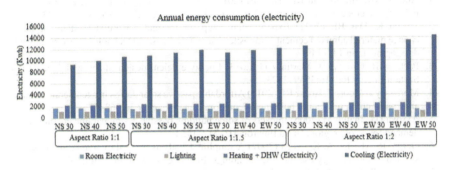

**Fig. 19** Annual energy consumptions for all the models

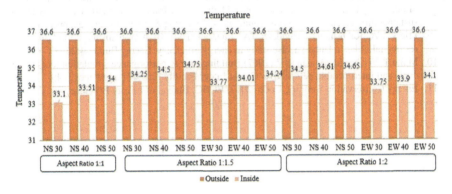

**Fig. 20** Summary of temperature variations between inside and outside environments

**Table 12** Annual energy consumption for NE and EW model-03

| Window to wall ratio (%) | Annual energy consumption (kWh) | | Difference in % |
|---|---|---|---|
| | NS | EW | |
| 30 | 17,674.4 | 18,029 | 2.01 |
| 40 | 18,304.9 | 18,619.6 | 1.72 |
| 50 | 19,244.5 | 19,548.3 | 1.58 |

# References

1. Aneesh NR, Shivaprasad KN, Das BB (2018) Life cycle energy analysis of a metro station building envelope through computer based simulation. Sustain Cities Soc 39:135–143
2. Snehal K, Das BB (2021) Experimental setup for thermal performance study of phase change material admixed cement composites—a review. In: Smart technologies for sustainable development, pp 137–149
3. Pacheco R, Ordonez J, Martinez G (2013) Energy-efficient design of the building. Renew Sustain Energy Rev 16:3559–3573
4. Morrissey J, Moore T, Horne RE (2011) Affordable passive solar design in a temperate climate: an experiment in residential building orientation. Renew Energy 36:568–577
5. Aksoya UT, Inallib M (2006) Impacts of some building passive design parameters on heating demand for a cold region. Build Environ 41:1742–1754
6. Ourghi R, Al-Anzi A, KrartI M (2007) A simplified analysis method to predict the impact of shape on annual energy use for office buildings. Energy Convers Manage 48:300–305
7. Eskin N, Turkmen H (2008) Analysis of annual heating and cooling energy requirements for office buildings in different climates in Turkey. Energy Build 40:763–773
8. Jazayeri A, Aliabadi M (2018) The effect of building aspect ratio on the energy performance of dormitory buildings in cold and semi-arid climates of Iran. In: International conference on sustainability, green buildings, environmental engineering & renewable energy, pp 1–6
9. Friess WA, Rakhshan K, Hendawi TA, Tajerzadeh S (2012) Wall insulation measures for residential villas in Dubai: a case study in energy efficiency. Energy Build 44:26–32
10. Maile T, Fischer M, Bazjanac V (2007) Building energy performance simulation tools a lifecycle and interoperable perspective. Center Integr Facility Eng Working Paper 1–49
11. Energy Conservation Building Code (2017) ECBC, Bureau of Energy Efficiency

# Recent Trends in Environmental and Water Resources Engineering

# Spatial Variability of Organic Carbon and Soil pH by Geostatistical Approach in Deccan Plateau of India

**N. T. Vinod, Amba Shetty, and S. Shrihari**

**Abstract** Proper soil nutrient management is necessary to meet India's rising population without degrading the environment. However, the state of soil organic carbon (SOC) and soil pH is a concern, especially in Indian vertisols that are productive when well managed. Due to a lack of scientific knowledge and poor soil management among the small-scale farmers (most Indian farmers hold less than 2 ha), the crop yield has declined. The current study examines the correlation between soil pH and SOM and their spatial variability in vertisols (Black-Cotton soil). Geostatistics and conventional statistics are used to produce the spatial distribution maps, with the R software and the SpaceStat. Sixty-eight soil samples at the root zone level (0–15 cm depth) are collected from Gulbarga Taluk, Karnataka, India. The random sampling method is adopted according to the agriculture fields distribution, and each sample consists of five subsamples. The soil pH was estimated by pH meter and SOC by Walkley and black method. The violin plots indicate that most soil pH samples range from 8.5 to 7.5 and SOM 0.20–0.50%. The Pearson correlation indicated a negative correlation between the two parameters ($r = -0.21$). In semivariogram analysis, the spherical and exponential models were best fitted for soil pH and SOM, respectively. The ordinary kriging accomplished by a traditional estimator is adapted for generating spatial distribution maps. In line with the negative correlation of soil pH and SOC, the predictable maps are the mirror images. The spatial variability maps give an overview of how extrinsic and intrinsic factors affect the availability of soil pH and SOC. In this region, the parent materials, fertilizers application, and agricultural practices are affecting the soil variability. Small scale farmers should assess these spatial variability maps before applying the fertilizers.

---

N. T. Vinod (✉)
School of Civil Engineering, Reva University, Bengaluru, Karnataka, India
e-mail: vinodnayak2448@gmail.com

A. Shetty
Department of Water Resources and Ocean Engineering, National Institute of Technology Karnataka, Surathkal, India

S. Shrihari
Department of Civil Engineering, National Institute of Technology Karnataka, Surathkal, India

**Keywords** Vertisols · Sustainable environment · Spatial variability · Ordinary kriging

## 1 Introduction

Soil is an important parameter for agricultural practices. The main source of available soil nutrients. Soil health plays a vital role in agricultural practices. Soil is a renewable natural resource which is influenced by many factors that inexorably shift the soil towards future degradation, and it is difficult to reverse these processes without the start of a new geological erosion cycle. It is estimated that the soil depletion in India occurred on 147 Mha of land including 16 Mha of acidification and rest for water erosion, inundation, wind erosion and salinity. The acidification is due to inappropriate agricultural practices. Inappropriate practices by farmers incorporate extreme tillage, unbalanced usage of fertilizers and pesticide, poor irrigation, and poor crop cycle preparation [2]. To understand the variability in soil quality, the quantification of SOC content and soil pH is crucial. The most frequently reported attribute is SOM, and it is chosen as the most essential factor of soil quality and agricultural sustainability [10]. Understanding the complex existence of soil systems (such as SOC and pH) could help clarify soil quality changes that occur in the farmland.

Soil pH is a measurement of soil alkalinity or acidity, indicating the essential chemical and physical properties of soil quality. The alkalinity and acidity alter the presence of soil available nutrients. The SOC is a lively combination affecting a pattern of soil and nutrient properties. SOC content and consistency are known to be a critical factor in the soil's ability to sustain ecosystem services, maintain ecological quality, and improve plant health [13]. If SOC soil concentrations are too limited, then soil mechanics and disrupted soil nutrient cycling pathways can undermine the productive potential of agriculture [12]. With population rise and economy growth, land uses in rural areas have shifted dramatically, which has a big effect on the natural climate. PH and SOC are particularly sensitive to adverse agricultural practices.

In India, the Deccan Plateau is a large plateau in southern India, largely covered with vertisols. Vertisols are a type of soil noticeable by their clay texture and dark colour. Because of stickiness and swelling properties, vertisols are not suitable for agricultural practices; if properly handled, they are more productive. It contributes to disproportion of nutrients when intensively cultivated without understanding the soil state. It induces antagonism of nutrients under these conditions [5].

Proper land supervision is an important tool for increasing agricultural production in sustainable agriculture. Soil management also depends on knowledge of spatial soil variation, particularly soil nutrients. In countries where farms are operated by companies, the scale of the farm is comparatively high and precision farming methods are readily used for sustainable agriculture. In several developing nations, particularly India, a number of aspects, for instance small-scale (below two acres) agricultural land, inadequate maintenance of innovative technical equipment attributed to financial constraints, and lack of technical information to farmers, are contributing to the decline in understanding of sustainable agriculture [16].

An appropriate strategy for assessing the spatial variability of soil properties is done through geostatistics. In India, the geostatistics are to be used for sustainable cultivation, although it has a long background in soil science activities [7, 19]. The spatial variability of soil nutrients in vertisol is vital to recognize. In this regard, it is recognized that soil pH and SOC play vital roles in agriculture, however determining their spatial variability is essential. There are several experiments carried using geostatistics around the world on spatial variation of soil pH and SOC [3, 4, 11] and few research on different areas of Indian farms [1, 6, 15, 17]. In specific, efforts to understand the spatial variations of soil properties on the Indian vertisol continents are marginal, despite poor crop productivity. In this analysis, certain soil samples were taken from small farms, and statistical and geostatistical techniques have been used to analyse soil pH values and SOC content data. The main objective of this work is to determine the spatial variability of soil pH and soil organic carbon, to characterize and recognize the spatial heterogeneity of the state of soil property, and to analyse the factors that influence the variability of soil properties.

## 2 Materials and Methodology

The study area, Kalaburagi taluk, is situated in north part of Karnataka, India, between 76°43′ E and 77°4′ E longitude and 17°27′ N and 17°37′ N latitude. The area of study is recognized as the tur bowel (pigeon pea vessel) of Karnataka (Fig. 1). The Hyperion satellite pass-over is the area selected for the collection of soil samples. The geographical area is 183.8 km$^2$. The region is completely covered with black cotton soil (vertisols). Pigeon pea and jowar are mostly cultivated in this region. The Hyperion satellite pass-over is the area selected for the collection of soil samples.

The sampling of the soil began during the harvest time on the second week of November 2016. Although the fields are distributed over a wide region and are random, the uniform sampling approach was difficult to implement. Topsoil samples (0–15 cm) are obtained at sixty-eight locations in the study area. These points are selected from the cultivated fields of pigeon pea. It was observed during sampling that few farmlands were waterlogged, and residual crops were burnt, as they prepare for another crops, those farmlands are not considered for sampling. The geographical coordinates of the sample positions are registered with Trimble Juno GPS with an accuracy of 2 m. Every soil sample comprises a mixture of five sub-samples with a range of 10 m, collected in plastic carry bag with a shipping label. The soil then air dried for chemical analysis and grounded through a 200-micron sieve. Soil pH is accordingly measured by the pH meter and SOC by Walkley–Black analysis [18].

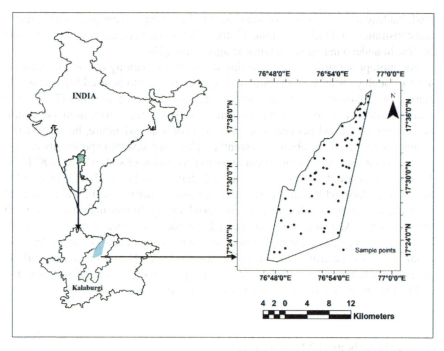

**Fig. 1** Study area and sampling locations

## 2.1 Data Analysis

Descriptive analysis and soil data normality test is performed in R software. The violin plots are plotted. Parameters such as minimum, maximum standard deviation, mean and coefficient of variation are measured. The data were also tested for skewness and kurtosis. Geostatistical study is performed in SpaceStat 4.0® program. The semivariogram is determined using Eq. (1), the semivariance $\gamma(h)$ is calculated at various lags of $h$ [7]. Semivariogram assesses the spatial correlation between data sets. The best semivariogram model is chosen based on the initial flatness of the curve and the lowest MSS error (mean sum of square error). Spatial dependence is determined by nugget to sill ratio.

$$\gamma(h) = \frac{1}{2N(h)} \sum_{\alpha=1}^{N(h)} [z(Y\alpha) - z(Y\alpha + h)]2 \qquad (1)$$

where $z(Y\alpha)$ and $z(Y\alpha + h)$ is the calculated value at $\alpha$th sample location and at the point $(\alpha + h)$th sample location, respectively [4]. Ordinary kriging process is used to obtain spatial heterogeneity maps, since it generates a non-sampled estimation with a low or no influence of outliers [11].

Spatial variability precision is verified by using Eqs. (2) and (3) respectively with a cross-validation procedure using Mean Absolute Error (MAE) and a Root Squared Error (RMSE) [16].

$$\text{MAE} = \frac{1}{N} \sum_{\alpha=1}^{N} z(y_\alpha) - \hat{z}(y_\alpha) \qquad (2)$$

$$\text{RMSE} = \sqrt{\frac{\sum_{\alpha=1}^{N}\left[z(y_\alpha) - \hat{z}(y_\alpha)\right]^2}{N}} \qquad (3)$$

In above equations $\hat{z}(y_\alpha)$ is the predicted value at location $\alpha$.

## 3 Results

### 3.1 Descriptive Statistics of Soil pH and SOC

The descriptive statistics are presented in Table 1. The soil pH in the study region varied from 6.52 to 8.82 with mean value of 7.98. It is noted that soil pH varied from neutral to alkaline in nature. The violin plot of soil pH also indicates most of the data are densely accumulated towards alkalinity (Fig. 2). The alkalinity in soil is inherited due to presence of basalt rock as parent material in study area. The SOC varied from 0.03 to 0.86% with mean of 0.34%. The SOC in the study region is

**Table 1** Descriptive statistics of SOC and soil pH

| Soil property | Mean | Maximum | Minimum | Standard deviation | Coefficient of variation | Kurtosis | Skewness |
|---|---|---|---|---|---|---|---|
| Soil pH | 7.98 | 8.82 | 6.52 | 0.48 | 6 | 0.59 | −0.99 |
| SOC (%) | 0.34 | 0.86 | 0.03 | 0.203 | 58.2 | −0.30 | −0.88 |

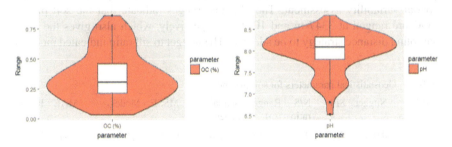

**Fig. 2** Violin plots for soil properties

**Table 2** Persons correlation for soil pH and SOC

| Soil property | Soil pH | SOC |
|---|---|---|
| Soil pH | 1 | |
| SOC | −0.21 | 1 |

below the permissible limits (less than 0.5%), even violine plot indicates the data are accumulated in the lower range. is due to the extensive use of chemical fertilizers by farmers.

The CV is categorized as low (if CV is less than 15%,), moderate (if CV ranges between 15 and 35%) and high (if CV is more than 35%) variability [15]. In the study region, soil pH was low and SOC was highly variable. The data were log transformed to reduce the negative skewness effect. The soil formation processes are influenced by topographical changes effects the variability of soil properties. The Pearson correlation analysis is used to identify the relationship between soil pH and SOC. The negative correlation indicates ($r = -0.21$) (Table 2) that maintaining the soil pH is essential to increase the SOC content. High pH soils also have poor chemical and physical conditions which are unfavourable for growth of crop and the production of the root system, resulting in a smaller amount of organic carbon input to the soil [8].

## 3.2 Geostatistics of Soil pH and SOC

The semivariogram tool is used to analyse the spatial structure for the soil properties. The well-established variograms (Gaussian, Exponential, Spherical, and Linear) are elevated for each property. The parameters included in the semivariogram model are nugget, sill, and range [13]. The degree of spatial dependence is calculated by taking nugget to sill ratio, it can be classified as strong (less than 25%), moderate (25–75%), and weak (more than 75%) [4]. The best semivariogram model is chosen based on the initial flatness of the curve and the lowest MSS error.

The results are tabulated in Table 3. The range is the maximal distance at which parameters are spatially associated. It shows the optimal sampling interval for correct spatial variability assessment. The obtained ranges indicate that the soil pH and SOC are ranged at 3347 m and 1051 m respectively, which also gives the future sampling distance strategy to be adopted. The nugget to sill ratio indicated moderate

**Table 3** Geostatistical parameters for soil pH and SOC

| Soil property | Nugget | Sill | N/S ratio | Range (m) | Spatial dependence | MSSE | Model | MAE | RMSE |
|---|---|---|---|---|---|---|---|---|---|
| Soil pH | 0.089 | 0.17 | 0.52 | 3347.8 | Moderate | 0.32 | Spherical | 0.34 | 0.96 |
| SOC (%) | 0.004 | 0.082 | 0.04 | 1051.6 | Strong | 0.26 | Exponential | 0.24 | 0.83 |

Spatial Variability of Organic Carbon and Soil pH ... 357

and strong spatial dependency for spherical and exponential models, respectively. The different ranges, which are attributed to sampling intensity and study area size, have been compared by Laekemariam et al. [9]. The range recorded differs due to the cumulative impact of agricultural activities, environmental factors, and parental content. In our field of study, the range is varied due to agricultural activities.

The Ordinary kriging is carried for spherical and exponential models as they are the best-fitted models for generating spatial viability maps. The spatial variability maps show that soil pH usually ranged from neutral to alkaline in nature, the southern part of the region is mainly alkaline (Fig. 3), which may be due to the consistency of irrigation water, as well as the rich presence of calcium carbonate [14]. The SOC was

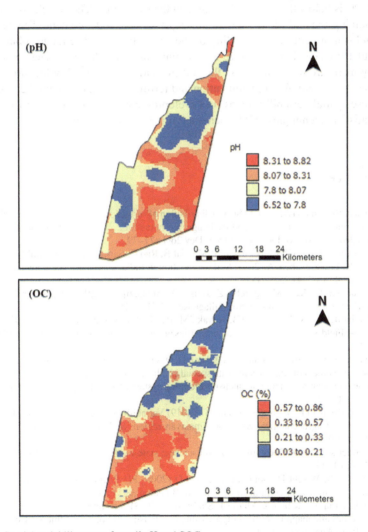

**Fig. 3** Spatial variability maps for soil pH and SOC

comparatively higher on the south side of the study area, even though it is below the permissible limit (Fig. 3). This variability may be due to the use of organic manure and fertilizer application.

## 4 Conclusion

The Soil pH and the SOC are important soil properties to determine soil health. The soil in the study region is basically alkaline in nature and some parts are neutral. The SOC is below the permissible range in the study area. The correlation indicates maintaining the soil pH is vital, which regulates the SOC content. The geostatistics helps in a better understanding of the soil behaviour. The ranges obtained are helpful for researchers in future to maintain sampling distance. The spatial variability maps are useful to assess the soil pH and SOC variability for farmers. The status of soil health for crop cultivation and fertilizer applications will be monitored by these spatial variability maps. In developing countries like India practicing the geostatistical techniques will be benefited for small scale farmers.

## References

1. Behera SK, Shukla AK (2015) Spatial distribution of surface soil acidity, electrical conductivity, soil organic carbon content and exchangeable potassium, calcium and magnesium in some cropped acid soils of India. Land Degr Dev 26(1):71–79
2. Bhattacharyya R, Ghosh B, Mishra P, Mandal B, Rao C, Sarkar D, Das K, Anil K, Lalitha M, Hati K, Franzluebbers A (2015) Soil degradation in India: challenges and potential solutions. Sustainability 7(4):3528–3570
3. Bogunovic I, Mesic M, Zgorelec Z, Jurisic A, Bilandzija D (2014) Spatial variation of soil nutrients on sandy-loam soil. Soil Tillage Res 144:174–183
4. Cambardella CA, Moorman TB, Novak JM, Parkin TB, Karlen DL, Turco RF, Konopa AE (1994) Field-scale variability of soil properties in central iowa soils. Soil Sci Soc Am J 58:1501–1511
5. Chatterjee S, Santra P, Majumdar K, Ghosh D, Das I, Sanyal SK (2015) Geostatistical approach for management of soil nutrients with special emphasis on different forms of potassium considering their spatial variation in intensive cropping system of West Bengal, India. Environ Monit Assess 187(4):1–17
6. Deb P, Debnath P, Denis AF, Lepcha OT (2019) Variability of soil physicochemical properties at different agroecological zones of himalayan region: Sikkim, India. Environ Dev Sustain 21(5):2321–2339
7. Goovaerts P (1999) Geostatistics in soil science: state-of-the-art and perspectives. Geoderma 89(1–2):1–45
8. Kemmitt S, Wright D, Goulding K, Jones D (2006) PH Regulation of carbon and nitrogen dynamics in two agricultural soils. Soil Biol Biochem 38(5):898–911
9. Laekemariam F, Kibret K, Mamo T, Shiferaw H (2018) Accounting spatial variability of soil properties and mapping fertilizer types using geostatistics in Southern Ethiopia. Commun Soil Sci Plant Anal 49(1):124–137

10. Liu X, Herbert SJ, Hashemi AM, Zhang X, Ding G (2011) Effects of agricultural management on soil organic matter and carbon transformation—a review. Plant, Soil Environ 52(12):531–543
11. Liu Z, Zhou W, Shen J, He P, Lei Q, Liang G (2014) A simple assessment on spatial variability of rice yield and selected soil chemical properties of paddy fields in South China. Geoderma 235:39–47
12. Loveland P, Webb J (2003) Is there a critical level of organic matter in the agricultural soils of temperate regions: a review. Soil Tillage Res 70(1):1–18
13. Mao Y, Sang S, Liu S, Jia J (2014) Spatial distribution of PH and organic matter in urban soils and its implications on site-specific land uses in Xuzhou, China. CR Biol 337(5):332–337
14. Srivastava P, Bhattacharyya T, Pal DK (2002) Significance of the formation of calcium carbonate minerals in the pedogenesis and management of cracking clay soils (Vertisols) of India. Clays Clay Min 50(1):111–126
15. Tamburi V, Shetty A, Shrihari S (2020) Characterization of spatial variability of vertisol micronutrients by geostatistical techniques in Deccan plateau of India. Model Earth Syst Environ 6(1):173–182
16. Tamburi V, Shetty A, Shrihari S (2020) Spatial variability of vertisols nutrients in the Deccan plateau region of North Karnataka, India. Environ Dev Sustain 23(2):1–14
17. Vasu D, Singh SK, Sahu N, Tiwary P, Chandran P, Duraisami VP, Ramamurthy V, Lalitha M, Kalaiselvi B (2017) Assessment of spatial variability of soil properties using geospatial techniques for farm level nutrient management. Soil Tillage Res 169:25–34
18. Walkley A, Black IA (1934) An examination of the Degtjareff method for determining soil organic matter, and a proposed modification of the chromic acid titration method. Soil Sci 37(1):29–38
19. Webster R, Oliver MA (2008) Geostatistics for environmental scientists, 2nd edn. John Wiley & Sons

# Hydro-geo Chemical Analysis of Groundwater and Surface Water Near Bhima River Basin Jewargi Taluka Kalburgi, Karnataka

**Prema and Shivasharanappa Patil**

**Abstract** River Bhima is one of the major sources of water in the study area of the Jewargi taluka, Kalburgi district of Karnataka. The rural regions' major livelihood depends on agriculture, and the main water source for drinking irrigation purposes is the surface water of the Bhima River. The Hydro-geo chemical parameters are an important aspect of classifying the quality of water for irrigation and domestic purposes. Surface water analysis from the Bhima river basin near Jewargi taluka Kalburgi district Karnataka was carried out from five sampling points along the river for a five-meter stretches of the river. The samples were collected once a week at every 1 m distance approximately during February to May, which is pre-monsoon, and October to January, which is the post-monsoon season in the year 2019. All these five samples were analysed for pH, TDS, TH, $Ca^{2+}$, $Mg^{2+}$, $Cl^-$, Nitrate, Sulphate, dissolved oxygen (DO), Alkalinity, Sodium, Potassium as per the APHA manual. The concentration of ions of all the samples was determined by using hydro-geochemical analysis of Piper diagram, Gibb's diagram, and USSL diagram. From the analysis, it is noticed that all the samples were moderately hard water and rock dominance during both pre- and post-monsoon season. According to Piper's plots, the water samples belong to the permanent hardness that is $Ca^{2+}$–$Mg^{2+}$–Cl–$SO_4^{2-}$ types of water. As per USSL diagram, the surface water samples are of $C_3S_1$ type, which are not suitable for agriculture purposes because of its high salinity and low sodium water quality. Gibb's diagram shows that the samples belong to rock dominance in both the seasons. Due to this rock dominance, the surface water is also influencing the characteristics of groundwater. Rock Weathering is the main source for this chemical composition of the Bhima river basin in the study area. The results showed that water is having a medium hardness in both pre-monsoon and post-monsoon seasons in all the five points of the study area. This study improves to know the present situation and study can be referred to improve the surface water quality of Bhima River in Jewargi taluka of Kalburgi district of Karnataka.

---

Prema (✉) · S. Patil
PDA College of Engineering, Kalburgi, Karnataka, India
e-mail: prematengli52@gmail.com

**Keywords** Surface water · Piper diagram · Gibb's diagram · Hydro-geochemical · Bhima river

## 1 Introduction

A rapid increase in population has increased the demand for water by industrialization and urbanization. As per WHO, about 80% of diseases in living organisms especially in human beings are caused by water. As the demand for water increased nowadays leading to scarcity of water worldwide. The groundwater reservoir is a reliable source of drinking water than surface water as it is available almost everywhere and also less contaminated than surface water. As a result, groundwater analysis is also assumed a top priority nowadays [1]. From 1123 BCM (Billion cubic metre) of usable water, about 38.55% is groundwater in India [2]. Graphical interpretation of chemical analysis makes a complex water system simpler and quicker. Chemistry of water is represented by methods like Piper diagram, US salinity diagram, Gibb's diagram, etc. [3, 4]. Surface water is in use for domestic use, industrial use, water supply for agriculture throughout the world. Surface water bodies are the major natural resources for human development, which are being polluted by improper disposal of sewage, industrial wastewater, and abundance of human activities, that affects physical and chemical composition of water bodies. The study area is a zone with a dense population and agricultural activities. There is no proper drainage system in this region, hence the sewage water reaches directly into the surface water of the Bhima river and some part of it will join the groundwater through infiltration hence it has a large impact on groundwater quality. The present studies have been undertaken to estimate the health of the ecosystem of Bhima River with reference to physical and chemical properties of river water. The main objectives are to characterize the Physico-chemical analysis of ground and surface water and the Hydro-Geochemical analysis for irrigation water quality.

### 1.1 Study Area

Sahyadri is the origin of the Bhima river which is on the Western side of Western Ghats of Bhimashanaka hills near Karjat, Maharashtra state India. The flow of the Bhima River is southeast through Maharashtra and Karnataka states. The study area considered is the Bhima river basin near the Jewargi taluka Kalburgi district 25 km away from the Kalburgi district on NH218 [1]. Kalburgi is the divisional Head-Quarters of 6 Districts of Bidar, Gulbarga, Yadgir, Raichur, Koppal, and Bellary. The location map of the study area is shown in Fig. 1. Geographically, Jewargi is located between 17.02A, North latitude, and 76.77° East longitude at a height of 393 m (1289 ft) above mean sea level [5]. The black cotton soil is predominantly available due to weathering to basalt rocks in the study area [6]. The study area is comprised of

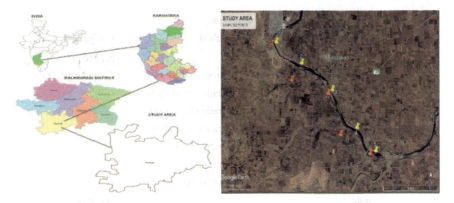

**Fig. 1** Location map of the Bhima river basin

villages nearby the Bhima river basin of Jewargi taluka which are Kolkur village, Raddewadgi village, and Katti Sangavi village. The sampling locations are selected in such a way that they cover the study area uniformly.

## 2 Material and Methodology

All the reagents used for analysis are prepared by using the standard method prescribed in the NEERI manual [7]. Physico-chemical parameters such as Hardness, Calcium, Magnesium, Alkalinity, $HCO_3$, DO, Cl, TDS, $SO_4$, $NO_3$, etc., are determined using standard methods prescribed in NEERI manual. The samples for the analysis of groundwater were collected from various points near the Bhima River basin. Figure 1 shows the location of sampling points A, B, C, D, and E are sampling points for groundwater and surface water sampling points are shown as point 1, 2, 3, 4 and 5. Table 1 shows the methods adopted and instruments used in the Physico-chemical analysis of samples. The samples were collected in pre-monsoon (Feb to May) and post-monsoon (Oct to Jan) seasons of the year 2019. Samples were collected using a grab sampling technique using 2.5liters of white polythene containers on every 6th day in the pre-monsoon and post-monsoon seasons. The samples were analysed for Physico-chemical analysis. Further, the analysed results are used to plot the hydro-geo chemical analysis like Piper diagram, US salinity diagram and Gibb's diagram with the help of AQQA software, Minitab software and MS Excel spreadsheet.

**Table 1** Parameters, methods and equipment employed

| Sl. No. | Parameters | Method | Equipment |
|---|---|---|---|
| 1 | pH | – | pH Meter |
| 2 | Total dissolved solids (TDS) | Gravimetric Method | China Dishes, Oven |
| 3 | Alkalinity | Titration by $H_2SO_4$ | – |
| 4 | Chloride | Titration by $AgNO_3$ | – |
| 5 | TH (total hardness) | Titration by EDTA | – |
| 6 | $Ca^{2+}$ (Calcium) | Titration by EDTA | – |
| 7 | $Mg^{2+}$ (magnesium) | Titration by EDTA | – |
| 8 | DO | Titrated by $Na_2S_2O_3$ | – |
| 9 | Nitrate | – | Spectrophotometer |
| 10 | Sulphate | Turbidimetric method | Spectrophotometer |
| 11 | Sodium | – | Flame photometer |
| 12 | Potassium | – | Flame photometer |

*All the values are in mg/l except pH

## 3 Results and Discussions

Maximum, minimum values were for all samples during pre-monsoon and post-monsoon were calculated and tabulated in Table 2. For surface water and Table 3 for groundwater samples. From the analysis, it is found that pH is alkaline during both the season, there is a slight increase in total hardness, chloride, alkalinity, nitrate,

**Table 2** Descriptive statistics of surface water samples during the year 2019

| Parameters | Pre-monsoon season 2019 | | Post-monsoon season 2019 | |
|---|---|---|---|---|
| | Max | Min | Max | Min |
| pH | 8.6 | 7.5 | 8.5 | 7.5 |
| TH | 223 | 130 | 243 | 195 |
| $Ca^{2+}$ | 156 | 124 | 168 | 130 |
| $Mg^{2+}$ | 69 | 61 | 78 | 63 |
| $Cl^-$ | 241 | 214 | 259 | 241 |
| TDS | 780 | 540 | 950 | 700 |
| $HCO_3^-$ | 247 | 220 | 269 | 231 |
| $Na^+$ | 99 | 74 | 88 | 52 |
| $K^+$ | 13 | 10 | 15 | 12 |
| $SO_4^{2-}$ | 99 | 90 | 109 | 95 |
| $NO_3^-$ | 36 | 30 | 42 | 32 |
| BOD | 3.1 | 1.8 | 3.8 | 2.01 |
| DO | 9.6 | 7.3 | 8.8 | 7 |

**Table 3** Descriptive statistics of groundwater samples during the year 2019

| Parameters | Pre-monsoon season 2019 | | Post-monsoon season 2019 | |
|---|---|---|---|---|
| | Max | Min | Max | Min |
| pH | 7.9 | 7.2 | 8.3 | 7.2 |
| TH | 112.3 | 90.2 | 124.5 | 97 |
| $Ca^{2+}$ | 46.9 | 39 | 55.9 | 41 |
| $Mg^{2+}$ | 65.4 | 51 | 69.2 | 56 |
| $Cl^-$ | 128 | 114 | 168 | 120 |
| TDS | 610 | 415 | 752 | 500 |
| Alk($HCO_3^-$) | 228 | 210 | 248 | 214 |
| $Na^+$ | 25.9 | 23 | 28 | 24.2 |
| $K^+$ | 12.65 | 9.5 | 14.1 | 10.5 |
| $SO_4^{2-}$ | 33.8 | 31 | 40.2 | 32 |
| $NO_3^-$ | 28.6 | 26.3 | 32.8 | 26.3 |
| DO | 7.9 | 6.4 | 8.5 | 6.66 |

sulphate, and potassium from pre-monsoon to post monsoon season for groundwater samples and all the Characteristics are within the acceptable limit of BIS standards for drinking water. Whereas for the surface water TDS of all the sampling point increases from pre-monsoon to post-monsoon, the pH is between 6.5 and 8.5 and it represents that the surface water is suitable for domestics purposes, and all the samples were within the desirable limit in the selected study location.

## 3.1 Hydro-Geochemical Analysis of Groundwater

### 3.1.1 Piper's Diagram

The results of the Physicochemical analysis of samples were plotted in a piper plot, which is a trilinear diagram. Percentage epm (equivalent per millions) of principle cations and anions proportions are plotted in this diagram [1, 4] as shown in Figs. 2 and 3 for both surface and Groundwater samples. Figure 4 shows the subdivisions of piper plots. By comparing the piper plots with subdivisions it is observed that the chemical composition of the sample points, comes in the sub-divisions of 05 and 9 as shown in Table 4, which represents Secondary alkalinity with Carbonate hardness exceeds 50% and No cat ion-anion pair exceeds 50%. On Piper plot, it is seen that all samples belong to $Ca^{2+}$–$Mg^{2+}$–$SO_4^{2-}$ type of water, which represents permanent hardness.

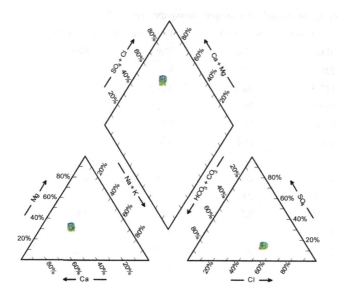

**Fig. 2** Piper diagram for surface water samples

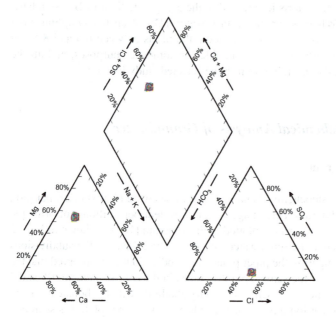

**Fig. 3** Piper Diagram for groundwater samples

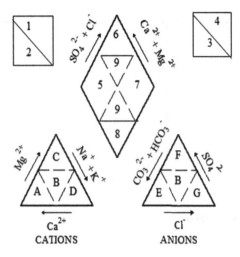

**Fig. 4** Sub-divisions of Piper diagram

**Table 4** Subdivisions of Piper's diagram

| Area | Subdivision |
|---|---|
| 1 | Alkaline earth exceeds alkalis |
| 2 | Alkalis exceed alkaline earths |
| 3 | Weak aids exceeds strong acids |
| 4 | Strong acids exceeds weak acids |
| 5 | Carbonate hardness (secondary alkalinity) exceeds 50% |
| 6 | Non carbonate hardness (secondary salinity) exceeds 50 |
| 7 | Non carbonate alkali (secondary salinity) exceeds 50% |
| 8 | Carbonate alkali (primary alkalinity) exceeds 50% |
| 9 | No one cation–anion pair exceeds 50% |

### 3.1.2 Gibb's Diagram

Gibb's diagram represent the ratio of ($Na^+/(Na^+ + Ca^{2+})$) and ($Cl^-/(Cl^- + HCO_3)$) as a function of Total Dissolved Solids, are used to determine the major source of dissolved chemical-constituent; those are precipitation, rock and evaporation-dominance [8]. Gibb's diagrams plotted for surface and groundwater are shown in Figs. 5 and 6. From the diagram, it is observed that most of the samples come under a rock- dominance type of the study area, and hence the rock types in all sampling points control water contamination in both seasons. Rock weathering is the main source for this chemical composition of the water in the study area.

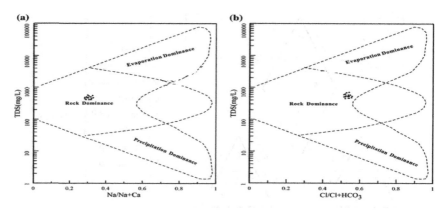

**Fig. 5** Gibb's diagrams for surface water samples

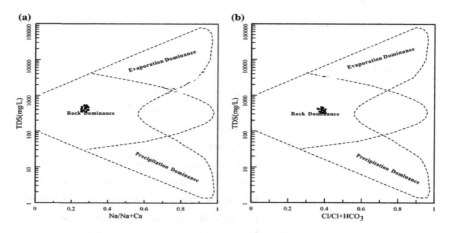

**Fig. 6** Gibb's diagrams for groundwater samples

### 3.1.3 U.S. Salinity Laboratory Classifications

The USSL diagrams plotted for pre-monsoon and post-monsoon season for groundwater are shown in Figs. 7 and 8 and USSL diagrams plotted for pre-monsoon and post-monsoon season for Surface water is shown in Figs. 9 and 10 for Pre-monsoon and Post-monsoon season. Which are plotted by taking SAR (Sodium Adsorption Ratio) verses EC (Electrical Conductivity). Groundwater sample of pre-monsoon belongs to C2S1 type, which represents 58.8% medium saline–low sodium water of the collected irrigated water samples, this type of water can be used if a moderate amount of leaching occurs with little danger of producing harmful levels of sodium. Where as in the post- monsoon 100% of samples belong to C3S1 type, that is high salinity and low sodium water. This class of water cannot be used on soils with restricted drainage otherwise, there is a possibility of development of harmful level

Hydro-geo Chemical Analysis of Groundwater and Surface ... 369

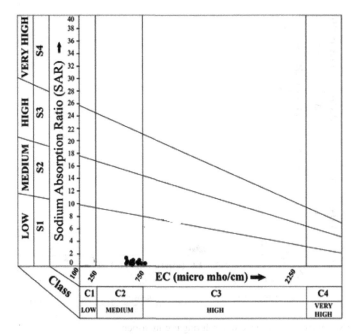

**Fig. 7** USSL diagram for groundwater during pre-monsoon

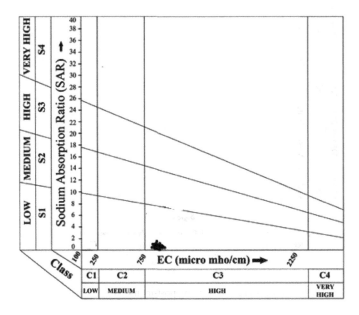

**Fig. 8** USSL Diagram for groundwater during post-monsoon

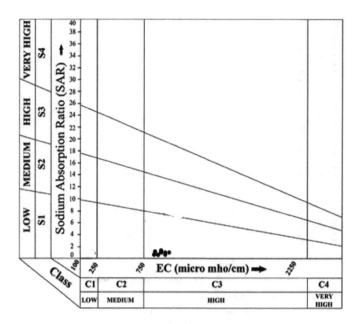

**Fig. 9** USSL Diagram for surface water during pre-monsoon

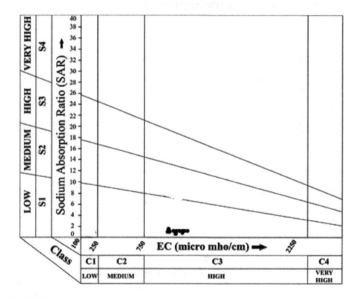

**Fig. 10** USSL Diagram for surface water during post-monsoon

of sodium in the soil. Similarly, all the samples collected from surface water points are of C3S1 type both during pre-monsoon and post-monsoon seasons.

## 4 Conclusion

After the careful analysis of physico-chemical characteristics and hydro-geochemical analysis, the conclusions were drawn. Water has a medium hardness in both pre-monsoon and post-monsoon seasons in all five points of the study area. The concentrations of sulphate, nitrates, potassium and sodium in both seasons are within the permissible limit of the BIS standard for drinking water. In the Piper plot, the characteristics of the sampling points fall in chloride and magnesium domains. Based on Gibb's diagram, samples under the study area fall under the classification of rock dominance. Hence, it can be recommended to use for agriculture (irrigation) purposes in the study location. The groundwater proves to be the best alternative for drinking and agriculture purpose of Jewargi taluka. However, the surface water in the study area is contaminated due to excessive use of fertilizers, chemicals, domestic effluents and dumping of solid waste and faecal matters, etc. due to the self-cleansing capacity of river Bhima, the characteristics of the surface water quality are well within the permissible-limit and suitable for domestic and agriculture purposes. As per the USSL diagram in the post-monsoon, 100% of samples belong to the C3S1 type of high salinity low sodium water. This type of irrigation water cannot be used on soils with restricted drainage facility or else, harmful level of sodium may develop. All the surface water samples belong to C3S1 type during pre-monsoon and post-monsoon season, high salinity, and low sodium water. The present study can be referred to improve the surface water quality of Bhima River in the Jewargi taluka of the Kalburgi district of Karnataka.

## References

1. Shivasharanappa, Yalkpalli A (2012) Hydro-geochemical analysis of Bhima River in Gulbarga district, Karnataka state, India. IOSR J Eng 2(4):862–882
2. Zolekar RB, Todmal RS (2020) Hydro-chemical characterization and geospatial analysis of groundwater for drinking and agricultural usage in Nashik district in Maharashtra, India. Environ Dev Sustain
3. Mirza A, Tanvir Rahman M, Saadat M, Islam MS, Al-mansur MA, Ahmed S (2017) Groundwater characterization and selection of suitable water type for irrigation in the western region of Bangladesh. Appl Water Sci 7(1):233–243
4. Lokhande PB, Mujawar HA (2016) Graphic interpretation and assessment of water quality in the Savitri River Basin. Int J Sci Eng Res 7(3):1113–1123
5. Shivasharanappa, Prema (2018) Ground water quality of Bhima River Basin near Jewargi Taluka Gulbarga district. Int J Res Eng Technol 7(8):42–44
6. Tamburi V, Shetty A, Shrihari S (2020) Spatial variability of vertisols nutrients in the Deccan plateau region of north Karnataka, India. Environ Dev Sustain 23(2):1–14

7. Laboratory manual on water analysis, National Environmental Engineering Research Institute (1987)
8. Chowdhury AK, Gupta S (2011) Evaluation of water quality, hydro-geochemistry of confined and unconfined aquifers and irrigation water quality in Digha Coast of West Bengal, India (a case study). Int J Environ Sci 2(2):576–589

# Conjunctive Use Modeling Using SWAT and GMS for Sustainable Irrigation in Khatav, India

**Ranjeet Sabale and Mathew K. Jose**

**Abstract** The present study engages two powerful models viz., Soil and Water Assessment Tool (SWAT) and Groundwater Modeling System (GMS) to simulate groundwater and surface water conditions to aid conjunctive use applications. This work elaborates a robust coupling of these two models for conjunctive use studies. The study area, Khatav in Maharashtra, is mostly drought-prone, semi-arid and has an average annual rainfall of about 560 mm. In order to have a sustainable and efficient water resource management in this semi-arid region, conjunctive use of groundwater and surface water is essential. Therefore, the SWAT and GMS models were calibrated and validated by using observed data for a period of twelve years, from 2000 to 2012. The basin area considered for the study purpose is 470 km$^2$. After the SWAT model has been processed, study area was divided into 51 hydrological response units (HRUs) and 5 sub-basins. The methodology used in this work was, the Soil Conservation Service (SCS) curve number was used to estimate surface runoff. As the SWAT model cannot be used to predict groundwater levels in the basin, the Groundwater Modeling System (GMS) has been used to model the groundwater processes. The sensitivity of work was assessed by using SWAT-CUP tool and with SUFI-2 algorithm. The results showed that the average annual flow from the basin is 3.059 × 10$^3$ m$^3$ and average curve number is 87.43. The study recommends recharging of the wells which are already available in the basin by operating them continuously to induce recharging by available surface water. The simulation shows that altogether due to reduction in evaporation and surface absorption, the groundwater level in basin is raised by 0.6 m. The validity of the simulations and acceptance of the SWAT model for the present study has been confirmed by values of RSR (0.60) and NSE (0.82) parameters.

**Keywords** Surface water · Groundwater · Conjunctive use · SWAT · GMS

---

R. Sabale (✉)
VTU, Belagavi, Karnataka, India
e-mail: ranjeetsabale123@gmail.com

M. K. Jose
National Institute of Hydrology, Regional Centre, Belgaum, India

© The Author(s), under exclusive license to Springer Nature Singapore Pte Ltd. 2023
M. S. Ranadive et al. (eds.), *Recent Trends in Construction Technology and Management*, Lecture Notes in Civil Engineering 260,
https://doi.org/10.1007/978-981-19-2145-2_29

## 1 Introduction

Conjunctive use of water means to use of surface and groundwater either in combination or separately to meet water demands. It is not a new technique but it is a less emphasized one. The conjunctive use term describes the integration of surface and groundwater through different hydrogeological processes like deep percolation, hydrological cycle, irrigation methods and water balance equation parameters [19]. The conjunctive use of water results in increase in water yield in low cost than different hydraulic structures like dam and reservoir. The conjunctive use projects are prominently categorized into (1) Stream diversion (2) Reservoir and Dams and (3) Total system [5]. The objectives of conjunctive use are to increase water yield, reliability of supply, improvement in water quality by dilution and overall efficiency of water management system. The conjunctive use of water resources can be used in space or in time. The strategies used in conjunctive use show that it is not necessary that all time groundwater and surface water has to use in conjunction but it can be used separately as per the command requirement. The conjunctive use technique was used to dilute the polluted water and to have systematic irrigation planning in Tehran plain, Iran [9]. The stormwater management model (SWMM) and Groundwater and Surface water Flow Model (GSFLOW) were together used to improve water resource management in water scare regions [23]. The water shortage problem can be minimized with the help of conjunctive use of groundwater and surface water [22]. As study area is semi-arid and water-stressed, the conjunctive use technique is suitable to alleviate the water shortage problem. Sabale and Jose [20] formulated a conjunctive use model using integration of SWAT and MODFLOW, the authors have adopted the artificial recharge technique. The study concludes that, conjunctive use of surface and groundwater leads to improvement in groundwater recharge, Moreover, the groundwater levels in study area were raised by 0.7 m.

In twenty-first century, it is difficult to build new dams and hydraulic structures because of environmental constraints and institutional issues. So, researchers and policymakers from around the globe are working on optimization/simulation of existing water resource systems. From the literature studies, the conjunctive use is the best tool which can be used to improve the hydraulic attributes of aquifers. Also it is able to reduce water logging, salination and pollutant contamination with the help of conjunctive use. So, in this study, an attempt was made to optimize water resources for sustainable water management through conjunctive use. To check feasibility and to optimize conjunctive use, the simulation model was developed by using SWAT and GMS software. The SWAT model works on data provided by remote sensing and GIS and model has been used since the last three decades for the betterment of water resource management [30]. The SWAT model was used to interpret the sediment yield and surface runoff for Lvohe river and according to the study the soil conservation management was adopted [13, 26, 29]. The evapotranspiration plays important role in surface water modeling [17] carried out the evaluation of evapotranspiration using SWAT. The authors concluded that SWAT is an efficient tool although less data is available [27]. Pereira et al. [18] worked on hydrologic simulation on Pomba river basin in Brazil. SWAT model was used to simulate soil scenarios and climate; model was used on daily time step basis, results show runoff of 13.6 and 6.5 mm for two scenarios considered for study. For sustainable conjunctive use, groundwater behavior study is more important, Lee et al. [12] and Zhang et al. [28] formulated a new approach to assess the groundwater behavior under the influence of climate change. The SWAT model has limitations in well-managed watershed, e.g. canal irrigation command area, because the canal seepage is not included in SWAT [25] have tried the SWAT model with modifications and performed on 732 km$^2$ area of Arkansas river. The SWAT model simulates the runoff and sediment yield very precisely; it can be used to manage/plan and to develop water resources sustainably [1].

In this work, an attempt was made to integrate surface and groundwater resources in the Khatav area in Maharashtra, India to aid sustainable irrigation. The significance of conjunctive use is that it consists of the interaction between groundwater and surface water through the recharging process, so it implies the reliable water resources. Moreover, it reduces the waterlogging and soil salination problems. This approach will alleviate the water scarcity issue, Moreover; the net returns from agricultural sectors will be increased so in turn, it will reinforce the nation's economy. The methodology adopted in this study was, the data obtained from SWAT model was used for groundwater modeling in GMS. The SWAT model was used to calculate surface runoff, percolation and evaporation. The GMS model was used to find out groundwater level and its variation according to the suction and recharge process.

## 2 Study Area

Khatav (Vaduj) taluka of Satara district in Maharashtra state, India, having total area 1358 sq.km was selected for study (Fig. 1). The area lies between 17°15′ and 17°45′ north latitudes and 74°15′ to 74°45′ east longitudes. The Ner dam and Yeralwadi dam are the two principal water sources of the study basin. The study area is mostly water-stressed and has annual rainfall about 560 mm. The area was categorized as semi-critical by the Ground Water Resource estimation carried out in 2013 by CGWB. The study basin receives a maximum portion of rainfall from south-west monsoon. The annual temperature varies from minimum 140 to maximum 400 and average wind speed for area is 7.4 km/hr [4]. Agriculture is a major land use aspect of study area. The traditional crops like Jowar, Wheat, Soyabin and Maize with some cash crops like sugarcane and pulses are grown on the area. The current study basin is geomorphologically a part of the Deccan plateau. The physiographical features of study area are older flood plain of Yerala River and basaltic soil. For study area, it is considered that aquifer is single layered and mostly unconfined, so horizontal hydraulic conductivity is estimated for the aquifers.

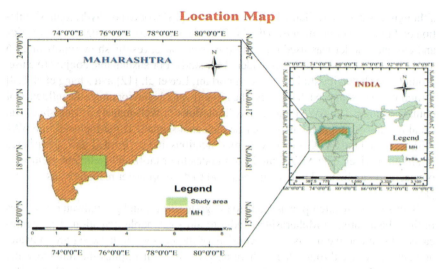

**Fig. 1** Location map of study area

## 3 Materials and Methods

The methodology used in this work, Soil and Water Assessment Tool was calibrated and validated for a period 2000 to 2010 and 2011–2012, respectively. The sequential Uncertainty Fitting-2 algorithm from SWAT-CUP tool was used for sensitivity analysis. The GMS model was used to estimate the groundwater parameters.

The SWAT model requires Digital Elevation Model (DEM), Land use/Land cover map and global weather data and those were taken from USGS earth data center. The number of water wells available in study area are marked and used in GMS model.

### 3.1 SWAT Model

The SWAT model has been developed to simulate the impact of land use on hydro-geological parameters [3]. It was developed by US-Agriculture department and is mostly used to simulate environmental and ecological process [14]. It is semi-distributed, physically based and the conceptual model. The SWAT model encompasses sediment transport, weather, water quality and quantity and agriculture management. The literature shows that SWAT model was used for climate change study and to estimate the impact of climate change on hydrology [21]. The SWAT model and GMS was used by Kareem [10] for simulation of surface and groundwater to aid conjunctive water use in Jolak basin, North Iraq. The SWAT model and

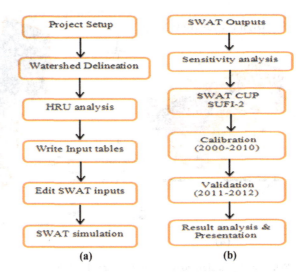

**Fig. 2** Flow chart of SWAT model: **a** run of model **b** model calibration

MODFLOW were successfully coupled and used for conjunctive water use simulation for Firoozabad watershed [6]. The suggestions of SWAT and MODFLOW study were, increase the rain gauge stations and temperature gauges in the basin.

The SWAT model is being semi-distributed, so it is unable to consider storage coefficient and hydraulic conductivity. Kim et al. [11] developed an integrated SWAT-MODFLOW model for sustainable groundwater management in Korea. The SWAT model was used by Venkatesh et al. [24] for hydrological modeling of Manimala river, in Kerala, India shows that the surface runoff was influenced by parameters like curve numbers and ESCO. The SWAT model uses input data such as digital elevation model (DEM), Soil map and land use/land cover (LULC) map of study area to delineate watershed into sub-basins. The sub-basins are again divided in to hydrologic response units (HRUs) that consist unique management of soil data and slope. The HRUs play a vital role in simulation and the data from each HRUs are added together to get basin parameters.

Based on the general rule and research studies, it is suggested to increase number of sub-basins than HRUs. Figure 2 shows the SWAT model flow chart including its run and Calibration–validation process.

### 3.1.1 DEM of Study Area

Digital elevation model explains the general topography and geometry of watershed. The Digital elevation model for current study area with 30 m resolution was downloaded from ASTER global dataset. The DEM data was available in two modules; by the mosaic tool the required DEM was created. The DEM was used to derive the parameters like percentage slope of basin, length of main channel, drainage pattern etc.

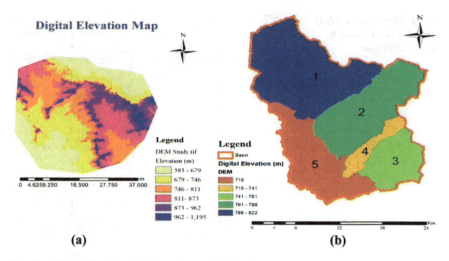

**Fig. 3** **a** Digital elevations (m) of study area **b** sub-basins

After delineation of DEM it is divided into number of sub-basins and HRUs. For current study, the DEM was projected into WGS_1984_UTM ZONE 43N. Figure 3a shows DEM of study area and Fig. 3b explains the number of sub-basins with elevation range.

### 3.1.2 LULC of Study Area

The land use and land cover map (LULC) of study area shows the physical use of area, e.g. forest, urban, bare land, etc., and reports how area is being used. The LULC data was taken from earth-explorer for period (2019–2020); data free from clouds were selected. The Landsat-8 images were used to prepare land use map for study area. For current study, LULC map of study area was developed with Arc SWAT (2012) version and classified according to maximum likelihood classification. Almost maximum study area was covered with high urban area (48.44%) and next to that with agriculture (26.64%), barren (17.73%), low urban (5.58%) and water bodies (1.60%). The maximum likelihood classification of LULC was verified manually from data available by local agriculture development authority. The verification showed good agreement between remote sensing data and manually available data. The authors reclassified the available LULC data into five groups (Fig. 4) for accurate analysis of curve numbers.

### 3.1.3 Soils in Study Area

The soil map for current study area was taken from NBSS and LUP, i.e. National Bureau of Soil survey and land use planning, department of India on the scale of

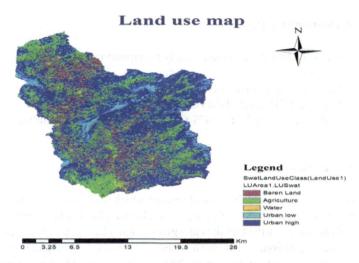

**Fig. 4** Land-use map of study area

1:250,000. The data was available in detailed manner showing the texture profile of soil and then it was digitized according to model use. Figure 5 Shows details of soils present in study area. The Loam soil (Nd51-2b-3820, 56.98%) was major soil in basin and next to that Clay (Vc43-3ab-3861, 22.55%), Clay loam (Hh11-2bc-3711, 15.73%) and Loam soil (I-Hh-3721, 4.74%) were present in basin.

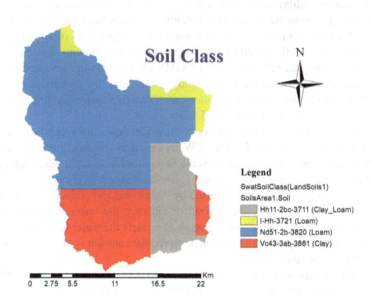

**Fig. 5** Soil classification of study area

### 3.1.4 Meteorological Data

The daily based meteorological data namely precipitation, solar radiation, temperature (max and min),

relative humidity and wind was collected from global weather data for period (2000–2014). The IMD gridded data and CFSR world data were updated in SWAT database. In the current study, the SCS-CN method was used to simulate runoff.

### 3.1.5 SWAT Simulation

SWAT is Agro-hydrological, basin-scale model which works on daily and sub-daily time step. In current work, SWAT model was calibrated for 470 km$^2$ area in Khatav. The model encompasses the GIS data and gives outputs like surface runoff, sediment yield, evapotranspiration, soil properties and many more hydrological parameters. The meteorological data for the period (2000–2014) was used to run the model.

## 3.2 Groundwater Modeling System (GMS)

The GMS is a very effective pre- and post-processor and it offers various groundwater operations [7]. It works in three steps like conceptual model design, data and construction and calibration of model. The polygon, arc and points are used to define features like pumping wells, streams, boundaries, rivers and piezometers. The calibration of model is a process used to compare the outputs and observed data. The calibration facilitates the modification in input data until it does not match to outputs. Marnani [15] coupled MODFLOW with GMS for groundwater resource study and management in Firozabad plain. The study reveals that the numerical model was helpful to simulate the groundwater flows and outcomes were used to predict water levels in Firozabad plain for the next 5 years.

Gurwin and Poprawski [8] developed a numerical model for groundwater renewal in the Odra river basin. The MODFLOW was used to simulate groundwater flows and the results shows that renewal of shallow aquifers were rapid than deep aquifer. The average time for aquifer to recharge and discharge was about 130 years. The MODFLOW and GMS model were used by Aghlmand and Abbasi [2] with limited available data for Birjand plain in Iran to enhance the water management. The outcomes of study show, due to integration of models the water tables in study area were raised and the agreement between observed data and simulated values indicates the acceptance of model.

The SWAT has semi-distributed nature and has limitations for groundwater modeling. The groundwater levels are not mapped with SWAT also canal seepage is not analyzed. Therefore GMS model was used in this work to simulate groundwater characteristics. The GMS was developed (1998) by the department of defense, to facilitate groundwater modeling. The GMS is 3D model and the flow is explained

Conjunctive Use Modeling Using SWAT and GMS ... 381

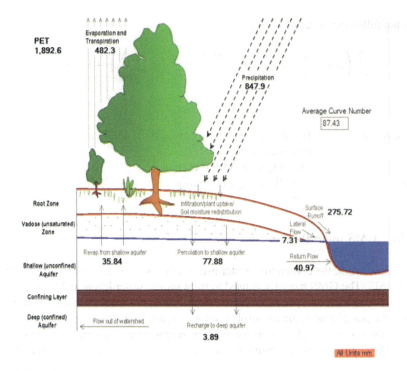

**Fig. 6** Hydrology of study area

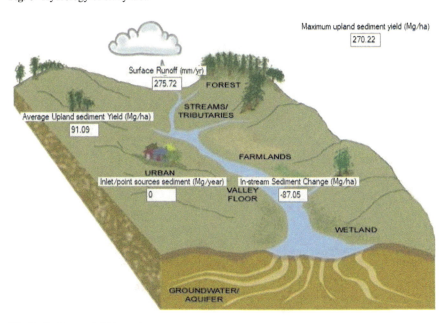

**Fig. 7** Sediment yield in study area

by the following equations [16],

$$\frac{\partial}{\partial x}\left(Kxx\frac{\partial h}{\partial x}\right) + \frac{\partial}{\partial x}\left(Kyy\frac{\partial h}{\partial y}\right) + \frac{\partial}{\partial x}\left(Kzz\frac{\partial h}{\partial z}\right) - W = Sc\frac{\partial h}{\partial t} \quad (1)$$

where

| | |
|---|---|
| Kxx, Kyy and Kzz | Hydraulic conductivity along X, Y and Z axes. |
| h | Hydraulic head (m). |
| W | Quantity of aquifer recharge. |
| Sc | Specific storage of aquifer. |
| T | Time. |

### 3.2.1 GMS Input Data

In current work, the conjunctive water use practice was tried with the help of SWAT and GMS. The GMS model was used to trace groundwater levels and to suggest the recharge pits. The study suggests some wells which are already present in basin as a recharge point to accumulate surplus runoff. Figure 8, shows the available wells in basin and suggests the locations for recharge points. In the current work the value of the coefficient of transmissibility and storage coefficient was taken from the CGWB

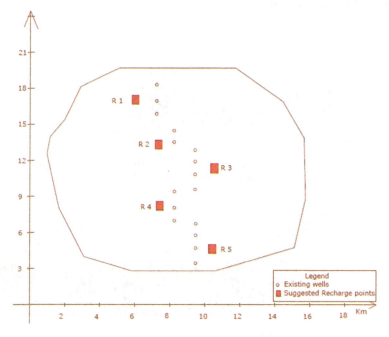

**Fig. 8** Location of wells

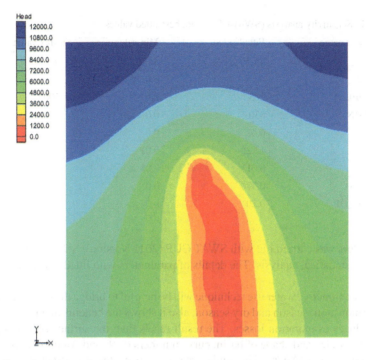

**Fig. 9** Groundwater contour map of area

report [4], the values were 1800 m²/d and 0.1 respectively. The hydraulic head of area was decreasing with length of site as shown in Fig. 9. The GMS model use input data as elevations of bottom and top layers of aquifers, coefficient of transmissibility and storage coefficient. The Yerala river was considered as constant head boundary which recharges the aquifers of study area. The 16 available wells, as shown in Fig. 8, were assumed to be operating at a rate (10 lit/s). The time duration for well operation was considered as 12 h/ day.

## 4 Results and Discussions

The surface water model, physical-based and semi-distributed SWAT was prepared in this study by using GIS data. The aim of this work was to imply the conjunctive use process for sustainable and efficient irrigation management in Khatav area. The methodology used was SWAT model was coupled with GMS model, also authors have tried to evaluate interrelationship between surface and ground water. The outcomes will be useful to balance, store the runoff in aquifer and that can be used in water deficit period (Fig. 6). This study also suggests the five recharge pits nearby 16 well which are already available in basin (Fig. 8). The sensitivity analysis for 9 hydrological

**Table 1** Sensitivity analysis (SWAT-CUP) with best-fitted values

| Parameter name | Fitted value | Min value | Max value |
|---|---|---|---|
| CN2 | 45.45 | 35 | 98 |
| ALPHA_BF | 0.55 | 0.0 | 1 |
| GW_DELAY | 203 | 0.0 | 500 |
| GWQMN | 330 | 0.0 | 5000 |
| ESCO | 0.41 | 0.0 | 1 |
| CH_N1 | 13.26 | 0.01 | 30 |
| CH_N2 | 0.017 | 0.01 | 0.30 |
| SOL_AWC | 0.37 | 0.0 | 1 |
| RCHRG_DP | 0.20 | 0.0 | 1 |

parameters was carried out with SWAT CUP (2012 Version), SUFI-2 algorithm was used for statistical analysis. The details of parameters with fitted values are given in Table 1

The conjunctive water use technique will be helpful to bridge the time gap between rainy (monsoon) season and dry season; also it shows the benefits of aquifer recharge and reduces evaporation losses. The result revels that, the surface runoff is greatly affected by size and shape of basin, curve numbers (CN) and Antecedent Soil moisture conditions (AMCs). The study area is being water stressed, the total runoff produced by basin was $3.059 \times 10^3$ cumec for (2003–2014), and that can be used to solve the water scarcity problem. The sediment yield from basin for time period (2003–2014) was $78.40 \times 10^4$ tones (Fig. 7). The average curve number (CN) of area was 87.43, annual average surface runoff was 275.72 mm and groundwater delay period was 203 days.

The land use aspect of present study area shows maximum proportion of urban and barren. Therefore curve number is more resulting sufficient runoff. The area suffers from critical soil erosion, so, SWAT will be useful to locate erosion-prone areas. The GMS simulates the groundwater flows, indeed, this conjunctive approach of SWAT and GMS will lead to plan and allocate water resource sustainably, to enhance crop production and also to mitigate environmental issues.

## 5 Conclusion

After the robust simulation work, this study concludes that the SWAT and GMS models can be efficiently used in data scare and water stressed watershed like Khatav for efficient and sustainable irrigation management. The outcomes of this work will be helpful to stakeholders, water managers and policy makers to reinforce the current irrigation and water resource management in Khatav and for similar regions. At low laying areas, it was seen that the water level rose by 0.6 m after the steady state is reached in one year. The sediment yield depicts the physical characters of basin as well as shows the critical basins for soil erosion. Altogether, this study was an attempt to alleviate the water scarcity issue in semi-arid region through conjunctive use modeling and to plan sustainable irrigation.

# References

1. Abebe T, Gebremariam B (2019) Modeling runoff and sediment yield of Kesem dam watershed, Awash basin, Ethiopia. SN Appl Sci 1(5):1–13
2. Aghlmand R, Abbasi A (2019) Application of MODFLOW with boundary conditions analyses based on limited available observations: a case study of Birjand plain in East Iran. Water (Switzerland) 11(9)
3. Arnold JG, Moriasi DN, Gassman PW, Abbaspour KC, White MJ, Srinivasan R, Santhi C, Harmel RD, Van Griensven A, Van Liew MW, Kannan N, Jha MK (2012) SWAT: model use, calibration, and validation. Trans ASABE 55(4):1491–1508
4. CGWB 2017 (2017) Report of Central Ground Water Board 2017 Aquifer maps and ground water management plan
5. Coe JJ (1990) Conjunctive use—advantages, constraints, and examples. J Irrig Drain Eng 116(3):427–443
6. Dowlatabadi S, Ali Zomorodian SM (2016) Conjunctive simulation of surface water and groundwater using SWAT and MODFLOW in Firoozabad watershed. KSCE J Civ Eng 20(1):485–496
7. Gogu RC, Carabin G, Hallet V, Peters V, Dassargues A (2001) GIS-based hydrogeological databases and groundwater modelling. Hydrogeol J 9(6):555–569
8. Gurwin J, Poprawski L (1999) Groundwater renewal in the Middle Odra River catchment based on flow model and isotopic data—the Lubin Glogow Copper Region, Poland. Acta Hydrochim Hydrobiol 27(5):383–391
9. Karamouz M, Kerachian R, Zahraie B (2004) Monthly water resources and irrigation planning: case study of conjunctive use of surface and groundwater resources. J Irrig Drain Eng 130(5):391–402
10. Kareem IR (2015) Conjunctive use modeling of surface water and groundwater in The Jolak Basin, North Iraq. J Kerbala Univ 13(1):236–246
11. Kim NW, Chung IM, Won YS, Arnold JG (2008) Development and application of the integrated SWAT-MODFLOW model. J Hydrol 356(1–2):1–16
12. Lee J, Jung C, Kim S, Kim S (2019) Assessment of climate change impact on future groundwater-level behavior using SWAT groundwater-consumption function in Geum River Basin of South Korea. Water (Switzerland) 11(5)
13. Lingling L, Renzhi Z, Zhuzhu L, Weili L, Junhong X, Liqun C (2014) Evolution of soil and water conservation in rain-fed areas of China. Int Soil Water Conserv Research 2(1):78–90
14. Luo Y, He C, Sophocleous M, Yin Z, Hongrui R, Ouyang Z (2008) Assessment of crop growth and soil water modules in SWAT2000 using extensive field experiment data in an irrigation district of the Yellow River Basin. J Hydrol 352(1–2):139–156
15. Marnani (2010) Groundwater resources management in various scenarios using numerical model. Am J Geosci 1(1):21–26
16. McDonald MG, Harbaugh A (1988) A modular three-dimensional finite-difference groundwater flow model
17. Parajuli PB, Jayakody P, Ouyang Y (2018) Evaluation of using remote sensing evapotranspiration data in SWAT. Water Resour Management 32(3):985–996
18. Pereira DR, Martinez MA, da Silva DD, Pruski FF (2016) Hydrological simulation in a basin of typical tropical climate and soil using the SWAT model part II: simulation of hydrological variables and soil use scenarios. J Hydrol: Reg Stud 5:149–163

19. Ramesh H, Mahesh A (2012) Conjunctive use of surface water and groundwater for sustainable water management. In: Sustainable development—energy, engineering and technologies—manufacturing and environment, July 2016
20. Sabale R, Jose M (2021) Hydrological modeling to study impact of conjunctive use on groundwater levels in command area. J Indian Water Works Asso 53(3):190–197
21. Saraf VR, Regulwar DG (2018) Impact of climate change on runoff generation in the upper Godavari river basin, India. J Hazar Toxic Radioactive Waste 22(4):04018021
22. Singh A (2014) Conjunctive use of water resources for sustainable irrigated agriculture. J Hydrol 519(PB):1688–1697
23. Tian Y, Zheng Y, Wu B, Wu X, Liu J, Zheng C (2015) Modeling surface water-groundwater interaction in arid and semi-arid regions with intensive agriculture. Environ Modell Softw 63:170–184
24. Venkatesh B, Chandramohan T, Purandara BK, Jose MK, Nayak PC (2018) Modeling of a river basin using SWAT model. Water science and technology library
25. Wei X, Bailey RT, Tasdighi A (2018) Using the SWAT model in intensively managed irrigated watersheds: model modification and application. J Hydrol Eng 23(10):04018044
26. Yesuf HM, Assen M, Alamirew T, Melesse AM (2015) Modeling of sediment yield in Maybar gauged watershed using SWAT, northeast Ethiopia. Catena 127:191–205
27. Yu M, Chen X, Li L, Bao A, de la Paix MJ (2011) Streamflow simulation by SWAT using different precipitation sources in large arid basins with scarce raingauges. Water Resour Manage 25(11):2669–2681
28. Zhang A, Zhang C, Fu G, Wang B, Bao Z, Zheng H (2012) Assessments of impacts of climate change and human activities on runoff with SWAT for the Huifa River Basin, Northeast China. Water Resour Manage 26(8):2199–2217
29. Zhang XS, Hao FH, Cheng HG, Li DF (2003) Application of swat model in the upstream watershed of the Luohe River. Chin Geogra Sci 13(4):334–339
30. Zhao WJ, Sun W, Li ZL, Fan YW, Song JS, Wang LR (2013) A review on SWAT model for stream flow simulation. Adv Mater Res 726–731:3792–3798

# Modelling and Simulation of Pollutant Transport in Porous Media—A Simulation and Validation Study

**M. R. Dhanraj and A. Ganesha**

**Abstract** In this paper, the transport of pollutants is simulated with COMSOL multi-physics in a transition-based cross-sectional model to trace the path and flow of pollutants after infiltration from the point of discharge of treated domestic sewage. The study is investigated using soil columns with known porosity and permeability and with various other boundary conditions. The solutions are tested using a finite-element numerical model built with COMSOL Multi-physics. It is found that the simulation model holds good in pollutant removal in comparison with the physical pilot scale model. It is also found that there is a strong correlation between the inlet and outlet parameters in removing the pollutants with naturally available media (soil). Also, the removal efficiency in BOD and COD in soil media is 54.54 mg/l and 50.13 mg/l respectively. The results of the reliability check also showed good agreement between the measured and the simulated value for both BOD and COD with ($R^2 = 0.83$), ($R^2 = 0.99$), respectively. Hence the results obtained from horizontal transport of pollutant is more significant.

**Keywords** Transport of diluted species (TDS) · Infiltration · Biochemical oxygen demand (BOD) · Chemical oxygen demand (COD)

## 1 Introduction

Water is the main source for any creatures to survive on the planet earth. In a country like India due to the diversified culture people are adopted to a different lifestyle. Due to the consistent increase in population growth, mostly in urban areas, the managing of wastewater has become a tedious process. The water demand is increasing day by day and generation of the wastewater is increasing, with the lack of proper management of wastewater disposal is a burning issue in almost all the cities in the present scenario. Adopting a systematic way to attain sustainability in the present scenario is very important due to the pollution, climate change, and deforestation the natural

---

M. R. Dhanraj (✉) · A. Ganesha
International Center for Applied Science, MAHE, Manipal, India
e-mail: dhanraj1117@gmail.com

© The Author(s), under exclusive license to Springer Nature Singapore Pte Ltd. 2023
M. S. Ranadive et al. (eds.), *Recent Trends in Construction Technology and Management*, Lecture Notes in Civil Engineering 260,
https://doi.org/10.1007/978-981-19-2145-2_30

existing water sources are deteriorating day by day. In India presently, as per the estimation 61,754 million liters of water per day (MLD) of sewage is generated per day, 22,963 MLD of sewage is treated and 62% of sewage is discharged directly into the water bodies without the treatment. By 2025 demand for industry and domestic usage may increase to 29.2 BCM population is expected to cross the 1.5 billion mark by 2050 [1]. So therefore the aim is to utilize the treated sewage as an alternative source of the main source to replenish and conserve it for future purposes using artificial recharge technique. This may increase the efficiency in water use by the approach of conjunctive use of groundwater thereby reducing the demand for freshwater sources. Though we have adopted the practice of stormwater recharge, the significant effects are still unknown after recharge for seasonal emerging pollutants. Water reuse plays a prominent role in future development, there are many forms of disposal of treated sewage the disposal is done either on surface or subsurface. Since because of the mandatory rules from the state pollution control board the disposal of the treated sewage has become a very big hectic planning. Zero effluent discharge for commercial, gated communities, institutions might be more troublesome to overcome such circumstances. Future this would not only be domestic, but even industry also treated waste would face problem is disposal of such a large quantity of water generated after treatment. A study was conducted to anticipate the reliability of the model using comsol to predict the outcome using the 2D model and its effectiveness for effective examination [2]. So, this research would give a solution to such problem of disposal which mega industries and domestic treatment plants are facing today. The amount of energy spent on treatment must reward back by some means, or in another way the treated sewage has to be utilized completely. There are many situations where the treated wastewater may be used completely, in developed countries after the tertiary treatment. But the stigma toward the developing countries for recycling and reuse is being present, due to a lack of concern. Therefore, adopting the groundwater recharge will overcome the disposal problem and also helps in replenishing as well as the withdrawal of water at any point in time for irrigation purposes [3]. For the present study, the water is being considered from the secondary clarifier, the considerations are being shown in the flowchart along with the treatment process (Fig. 1).

Analyzing the risk associated with the contamination of soils and identification of the sources of its propagation and fate processes is needed [4]. According to Jiang and Chen [5], their study suggested that both the physical (advection, diffusion,

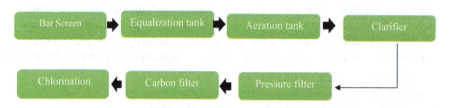

**Fig. 1** Flowchart of STP

dispersion) and reaction (adsorption and degradation) process controlling contaminant transport should be analyzed in the numerical model. Lie et al. [6] in their study solved the one-dimensional advection-dispersion equation in a heterogeneous porous media coupled linear and nonlinear sorption or decay using the generalized integral transforms technique. Gao et al. [7] studied one-dimensional model convective diffusion in the soil contaminant transport subjected to time-dependent boundary conditions. Comsol multi-physics software based on a partial differential equation. It is to solve the linear or nonlinear problem, and the steady-state or transient problem which was related to geometry mode of 1D, 2D, 3D problems. Comsol was mainly used to study at present in earth science research of groundwater flow and the soil water infiltration applied the software to simulate the migration of pollutant in soil and unverified the applicability of comsol to simulate soil solute transport. In this study based on comsol 5.5, 3d transient cross-sectional models were used to investigate water and solute transport in the soil column denoting variable sand column forms. These research gap findings with horizontal transport of pollutants would help generate the baseline data on removal of pollutant concentration after infiltration and serve the purpose of disposal after the treatment. Also, this will help to find the distance of pollutant transport from the point of discharge in the same directions.

## 2 Methodology

The working model is made up of concrete with a diameter of 1.2 m and a height of 1.5 m consists of a perforated pipe at the center with a diameter of 0.15 m. This center pipe acts as a well for recharge, and it also includes outlets for sampling. The water is pumped from the clarifier using a 0.5HP pump with a controlled flow to the recharge well. The water is being allowed to percolate and tried with two types of media. The water which comes out of the model is tested for various parameters such as BOD, COD and then checked for efficiency under inlet flow conditions, with the same existing media of sand, followed by the same parameters. The water sample is collected using polyethylene cans. Further, it is being transported and stored in a deep freezer at 4 °C for further analysis. A total of 50 samples were collected to check the consistency in the media used for the study. BOD and COD are the two important water quality parameters required to assess the waste assimilative capacity in the soil media [8]. BOD is employed as a gross measure of the oxygen demanding potential of the effluent. Assimilative capacity varies following variations in hydrodynamic conditions and other ecological processes. COD is employed as an indicative measure of the amount of oxygen that can be consumed by reactions in a measured solution (Fig. 2).

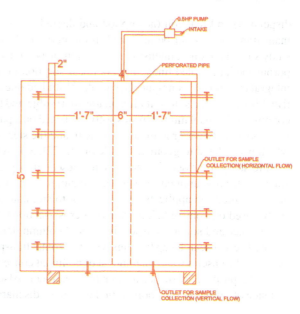

**Fig. 2** Pilot-scale model

## 2.1 BOD and COD Analysis

The analysis of water samples is done in the laboratory by collecting the water samples from both inlet and outlet. The collected water samples are stored in polyethylene bottles and stored at 5 °C while transportation. $BOD_5$ @20 °C is measured by the dilution method. Samples have undergone three trials and the best two values have been considered. The COD of the water is measured by the closed reflux method using potassium dichromate by colorimetric method (IS 10500:2012) (Fig. 3).

## 2.2 Model Characteristics

COMSOL Multi-physics modules were used to simulate and trace the pollutant transport in subsurface porous soil media. To verify the efficiency of the simulated model, experimental data is required. The COSMOL program, which simulates the velocity, pressure head, and the concentration distribution based on the boundary condition using an implicit iteration solution, has been used in developing and understanding the efficient working conditions of the model. A detailed description of the model is presented on the COMSOL website. In the following sections, a brief description of the main features of the model is presented (Tables 1 and 2).

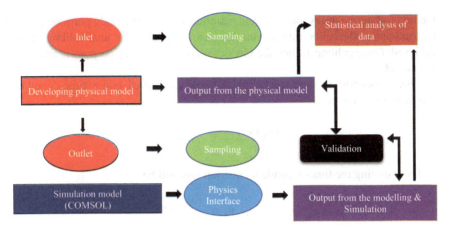

**Fig. 3** Methodological flowchart of the study

**Table 1** Boundary and initial conditions

| Parameters | Soil |
|---|---|
| Permeability (cm/sec) | $1.79 \times 10^{-5}$ |
| Density (gm/cm$^3$) | 1.46 |
| Specific gravity | 2.54 |
| Grain size (uniformity coefficient) | 1.44 |
| Porosity | 0.3 |

**Table 2** Flow measurements

| Parameters | Value |
|---|---|
| Velocity (m/s) | 0.28 |
| Discharge (m$^3$/s) | 0.044 |
| Hydraulic head (m) | 1.5 |
| Pressure head (m) | 1.38 |

## 2.3 Model Calibration and Governing Equations

Model calibration is a trial-and-error adjustment of parameters until the model solution matches the field physical model. In general, the calibration aims to design a steady state model for head distributions to be used as an initial condition for a transient state simulation after applying and trying different states of scenarios on it. Model calibration also includes a successive refinement for model input parameters from the initial condition to improve the fit between observed input and model predicted results. The calibration procedure typically begins with the definition of parameters also based on the availability of data with an initial conceptual model of the hydrogeology systems. The parameters chosen for the calibration process include soil permeability, porosity and materials used for the study purpose. Several adjustments to the calibration parameters were made until the final calibration was achieved. Once calibrated then the physics used for the study purpose is applied to the model.

Phase transport in porous media for study controlled under time-dependent conditions is given by

$$\frac{\partial \varepsilon_p \rho_{s_i} s_i}{\partial t} + \nabla \cdot \mathbf{N}_i = 0 \qquad (1)$$

By assuming the time-dependent equation this will be

$$\mathbf{N}_i = \rho_{s_i} \mathbf{u}_i, \ \mathbf{u}_i = -\frac{\kappa_{rs_i}}{\mu_{s_i}} \kappa \nabla \rho_{s_i}$$

When the outflow is set for the model $-\mathbf{n} \cdot \mathbf{N}_i = -\mathbf{n} \cdot \left(\rho_{s_i} \mathbf{u}_i\right)$.

## 3 Results and Discussion

See Tables 3 and 4.

### 3.1 COMSOL Multi-physics for Simulation and Modelling

This study traced the contaminant transport in a horizontal direction and analyzed it using a 2D-model transport of pollutant in porous media with controlled discharge using COMSOL multi-physics created with soil matrix for water recharge of treated domestic sewage. The study's objective is to understand the concentration distribution in the saturated porous media of the system. Based on the different time the variations in the concentration is observed at the Outlet. A study using COMSOL multi-physics software for the Artificial recharge in Jordon compares the measured and the software modeling (Azad et al. 2010). The validation work has applied the same approach. A study was conducted in the vadose zone to model the solute transport and their travel times toward groundwater bodies. The study to understand the impact of the unique state of soil in terms of spatial infiltration of solute-rich water [9] is needed in COMSOL to trace the pollutant transport in soil concerning time (Fig. 4).

Table 3 Summary of the BOD results from the physical model

| Parameter (mg/l) | Influent (mg/l) | Measuring range | Saturation time (min) | Effluent (mg/l) | Mean | Samples numbers | SD | SEM | Median |
|---|---|---|---|---|---|---|---|---|---|
| BOD$_5$ @20 °C | 30.06 | 30 | 30 | 15.96 | 10.26 | 10 | 7.34 | 2.32 | 6.20 |
| COD | 94.66 | 30 | 30 | 56.37 | 66.93 | 10 | 43.61 | 10.90 | 46.3 |

**Table 4** Summary of the BOD results from the simulation model

| Parameter (mg/l) | Influent (mg/l) | Measuring range | Saturation time (min) | Effluent (mg/l) | Mean | Samples number | SD | SEM | Median |
|---|---|---|---|---|---|---|---|---|---|
| $BOD_5$ @20 °C | 30.06 | 30 | 30 | 23 | 9.8 | 10 | 6.66 | 2.10 | 6.20 |
| COD | 94.66 | 30 | 30 | 39.06 | 39.05 | 10 | 22.62 | 5.19 | 26.2 |

# Modelling and Simulation of Pollutant Transport ...

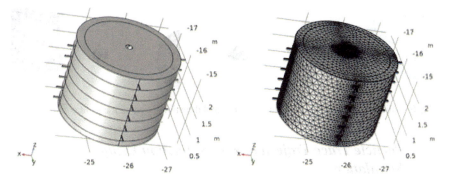

**Fig. 4** Physical model developed using COMSOL

## 3.2 Performance Measures Between Measured Values and Simulated Values for BOD and COD

There is a significant reduction in the output of the model after the horizontal movement. The presented graph shows the variations at the inlet and outlet from the physical and simulation model of the study.

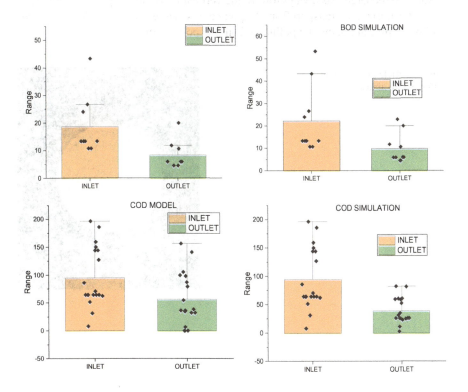

**Fig. 5** Validation of pollutant transport in horizontal direction based on transient condition

## 3.3 Particle Trace Trajectories for Validation Using Simulation

This Figure clearly shows the transport of pollutant in the horizontal direction. The distribution of concentration is observed at a time interval of $t = 30$ min (Fig. 5).

## 3.4 Dispersion of Pollutants

Horizontal displacement of the pollutant is not introspected well because the focus toward the recharge was always toward the vertical transport of species. Here Fig. 6 shows the horizontal transport from the point of discharge in different directions, this will be helpful in tracing the pollutant from the point of discharge.

**Fig. 6** Particle dispersion in different directions

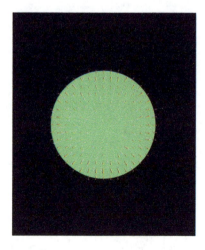

## 3.5 Numerical Analysis of BOD and COD Profiles

For data consisting of $x$ and $y$ subjected to each $x$ corresponds to $y$ and vice-versa which causes at least two distinct straight lines, correlation analysis, and various significance tests. The curve fitting is standardized with 95% confidence level. Therefore, from this analysis, the results obtained from the physical model and simulation model have a strong correlation with an $R^2$ value of 0.8 or more with good significance. Also, for all the measured value between the inlet and outlet samples has an $R^2$ value of 0.8 or more for the simulation value with good significance. The below-mentioned chart represents the graph drawn between the Inlet (influent) and Outlet (effluent) of measured value/Simulated value for soil media (Graph 1).

## 4 Statistical Analysis of Models

See Table 5.

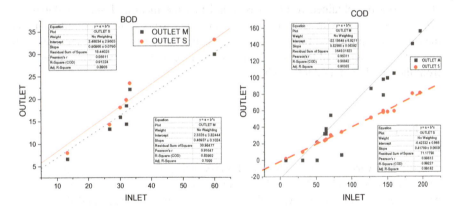

**Graph 1** Regression analysis of BOD and COD

**Table 5** Regression

| Models | Parameter | $R^2$ |
|---|---|---|
| Measured value | $BOD_5$ | 0.91 |
| Simulated value | $BOD_5$ | 0.83 |
| Measured value | COD | 0.90 |
| Simulated value | COD | 0.99 |

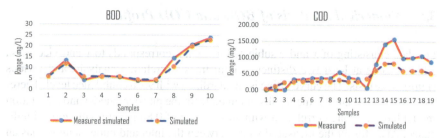

**Graph 2** Validation between measured and simulated

## 4.1 Validation of the Results

In this paper, to tackle the common issues in the prediction process of simulation models the COMSOL multi-physics is used and validated with the measured values which are done on a pilot scale in the college campus. The main objective of the study was to prove the lateral movement in soil porous media. The forecasted data of pollutants for BOD and COD was measured for a year from January to December 2020. From the obtained results it is found that the measured and simulated are in good agreement with less than 10% error. The results from the simulation and measure are represented further using statistical analysis. Also, from the statistical analysis, it is found that there is a strong acceptance of data between measured data from the simulation model (Graph 2).

## 5 Discussion

The Effluent variations were closely observed and it is found that the influent has an impact and primly effected the target value at the Effluent. The comparison pertaining to pollutant removal efficiency between the measured and the Simulation holds good is contributing a good amount of pollutant removal after infiltration. The BOD removal concept remains the same as the water gets recharged. The retention time attained during transport reduces the pollutant concentration in the outlet and also in the simulation model. The chemical oxidations also go well during the horizontal infiltration after the recharge, which is also closely observed in BOD due to the relationship between BOD and COD. Therefore, the organic matter removal also takes place in the horizontal direction resulting in the impact of artificial recharge.

## 6 Conclusion

Artificial recharge and its removal of pollutants is the simplest form of disposal of treated domestic sewage. Where a high-quality consistent value is required. It is found that the permeability of the soil medium has a direct influence on the removal efficiency. The BOD removal is found to be 54.54 mg/l and COD removal is 50.13 mg/l. Also, the results obtained from the physical model and simulation model have a very strong correlation, i.e., $R$-square $= (0.7 < |r| \leq 1)$. The model can also be used to understand the pollutant travel distance for zero transport from the point of discharge. Therefore, artificial recharge of treated domestic sewage has an impact on unconfined aquifer with reduction of pollutant concentration from influent to effluent and hence artificial recharge can be used as a method of disposal for treated domestic sewage (used after secondary treatment). However, as a safety measure, regular monitoring of the recharge water quality and ground water quality in the surrounding area of recharge site is essential.

**Acknowledgements** The author and co-author are extremely thankful to the Civil engineering department for providing the software and also extremely grateful to the director, ICAS, and management for their perpetual support, encouragement, and inspiration along with the excellent library facilities, and accessibility for top indexed journals provided to the authors during this work.

## References

1. Bhawan P, Nagar EA (2020) Central Pollution Control Board
2. Al-mansori NJ, Al-kizwini RS, Al-husseini FK (2020) Modeling and simulation of pollutants dispersion in natural rivers using comsol multiphysics. J Eng Sci Technol 15(2):1167–1185
3. Dhanraj AG (2019) Artificial recharge practice of treated domestic sewage—its challenges and opportunities in Indian perspective: a Mini systematic review. Indian J Environ Protect 39(12):1158–1165
4. Viccione G, Stoppiello MG, Lauria S, Cascini L (2020) Numerical modelling on fate and transport of pollutants in the Vadose Zone. Environ Sci Proc 2(34). https://doi.org/10.3390/environsciproc2020002034
5. Jiang WQ, Chen GQ (2019) Environmental dispersion in layered wetland: moment based asymptotic analysis. J Hydrol 569:252–264
6. Lei Y, Li Y, Su F, Li H (2019) Metabolomics study of subsurface wastewater infiltration system under fluctuation of organic load. In: Current microbiology. Springer
7. Gao G, Zhan H, Feng S, Fu B, Ma Y, Huang G (2010) A new mobile-immobile model for reactive solute transport with scale-dependent dispersion. Water Resour Res 46(8)
8. Thomann RV, Mueller JA (1987) Principles of surface water quality modeling and control. Harper and Row, New York
9. Cueto-Felgueroso L, Suarez-Navarro MJ, Fu X, Juanes R (2020) Interplay between fingering instabilities and initial soil moisture insolute transport through the Vadose Zone. Water (Switzerland) 12(3). https://doi.org/10.3390/w12030917
10. Indian standard code book for analysis of water, IS 10500:2012
11. Omar K (2015) Treated wastewater use in Saudi Arabia: challenges and initiatives. Int J Water Resour Dev:799–809

# Adsorptive Removal of Malachite Green Using Water Hyacinth from Aqueous Solution

**Sayali S. Udakwar, Moni U. Khobragade, and Chirag Y. Chaware**

**Abstract** Effluent from dye industries contains the reactive dye in large amount, which imparts color to the water and causes harm to the wellbeing of humans and aquatic animals. Malachite green (MG) is one of the dyes used in the dye industry, which is mutagenic, carcinogenic and therefore toxic. Water hyacinth (WH) is an aquatic weed, which is known for its proliferating growth. This leads to the eutrophication of water bodies. Water hyacinth (Eichhornia crassipes) powder is used as bio adsorbent to remove malachite green dye from aqueous solution. The adsorptive removal capacity of malachite green using water hyacinth powder was studied by carrying out the batch study on synthetic samples to assess malachite green removal efficiency. The effect on the removal was a function of various parameters like agitation speed, adsorbent dose, contact time, initial concentration, particle size and pH. At pH 7 adsorption reaches equilibrium at an agitation speed of 130 rpm at a contact time of 30 min at temperature of 27 °C. The pseudo-second-order kinetics and both Freundlich and Langmuir isotherms with $R^2$ 0.9987, 0.9970 and 0.9923, respectively, fit the kinetics and isotherm models. The water hyacinth characterization was carried out with SEM and EDX, which suggests adsorption of malachite green. From the results obtained it can be observed that water hyacinth powder can adsorb the malachite green dye under optimum conditions which are easy to maintain. Therefore, water hyacinth can be potentially applied for the removal of malachite green dye from dye wastewater.

**Keywords** Bio adsorption · Malachite green dye · Water hyacinth · Kinetic study · Material characterization

---

S. S. Udakwar · M. U. Khobragade · C. Y. Chaware (✉)
Department of Civil Engineering, College of Engineering Pune, Pune 411005, India
e-mail: chawarecy19.civil@coep.ac.in

S. S. Udakwar
e-mail: udakwarss18.civil@coep.ac.in

M. U. Khobragade
e-mail: muk.civil@coep.ac.in

© The Author(s), under exclusive license to Springer Nature Singapore Pte Ltd. 2023
M. S. Ranadive et al. (eds.), *Recent Trends in Construction Technology and Management*, Lecture Notes in Civil Engineering 260,
https://doi.org/10.1007/978-981-19-2145-2_31

# 1 Introduction

One of the main sources of water on the earth is river water. About 97% water on the earth is saline in nature, 3% is fresh water. Out of 3, 69% is in the form of glaciers and ice cap; 30% is ground water and only 1% is surface water which is in the form of rivers, streams and lakes. The importance of water is known to one and all. However, despite being the limited resource on the earth and basic human need, this vital resource is being wasted, contaminated, and depleted [1, 2]. Therefore, it is necessary to conserve water and prevent pollution of water. Many cities in India and worldwide are already facing severe water shortage due to reduced rainfall, human-made climate change, reduced groundwater levels, pollution explosion, industrialization, and staggering water wastage. Water contamination, scarcity of drinking water, poor sanitation, open discarding of waste, and forest cover loss are some of India's problems. Fatal water borne-diseases leading to a higher mortality rate in infants and organs impairment is caused due to water pollution [3]. Animals drink water directly from the surface of water bodies, polluted water causes harm to the animal's health so, it is necessary to take care of this dye wastewater [4, 5].

Pollution of water due to dumping of dye waste water in the natural stream is a big concern nowadays as it imparts the color and toxicity to the water [6–8]. Many dying, fabrication, paper and textile industries use reactive dyes of imparting color to the products. These dyes are highly reactive and toxic [9]. If it is present in large amount causes damage to the nervous system of human beings. Textile industries dump this dye waste water directly into natural water bodies which imparts color to the water and forms toxic compounds [10]. This surface water in the natural stream is used for various purposes by human being and other living beings. So, treatment of dye wastewater is necessary before dumping into the natural water bodies [3]. Conventional filtration methods are less efficient in treating dye wastewater. Therefore, removal of dye from dye wastewater was studied using absorption technique. The abilities of water hyacinth are higher growth rate, pollutants absorption efficiency [11, 12], low operation cost and renewability which shows using this plant can be considered as a suitable technology for the treatment of dye waste water.

# 2 Materials and Methods

## 2.1 Adsorbate

Malachite green a cationic reactive dye was used as the model pollutant or reagent in the study. It was purchased from the market. This dye was selected because it was the most commonly used in cotton, acrylic, nylon, silk, wool, leather, plastic and paper dyeing textile industries. Considering this there is a possibility of bulk release of dye in effluents from the textile industry [13]. For the preparation of the stock solution of 1000 mg/l 1 g malachite green as adsorbate was dissolved in 1000 ml of distilled water. By diluting the stock solution with adequate distilled water experimental solutions were prepared.

## 2.2 Preparation of Bio Adsorbent

In this study, water hyacinth leaves and stem powder was used as adsorbent, which is collected from Pashan Lake situated in Bavdhan, Pune. The collected adsorbents were washed thoroughly 2–3 times to clean the adsorbent from dust and mud. Leaves and stems were separated after washing. It was kept in 0.25 M EDTA solution for 24 h. at room temperature to removal, the metal and other surface impurities. It was rinsed with distilled water after 24 h. After washing with distilled water, it was soaked in sunlight for 2 h. After that keeping it in a hot air oven for 48 h at 110 °C which is an optimized temperature.

Crushed and dried water hyacinth leaves and stems were grinded in mixer grinder into a fine powder. Grinded powder is then sieved through 75 μm sieve to get fine powder. After sieving, water hyacinth leaves and stem powder were stored in airtight plastic bottles at room temperature 27 ± 2 °C and further used as the adsorbent in batch study.

## 2.3 Batch Bio Adsorption Experiments

The batch study experimentation was used to study the adsorption of malachite green using water hyacinth powder. Conical flask of 250 ml was used to perform batch study experiments and round bottom volumetric flask of 100 ml. A fixed volume and concentration of adsorbate solution were added in conical flask with the fixed amount of adsorbent powder. It was placed in orbital shaking incubator and temperature and agitation speed in orbital shaking incubator were fixed. After shaking in the orbital shaking incubator for a fixed time, conical flasks were removed from the orbital shaker and kept on a flat platform to rest so that suspended particles could settle down. As the suspended particles in the solution settled, the solution above the settled particles was removed to other conical flask. The sample was taken for analysis in the UV-spectrophotometer. By knowing absorbance, the concentration of the chemical substance in the solution was calculated. The results obtained from the spectrophotometer were used to find the removal percentage of malachite green and adsorption capacity of the adsorbent. This process was repeated by varying parameters such as contact time, pH, initial dye concentration, adsorbent dose, agitation speed and particle size. The calculation of adsorption efficiency $\eta\%$ and adsorption capacity $q_e$ of the adsorbent was calculated by Eqs. 1 and 2.

$$q_e = (C_o - C_e)/m \tag{1}$$

$$\eta\% = [(C_o - C_e)/C_o] * 100 \qquad (2)$$

where,

$C_o$   Initial dye concentration in mg/L
$C_e$   Final dye concentration in mg/L
$m$   Mass of adsorbent in mg/L.

Adsorption efficiency and adsorption capacity are important parameters for evaluating the performance of adsorbents. $q_e$ gives us the idea about the quantity of dye, adsorbed on the unit mass adsorbent, and the adsorption efficiency is just the removal efficiency of adsorbent irrespective of amount. The results were obtained by following the same procedure and were plotted to get the calibration curve that is used as basis for further calculations in the study.

## 2.4 Surface Characterization

### SEM and EDX analysis

The removal of malachite green using water hyacinth leaves powder was examined a Scanning Electron Microscope (SEM) and Energy dispersive X-ray (EDX). A working solution of a malachite green dye having concentration 12 mg/L was prepared. The working solution was kept in orbital shaking instrument at temperature 27 °C and agitation speed of 130 rpm. It was then taken out and allowed to settle. The solution above the settled particles was removed from the top and solution at the bottom of the flask was allowed to dry. This dried powder was tested under a scanning electron microscope with 15 kV and 10,000 times magnification.

## 3 Results and Discussion

## 3.1 Adsorbent Material Testing

### SEM and EDX analysis of water hyacinth water hyacinth leaves powder (Figs. 1, 2, 3 and 4):

The above graphs, images and table indicate the adsorption of malachite green on water hyacinth leave powder. Dye molecules are visible on the adsorbent material surface in SEM images. The main constituent of malachite green oxalate is oxygen; there is a significant increase in oxygen percentage on the adsorbent surface after dye removal. Oxygen is increased by approximately 8%. This implies that adsorption takes place and water hyacinth leaves powder adsorbed the malachite green oxalate dye.

**Fig. 1** SEM photograph of water hyacinth leaves powder before malachite green dye removal

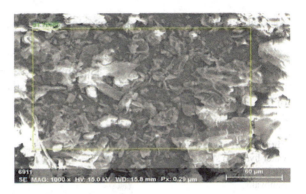

**Fig. 2** SEM photograph of water hyacinth leaves powder after malachite green dye removal

**Fig. 3** EDX graph before removal using leaves powder

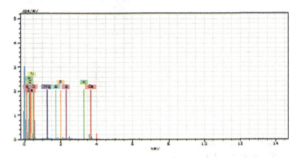

**Fig. 4** EDX graph after removal using leaves powder

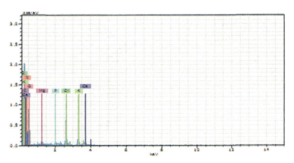

## SEM and EDX analysis of water hyacinth water hyacinth stem powder (Figs. 5, 6, 7 and 8)

Water hyacinth stem powder shows similar behavior as that of water hyacinth leaves powder. The images, graphs, and tables proved that malachite green oxalate's adsorption takes place on the surface water hyacinth stem powder. There is an increase of 5% in the amount of oxygen after the removal of malachite green oxalate dye.

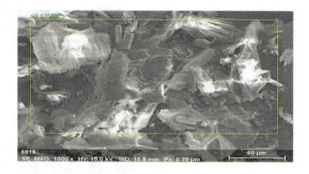

**Fig. 5** SEM photograph of water hyacinth stem powder after malachite green dye removal

**Fig. 6** SEM photograph of water hyacinth stem powder before malachite green dye removal

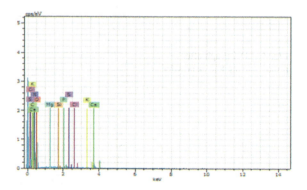

**Fig. 7** EDX graph before removal using stem powder

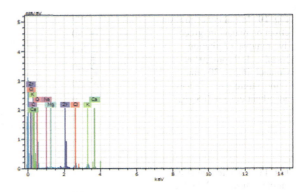

**Fig. 8** EDX graph after removal using stem powder

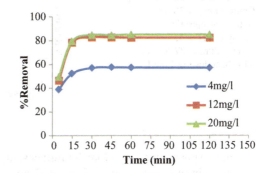

**Fig. 9** Contact time effect on MG dye removal (Agitation speed = 150 rpm, PH = 7, Temperature = 27 °C, and adsorbent dose = 1.0 g/L)

## 4 Batch Adsorption Study for Water Hyacinth Leaves Powder

### 4.1 Effect of Contact Time

See Fig. 9.

From the above graph, it can be concluded that the contact time of 30 min is optimum for the adsorption of malachite green oxalate dye using WH leaves powder from the aqueous solution. Therefore, for the further experimental process, the optimum contact time was taken as 30 min.

### 4.2 Effect of Initial Dye Concentration

See Fig. 10.

From the above graph, it can be concluded that the initial dye concentration of 12 mg/L is an equilibrium concentration after which the rate of increase in removal

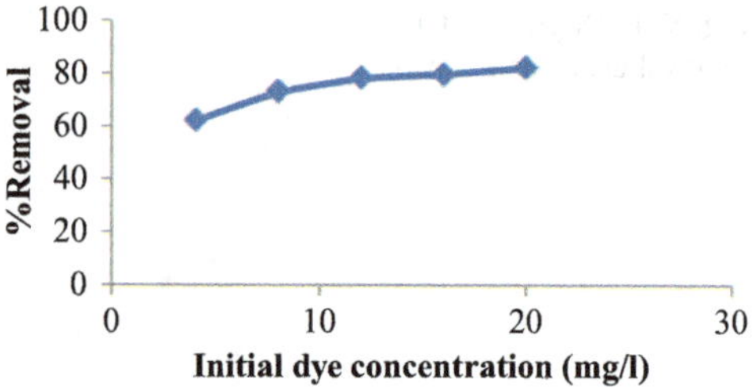

**Fig. 10** Initial concentration effect on malachite green dye removal (Agitation speed = 150 rpm, pH = 7, Contact time = 30 min, Temperature = 27 °C and adsorbent dose = 0.5 g/L)

**Fig. 11** Particle size effect on the removal of MG dye (Agitation speed = 130 rpm, Contact time = 30 min Initial concentration = 12 mg/L, PH = 7, Temperature = 27 °C and adsorbent dose = 1.0 g/L)

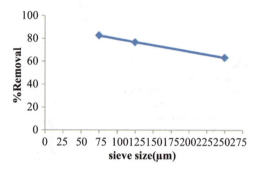

efficiency slows down. Therefore, for further study to get the optimum results solution having an initial concentration of dye at 12 mg/L was taken.

### 4.3 Effect of Particle Size

Maximum removal efficiency of 82.73% is obtained at the adsorbent particle size of 75 μm. Smaller the particle more will be surface area and more will be the adsorption. Therefore, water hyacinth powder having 75 μm of particle size was used for the experimental work (Fig. 11).

### 4.4 PH Effect

At pH 7.5 the peak removal effectiveness of malachite green dye using water hyacinth powder was observed. From the above results, it can be concluded that neutral pH, i.e., 7 is an optimum pH for removal of malachite green oxalate dye from the aqueous solution. Therefore, neutral pH, i.e., 7 is adopted for further experimental procedures (Fig. 12).

**Fig. 12** pH effect on malachite green dye removal (Agitation speed = 150 rpm, Initial concentration = 12 mg/L, Temperature = 27 °C, Contact time = 30 min and adsorbent dose = 1.0 g/L)

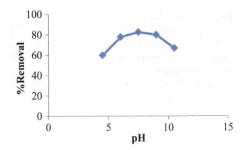

**Fig. 13** Adsorbent dose effect on removal of MG dye (Initial concentration = 5 mg/L, pH = 7, Temperature = 27 °C, Contact time = 30 min and Agitation speed = 150 rpm)

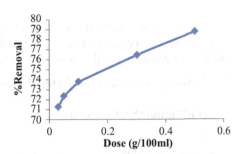

### Influence of Adsorbent Dose

See Fig. 13.

From the above graph, it can be concluded that 1 g/L of the adsorbent dose is required for optimal removal dye from an aqueous solution having initial dye concentration constant at 5 mg/L. Above dose 1 g/L or 0.1 g/100 ml removal efficiency do not increase by considerable amount. Therefore, for further study, adsorbent dose of 1 g/L was adopted.

#### 4.4.1 Effect of Agitation Speed

See Fig. 14.

**Fig. 14** Agitation speed effect on the removal of MG dye (Contact time = 30 min, PH = 7, Initial concentration = 12 mg/L, Temperature = 27 °C and adsorbent dose = 1.0 g/L)

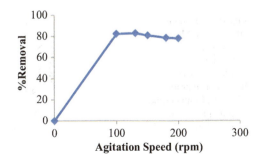

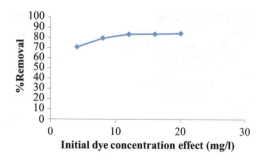

**Fig. 15** Initial dye concentration effect on the malachite green dye removal using water hyacinth stem powder(Agitation speed = 130 rpm, pH = 7, Contact time = 30 min, Temperature = 27 °C and adsorbent dose = 1.0 g/L)

From the above graph, it can be concluded that as speed increases, detachment of dye molecules occurs therefore, with further increase in agitation speed removal efficiency decreases. For this study agitation speed of 130 rpm was optimum.

## Batch Adsorption Study for Water Hyacinth Stem Powder

### Effect of Initial Dye Concentration

See Fig. 15.

For this batch study adsorbent shows maximum removal efficiency 83.55% at an initial dye concentration of 12 mg/L like water hyacinth leaves powder. Therefore, the initial concentration of 12 mg/L is an equilibrium concentration.

### Effect of Adsorbent Dose

See Fig. 16.

From the study, the optimum removal efficiency of 81.45% is obtained for an adsorbent dose of 1.5 g/L. Therefore, it can be concluded that water hyacinth stem powder is required in more quantity than that of leaves powder to remove malachite green dye from aqueous solution.

### Adsorption Isotherms

The adsorption isotherms such as Langmuir isotherm and Freundlich isotherm model, depict the relationship between adsorbent and adsorbate. To carry out the analysis of these model's batch experiment was carried out by varying initial dye concentration

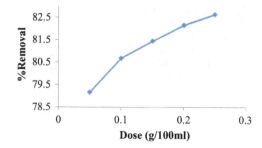

**Fig. 16** Adsorbent dose effect on the MG dye removal using water hyacinth stem powder (Agitation speed = 150 rpm, Initial concentration = 12 mg/L, Contact time = 30 min, pH = 7 and Temperature = 27 °C)

**Table 1** Langmuir isotherm constants

| Langmuir isotherm | | |
|---|---|---|
| $K_L$ (L/mg) | $q_m$ (mg/g) | $R^2$ |
| 0.1113 | 16.9690 | 0.9923 |

**Table 2** Freundlich isotherm constants

| Freundlich isotherm | | |
|---|---|---|
| $K_F$ (L/g) | $n$ | $R^2$ |
| 8.5729 | 20.03 | 0.9970 |

4, 8, 12, 16 and 20 mg/L and keeping constant adsorbent dose 1 g/L, contact time 30 min, agitation speed 130 rpm, pH 7 and temperature 27 °C. The values of constants in Langmuir and Freundlich isotherm are given in Tables 1 and 2,, respectively. The graph of Langmuir and Freundlich adsorption isotherm is shown in Figs. 17 and 18, respectively.

From Tables 1 and 2, it is clearly shown that the correlation coefficient ($R^2$) is almost similar in both isotherms Langmuir and Freundlich, which is 0.9923 and 0.9970, respectively. This depicts that the adsorption of malachite green dye using water hyacinth powder as adsorbent occurs in monolayer as well as a multilayer. Therefore, the adsorption of malachite green follows both Langmuir and Freundlich isotherm and adsorption technique is suitable for removing malachite green using water hyacinth powder from aqueous solution.

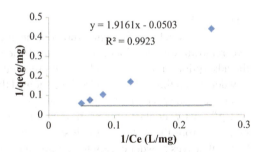

**Fig. 17** Langmuir adsorption isotherm

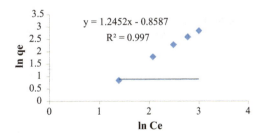

**Fig. 18** Freundlich adsorption isotherm

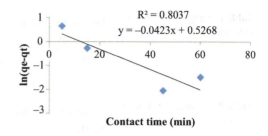

**Fig. 19** Kinetic model Pseudo-first-order

## 4.5 Adsorption Kinetics

Pseudo-first-order and pseudo-second-order two kinetic models are utilized to analyze the dynamics of malachite green oxalate dye adsorption on the water hyacinth. The best fit for adsorption between pseudo-first-order and pseudo-second-order is examined using the coefficient of correlation. The correlation coefficient ($R^2$) for the pseudo-first-order is 0.8037 and for pseudo-second-order is 0.9989. It was observed that the adsorption process of malachite green oxalate dye using water hyacinth powder follows pseudo 2nd order kinetics. Therefore, for the removal of malachite green dye adsorption technique is water hyacinth powder (Fig. 19).

## 5 Conclusions

See Fig. 20.

It can be concluded from the study that an efficient adsorbent for malachite green dye removal was developed from water hyacinth plant and successfully used in the adsorption process. The adsorbent prepared from water hyacinth plant's leaves powder sieved through 75 μm IS sieve showed the optimum removal efficiency 81% at an agitation speed of 150 rpm, initial dye concentration 12 mg/L, adsorbent dose of 1 g/L, pH 7, contact time of 30 min, and temperature 27 ± 2 °C. The adsorbent prepared from the water hyacinth stem powder showed the same as that of water hyacinth leaves powder except the optimum removal efficiency of 83% at initial dye

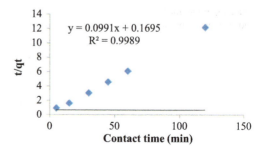

**Fig. 20** Kinetic model Pseudo-second-order

concentration of 12 mg/L and adsorbent dose of 1.5 g/L. Therefore, both leaves and stem powder shall be utilized for malachite green dye removal as a bio adsorbent. The study concludes that water hyacinth is useful in removing dye solution Adsorption process follows both Langmuir and Freundlich adsorption isotherm as the value of the correlation coefficient of isotherms is 0.9923 and 0.997, respectively. In the kinetic study, this adsorption process follows a pseudo-second-order kinetic model with an $R^2$ value of 0.9989. which shows the significant amount of adsorption that occurred in the experiments. Therefore, the adsorption process is best suitable for the removal of malachite green dye.

# 6 Recommendation

The water hyacinth plant from the surface water bodies should be used for removing malachite green dye contaminants from dye wastewater with high efficiency. The separation of malachite green dye from the solution was observed between 80 and 83%. Thus, it is, recommended to modify the surface of water hyacinth powder chemically to improve the removal efficiency of malachite green.

# References

1. Kumar B, Smita K, Flores LC (2017) Plant mediated detoxification of mercury and lead. Arab J Chem 10:S2335–S2342
2. Shokry H, Elkady M, Hamad H (2019) Nano activated carbon from industrial mine coal as adsorbents for removal of dye from simulated textile wastewater: operational parameters and mechanism study. J Mater Res Technol 8:4477–4488
3. Sukasem N, Khanthi K, Prayoonkham S (2017) Biomethane recovery from fresh and dry water hyacinth anaerobic co-digestion with pig dung, elephant dung and bat dung with different alkali pretreatments. Energy Procedia 138:294–300
4. Kulkarni MR, Revanth T, Acharya A, Bhat P (2017) Removal of crystal violet dye from aqueous solution using water hyacinth equilibrium, kinetics and thermodynamic study. Resource-Efficient Technol 3:71–77
5. Wanyonyia WC, Onyarib JM, Shiunduc PM, Mulaad FJ (2017) Biodegradation and detoxification of malachite green dye using novel enzymes from bacillus cereus strain KM201428: kinetic and metabolite analysis. Energy Procedia 119:38–51
6. Elhalil A, Tounsadi H, Elmoubarki R, Mahjoubi FZ, Farnane M, Sadiq M, Abdennouri M, Qourzal S, Barka N (2016) Factorial experimental design for the optimization of catalytic degradation of malachite green dye in aqueous solution by Fenton process. Water Resour Ind 15:41–48
7. Sinha A, Lulu S, Vino S, Osborne WJ (2019) Reactive green dye remediation by Alternanthera philoxeroides in association with plant growth promoting Klebsiella sp. VITAJ23: a pot culture study. Microbiol Res 220:42–52
8. Abdelrahman EA, Hegazey RM, El-Azabawy RE (2019) Efficient removal of methylene blue dye from aqueous media using Fe/Si, Cr/Si, Ni/Si, and Zn/Si amorphous novel adsorbents. J Mater Res Technol 8:5301–5313
9. Jarusiripot C (2014) Removal of reactive Dye by adsorption over chemical pretreatment coal based bottom ash. Procedia Chem 9:121–130

10. Sousa HR, Silva LS, Sousa PAA, Sousa RRM, Fonseca MG, Osajima JA, Silva-Filho EC (2019) Evaluation of methylene blue removal by plasma activated palygorskites. J Mater Res Technol 8:5432–5442
11. Kanawade SM, Gaikwad RW (2011) Removal of methylene blue from effluent by using activated carbon and water hyacinth as adsorbent. Int J Chem Eng Appl 2(5):317–319
12. Sarkar M, Rahman AKML, Bhoumik NC (2017) Remediation of chromium and copper on water hyacinth (E. crassipes) shoot powder. Water Resour Ind 17:1–6
13. Rajabi M, Mahanpoor K, Moradi O (2019) Preparation of PMMA/GO and PMMA/GO-Fe$_3$O$_4$ Nano composites for malachite green dye adsorption: Kinetic and thermodynamic studies. Compos B 167:544–555

# Sediment Yield Assessment of a Watershed Area Using SWAT

**Prachi A. Bagul and Nitin M. Mohite**

**Abstract** Soil erosion is a major concern and severe challenge for the sustainability and productivity of agricultural systems, land all over the world. It is directly and majorly impacting on storage capacity and life span of reservoir, dam. It is important to reduce soil loss from cultivated lands and minimize degradation in water quality. Sediment yield is the volume of sediment eroded from the ground surface by runoff and transferred to a stream system or basin outlet over time. Estimation of sedimentation yield is needed for studies of its impact on reservoir storage capacity, river morphology and planning of soil and water conservation measures. The present study has been carried out by using Soil Water Assessment Tool (SWAT), to develop model for watershed area which comes under Pawana Watershed, Maharashtra state, India. SWAT requires large number of input parameters which include digital elevation model, soil map, Land use land cover map, slope map, weather data. These inputs of SWAT have been prepared under the QGIS2.6.1 environment. All these inputs- thematic maps and attribute information of study area have been collected from various Government agencies and sources. The simulated result is obtained after successful run of SWAT which can be visualized statically, graphically, and numerically.

**Keywords** Pawana watershed · Sediment yield · Quantum geographic information system · Soil water assessment tool

---

P. A. Bagul (✉) · N. M. Mohite
Department of Civil Engineering, College of Engineering Pune, Pune, MH, India
e-mail: bagulaprachi@gmail.com

## 1 Introduction

Accelerated soil and water loss, seriously threaten land and water resources and ecological environment, by posing a severe challenge to the productivity of land by the loss of fertile soil and to the life of reservoirs by the deposition of sediment. If we see, Sedimentation is an important parameter to assess the life of a reservoir, sediments deposited in reservoir may affect the safety of reservoir and it also effect on hydropower production. For best watershed management practices, it might be to minimize the amount of sediment, to divert sediment around or through the reservoir, or to remove deposited sediments it is necessary to quantify and analyse the sedimentation yield.

In recent years, the study of soil erosion is more common by using geographic information system technology, remote sensing and Universal Soil Loss Equation (USLE) models like Revised USLE (RUSLE) and Modified USLE (MUSLE) are increasingly being used [1]. SWAT shows the efficient results in different scenarios like- to quantify the special and temporal runoff estimation from gridded rainfall data [2], Study areas of ungagged watersheds and which are closely located, having similar meteorological and hydrological characteristics [1]. SWAT shows efficient application of sediment yield in different scenario like Urbanized basin, highly urbanized basin, a temporary large basins management [3], in zone of semi-arid and arid catchments, though it is challenging due to unavailability of sufficient data, with gridded rainfall data and even in ungauged watershed area [4]. SWAT provides effective ways and methods for runoff and sediment simulation in different watersheds.

## 2 Study Area

The study area—Pawana dam lies between 730 40′ 30″ E longitudes and 180 21′ 30″ N latitudes and has an area of Pawana catchment is approximately 113.36 km$^2$. Pawana River (Pawana) is located in the western part of the state of Maharashtra, India, in the Pune district. It originates in the ghats of the West. It is a tributary of the Bhima River and merges into the Mula River in the city of Pune. Average annual rainfall in the catchment is 2800 mm. Figure 1 shows the location of study area.

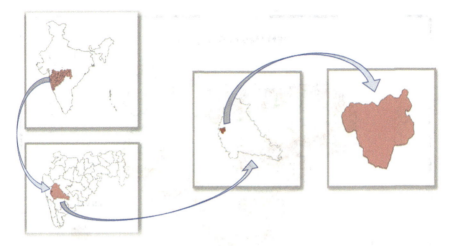

**Fig. 1** Location of study area Pawana watershed

## 3 Model Description and Methodology

The hydrological modelling of study area Pawana watershed is carried out by using SWAT. Digital Elevation Model (DEM), land use land cover (LULC), soil and slope maps are provided as inputs data to run the SWAT model. DEM of study area is extracted under QGIS environment, then delineation of the watershed is carried out. Soil map, slope map and land use land cover map are used to create hydrological response units. Figure 6 shows the Pawana watershed with 63 sub-basins by using stream network, where outlet point present at Pawana dam. Then using weather data as an input with hydrological response unit run the SWAT model.

**Database preparation**—For the hydrological runoff simulation, the SWAT model needs the following data which will work as input.

## 4 Digital Elevation Model (DEM)

From the Bhuvan website (http://bhuvan.nrsc.gov.in/data/download/index.php) a digital elevation model (DEM) was downloaded with a resolution of 30 m. Figure 2 depicts the various elevation bands, with high values suggesting higher elevations in the Western Ghats.

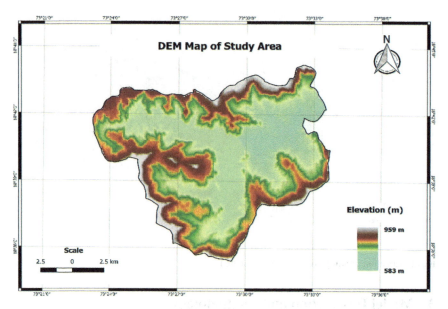

**Fig. 2** Digital elevation module of study area

## 5 Weather Data

SWAT requires weather data which includes value of daily precipitation (mm), maximum and minimum air temperature (o C), solar, wind speed and relative humidity data with information on the location of the weather station (latitude and longitude). For the period 1978 to 2014 From India Meteorological Department (IMD) gridded precipitation data and from www.swat.tamu.edu. Maximum and minimum air temperature (°C), relative humidity, solar, wind data is collected and used.

## 6 Land Use Land Cover (LULC) Map

The land use land cover map of Pawana watershed was digitized and rasterized in QGIS, which was developed by adding web map service (WMS) layer provided by www.bhuvan.nrsc.gov. in QGIS. Figure 3 shows the type of land cover presented in the watershed area.

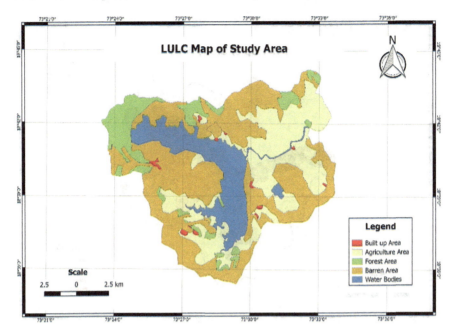

**Fig. 3** LULC map of study area

## 7 Soil Map

Soil map of Maharashtra developed by NBSS, Nagpur was geo-referenced and used for digitized and rasterize soil map. The Pawana river basin is made up of three separate soil types: loamy soils (77), drained loamy soils (118), and fine soil (266), but the majority of the basin is made up of rocky/sandy loam. Figure 4 shows a soil map for study area.

## 8 Hydrological Response Units (HRU)

The regions with homogeneous soil, land use and topographical characteristics are known as hydrological response units (HRU). SWAT requires the soil map, land use and slope classification map for creating HRU, which are areas within a watershed that respond similar to the given input. It is a method of representing spatial heterogeneity of watershed. Nearly 292 HRUs were created from SWAT overlay analysis, shown in Fig. 5.

QSWAT Ref 2012.mdb file SWAT Code has been modified by copying useroil.xls to excel. Thirty-three years of daily weather data such as precipitation (mm), wind speed (m/s), solar radiation (MJ/m$^2$) and temperature (°C) for the study area

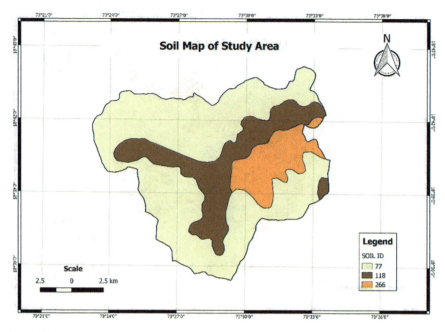

**Fig. 4** Soil map of study area

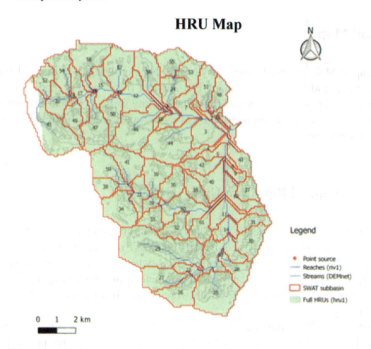

**Fig. 5** HRU map of study area

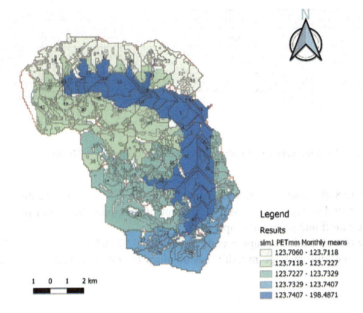

**Fig. 6** Monthly evapotranspiration in mm

have been downloaded from the http://globalweather.tamu.edu site. Calculation of TMPMX (mean maximum temperature), TMPMN (mean minimum temperature), TMPSTDMX (deviation of max temperature), SOLARAV (mean solar radiation), WNDAV (mean wind speed) was carried out with the help of Microsoft Excel. Precipitation-related parameters and dew point have been computed using custom software called pcpSTAT.exe and dewpoint.exe. All results are then copied to the WGEN WatershedGan.xls file. The SWAT model divides the watershed delineation into several sub-basins, and then determines the water balance components for each sub-basin.

## 9 Results

SWAT was run using the data obtained, analysed, and developed as mentioned above. The simulated result is obtained after successfully running SWAT. The outcomes can be presented on an overall, annual daily, monthly, and yearly basis. For each subbasin, it offers all types of hydrological components (Fig. 6).

The 63 subbasins were obtained in this analysis. Subbasins cover 10,763.73 km$^2$ of the total watershed area. Graphically, statically, and numerically, the output can be visualized. Figure 7 provides a graphic representation of the different parameters of subbasin 1.

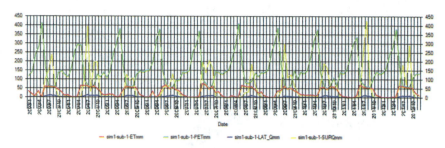

**Fig. 7** Graphical presentation of variations of all the parameters for subbasin 1

Figure 8 illustrates the groundwater contribution of subbasins on a monthly basis which includes water yield average amount of precipitation, actual precipitation, surface runoff and potential evapotranspiration.

The 63 sub basins occupy a combined area of 10,763.73 km$^2$. SWAT Check is used to measure average monthly basin values for rainfall, water yield, sediment

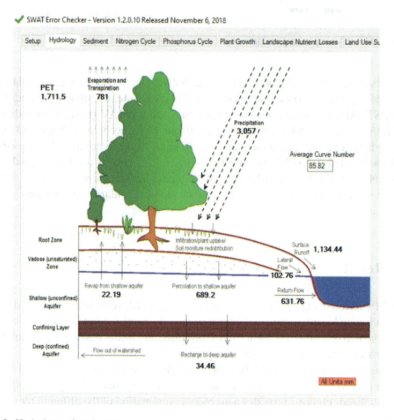

**Fig. 8** Hydrology of study area

# Sediment Yield Assessment of a Watershed Area Using SWAT

**Table 1** Hydrological response unit analysis output

| Land Use Pattern | Area (Ha) | % Watershed Area |
|---|---|---|
| Watershed (WATR) | 2721.35 | 25.28 |
| Agricultural Land (AGRL) | 1581.59 | 14.69 |
| Forest Area (FRST) | 1380.65 | 12.83 |
| Pasture (PAST) | 5080.14 | 47.2 |

yield, surface runoff, evapotranspiration, and potential evapotranspiration (PET). The SWAT measures the amount of various land use areas covered by the watershed, which is shown in percentage in the following Tables 1 and 2 (Fig. 9).

**Table 2** Average monthly basin values

| MONTH | RAIN (mm) | SNOWFALL (mm) | SURFQ (mm) | LATQ (mm) | WATER YIELD (mm) | ET (mm) | SED. YIELD (mm) | PET (mm) |
|---|---|---|---|---|---|---|---|---|
| 1 | 0.62 | 0 | 0 | 2.01 | 14.59 | 46.71 | 0 | 152.18 |
| 2 | 0.88 | 0 | 0 | 1 | 4.08 | 43.66 | 0 | 151.78 |
| 3 | 2.38 | 0 | 0 | 0.66 | 2.31 | 86.27 | 0 | 194.67 |
| 4 | 0.89 | 0 | 0 | 0.38 | 1.53 | 69.74 | 0 | 203.88 |
| 5 | 29.21 | 0 | 0.99 | 0.4 | 2.28 | 66.48 | 0.06 | 220.9 |
| 6 | 628.72 | 0 | 203.91 | 4.68 | 212.14 | 81.74 | 7.79 | 113.19 |
| 7 | 967.77 | 0 | 425.52 | 17.9 | 509.24 | 53.18 | 25.16 | 62.21 |
| 8 | 774.51 | 0 | 307.23 | 24.64 | 478.31 | 57.46 | 65.95 | 67.21 |
| 9 | 451.25 | 0 | 142.09 | 22.4 | 329.48 | 83.29 | 45.8 | 101.58 |
| 10 | 151.13 | 0 | 48.85 | 16.03 | 205.97 | 80.36 | 18.49 | 142.83 |
| 11 | 45.53 | 0 | 5.78 | 8.22 | 95.06 | 60.99 | 1.98 | 146.71 |
| 12 | 4.15 | 0 | 0.05 | 4.43 | 47.12 | 50.68 | 0.02 | 152.89 |

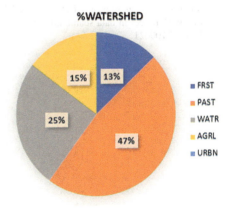

**Fig. 9** Pie chart of presenting watershed area in percentage

From Table number 2, shows the average annual monthly basin values of rain, water yield and sediment yield, evapotranspiration (ET), potential evapotranspiration (PET). The results show that during wet months, discharge and sediment yield increase, while during dry months, they decrease.

## 10 Conclusions

The best way to assess the availability of water in various components of the hydrological cycle, as well as adjustments between these components, is to use water balance. The SWAT model has proved to be a valuable method for simulating the hydrology of large basins at the watershed scale. This shows the simulated outcomes for each parameter. This water balance study reduces the possibility of drought and mismanagement, resulting in more effective use of available water supplies. The best way to assess the availability of water in various components of the hydrological cycle, as well as variations between these components, is to use water balance. Open-source geospatial techniques like Q-GIS were used to construct various thematic maps of the study area's impact on land use, soil and slope, which were then fed into the SWAT model. This water balance study lessens the chances of drought and poor management, resulting in more effective use of available water supplies.

## References

1. Prabhanjan A, Rao EP, Eldho TI (2015) Application of SWAT model and geospatial techniques for sediment-yield modeling in ungauged watersheds. J Hydrol Eng 20(6):C6014005
2. Jothiprakash V, Praveenkumar C, Manasa M (2017) Daily runoff estimation in Musi river basin, India, from gridded rainfall using SWAT model. Eur Water 57:63–69
3. Busico G et al (2020) Evaluating SWAT performance to quantify the streamflow sediment yield in a highly Urbanized Basin. Environ Sci Proc 2(1). Multidisciplinary Digital Publishing Institute
4. Mengistu AG, van Rensburg LD, Woyessa YE (2019) Techniques for calibration and validation of SWAT model in data scarce arid and semi-arid catchments in South Africa. J Hydrol Reg Stud 25:100621

# "NDVI: Vegetation Performance Evaluation Using RS and GIS"

### A. Khillare and K. A. Patil

**Abstract** Vegetation as an ecosystem's crucial part, plays a key role in soothing global environment. Normalized Difference Vegetation Index (NDVI) is one such remote sensing technique that is widely used to compute vegetation cover change. Remote sensing and Geographical Information System methods are used often in examining natural resources, determination of land changes and related planning work. The methodology discussed in this study is based on association with vegetation remote sensed data in the form of Normalized Difference Vegetation Index (NDVI). The major application of this index is to monitor the vegetative cover. NDVI is a function of reflected Near Infrared (NIR) and Visible (VIS) radiance's spectral contrast from a surface. A further study is made on the calculated NDVI to evaluate the agricultural drought index in the form of Vegetation Health Index. This index comprises of Vegetation Condition Index (VCI) and Land Surface Temperature (LST). Vegetation health is assessed based on VHI, which is suitable indicator of agricultural drought extent. A correlation is studied statistically between NDVI, VHI, precipitation and temperature. The present study is focussed on the Shirur and Khed talukas of Pune district for the years 2000, 2003, 2009, 2012, 2015 and 2018 for particular months. The use of data Landsat 7 ETM+ for the year till 2012 and data Landsat 8 OLI for 2015 and 2018 was made. Data was obtained from U. S Geological Survey. The precipitation data was taken from maharain.gov.in. Thus, vegetative cover over the specified area was studied including the drought severity. A liner regression analysis is performed using the evaluated data which can be used to forecast the vegetation condition as an early warning system for agricultural drought.

**Keywords** NDVI · VHI · Agricultural drought · Landsat · VCI

A. Khillare (✉) · K. A. Patil
Civil Engineering, College of Engineering, Pune, India
e-mail: khillareanjali21@gmail.com

K. A. Patil
e-mail: kap.civil@coep.ac.in

© The Author(s), under exclusive license to Springer Nature Singapore Pte Ltd. 2023
M. S. Ranadive et al. (eds.), *Recent Trends in Construction Technology and Management*, Lecture Notes in Civil Engineering 260,
https://doi.org/10.1007/978-981-19-2145-2_33

## 1 Introduction

Indian economy principally depends upon agriculture sector. Thus, agriculture can be entitled as a back bone of India. To develop these valuable asset-like landscapes, green open spaces is of a great concern. The way to advance in any field is to gain knowledge of advanced technology. With the help of advanced tools and a combination of educated and skilled employees, it is possible to perform better in various fields. Remote sensing and Geographical Information System are ways to enhance in the monitoring of vegetation, determination of land changes and planning work. The satellite imagery is used for yield and production forecasting, green cover inventory and assessment of drought-like catastrophe. By studying the temporal and spatial variations in the vegetative structure, monitoring and analysis can be performed. Vegetation indices are widely used in this field. There are numerous indices in use to study the changing pattern of the vegetation. These indices are the indicators of health and greenness of vegetation and a measure of density. To compute the vegetation indices the band combination is used which mainly comprises of red, green and infrared spectral bands.

One of the approved index is Normalized Difference Vegetation Index (NDVI) to monitor vegetation stress. Normalized Difference Vegetation Index (NDVI) offers better results to understand the vegetation change. It converts multi spectral data into a single image band which displays vegetation distribution. To quantify the healthy green vegetation based on satellite images, NDVI a graphical indicator makes use of the distinctive reflection of green vegetation in the visible and near infrared portions (NIR) portion of spectrum thus providing condition of the vegetation. NDVI value ranges from $-1$ to $+1$, value leaning towards $+1$ denoting healthier vegetation. Further, the remotely sensed data from satellite is adopted to analyse drought risk and has become widespread these days. Drought, a natural hazard causes noteworthy loss in the field of crop production, water supply and harness the well-being of humans. Recognition of such calamity becomes important to decrease its impact and severity in future. The way to mitigate drought effectively is to monitor such risk in advance with the help of remote sensing technology. Drought indices have been developed which comprise of spatial extent of vegetation, duration, intensity of meteorological factors. NDVI has an effective approach to monitor drought. A combination of NDVI in the form of Vegetation Condition Index (VCI) along with land surface temperature (LST) delivers a strong correlation thus providing information about agricultural drought beforehand.

Study aims to examine vegetation extent over the Shirur and Khed talukas of Pune District. The relationship between rainfall and NDVI in the context of these talukas Shirur and Khed is analysed for drought monitoring, rainfall being key factor in vegetative health. Use of Landsat dataset for examination of these indices is made and is explained in the methodology section.

## 2 Study Area

See Fig. 1.

Study is focussed on Shirur and Khed talukas of Pune district. Pune district is situated at 17.54–19.24-degree North latitude and 73.19–75.10-degree eastern longitude, located in western part of Maharashtra state, India with Shirur located at18.8250° N, 74.3776° E and Khed at 18.8405° N, 73.9072° E. In arrears to the geographical conditions, Pune district has uneven rainfall distribution. The Western part of district adjoining West coast having mountainous area with forest cover. Hence, rainfall intensity is more in this region than the eastern parts. Mainly, southwest monsoon winds are responsible for rain in summer. About 87% of rainfall is during the monsoon months June to September with maximum intensity in the month of July and August. Around 73% of cropped area is cultivated under rainfed conditions. Agriculture is Rabi crop dominated in these areas and the second being the Kharif Crops. Summer crop production is comparatively less. Shirur is on the banks of the Ghod River on the eastern boundary of Pune District. Shirur is influenced by the semi-arid climate. Khed lies in the western region of Pune district with river Indrayani flowing through this taluka with a somewhat cool climate.

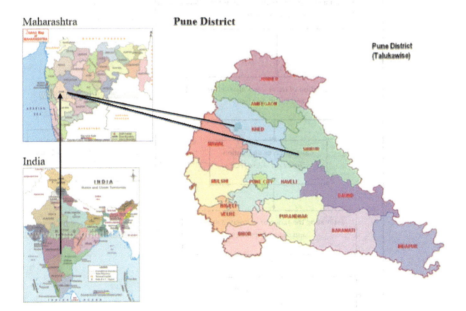

**Fig. 1** Location map of study area

## 3 Methodology

In the following research, Vegetation condition is identified for long-term sequence for the year 2000, 2003, 2006, 2009, 2012, 2015 and 2018 of months January, May, September and December. For the analysis of NDVI and land surface temperature (LST) multispectral and thermal data from Landsat are used. The dataset used for the year 2000, 2003, 2006, 2009, 2012 is Landsat 7 ETM+ and for 2015 and 2018 is Landsat 8 OLI. The following flowchart describes the methodology briefly (Fig. 2).

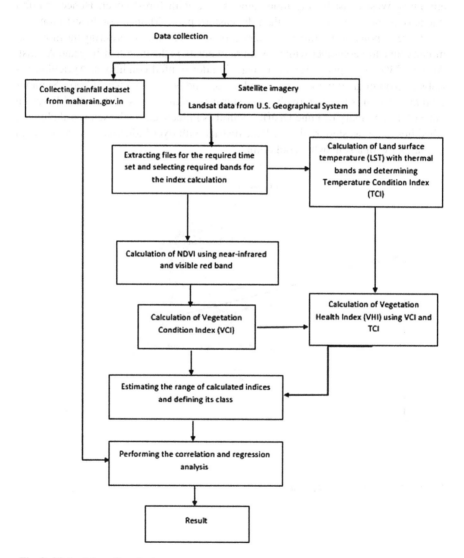

Fig. 2 Methodology flowchart

NDVI is expressed by the formula:

$$NDVI = \frac{NIR - VIS}{NIR + VIS} \quad (1)$$

NDVI calculation for the Landsat 7 ETM+ is calculated using relation:

$$NDVI = \frac{band4 - band3}{band4 + band3} \quad (2)$$

NDVI calculation for the Landsat 8 OLI is calculated using relation:

$$NDVI = \frac{band5 - band4}{band5 + band4} \quad (3)$$

where, NIR = Near Infrared Light, VIS = Visible Light.

Provided, bands are having their usual meaning according to the data set used. The classification of vegetative cover based on the NDVI values is according to the following Table 1 [1].

NDVI and LST time series have the capability to set out the various dynamics of dry conditions [2]. Landsat data sets are adopted for the calculation of vegetation condition index (VCI) and temperature condition (TCI). TCI is obtained from the calculated Land surface temperature which is calculated following the procedure in Landsat handbook provided by U.S Geological Survey. Combined VCI, TCI data is then used to compute VHI. The calculation of VHI is performed based on the relation shown below. VCI is determined from Normalized Difference Vegetation Index (NDVI) to monitor vegetation conditions [3]. The VCI data is obtained by the following equation.

$$VCI = \frac{(NDVIa - NDVImin)}{NDVImax - NDVImin} \times 100 \quad (4)$$

Here, NDVIa denotes the value of NDVI for current month, NDVImin, NDVImax represents the minimum, maximum NDVI values sequentially, during span of observation. VCI has been commended as drought tool. But, solely utilizing VCI is not sufficient to relate to drought analysis precisely. Thus, TCI is used to seize

Table 1 NDVI classification range

| Cover type | NDVI range |
|---|---|
| Dense green leafy vegetation | 0.500–1.000 |
| Medium green leafy vegetation | 0.140–0.500 |
| Light green leafy vegetation | 0.090–0.140 |
| Bare soil | 0.025–0.090 |
| Swampy areas/wet lands | −0.046 to 0.025 |
| Water bodies | −1 to 0.046 |

**Table 2** VHI classification range

| Drought class | VHI |
|---|---|
| Extreme drought | <10 |
| Severe drought | 10–20 |
| Moderate drought | 20–30 |
| Mild drought | 30–40 |
| No drought | >40 |

various replication of vegetation together with in-situ temperature as supplementary data. This can be attained by using thermal channels for drought monitoring. TCI calculation is carried out using the following relation

$$TCI = \frac{(LSTmax - LSTa)}{(LSTmax + LSTmin)} \times 100 \qquad (5)$$

where LSTa is the LST value for the current month, LSTmax and LSTmin represent maximum and minimum LST values respectively. VHI is calculated to analyse vegetation stress to define the drought severity. VHI is expressed as shown below

$$VHI = \alpha \, VCI + (1 - \alpha)TCI \qquad (6)$$

VHI is expressed as a composition of VCI and TCI by parameter $\alpha$. Here, parameter $\alpha$ in equation takes a value between 0 and 1. Since there is no foregoing information of its contribution for moisture and temperature conditions to the vegetation health, the value of $\alpha$ is generally taken as 0.5 [4, 5]. Following the classification of drought monitoring is used on the basis of following Table 2 [6]

Further, rainfall data is obtained from the Department of Agriculture, Maharashtra state for rainfall data. The annual precipitation was considered for the analysis. With the calculated data for NDVI, VHI, LST and rainfall data collected, correlation and linear regression analysis is carried out and the results of which are discussed below.

## 4 Results and Discussion

### 4.1 Results of NDVI

The NDVI image of Khed and Shirur are tabulated for month of January, May, September and December of the year 2000, 2006, 2009, 2012, 2015 and 2018 (Fig. 3).

"NDVI: Vegetation Performance Evaluation Using RS and GIS" 431

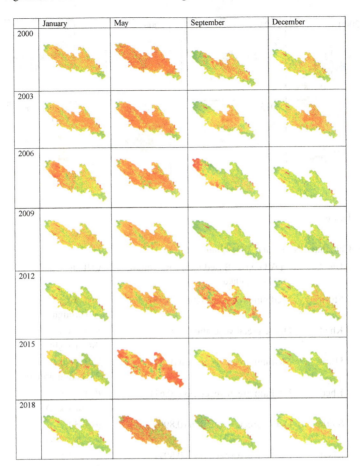

**Fig. 3** NDVI images

## 4.2 Results of VHI

See Fig. 4.

## 4.3 Classification

See Table 3.

Here, the values tabulated clearly illustrate the class and severity of NDVI and VHI. Temperature being an important factor in the calculation of VHI, it is to be noted that in fact of having high NDVI values, its VHI value obtained can be less. In the year 2012, Shirur with NDVI value is having VHI in class of mild drought but

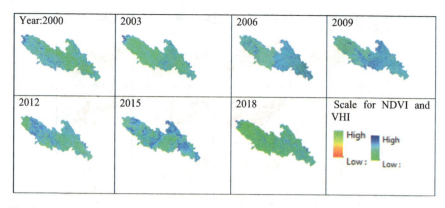

**Fig. 4** VHI images

**Table 3** Calculated NVDI and VHI values

| Year | Block | NDVI Class (Based on the maximum values obtained) | VHI class |
|---|---|---|---|
| 2000 | Shirur | Medium green vegetation (0.55) | 24.3–49.3 (No drought) |
|  | Khed | Dense green vegetation (0.83) | 23.6–74.32 (No drought) |
| 2003 | Shirur | Medium green vegetation (0.31) | 25.4–30.4 (Moderate drought) |
|  | Khed | Medium green vegetation (0.61) | 29.4–53.55 (No drought) |
| 2006 | Shirur | Dense green vegetation (0.890) | 26.3–52.3 (No drought) |
|  | Khed | Dense green vegetation (0.897) | 25.1–89.1 (No drought) |
| 2009 | Shirur | Dense green vegetation (0.942) | 25.8–54.8 (No drought) |
|  | Khed | Dense green vegetation (0.73) | 25.16—68.96 (No drought) |
| 2012 | Shirur | Medium green vegetation (0.71) | 10.42–36.6 (Mild drought) |
|  | Khed | Dense green vegetation (0.69) | 13.9–49.14 (No drought) |
| 2015 | Shirur | Medium green vegetation (0.579) | 25.90–39.70 (Mild drought) |
|  | Khed | Dense green vegetation (0.81) | 25.6–80.19 (No drought) |
| 2018 | Shirur | Light green vegetation (0.398) | 29.32–30.9 (Moderate drought) |
|  | Khed | Dense green vegetation (0.77) | 31.76–75.68 (No drought) |

**Table 4** Correlation values

| Block | NDVI and VHI | NDVI and Rainfall | VHI and Rainfall |
|---|---|---|---|
| Shirur | 0.838127 | 0.890172 | 0.889107 |
| Khed | 0.909356 | 0.778066 | 0.8072 |

for Khed NDVI is 0.69 and having VHI in the class of no drought category, as Shirur has a hot and dry climate with Khed having somewhat cool climate.

## 4.4 Correlation and Liner Regression Analysis Results

Correlation between NDVI, VHI and annual Rainfall (in mm) for Shirur and Khed (Table 4).

It is found that these indices proficiently be used to identify the spatio-temporal extent of agricultural drought. Furthermore, it explains classes of drought severity in the research areas through combined examination of vegetation health through vegetation condition along with temperature condition of vegetation with the help of VHI. Thus, the variables under the study are correlated with the strength of their association with Shirur and Khed as shown above.

Correlation between NDVI and Temperature.

The result obtained for the correlation of NDVI and temperature is negative which is Shirur corr. factor $= -0.396$ and Khed corr. factor $= -0.637$ which denotes that the rise in temperature lowers the value of NDVI.

(1) Regression analysis for NDVI and annual rainfall (in mm) (Table 5).

(2) Regression analysis for VHI and annual rainfall (in mm) (Table 6).

**Table 5** Regression analysis for NDVI and annual rainfall (in mm)

| Block | $r^2$ | Significance value | Equation |
|---|---|---|---|
| Shirur | 0.7924 | 0.0072 | $Y = 0.2428 + 0.000821 \times (X)$ |
| Khed | 0.60538 | 0.0398 | $Y = 0.62179 + 0.0001998 \times (X)$ |

**Table 6** Regression analysis for VHI and annual rainfall (in mm)

| Block | $r^2$ | Significance value | Equation |
|---|---|---|---|
| Shirur | 0.7905 | 0.0074 | $Y = 25.65103 + 0.03508 \times (X)$ |
| Khed | 0.6515 | 0.0281 | $Y = 48.0341 + 0.03163 \times (X)$ |

(a) $r^2$ value: The value of $r^2$ obtained is 0.7924(for Shirur NDVI &rainfall), the value closer to the 1, better the regression line fits the data. To check the significance of the result obtained statistically, significance value is observed which is 0.0072 (<0.05 Pearson coefficient) and is ok.
(b) Linear Regression equation: $Y = 0.2428 + 0.000821 \times (X)$ (for Shirur case). Here, $Y =$ Shirur annual rainfall and $X =$ NDVI values. The equation illustrates that 1 unit change in rainfall, changes NDVI value by 0.00821 units. Thus, these coefficients can be used to forecast the change in vegetative cover. The equations developed are useful in forecasting, crop insurance design and drought monitoring.

## 5 Conclusion

Study endeavours to identify the spatio-temporal extent of vegetative cover and agricultural drought for Shirur and Khed talukas with the use of remote sensing data based on NDVI, VHI and thus can be successfully used to identify the same. It explains the severity of drought with its class in the research areas by analysing both vegetation health by vegetation condition and temperature condition. Based on VHI values, Shirur area falls under medium green vegetation with moderate drought and Khed area with dense green vegetation with no drought. NDVI, VHI, rainfall and temperature under the study are correlated with the strength of their association for Shirur area and Khed area in the range of 77% to 90%. The estimated results of VHI lead to monitor the commencement of agricultural drought as an early warning system. The correlation and linear regression analysis can be made useful to forecast the vegetative cover condition. Results can further be useful for land use/ land cover database creation, identification of multiple crop and soil type.

**Acknowledgements** Authors are appreciatively grateful to US Geological Survey (USGS) for provision of Landsat dataset and Department of Agriculture, Maharashtra state for rainfall data used in the research.

## References

1. Bisrat E, Berhanu (2018) Identification of surface water storing sites using Topographic Wetness Index (TWI) and Normalized Difference Vegetation Index (NDVI). J Nat Resour Develop 08:91–100. https://doi.org/10.5027/jnrd.v8i0.09
2. Wang H, Lin H, Liu D (2014) Remotely sensed drought index and its responses to meteorological drought in Southwest China. Remote Sens Lett 5(5):413–422
3. Kogan FN (1995) Application of vegetation index and brightness temperature for drought detection. Adv Space Res 15(11):91100
4. Rojas O, Vrieling A, Rembold F (2011) Assessing drought probability for agricultural areas in Africa with coarse resolution remote sensing imagery. Remote Sens Environ 115:343–352

5. Kogan FN (2001) Operational space technology for global vegetation assessment. Bull Am Meterological Soc 82(9):194964
6. Sholihah RI (2016) Identification of agricultural drought extent based on vegetation health indices of Landsat data. Procedia Environ Sci 33:14–20

# Comparison of Suspended Growth and IFAS Process for Textile Wastewater Treatment

Sharon Sudhakar, Nandini Moondra, and R. A. Christian

**Abstract** Water is an essential commodity that is getting scarce due to its overuse. The primary consumers of water resources are the industries, and due to industrialization, the generation of wastewater is increasing. When discharged into water bodies, this wastewater contaminates the water bodies greatly, and thus its treatment is of utmost importance. The treatment facility is not economical for small and medium-scale industries. Hence, for such industries, a Common Effluent Treatment facility (CETP) is provided where the wastewater from a cluster of industries is collected, treated and finally discharged. The present study treats CETP wastewater using a sequencing batch reactor (SBR) with suspended biomass and an SBR-IFAS, which uses a combined suspended and attached growth process. The study was done on homogeneous textile wastewater from a CETP. HRT for the system was 10 h, with 9 h aeration and 1 h settling. The pre-treated wastewater was used as an influent in the study with an inlet COD concentration varying between 600 and 1160 mg/L. After the treatment in the SBR reactor, the maximum COD removal was 65.7% for homogeneous textile wastewater. Chromium removal was 48%, at a maximum suspended MLSS of 2852 mg/L with an SVI of 30 mL/g. In SBR- IFAS, maximum COD removal was 74.28%, chromium removal was 53.33% with a suspended MLSS of 2600 mg/L and attached biomass of 6.72 mg/m with SVI of 96 mL/g. Thus, the study indicates that the IFAS system incorporated in SBR will enhance the treatment efficiency while reducing the effluent concentration and help treat a larger volume of industrial wastewater for the same footprint.

**Keywords** Textile wastewater · SBR · SBR-IFAS · Ring lace · COD

## 1 Introduction

Water resources are getting reduced daily, so its sustainable use is considered one of the world's most critical environmental issues. Wastewater produced contains

---

S. Sudhakar (✉) · N. Moondra · R. A. Christian
Civil Engineering Department, SVNIT Surat, Surat, India
e-mail: sharonsudhakar6594@gmail.com

© The Author(s), under exclusive license to Springer Nature Singapore Pte Ltd. 2023
M. S. Ranadive et al. (eds.), *Recent Trends in Construction Technology and Management*, Lecture Notes in Civil Engineering 260,
https://doi.org/10.1007/978-981-19-2145-2_34

many pollutants like COD, phenols, heavy metals, etc., 12. The scarcity of water is increasing. Hence, the treatment of wastewater generated from various sources such as industries, agriculture, etc., is considered an effective solution for it to be reused or disposed of. Due to the growing industries globally, pollution from industries has become a significant concern that threatens the environment. It was seen that nearly 50% of small and medium scale industries are significant producers of industrial wastewater. In India, small-scale industries are very high, and these industries produce wastewater with very toxic pollutants, which makes treatment necessary before its disposal. Small scale industries cannot afford treatment systems because of space constraints, the need for skilled laborers for operation and maintenance, capital cost, etc. Hence, many other facilities were considered for the centralized treatment of such wastewater.

The aim of establishing CETP is to bring effluents of small and medium scale industries to a centralized treatment plant. CETP can be a better and economical option as equalization and neutralization are done more often. The effluent fluctuates in quality and quantity from different industries such as pharmaceutical, chemical, textile etc. The inlet characteristics of the wastewater vary depending on the industry and its processes. Therefore, this mixed wastewater treatment has gained more attention [1].

The effluent at CETP undergoes primary, secondary and tertiary treatment before being disposed. The CETP works continuously and has a series of operations like screening, equalization, primary clarifier for removing suspended solids present, biological treatment, secondary clarifier, advanced treatments, etc.

A wide variety of industries are present in India and along with the world. Industries produce wastewater based on the manufacturing process involved. Some major industries that use a significant amount of water for processing generate massive wastewater, including textiles, diaries, paper and pulp, chemicals, distillery, etc. The wastewater from the textile industry has high color, high organic and TDS load. The textile wastewater generated is highly polluted due to the presence of dyes, which are not readily amenable to biological treatment; hence primary treatment should be provided before secondary treatment. The color present in water causes scarcity of light, which is essential for developing aquatic organisms. Hence for the discharge of textile wastewater into the river, many treatment processes, including physical, chemical, biochemical, and hybrid treatment processes, have been developed to treat it economically and efficiently [2].

## 1.1 Integrated Fixed-Film Activated Sludge (IFAS) Process

The integrated fixed-film activated sludge (IFAS) process combines suspended and the attached growth systems. The biofilm in IFAS provides many advantages over conventional methods such as improved nutrient removal, larger solid retention time and high removal efficiency of anthropogenic pollutants. IFAS effectively removes dissolved organic matter as carbon and offers high nitrification and de-nitrification [3].

The mode of operation for IFAS is similar to the conventional activated sludge process. The IFAS media is available in many types such as plastic elements, webs, sponges, string type, lace type, etc., and are manufactured by many companies. The biomass is attached to the media, which helps in reducing the loading on the secondary clarifier. Loading on secondary clarifiers is reduced due to biomass attachment on the media [4].

This present study compares the sequencing batch reactor (SBR) and sequencing batch reactor with an integrated fixed-film activated sludge process (SBR-IFAS) for pre-treated wastewater from CETP.

The study involves analysing pH, dissolved oxygen, COD and chromium for homogeneous textile wastewater in both reactors and analysing the mixed liquor suspended solids grown in suspended and attached form.

## 2 Materials and Methodology

### 2.1 Plant Setup

The setup of the experiment consisted of two reactors of 15 L each. One reactor is an SBR with suspended biomass and another reactor is with IFAS strips, as shown in Fig. 1. Both reactors were provided with outlets at 40, and 50% of the volume and 40% is utilized in the present study. Aeration in the reactor was provided with the help of aquarium pumps. IFAS reactor contained ring lace as fixed media made of PVC. Five media of 14.5 cm height were inserted such that they remained suspended and fixed. The reactors were covered with nets to prevent the breeding of mosquitoes. Influent feeding was done as a batch process daily to maintain the HRT of 10 h during the study.

### 2.2 Raw Wastewater Characteristics

The wastewater was analysed for parameters like pH, Dissolved Oxygen (DO), mixed liquor suspended solids (MLSS), chemical oxygen demand (COD) and chromium. Table 1 illustrates the initial characteristics of wastewater. The wastewater for the study was collected from CETP containing homogeneous textile wastewater coming from 150 dyeing and printing industry. The wastewater collected from the CETP used for the present study has undergone primary treatment such as screening and coagulation using lime, PAC, polyelectrolyte, etc.

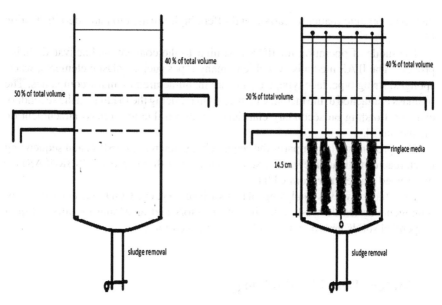

**Fig. 1** Reactor setup for SBR and SBR-IFAS

**Table 1** Initial characteristics of wastewater

| Wastewater Sample | pH | TSS (mg/L) | DO (mg/L) | COD (mg/L) | Chromium (mg/L) |
|---|---|---|---|---|---|
| Textile wastewater | 7.1–8.1 | 480–2000 | 0 | 600–1160 | 0.5–1 |

## 2.3 IFAS and Seeding

Seeding was done to acclimatize the bacteria in the new environment by aerating the reactor for a day followed by 1 h of settling. After settling, the supernatant was replaced by a fresh influent. MLSS of the seeded sample was around 1200 mg/L. Ring lace media is a fixed type of IFAS media used in biological wastewater treatment that will be submerged entirely even after decanting. It is a flexible PVC rope type of material attached to strands, as shown in Fig. 2. The strands contain loops that protrude out of them. Each loop has an approximate diameter of 5 mm, and aluminum frames were used to hold these media at the top and bottom of modules [5]. The strips were arranged so that it remained submerged after decanting to maintain the biomass grown on them.

**Fig. 2** Ring lace media used before the growth of biomass and after the growth of biomass

## 2.4 Methodology

The biomass provided as seeding from aeration tank of CETP was aerated for 24 h, and 40% of the effluent decanted after 1 h settling and filled with homogeneous textile wastewater as an influent. The reactors were aerated for 9 h, and 1 h settled, and 40% of the reactor volume was collected as an effluent after settling. The treated effluent was collected and analysed for pH, DO, COD, MLSS and chromium. The reactors were operated for 35 days. All the parameters were analysed according to standard procedure as per APHA. The attached growth on Ring lace media, as shown in Fig. 2, was analysed for MLSS by oven drying method.

## 3 Result and Discussion

### 3.1 Variation of pH

pH is an essential factor in the wastewater treatment process. Still, it to be in the optimum range (6.5–8.5) is necessary to grow microorganisms (Metcalf and [6]. The inlet sample's pH is of the range 7–8.3, and after treatment in both SBR-IFAS and SBR reactor, pH is noted to have decreased as given in Fig. 3. Treatment did not impart considerable change in pH. A decrease in the pH of wastewater after the treatment process was probably due to the release of carbon dioxide on organic matter's biodegradation.

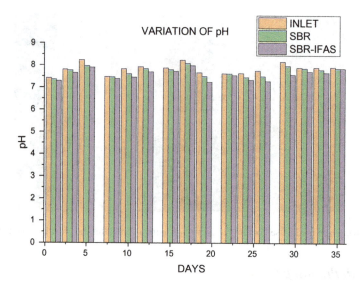

**Fig. 3** Variation of pH

## 3.2 Variation of Dissolved Oxygen (DO)

DO being a vital parameter is necessary for the effluent to be discharged into water bodies as it affects their properties. It is seen that initially, DO in both SBR-IFAS and SBR reactors was nearly zero as the removal of organic matter in wastewater was significantly less due to the acclimatizing period of biomass. After treatment, DO in both reactors increased, which might be due to the increased degradation of organic matter present in wastewater by the biomass. But with further increase in MLSS, the COD removal reduced as the reactor has high biomass, and this biomass with less food can contribute to organics and start to degrade. Effluent DO was higher; this is basically due to a decrease in COD due to biological treatment. DO in the SBR-IFAS reactor varied from 0 to 2.8 mg/L, and in SBR, it varied from 0 to 2.4 mg/L. An increase of DO to 2.4 mg/L in IFAS caused good removal, and the decrease towards the end may have been due to the reduction in active biomass due to high MLSS, causing the reduction in F/M ratio. The excess biomass contributes to organic matter, which will degrade the quality and hence some amount of wasting of active biomass will be necessary. The variation in DO for both reactors is as given in Fig. 4.

## 3.3 Variation of Chemical Oxygen Demand (COD)

The wastewater, after undergoing pre-treatment, has an inlet COD in the range 600–1160 mg/L. The SBR-IFAS reactor showed a reduction in effluent COD varying from

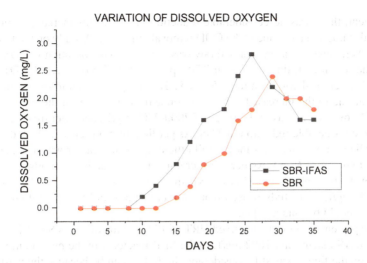

**Fig. 4** Variation of dissolved oxygen

180 to 720 mg/L with a removal varying from 12.5% initially to maximum removal of 74% with the biological treatment provided for an HRT of 10 h, and SBR showed an effluent COD of 240–780 mg/L with a percentage removal of 10–65% as shown in Fig. 5. The biomass grown on the media might also contribute to the removal of organic matter in wastewater. After maximum removal of 74%, the COD removal decreased; this may be due to increased biomass, thus overloading the reactor. In the present study, the TDS of influent was in the range of 2000-2500 mg/L, and after

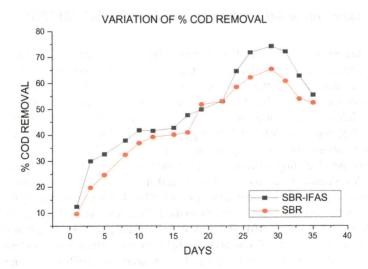

**Fig. 5** Variation of % COD removal

treatment, there was no considerable reduction in TDS. This increased amount of TDS also might have reduced the COD removal efficiency. A study on the treatment of synthetic textile wastewater and raw was done based on variation in TDS, it was seen that a 24 h cycle time increase in TDS up to 5000 mg/L reduced the COD removal and COD removal was seen to be 59.44%. The study was seen to reduce the COD removal due to the increased TDS effect on activated sludge. In the present study, the TDS of influent was in the range of 2000-2500 mg/L, and after treatment, there was no considerable reduction in TDS. As per literature presence of TDS decreases the COD removal, and thus the overall efficiency of the reactor was less than 75% in the removal of COD. As per the literature study on activated sludge for textile wastewater treatment for an HRT of 18 h, the removal efficiency was 77% at MLVSS of 5000 mg/L. The study proved that with an increase in MLVSS and HRT, more removal could be obtained [7].

The present study was on a fixed HRT of 10 h at a maximum VSS of 1430 mg/L in the IFAS reactor and 1602 mg/L in the SBR reactor. But the maximum removal based on the figure was still considerably high; this can be because the wastewater obtained was already pretreated. The biomass on the IFAS media might be more volatile as more removal was obtained in the IFAS reactor. With a further increase in HRT and MLVSS, more removal can be achieved. Previous studies showed that for treatment using ASP, COD removal for textile wastewater was seen to be 89% at an HRT of 2 days [8]. The presence of dye can reduce removal efficiency and hence requires more HRT. HRT being 10 h gave maximum removal of 74% in SBR-IFAS and 65% in SBR. But in the present study, the presence of dye was not heavily visible and also the presence of heavy metals like chromium was less hence more removal.

## 3.4 Variation of Mixed Liquor Suspended Solids (MLSS)

In the present study, the variation of mixed liquor suspended solids (MLSS) was studied four times a week during the study period. It was seen that MLSS kept on increasing daily in both the reactors. This increase in MLSS caused an increase in COD removal. MLSS in the SBR-IFAS reactor increased from 660 to 2600 mg/L with a final MLVSS concentration of 1430 mg/L, and in the SBR reactor, it increased from 610 to 2852 mg/L and VSS of 1602 mg/L in suspended form and IFAS as attached growth it was seen to be 6.72 mg/m. The variation of MLSS in both reactors is given in Fig. 6. Initial seeding to the reactor was done from the CETP aeration basin, which had an MLSS concentration of 1200 mg/L, and after 24 h and 1 h, the settling sample from the aeration tank was replaced with 6 L of effluent. In the present study for MLSS of 2852 mg/L, the removal was 65.7% in the SBR reactor and 74% for a suspended MLSS of 2600 mg/L in SBR-IFAS. A literature study on textile wastewater also showed that COD removal was 60% for MLVSS concentration of 3500 mg/L [9]. The increased removal can also be due to the non-interference of harmful components like chromium that affect the efficiency of biomass in the removal of organics. In the IFAS reactor, some biomass attached to the media may have contributed to the

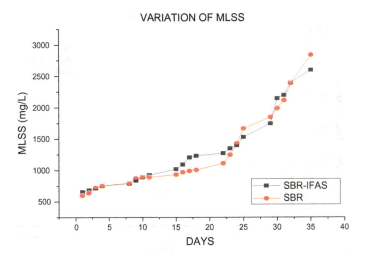

**Fig. 6** Variation of mixed liquor suspended solids (MLSS)

removal. Analysis of the variation of MLSS in previous studies was compared with the present study. A study on textile CETP in Rajasthan showed removal of COD was at 75–85% at an MLSS concentration of 3500 mg/L for an HRT of 12–15 h in an aeration tank after primary treatment [10], which is similar to the removal efficiency in the present study.

### 3.4.1 Variation of Chromium ($Cr^{6+}$)

If discharged into the river, chromium being a heavy metal, can damage the aquatic flora and fauna, and chromium consumption is harmful to humans. Textile wastewater supplied was colored. Thus, chromium was analysed as textile wastewater contains chromium due to dyeing. The wastewater was seen to have chromium in the range 0.5–1.02 mg/L, which is within limits per standards; this chromium reduction may be due to coagulants added. The effect of biomass in SBR-IFAS and SBR was analysed for the removal of chromium. After treatment using IFAS, chromium removal varied from 36.67% to 50%, and SBR reduction varied from 30 to 48.03%, as shown in Fig. 7. Studies have shown that activated sludge can remove chromium ($Cr^{6+}$) by 40% but is dependent on sludge acclimatization and a longer hydraulic retention time. According to the literature, as inlet chromium is less than 1 mg/L, it is proven that Chromium less than 5 mg/L will not affect COD removal 11. The minimum value of effluent through IFAS treatment was 0.3 mg/L, limiting the discharge limit to 2 mg/L. The amount of chromium present is mainly due to the usage of dye for the coloring of textiles.

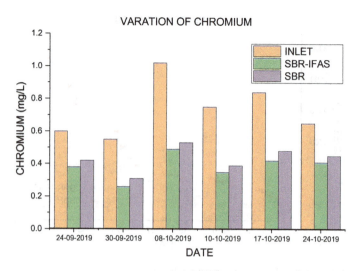

**Fig. 7** Variation of chromium ($Cr^{6+}$)

## 3.5 Sludge Volume Index

Sludge volume index (SVI) is used to analyse the sludge's settling characteristics in the present study. The sludge produced after treatment of CETP wastewater containing wastewater from textile industries was found in both SBR and SBR-IFAS. It was seen that SBR produced a sludge volume of 8 mL per 100 mL, thus giving an approximate sludge volume index (SVI) of 28 mL/g. For IFAS, a sludge volume of 25 mL per 100 mL, maintaining an SVI of 96 mL/g considering only the MLSS present in the suspended form for MLSS of 2852 and 2600 mg/L. The settling of sludge from the SBR-IFAS reactor was seen to be better than the SBR reactor.

## 4 Conclusion

Based on the study, the conclusions made were:

- A study was done to analyse the removal efficiency of SBR and SBR IFAS for textile wastewater.
- Based on the study, the SBR-IFAS treatment DO varied from 0 to 2.8 mg/L. The MLSS in suspended form was 2600 mg/L with a maximum COD removal of 74.28%, chromium removal of 53.33% and pH of 7.3–8; attached growth was measured by weight difference method obtained as 6.72 mg/m with an SVI of 96 mL/gm.

- For SBR, DO was in the range 0–2.4 mg/L with a maximum MLSS of 2852 mg/L and a maximum COD removal of 65.7%. Chromium removal was 48.03%, pH was maintained at 7.4–8.2, SVI was 30 mL/gm.
- Based on the removal efficiency of the COD SBR –IFAS reactor performed better than the SBR reactor, additional biomass grown on the Ring lace helped in improving the performance.
- The study was done without addition of nutrients; the IFAS reactor might have performed well if nutrients are supplied and also by maintaining the F/M ratio.

# References

1. Popat A, Nidheesh PV, Singh TSA, Kumar MS (2019) Mixed industrial wastewater treatment by combined electrochemical advanced oxidation and biological processes. Chemosphere 237:124419
2. Holkar CR, Jadhav AJ, Pinjari DV, Mahamuni NM, Pandit AB (2016) A critical review on textile wastewater treatments: possible approaches. J Environ Manage 182:351–366
3. Waqas S, Bilad MR, Man Z, Wibisono Y, Jaafar J, Mahlia TMI, Khan AL, Aslam M (2020) Recent progress in integrated fixed-film activated sludge process for wastewater treatment: a review. Journal Environ Manage 268:110718
4. Hait S, Mazumder D (2008) Scope of improvement of treatment capacity of activated sludge process by hybrid modification. J Environ Eng 7:147–158
5. Sen D, Farren G, Sturdevant J, Copithorn RR (2006) Case study of an IFAS system—Over 10 years of experience 4309–4324
6. Metcalf & Eddy, Inc. (2017) Wastewater engineering: treatment and reuse. Boston :McGraw-Hill
7. Mirbolooki H, Amirnezhad R, Pendashteh AR (2017) Treatment of high saline textile wastewater by activated sludge microorganisms. J Appl Res Technol 15:167–172
8. Maddad M, Abid S, Hamdi M, Boullagui H (2018) Reduction of adsorbed dyes content in the discharged sludge coming from an industrial textile wastewater treatment plant using aerobic activated sludge process. J Environ Eng 223:936–946
9. Kumar K, Singh GK, Dastidar MG, Sreekrishnan TR (2014) Effect of mixed liquor volatile suspended solids (MLVSS) and hydraulic retention time (HRT) on the performance of activated sludge process during the biotreatment of real textile wastewater. Water Resour Ind 5:1–8
10. Pophali GR, Kaul SN, Mathur S (2003) Influence of hydraulic shock loads and TDS on the performance of large-scale CETPs treating textile effluents in India. Water Res 37:353–361
11. Vaiopoulou E, Gikas P (2012) Effects of chromium on activated sludge and on the performance of wastewater treatment plants: a review. Water Res 46:549–570
12. Mojiri A, Ohashi A, Ozaki N, Kindaichi T (2018) Pollutants removal from synthetic wastewater by the combined electrochemical, adsorption and sequencing batch reactor (SBR). Ecotoxicol Environ Saf 137–144

# Innovative Arch Type Bridge Cum Bandhara for Economical and Quick Implementation of Jal Jeevan Mission

**P. L. Bongirwar and Sanjay Dahasahasra**

**Abstract** Total number of villages in India are 6,40,932, out of which 5,97,608 are inhabited villages having total 19.12 crore households. Jal Jeevan Mission (JJM) has been established by the government of India (GoI) with the provision of Rs 3.5 lakh crores to supply piped drinking water by functional household Tap connection (FHTC) by 2024 to all the rural population at the rate of 55 L per capita day. It can be noted that 75% of villages in India are below 1500 population. Due to inadequate water, especially in summer, tankers are required to be deployed at a very heavy cost. Generally, the water source at the village level is not reliable and assured, hence Regional Water Supply Schemes (RWSS) are planned at a cost of 150 lakhs to 200 lakhs per village and are executed to meet the need of cluster of villages. Each village is located near a natural stream, called Nalla which generally has a steep slope, hence rainwater drains off fast once the rain stops. They, therefore, do not help the ground recharging. Such villages are always connected by a small bridge which is proposed to be converted into bridge cum bandhara to act as a storage and crossing structure. After extensive model analysis and research, a new innovative affordable elliptical concrete arch is evolved which is permanently fixed to piers to create water storage up to a depth of 2.4 m. This water is proposed to be stored in nearby "Shet Tale" for meeting the need of drinking water and irrigation. Artificial recharging also can be adopted. Several structures have been constructed and it has revealed that besides meeting the drinking water need, we can have irrigation for 150–200 acres per village. Construction techniques are developed to complete structures in just 20 days and the cost of storage is 60% of norms as well 60% compared to the cost of cement weir/plug. The total cost per village (including BCB, solar pumps, Shet Tale, and filtering system) is on an average Rs 50 Lakhs per village against 150 lakhs to 200 lakhs in regional water supply scheme. Government intends to double the farmer's income and this objective will be fulfilled with this innovative design. Objective of this paper is to discuss this innovative technique. It is recommended

P. L. Bongirwar (✉)
Public Works Department, Govt. of Maharashtra, Mumbai, India
e-mail: plbongirwar@rediffmail.com

S. Dahasahasra
Maharashtra Jeevan Pradhikaran, Mumbai, India

© The Author(s), under exclusive license to Springer Nature Singapore Pte Ltd. 2023
M. S. Ranadive et al. (eds.), *Recent Trends in Construction Technology and Management*, Lecture Notes in Civil Engineering 260,
https://doi.org/10.1007/978-981-19-2145-2_35

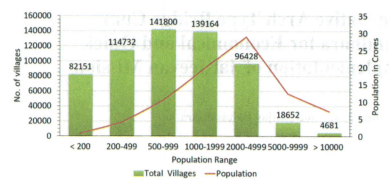

**Fig. 1** Population wise grouping of villages

that this model be adopted by JJM to achieve FHTC at minimum cost and earlier than 2024.

**Keywords** Jal Jivan Mission · Bandhara · Drinking water · Irrigation

## 1 Introduction

Total number of villages in India are 6, 40, 932, out of which 5, 97,608 are inhabited villages having total of 19.12 crore households. Population-wise grouping of villages is shown in Fig. 1.

Jal Jeevan Mission (JJM) has been set by the GoI to supply piped water to all the rural population at the rate of 55 L per day. Out of 19.12 crores, as of 18 Jan 2021, there still remains 12.72 crore households to be supplied with Functional Household Tap Connections (FHTC). It can be noted that 75% of villages are below 1,500 populations. If water supply is provided to these villages, then the goal of JJM can be achieved.

Objective of innovative and cost-effective initiative is to make water supply schemes with sustainable water sources at village level using Arch Type Bridge cum Bandhara (BCB).

## 2 Scenario in Rural Water Supply

### 2.1 Supply of Water by Tanker

Due to inadequate water especially in summer, tankers are required to be deployed at a very heavy cost. In Nagpur district, Rs 175 lakhs was spent to supply drinking water by tankers to 45 thirsty towns/villages, out of which 15 are below 1500 population.

**Fig. 2** Long queue for getting drinking water

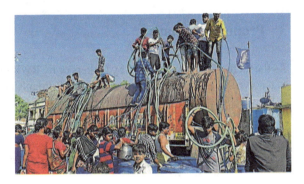

**Fig. 3** Rush for getting their buckets filled, when the tanker comes in the village

In Amravati, Rs 180 lakhs was spent for 65 villages, out of which 25 villages are below 1500 population. The supply by tanker creates a pathetic situation as can be seen from Figs. 2 and 3.

## 2.2 Regional Water Scheme

If the water source at the village level is not reliable and assured, Regional Water Supply Schemes (RWSS) are generally planned to meet the need of a cluster of villages. The source of RWSS is surface water from the reservoir of dam, major rivers, or natural lakes. Average cost per village is Rs 1.5–2 crore and per person the cost is Rs 5500–6000. Data of a few Regional Schemes are shown in Table 1.

**Table 1** Data of a few regional scheme

| SN | Name | No. of villages | Population rage | Below Population of 1500 | Cost per village (Rs lakhs) | Per person (Rs) | Source |
|---|---|---|---|---|---|---|---|
| 1 | Pompbhurna | 15 | 500–3200 | 10 | 192 | 5911 | Wainganga river |
| 2 | Khambhora | 60 | 140–3150 | 44 | 85 | 5654 | Pick-up weir on Katepurna |
| 3 | Mul | 24 | 120–4360 | 17 | 192 | 14,577 | Wainganga river |
| 4 | Pombhurna Grid | 15 | 471–2897 | 13 | 200 | 15,456 | Wainganga river |

## 3 Proposed Initiative

As the methods of water supply by tankers or by RWSS are too costly, there is a need to resolve this problem indigenously at an affordable cost to the villagers preferably at the village level.

### 3.1 Source of Water for Drinking Purposes

Groundwater sources of rural village schemes are mostly unsustainable due to dwindling of the water table (Fig. 4). To make it sustainable, recharging of groundwater should be carried out. This can be done by storing the rainwater (which otherwise goes waste) against the weir/bandhara and then allow it to percolate into the ground by various methods. But this action requires the first task to make water storages. The proposed initiative does the same (making of water storages) by constructing the *Bridge cum Bandhara* (BCB) across a small stream flowing near the village.

Source of water should preferably be developed near the village if feasible and possible as it proves to be a least cost solution. Generally, each village is situated

**Fig. 4** Dried river near Piplagaon-14th August 18

**Fig. 5** Typical innovative structures of BCB

near a small river called, Nalla. Such rivers have a steep slope of 1:250 and hence drain off once the rain stops. They, therefore, do not help ground recharging.

## 3.2 Techniques of Water Storage Structures Within Banks of Rivers

### 3.2.1 Arch Type Bridge Cum Bandhara (BCB)

A new model of constructing elliptical concrete arch (Fig. 5) abutting the bridge has now been evolved so that bridge serves dual purposes, i.e., crossing and storage. This type of arch type Bandhara is an almost permanent structure having a height up to 2.4 m in the form of an ellipse to create storage of water upstream. The design is evolved after extensive literature survey and model analysis.

### 3.2.2 Cement Plug/Cement Weir

It is the practice of agriculture/irrigation department to construct a permanent small weir known as cement plug/cement weir to store water within riverbank. Generally, storage planned is up to 2.5 m. These structures also create a pool of water which helps in increasing groundwater. The water is used as a source for drinking water, limited irrigation, etc. The typical structures are shown in the photographs in Figs. 6 and 7. Comparison is made in Table 2.

**Fig. 6** Cement plug/Cement weir

(a)  (b)

(a)  (b)

**Fig. 7** Anicuts/Cement plugs

## 4 Special Feature of Innovative Arch

### 4.1 Discharge Over Weir

IIT Roorkee [1] has done extensive research to find the discharge capacity of weir. This has helped us to evolve the most cost-effective shape to ensure maximum discharge with minimum afflux so that adjoining fields are not affected.

### 4.2 Obstruction to Waterway of Bridge

Basic question which is being raised is whether by creating a permanent storage, we are encroaching on waterway. This paper gives an explanation that we are just changing the direction of flow and not creating an obstruction to flow. By using the standard equations for discharge over weir after correcting for equivalent straight length as per the above findings, we can calculate afflux, highest flood level (HFL), and check whether it is likely to enter adjoining fields or touching the slab causing additional force for a high-level bridge. Storage depth has to be then readjusted.

### 4.3 Plan Area of Arch and Shape

Extensive model analysis was carried out at various places. The visual observation has revealed that if the plan area of the arch is 1.5 times the area of obstructed flow, then in that case the disturbance due to vortex formation is minimum.

### 4.4 Vortex Formation

During floods, water falls down from all sides of the arch and tries to rush out through the bridge opening. In this process, vortex is formed within the arch which kills all the

Innovative Arch Type Bridge Cum Bandhara for Economical … 455

**Table 2** Comparison

| Sr. No | Element | Cement weir | Curved weir | Remarks |
|---|---|---|---|---|
| 1 | Shape | | | Being gravity structure, requires more materials while in arch being closed RCC structure, structurally efficient and requires much less materials |
| 2 | Length | More than width of minor river | Same width | To keep afflux low, length has to be increased in weir but due to curved shape same length is adequate |
| 3 | Wing wall | To channelize flow and to contain within banks | No need | Other elements of bridge channelize flow |
| 4 | Protection | Excessive protection required | No need | In cement weir, energy is killed by hydraulic jump requiring protection while in arch; it is killed by vortex formed inside the arch |
| 5 | Stability | By gravity | Shape makes it safe | Lever arm at base being more, dead load makes it safe. Cut off add resistance against sliding Virtually very minor force transferred to pier |

energy. In conventional weir, the energy is killed by hydraulic jump requiring extensive protection measures while in the present case, virtually no protection measures are required.

### 4.5 Common Shape

To fulfill the condition of plan area of 1.5 times storage area, an ellipse shape is found to be ideal. With an objective of evolving common formwork, the shape for different spans is adjusted by keeping central straight portion.

### 4.6 Flow Characters of Small Rivers

Technology is developed for small villages to make them self-sufficient for drinking water by converting bridge as bridge cum bandhara to act as storage structures Such small villages generally have nearby stream(nalla) having a width of 40–60 m and maximum flood depth of 5 m. We have already checked the stability for this condition, i.e., when stored water is 2.4 m and flood depth up to 8 m and found to be safe. Several completed BCB structures have been already subjected to maximum flood.

## 5 Cost of Structure

The structure is innovative cost-effective and affordable compared to the norms prescribed by the Government of Maharashtra (GoM). For storage structure, the cost is 50–60% only and similarly compared to cement weir/plug, the cost is 60–70%. Cost per running meter comes to about Rs 90,000 per running meter. As mentioned above, the cost of supplying drinking water under RWSS per village is 150 lakh to 200 lakh while in this innovative design, it can be brought down to 50 lakhs.

## 6 Model Study

Extensive model studies have been carried out at (1) Pune Engineering College [2], (2) Amravati Engineering College, and (3) "Chandrabhaga" Canal, at Amravati. To get a better understanding of the behavior of the flow, the arrangement is shown in Fig. 8, 9 and 10.

All these model studies have revealed that there are no adverse flow conditions such as heavy turbulence, hydraulic jump, etc. On the contrary, the vortex formed due to water falling on all sides creates a vortex and thereby kills energy. There is

# Innovative Arch Type Bridge Cum Bandhara for Economical ...

**Fig. 8** Lab study at Pune Engineering College

**Fig. 9** Study at canal

**Fig. 10** Study at Chandrabhaga canal

also a water bath within a span of bridge and arch which also absorbs the energy of falling water. As such, no separate protection measures are required. The photographs of the actual structure are as shown, and on the contrary, the flow is smooth. The model study also confirmed that afflux as calculated by the curved weir formula is reasonably correct. For multiple buried Hume pipe culverts with high face walls or buried boxes with high face well, we can plan storage even above the top of the pipe or top of the slab as seen in the model.

Table 3 Details of various villages

| SN | Details of Village | | |
|---|---|---|---|
| | Name | Population | Bridge details |
| 1 | Bothali | 745 | 3 * 7 = 21 |
| 2 | Peryl | 1157 | 3 * 5 = 15 |
| 3 | Gaurla | 233 | 3 * 6 = 18 |
| 4 | Jamkrdi | 1226 | 5 * 2 = 10 |
| 5 | Karambi | 342 | 3 * 5 = 15 |
| 6 | Dudha | 320 | 1 * 7 = 7 |
| 7 | Korgaon | 1250 | 3 * 5 = 16 |
| 8 | Umra Lahura | 426 | 3 * 6 = 18 |

## 7 Water Required for Drinking Purpose

As per norms of JJM, each household should get 55 L of tap water and assuming 20% more the requirement of water for 1500 population is 42,000 cubic meters. As per the Irrigation department, 1 km$^2$ can give water of 14,197 cubic meters for 50% dependability and a 10 m waterway bridge can give at least 72,000 cubic meters minimum in the worst year. Table 3 indicates the quantity of water required for different population.

## 8 A Model for Supplying Drinking Water Through Water Stored in BCB

The overall water supply arrangement right from BCB to villages is shown in Fig. 11.

The total cost covering bandhara, Shet Tale, filtering system, solar, panels, pipe, pump, etc., works out to 50 lakhs. Solar panels are used to generate power to pump water to Shet Tale and later on to the elevated reservoir to distribute water. Excavation

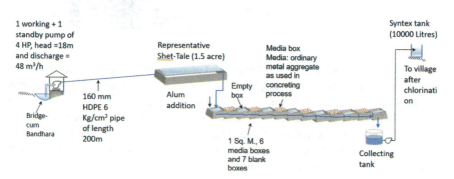

Fig. 11 Water supply arrangement from BCB to village

work of Shet Tale can be taken up under National Rural Employment Guarantee Act (NAREGA). Specially designed horizontal filter is used for designing filtering the water. Horizontal filter is easy to maintain. Water from the Bridge-Cum-Bandhara (BCB) as shown in Fig. 11 will be pumped by a pump of 5 HP which will be energized by a solar panel (Fig. 21). The rising main is 225 mm HDPE, PE100, 6 kg/cm2 with length of 200 m.

## 9 Success Stories

Based on the proposed initiative, several structures have been constructed. Benefits of a few are described in Table 4.

## 10 Design of Structure

**Guidelines for selection of site and data to be collected**

The elliptical concrete cost-effective technology is developed to convert existing minor bridges into storage structures; standard designs are prepared for bridges of span up to 10 m. Drawings are now available to cover the following typical structures to BCB. The typical bridge shown in photographs can be converted into storage structures. Standard drawings are available for various spans and foundation condition. Tentative sites are shown in Figs. 12a–c.

## 11 Construction Technologies

Work is being executed in a remote area and as cost is low, big contractors are not expected to come forward for executions. We intend to construct a water-retaining structure, hence, the quality of concrete has to be good and the same should be dense without any honeycombing. This is only possible if good quality formwork is adopted which means higher cost. Experience has shown that the cost of formwork is comparable to that of concrete cost. It has also been decided to use M-30 self-compact concrete which may not require much vibration and thin sections can be casted having dense concrete. We have developed the precast construction technology which ensures good and dense concrete. A typical structure can be completed in 20 days using common equipment as it is now available even at talk level.

**Table 4** Benefits of proposed BCB structures

| SN | Village and Population | Bridge details | Slope | Storage in Bank (CuM) | Benefits/ remark |
|---|---|---|---|---|---|
| 1 | Kathrgaon, population = 600 | 5 Spans of 6 m | 1:270 | 30,000 | Irrigation for Kharip for the first time for 150 acres, Water level in wells rose by 25 m** |
| 2 | Ellora, population = 1500 | 6 spans of 5 m | 1:370 | 35,000 | Water level rose by 16 m in wells, more than 150 acres Kharip irrigation expected** |
| 3 | Kinh, population = 1500 | 3 spans of 6 m | 1:500 | 20,000 | At least 100 acres irrigation expected, totally in tribal area |
| 4 | Piri-kalga, population = 2250 | 3 spans of 7 m | 1:260 | 20,000 | Additional 25 acres irrigation in Kharip Fishing has an additional source of revenue |
| 5 | Mull Marodi, population = 1500 | 3 spans of 5 m | 1:500 | 20,000 | Expected 100 acres in this year |
| 6 | Sipna in Dharni Backwater of 2.5 km | Major bridge | 1:1500 | storage of 3 lakhs Cum | New Shet Tale constructed filled by solar pump to create irrigation for 60 acres. All tribal population 80% below BPL |
| 7 | Anhdari near chiroli | Major bridge | 1:1400 | Backwater of 4 km, storage of 4.5 lakh Cum | Four natural lakes which will be filled using solar panels to create storage of 16 lakh cubic meters to irrigate 1600 acres; 60% population is tribal and BPL; 70 acres irrigation in Kharip done from backwater. Average cost of irrigation is Rs 20,000 Rs per acre |

(continued)

Innovative Arch Type Bridge Cum Bandhara for Economical ...    461

**Table 4** (continued)

| SN | Village and Population | Bridge details | Slope | Storage in Bank (CuM) | Benefits/ remark |
|----|------------------------|----------------|-------|----------------------|------------------|
| 8  | Apoti, Population = 2500 | 3 spans of 6 m | 1:600 | 30000 m$^3$ | Saline track in Black cotton soil, groundwater not suitable for irrigation or drinking, 150 acres irrigation done in backwater for the first time, a natural tank of 54 acres is proposed to be filled using solar panels to create irrigation for 500 acres at an average cost of Rs 12,000 per acre. Sand filter being installed to supply drinking water at the rate of Rs one for 20 L |

** The water supply for these two villages is part of RWSS for 114 villages at a cost of 221 crores. We can now supply water from our bandhara

(a) A typical bridge1      (b) A typical bridge2      (c) A typical bridge3

**Fig. 12** Typical bridges

## *11.1 Precast Cut-off*

Some of the precast cut off are shown in Figs. 13, 14, 15.

**Fig. 13** Precast cut off

**Fig. 14** Placing precast arch

**Fig. 15** Completed structure

## 12 Details of Successful Projects

### 12.1 Success stories

The initiative was conducted at the following places:

(1)  Saline track village "Apoti"

(a) BCB  (b) Imponded water  (c) ARCH BANDHARA  (d) WOMEN COLLECTING WATER  (e) DISINFECTION ROOM  (F) SHET TALE

**Fig. 16** Village Apoti

The scheme is shown in Fig. 16. The benefits are: (a) 150 acres irrigation with sprinkler, (b) Pure drinking water—20 L per family at Rs one, (c) 50-acre lake to create irrigation for 500 acres.

(2) Village "Mul Marodi"

The scheme is shown in Fig. 17.

(3) Other Villages

The schemes are shown in Fig. 18, 19, 20.

(4) Economical Irrigation

Solar pumps are used in this project as shown in Fig. 21. BCB at Major Bridge Andhari Irrigation is shown in Fig. 22.

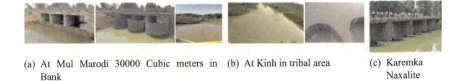

(a) At Mul Marodi 30000 Cubic meters in Bank  (b) At Kinh in tribal area  (c) Karemka Naxalite area

**Fig. 17** Village Mul-Marodi

**Fig. 18** BCB at Ellora and Kathargaon

**Fig. 19** BCB at Pipri Kalga

**Fig. 20** BCB at Apoti

**Fig. 21** At Dharani Irrigation for 60 acres using solar panels

## 13 Comparison of Cost Economics

Comparison of cost, of this novel model prepared in this paper and discussed above, has been made with the Regional water supply scheme of "Pomphurna whose source for supply is the Wainganga River.

"Pombhurna" regional water supply scheme—The scheme is planned to supply drinking water to 15 villages having a total population of 20,000 at a cost of 28.82 crores, out of which, the cost of bringing filtered water up to the village is 17.30.

Innovative Arch Type Bridge Cum Bandhara for Economical …

**Fig. 22** BCB at Major Bridge Andhari Irrigation for 1600 acre by filling existing lakes for 16 lakh cub m using solar panels, cost 20,000 Rs per acre

**Fig. 23** Photographs of BCBs near villages

Out of 15 villages, 11 villages have a population less than 1500, 3 villages have a population less than 1800, and balance one village has a population of 3200. The cost of the new design is 14.00 cores including irrigation for 1600 acres. Few photographs of bridges near these villages are attached in Fig. 23.

## 14 Conclusions

Jal Jeevan Mission (JJM) has a vision—"Har Guar Jal, every rural household has drinking water supply in adequate quantity of prescribed quality on regular and long-term basis at affordable service delivery charges leading to improvement in living standards of rural communities." Main task of JJM is to make functional household tap connection (FHTC) to all these households. A provision of Rs 3.5 lakh crores is made to do this job of providing FHTC at the rate of 55 L per day per person and target is set to be achieved by the year 2024.

Suggestion for meeting the above mission.

(1) The new design of elliptical concrete arch to convent bridges into storage structure, i.e., Bridge Cum Bandhara should be made a part of JMM.
(2) The above innovative design is cost-effective, and can be completed in 20 days through local contractor. It does not require land acquisition and is cost-effective
(3) Funds from JMM may be made available to water supply department for planning and execution.
(4) As construction can be completed in a short period, cost of escalation is ruled out. The technology should be given first preference.
(5) The additional benefit of irrigation in villages will help in doubling the income of the farmer.
(6) Target to achieve 175 GW by 2022 through non-conventional energy sources has been planned and large-scale adoption of solar pumps (additional electricity in the national grid) in this innovative design for villages would help to achieve this target.
(7) Draft national water policy 2020 gives importance to cost-effective sustainable option and discourage large-scale off-site storage and long-distance transport of water. As we are using rainwater near the villages, the objective of clean and safe water also will be fulfilled.
(8) Advantage of NAREGA can be taken for constructing the Shet Tale.
(9) More solar panels can be fixed to meet the power need of street lights, Gram Panchayat office, village health centers, and other goverment offices.
(10) Government of India has taken a decision that on new National Highway and PMGSY works, BCBs can be constructed at their cost hence these funds also are available for JMM
(11) Technology for water filtration being simple, the training classes can be arranged at village levels.

## References

1. Kumar S et al (2012) Discharge characteristics of sharp crested weir of curved plan-form. Res J Eng Sci 1(4):16–20
2. Nilgunde P Study of Arch Type Bridge cum Bandhara. A mini Project Report Guided by Dr. A. D. Thube, COE, Pune
3. PMGSY guidelines P -1701 /S/2011-RC (part) (FMS-337815), dated 16 -10-4, "Provision of Bandhara on the bridges under construction under PMGSY roads", Amendment to Para 8.5 (V)
4. MORTH RW Construction of BCB Barrage Structure on NH serving dual purpose i.e. to cross water body as well as to store water
5. R&D Project on Development of standard design and drawings of Bridge Cum Bandhara System-Department of applied Mechanics ,VNIT, Nagpur

# Green Synthesis of Zinc Oxide Nanoparticles and Study of Its Adsorptive Property in Azo Dye Removal

C. Anupama and S. Shrihari

**Abstract** Dyes are important colouring chemical compounds widely used in textile, paper, food, cosmetics and pharmaceutical industries. Most of the dyes are highly toxic and potentially carcinogenic. Exposure to high doses of these compounds can have severe effects in the human body. When untreated dye effluents are directly discharged into water bodies it leads to water contamination. Adsorption is an efficient method for the dye removal. It is also an economical technique. Green synthesis helps in the production of cost-effective and ecofriendly adsorbent. In this paper, green-synthesized zinc oxide nanoparticles were widely used for the removal of azo dye; methyl orange. Zinc oxide nanoparticles (ZnO NPs) were synthesized using Plectranthus Amboinicus leaf extract. The structure of nanoparticles was visualized using Field Emission Scanning Electron Microscope (FESEM). Nanoparticles due to their small size and thus large specific area are strong adsorbents. The effect of various parameters like dye dosage, zinc oxide nanoparticle loading, pH and contact time was evaluated. In batch adsorption study, maximum removal efficiency of 85.41% was obtained with zinc oxide nanoparticle dosage of 3 g initial dye concentration of 10 mg/l, pH 6 and contact time of 3 h. Adsorption parameters fitted well in the Langmuir isotherm model with a correlation coefficient ($R^2$) of 0.975. Adsorption kinetic studies revealed that adsorption strongly followed pseudo-second order. This paper suggests a safe, economical and environment-friendly technique for the removal of azo dyes.

**Keywords** Azo dye · Methyl orange · Zinc oxide nanoparticles · Adsorption

## 1 Introduction

Colour has always attracted human minds for aesthetic and social reasons. Dyes are colouring compounds that are used widely in paper, cosmetics, textile, pharmaceutical and food industries. To satisfy the demands and for marketing purpose

---

C. Anupama (✉) · S. Shrihari
Department of Civil Engineering, National Institute of Technology Karnataka, Mangaluru, India
e-mail: anupama23495@gmail.com

large variety of dyes are used to make different shades of fabrics. With the increasing demands, the environmental issues related to the production and applications of dyes are also increasing. Large quantities of wastewater with intensive colour and toxicity are being released in the environment during the production and manufacturing of dyes. About 200,000 tonnes of dyes are being released in textile effluents yearly due to the inefficiency in dyeing process. These dyes can be found in both suspended as well as dissolved state. They are very difficult to get degrade and are toxic, making the water fail to meet its purpose.

Coloured pollutants can be easily recognized in the environment. Textile industries in discharging effluents without any treatment contain chemicals that are toxic and which can lead to environmental issues like water pollution. It decreases the penetration of light into the water bodies which leads to decrease in the photosynthetic activities of aquatic plants. It also reduces the reoxygenation capacity of water bodies. Some of the synthetic dyes are carcinogenic. Dyes are resistant to biodegradation and are stable against temperature and light. Textile effluents also contain heavy metals that get accumulated in the body of organisms. It may affect the ecosystem adversely. Dyes are teratogenic or mutagenic in various species [8]. It may also damage the reproductive system, kidney, liver, nervous system and brain of human beings [9].

Dyes can be classified based on their chemical structure. Based on the chromophore present, they are classified as Nitro dyes, Azo dyes, Anthraquinone dyes, Indigoid dyes, Phthalein dyes, Nitroso dyes and Triphenyl methyl dyes [6]. Azo dyes are widely used dyes due to its productiveness, firmness and availability in a variety of colours. About 70% of total dyes used in industries are Azo dyes. The common structural feature of these dyes is the azo linkage (–N = N–) [2]. Azo dyes contain abundant organic compounds that can lead to various environmental issues. Effluents from various industries contain traces of dyes which may affect the water quality as well as human health because of its toxicity. Azo dyes are potentially carcinogenic and mutagenic. It is very difficult to remove these dyes by conventional treatment technologies [4]. In this paper, we study mainly about the azo dye; methyl orange.

Methyl orange is an anionic azo dye which is mostly used in pharmaceutical, textile industries, food, paper and printing industries and some research laboratories. It is a heterocyclic aromatic compound with molecular formula $C_{14}H_{14}N_3NaO_3S$ [1] (Fig. 1).

These dyes can cause eye irritation, gastrointestinal irritation with vomiting, nausea and diarrhoea. It may also cause skin irritation and respiratory tract irritation.

There are chemical, biological and physical methods for the removal of dyes from industrial effluents. Chemical method includes oxidation, electrocoagulation, photochemical oxidation, photocatalytic degradation, ozonation, etc. Biological method

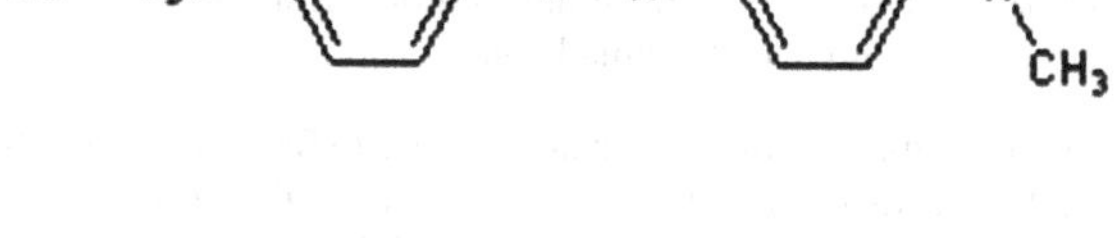

Fig. 1 Methyl orange [1]

includes decolouration and degradation of azo dye either by biodegradation or by adsorption on to biomass. Physical methods of removal include filtration, flocculation and coagulation, ion exchange, reverse osmosis and adsorption.

Adsorption due to its high removal efficiency of pollutants had become a trending removal technique. Adsorption results in high-quality products and is also economically feasible [9]. In adsorption, the process involved is the transfer of soluble organic dye from wastewater to highly porous solid materials. Substances that get collected on the solid interface is referred as adsorbate and adsorbent is the solid on which adsorption occurs. Factors that influence the adsorption efficiency are particle size of adsorbent, surface area, temperature, pH, time of contact and the interaction between adsorbent and adsorbate. Optimizing these parameters will greatly help in improving the removal efficiency. Activated carbon is the most commonly used adsorbent.

Nanoparticles have good potential in wastewater treatment. Its characteristics like high surface area, catalysis activity and high reactivity make it a good treatment method. Nanoparticles due to its small size and large surface area act as a good adsorbent. Various organic and inorganic pollutants, heavy metals and bacteria are removed using varieties of nanoparticles.

Zinc oxide has been most frequently used for the degradation of many organic pollutants. Zinc oxides have strong activity even in small amounts. They have great heat resistance and durability [3]. Due to its high surface area to volume ratio, they have large number of active sites for the interaction of pollutant species for increased adsorption capacity. They also have antibacterial properties [7]. They also have potential application in heavy metal removal also. There are many methods for the synthesis of ZnO nanoparticles.

Green Synthesis of nanoparticles has been widely increased now since they are economical, safe and environment friendly. Bacterial extracts, plant extracts, certain enzymes and other organic sources are used for green synthesis. Anisochilus carnosus, Tamarindus indica, Solanum nigrum, Vitex negundo, Plectranthus amboinicus and Hibiscus rosasinensis are some of the plant extracts used for the synthesis of ZnO Nps [5]. Green synthesis can also help in the reduction of use of strong chemicals as reducing agents.

The main objectives of this paper include green synthesis of ZnO nanoparticles, study the effective removal of methyl orange dye by adsorption onto ZnO nanoparticles, understand the effect of various parameters like ZnO dosage, initial concentration of dyes, pH and contact time on adsorption.

## 2 Materials and Methodology

### 2.1 Materials

Methyl orange (MO), Zinc nitrate hexahydrate with purity of 98% for ZnO NP synthesis and ethanol were purchased from Gennext. Plectoranthus amboinicus leaves were collected from Kozhikode, Kerala.

### 2.2 Adsorbent Preparation and Analysis

Zinc oxide nanoparticles were selected as the adsorbent. It was synthesised biologically using plectoranthus ambinicus leaves.

#### 2.2.1 Collection and Preparation of Plectoranthus Amboinicus Leaf Extract

Plectoanthus amboinicus leaves were collected and cleaned in distilled water. Then, 5 g of leaves were taken and cut into smaller pieces. The leaves were then boiled in 30 ml of distilled water for about 15 min. Then the cooled leaf extract was filtered through Whatman No. 41 filter paper and stored in the refrigerator for further use.

#### 2.2.2 Biosynthesis of Adsorbent ZnO Nanoparticles

Zinc nitrate solution (0.1 M) was prepared using 30 mL of distilled water. To this solution 10 mL of leaf extract was added. It is continuously stirred at 80 °C for 4 h. Yellow colour paste will be obtained through centrifugation. This paste was collected in a crucible and kept in a hot air oven at 100 °C for 3 h. A pale yellow colour powder is obtained and it should be washed with ethanol and water. Zinc oxide Nanoparticles will be obtained after annealing the sample in a furnace for 2 h at 400 °C.

#### 2.2.3 Characterization of Biosynthesized Zinc Oxide Nanoparticles

The biosynthesized ZnO NPs were characterized with the help of Field Emission Scanning Electron Microscope (FESEM). Instead of light, these microscope works with electrons. These electrons are liberated by a field emission source. It helps to obtain the surface morphology of the nanoparticles. The nanoparticles should be washed, dried and made free from any impurities before doing characterization.

## 2.3 Preparation of Adsorbate and Analysis

Methyl orange was the azo dye selected as adsorbate.

### 2.3.1 Preparation of Azo Dye

Synthetically prepared MO dye solution was used for the study. 100 mg/l of stock solution of the dye was prepared with 500 ml water. Sample solution was prepared by diluting this stock solution just before the study.

## 2.4 Adsorption Studies

### 2.4.1 Batch Adsorption Studies

Batch adsorption study was done by adding the required amount of ZnO nanoparticles to the prepared dye solution and mixed well on an orbital shaker for preferred time. Then the solution will be filtered and the residual concentration of dye will be monitored using UV–visible spectrophotometer. Effect of dosage of ZnO NPs, effect of initial concentration of dyes, effect of pH and effect of contact time on the adsorption rate was analysed.

The percentage removal of dye ZnO nanoparticles is calculated using,

$$\% \text{ Dye removal} = \frac{C_0 - C_e}{C_e} \times 100$$

$C_0$ is the initial concentration of dye and $C_e$ is the concentration of dye at equilibrium.

## 3 Results and Discussions

### 3.1 Characterization of Zinc Oxide Nanoparticles

The morphology of the ZnO NPs synthesized was examined using Field Emission Scanning Electron Microscope (FESEM). The analysis gave a lotus petal-like flake-shaped structure. The average size of the ZnO NPs was 1 μm. Large surface area of ZnO NPs enables it to adsorb large amount of pollutants making it a good adsorbent (Fig. 2).

**Fig. 2** FESEM image of ZnO NPs synthesised

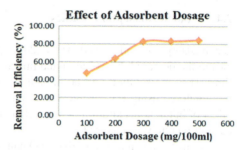

**Fig. 3** Effect of ZnO NPs dosages on removal of methyl orange dye

## 3.2 Methyl Orange Dye Adsorption Analysis

### 3.2.1 Effect of Adsorbent Dosage

For a fixed concentration of 10 mg/l and a volume of 100 ml, the percentage removal of methyl orange dye was examined by changing the dosage of ZnO NPs from 100 to 500 mg. Figure 3 shows that there was a good increase in the removal of dye from 100 to 300 mg dosage of ZnO NPs. This increase might be due to the availability of more sites and large surface area for adsorption with the increase in adsorbent dosage. But after 300 mg further increase in the adsorbent dosage didn't show much increase in the removal. This may be the result of formation of aggregation of particles. Since there was no significant increase in the removal efficiency with the further increase in adsorbent dosage, the optimum dosage was taken as 300 mg.

### 3.2.2 Effect of Initial Adsorbate Concentration

The initial concentration of methyl orange dye was varied from 5 to 25 mg/l at a constant pH and contact time. Then the efficiency in the removal of dye

Fig. 4 Effect of Initial adsorbate concentrations on removal of methyl orange dye

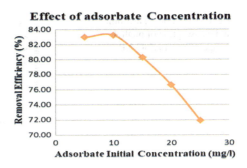

with the optimum dosage of ZnO NPs and was studied. Initially, there was good removal efficiency but with increase in concentration, the removal efficiency also decreased (Fig. 4). The initial increase in the removal might be due to the availability of large number of vacant active sites at lower concentrations. But when concentration increased the sites may have got saturated and difficult to capture the dye molecules.

### 3.2.3 Effect of pH

Optimum concentration of methyl orange dye was prepared at a varying pH of 2 to 12 and was allowed to react with ZnO NPs at a contact time of 3 h. Figure 5 shows the result of this adsorption study. Initially, there was an increase in the removal efficiency till a pH of 6 then the removal efficiency decreased. Maximum adsorption at pH 6 might be the result of good electrostatic attraction between the positively charged ZnO NPs and the anionic dye molecules. The low removal efficiency at very low pH may be because of the larger quantity of protons getting attached with N = N and making methyl orange dye protonated and as a result repelling the positively charged ZnO NPs. When pH was increased there will be more negatively charged sites. They may repel the anionic dyes and reduce the adsorption. Also more OH$^-$ ions may be available at higher pH which competes with the anionic dyes to occupy the positive sites which in fact reduces the removal efficiency.

Fig. 5 Effect of pH on adsorption of methyl orange dyes on to ZnO NPs

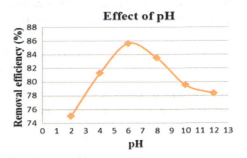

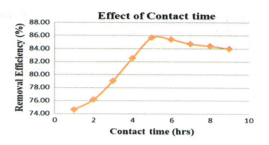

**Fig. 6** Effect of contact time in the adsorption of methyl orange dye on to ZnO NPs

### 3.2.4 Effect of Contact Time

Optimum concentration of methyl orange dye with optimum pH was prepared and was allowed to react with a varying time interval of 1–5 h. The result of this adsorption study is shown in Fig. 6. The result shows a good increase in the removal efficiency up to 3 h, but later there was no considered change in the removal efficiency. This probably resulted from the saturation of active sites at 3 h and followed by adsorption and desorption process.

## 3.3 Adsorption Isotherms

At a constant temperature, the quantity of adsorbate adsorbed is a function of concentration of adsorbent. This function is called the adsorption isotherm. It is important to understand the adsorption behaviour to identify isotherm model. Isotherm models help us to identify the adsorption mechanism, surface properties and adsorption capacity. It helps to assess the efficiency of the synthesized adsorbent. The adsorption equilibrium data can be analysed using Langmuir and Freundlich isotherms.

### 3.3.1 Langmuir Isotherm

Langmuir isotherm was developed by some assumptions like the monolayer adsorption on a uniform surface with finite number of adsorption sites all having the same energy and the adsorption process is reversible. It is expressed as shown below.

$$Q_e = \frac{Q \max K_L C_e}{1 + K_L C_e}$$

Langmuir equation is rearranged and isotherms were plotted between $C_e/Q_e$ versus $C_e$ and the Langmuir constants were found out from the slope and intercept.

$$\frac{C_e}{Q_e} = \frac{1}{Q \max K_L} + \frac{C_e}{Q \max}$$

**Fig. 7** Langmuir isotherm

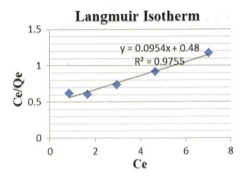

The important feature of Langmuir model can be defined by a dimensionless constant $R_L$ (Seperation factor).

$$R_L = \frac{1}{1 + K_L C_0}.$$

Figure 7 shows the Langmuir isotherm plotted.
Langmuir Constants,

$$Q_m, \text{ Maximum adsorption Capacity} = 10.482 \text{mg/g}$$

$$K_L, \text{ Langmuir Isotherm Constant} = 0.19875 \text{l/mg}$$

$$R_L, \text{ Separation factor} = 0.1675$$

$$R^2, \text{ Coefficient of Determination} = 0.9755$$

The maximum adsorption capacity obtained is 10.482 mg/g which is more than many other studies. The determination coefficient value ($R^2 = 0.9755$) shows strong positive correlation. Separation factor indicates whether the isotherm is irreversible ($R_L = 0$), favourable ($0 < R_L < 1$) or unfavourable ($R_L > 1$). It can be seen that $R_L$ value obtained is less than 1 indicating the adsorption is favourable.

### 3.3.2 Freundlich Isotherm

It was developed based on some assumptions like heterogeneous surface energy. This isotherm defines reversible and non-ideal adsorption. It is not restricted to monolayer adsorption, multilayer adsorption can occur. It is expressed as below.

$$Q_e = K_f C_e^{1/n}$$

**Fig. 8** Freundlich isotherm

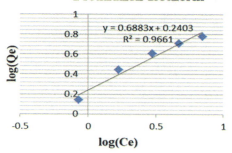

Freundlich equation Freundlich isotherms were plotted between $\log(Q_e)$ versus $\log(C_e)$.

$$\log(Q_e) = \log K_f + \log C_e$$

Figure 8 shows the plotted Freundlich isotherm.
Freundlich Constants,

$K_f$, Freundlich Capacity Factor $= 3.576$ mg/g

$1/n$, Freundlich Intensity Parameter $= 0.6883$

$n$, Freundlich constant $= 1.4528$

$R^2$, Coefficient of Determination $= 0.9661$

Freundlich Capacity Factor $K_F$ is a factor showing the adsorption capacity which is equal to 3.576 mg/g. 1/n indicates adsorption intensity or surface heterogeneity. When 1/n becomes closer to zero, it indicates more heterogeneous and when below 1, it shows a normal Langmuir isotherm and when greater than 1 indicates cooperative adsorption. n represents adsorption intensity. When $n \ll 1$, adsorbate was unfavourably adsorbed on the adsorbent and when $1 < n < 10$, adsorption was favourable. Here, $n = 1.4528$ which means adsorption is favourable. Coefficient of Determination obtained is 0.9661.

Comparing both the isotherms, $R^2$ value for Langmuir isotherm and Freundlich Isotherms are 0.9755 and 0.9661, respectively. This shows that the most appropriate isotherm that describes this adsorption study is the Langmuir isotherm. This shows that adsorption of methyl orange dye is predominantly by homogenous monolayer formation.

## 3.4 Adsorption Kinetics

The basics of kinetic studies are the kinetic isotherm obtained experimentally by tracking the adsorbate concentration against time. To study the mechanism of adsorption of the adsorbate onto the adsorbent kinetic studies are done.

### 3.4.1 Pseudo-First Order

The differential equation for pseudo first-order reaction is given by lagergren equation,

$$\frac{dQ}{dt} = k_1(Q_e - Q_t)$$

Its linearized form is obtained by integrating the above equation under the boundary condition $Q_t\ (t = 0) = 0$,

$$\log(Q_e - Q_t) = \log(Q_e) - \frac{k1}{2.303}t$$

The pseudo-first-order graph plotted between $\log(Q_e - Q_t)$ versus $t$ is given in Fig. 9. The pseudo-first-order parameters were found out from the slope and y intercept of the graph.

Pseudo-First-Order parameters,

$$\text{Rate constant, } K_1 = 0.0048\ \text{min}^{-1}$$

$$\text{Adsorption capacity at Equilibrium, } Q_e = 0.8977\text{mg/g}$$

$$\text{Coefficient of Determination. } R^2 = 0.3838$$

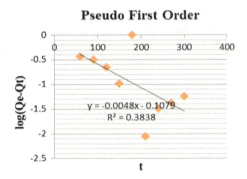

Fig. 9 Pseudo-first order

By analysis, it can be seen that the $R^2$ value is very less and also the experimental adsorption capacity at equilibrium did not match with the theoretical $Q_e$ value obtained from the graph. It shows that adsorption of methyl orange dye on to ZnO NPs does not elucidate pseudo-first-order kinetics.

### 3.4.2 Pseudo-Second Order

This model assumes that the uptake rate is second order with respect to available sites. That is when the adsorbate reacts with two adsorption sites, the rate of adsorption is as follows,

$$\frac{dQ}{dt} = k_1(Q_e - Q_t)^2$$

Its linearized form is,

$$\frac{t}{Qt} = \frac{1}{k_2 Q_e^2} + \frac{t}{Q_e}$$

The graph plotted between $t/Q_t$ versus t graph is shown in Fig. 10. The pseudo-first-order parameters were calculated from the slope and y intercept of the graph.

Pseudo-Second Order Parameters,

Rate constant, $K_2 = 0.03073$ min$^{-1}$

Adsorption capacity at Equilibrium, $Q_e = 2.942$ mg/g

Coefficient of Determination. $R^2 = 0.9984$

The correlation coefficient value of pseudo-second-order graph is very high and also the experimental $Q_e$ value shows a good agreement with the $Q_e$ value obtained

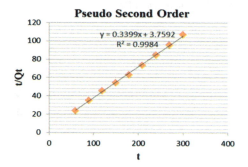

**Fig. 10** Pseudo-second order

from graph. This shows that adsorption of MO dye onto ZnO NPs follows the pseudo-second-order adsorbtion kinetics.

## 4 Conclusion

The main aim of this study was to develop an ecofriendly and economical method for the removal of azo dyes from the wastewater. ZnO NPs were green synthesized followed by batch adsorption study was done for removal of methyl orange dye. Characterization using FESEM was done on green synthesized ZnO NPs and it showed a large surface area which is important for the adsorption. The optimum dosage, initial concentration of azo dye, pH and contact time of 0.3 g, 10 mg/L, 6 and 3 h gave a maximum removal of 85.41%. Adsorption isotherm studies clearly show that the adsorption of methyl orange dye on to ZnO NPs had strongly followed Langmuir adsorption isotherm and the maximum adsorption capacity was 10.482 mg/g. Adsorption had followed pseudo-second-order adsorption kinetics. Green synthesized ZnO NPs can be used for efficient removal of azo dyes in a cost-effective and ecofriendly manner.

## References

1. Bazrafshan E, Zarei AA, Nadi H, Zazouli MA (2014) Adsorptive removal of methyl orange and reactive red 198 dyes by Moringa Peregrina ash. Indian J Chem Technol 21(2):105–113
2. Benkhaya S, El Harfi S, El Harfi A (2017) Classifications, properties and applications of textile dyes. A review
3. Dimapilis EAS, Hsu CS, Mendoza RMO, Lu MC (2018) Zinc oxide nanoparticles for water disinfection. Sustain Environ Res 28(2):47–56
4. Ecology A, Bhatnagar P (2014) Assessing mutagenicity of textile dyes from pali (Rajasthan) using ames bioassay
5. Fakhari S, Jamzad M, Kabiri Fard H (2019) Green synthesis of zinc oxide nanoparticles: a comparison. Green Chem Lett Rev 12(1):19–24
6. Mariselvam R, Ranjitsingh AJA, Thamaraiselvi C, Ignacimuthu S (2019) Degradation of AZO dye using plants based silver nanoparticles through ultraviolet radiation. J King Saud Univ Sci 31(4):1363–1365
7. Ngoepe NM, Mbita Z, Mathipa M, Mketo N, Ntsendwana B, Hintsho- Mbita NC (2018) Biogenic synthesis of ZnO nanoparticles using Monsonia burkeana for use in photocatalytic, antibacterial and anticancer applications. Ceram Int 44(14):16999–17006
8. Samchetshabam G, Hussan A, Choudhury TG (2017) Impact of textile dyes waste on aquatic environments and its treatment impact of textile dyes waste on aquatic environments and its treatment
9. Yagub MT, Sen TK, Afroze S, Ang HM (2014) Dye and its removal from aqueous solution by adsorption: a review. Adv Coll Interface Sci 55(209):172–184

# Anthropogenic Impacts on Forest Ecosystems Using Remotely Sensed Data

**Gaurav G. Gandhi and Kailas A. Patil**

**Abstract** Modernizations are beneficial to the comfortable condition of living; they also negatively impact the world of which we belong. Human activity that causes environmental harm (either directly or indirectly) on a global scale include human reproduction, overconsumption, overexploitation, pollution, and deforestation, to name just a few. Normalized Difference Vegetation Index (NDVI) is a band ratio technique that helps to determine the health of the vegetation. NDVI has been extensively used to study land use and land cover changes, agricultural drought analysis, forest fire analysis, and climate change detection. In this study, NDVI of three forest areas in the Vidarbha region of India, namely, Pench National Park, Tadoba Tiger Reserve, and Gadchiroli forest are studied to determine the anthropogenic effects of Nagpur, Chandrapur, and Gadchiroli on the forest areas respectively for past twenty-one years. In the process, the land use and land cover of the cities have been studied and change in weather conditions has been predicted. On the other hand, the meteorological parameters, like the temperature and rainfall have been studied for the three forests and have been related to NDVI of these forests and then compared with the weather predictions from the land use and the land cover maps of the cities. The anomalies in these two are justified with probable justifications. The study revealed the effects of local anthropogenic activities on these forests. It is concluded that the local human activities have an impact on the forest vegetation along with the locational aspect of an area.

**Keywords** NDVI · Land use/Land cover · Climate change · Precipitation · Temperature

---

G. G. Gandhi (✉)
Environmental and Water Resources Engineering, College of Engineering, Pune, India
e-mail: gandhigg19.civil@coep.ac.in

K. A. Patil
Civil Engineering, College of Engineering, Pune, India

© The Author(s), under exclusive license to Springer Nature Singapore Pte Ltd. 2023
M. S. Ranadive et al. (eds.), *Recent Trends in Construction Technology and Management*, Lecture Notes in Civil Engineering 260,
https://doi.org/10.1007/978-981-19-2145-2_37

## 1 Introduction

Human beings have experienced numerous changes in the course of their culture since history has been documented. Although these modernizations are beneficial to the comfortable condition of living, they also negatively impact the human life. With great benefits that the advancement of technology has given us, it has also caused some environmental harm.

Anthropogenic practices have injected many toxins into the biosphere in the form of solids, liquids, and gases, which in turn degrade the environment in the short or long term. Biodiversity consists of several species of flora and fauna that are highly vulnerable to the slightest alteration in the nature of the biosphere and are in turn at risk of endangerment and destruction in a developed world subject to human activities. Certain human actions that directly or indirectly cause environmental damage on a global scale include human reproduction, overconsumption, overexploration, deforestation, and waste. Natural resources actively or indirectly caused by humans include anthropogenic environmental effects, including global warming and the atmosphere, biodiversity, biological hazards, and environmental failure.

To locate vegetation index, ground cover classification, vegetation, open field, farm area, woodland, with few band variations of the remote-sensed data, NDVI uses the multispectral remote sensing data technique. (Meera Gandhi et al. 2015). The highest monthly temperature and precipitation have greatly influenced the NDVI value (Fanghua Hao et al. 2011). The Landsat archive shows six GDE study areas, i.e., Annual NDVI vegetation, runoff, evaporative demand, groundwater depth, and land and water management (CUI Linli et al. 2010). The LULC map can be used as a base map to support the advancement of mine maintenance plans and sustainable development planning in the nearby local communities (Justin Huntington et al. 2016). The variance in fire patterns is related to both vegetation and precipitation variations (Susana Burry et al. 2017).

The combination of RS with GIS offers an excellent platform for data capture, preparation, synthesis, calculation, and interpretation, all of which are important for environmental interpretation. Satellite photographs are highly useful because they include regularly updated maps of remote regions with constantly evolving landforms. Geographic Information Systems (GIS) and Remote Sensing (RS) have developed as important tools to support environmental monitoring assessments. Ecological surveillance is aimed to study the ecological structure of a given area in time and space, surveillance the environmental condition of the transition in key parameters, evaluating and forecasting the effect of human activities on the ecological system.

The current study intends to take an account of the effects on two different forest ecosystems which are affected and influenced by several anthropogenic activities in two cities located in close proximity, i.e., local influence is taken into consideration for these two forest ecosystems. The results are compared to another forest ecosystem which is so far not affected by any anthropogenic activities locally. Global effects of several human for activities are predominant in all of these forests, considered in the study. The current understanding from this project is only based on local activities.

The goal of this initiative is the effect of anthropogenic behavior on the forest environment. In the current study, RS and GIS are incorporated to study and understand the effects of human activities on Tadoba-Andhari Tiger Reserve, Pench National Park, and Gadchiroli forest. The study also focuses on evaluating whether anthropogenic behaviors in surrounding urban areas have a beneficial or detrimental impact on vegetation.

The objectives of the study are as given below:

1. To find positive or negative effects of anthropogenic impacts of nearby urban areas on the vegetation.
2. To determine the impact of temperature and rainfall on surrounding vegetation using NDVI data.
3. To check the effectiveness of management activities undertaken by the government and forest management agencies.
4. To study change detection in the vegetation distribution in the study area and its temporal variation during the period 2000–2020.
5. To study the long-term trend of rainfall, temperature, and NDVI.
6. To determine the correlation between NDVI and different climatic parameters.

## 2 Study Area

The study is based on two forest areas, Pench National Park and Tadoba-Andhari National Park, which are situated next to two major cities, Nagpur and Chandrapur, respectively, of Maharashtra state of India. Nagpur is an industrial area, whereas mining activities are predominant in Chandrapur district. The results are compared with Gadchiroli forest, which is comparatively unaffected by any such activities. The full description of the area of the analysis is given in Fig. 1.

### 2.1 Pench National Park:

The forest is situated on the southern border of Madhya Pradesh and the state of Maharashtra. Nagpur district lies between 20.35 and 21.44 N latitude and 78.15 to 79.40 E longitude extending over an area of 9892 km$^2$. It is the closest extremely industrial city in the area of the Pench forest. Totladoh dam, adjacent to Madhya Pradesh in India, is a gravity dam on the Pench river near Ramtek in Nagpur district. The tank submerges 54 km$^2$ of the park district (Table 1).

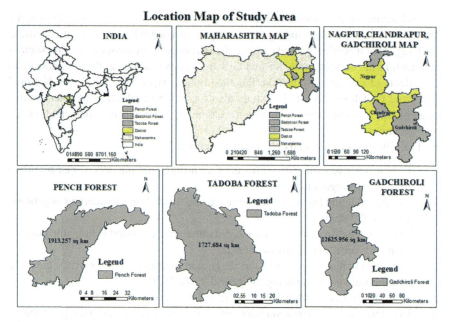

**Fig. 1** Location of the study area

| Table 1 Pench National Park Data | Total Pench National Park Area | 1913.257 km² |
|---|---|---|
| | Latitude | 21.64 N |
| | Longitude | 9.22 E |
| | Temperature | 6–31 °C in Winter to 25–48 °C in Summer |
| | Average rainfall | 1300 |

## 2.2 Tadoba Tiger Reserve

Chandrapur district is located between 19.30 °N and 20.45 °N latitude and at 78.46 °E longitude extending over an area of 11,443 km². Chandrapur district is known for its thermal power stations and its vast reserves of coal in Wardha Valley coalfields (Table 2).

## 2.3 Gadchiroli Forest

It lies between 18° 08′ and 20° 50′ N latitudes and 79° 45′ and 80° 54′ E longitude. The district has a geographical area of 14915.54 km². Forests cover more than 79.36%

**Table 2** Tadoba Tiger Reserve Data

| | |
|---|---|
| Total Tadoba Tiger Reserve Area | 1727.684 km$^2$ |
| Latitude | 20.16 N |
| Longitude | 79.40 E |
| Temperature | 25–30 °C in Winter to 47 °C in Summer |
| Average rainfall | 1388 |

**Table 3** Gadchiroli Forest Data

| | |
|---|---|
| Gadchiroli Forest Area | 12625.956 km$^2$ |
| Latitude | 20°50 N |
| Longitude | 80°54 E |
| Temperature | 14.6 °C in Winter to 42.1 °C in Summer |
| Average rainfall | 1750 |

of the hilly geographical of the district. Several hill ranges span across the region of Gadchiroli. These hills provide the catchment area for valleys formed by the tributaries of rivers Pranhita, Vainganga, and Indravati (Table 3).

## 2.4 Data and Software Used

The data used, materials gathered, and software used for the completion of the goals and analysis for the study have been tabulated (Table 4).

## 3 Methodology

The methodology consists of collecting all the required data for the study, carrying out all the necessary correction and processing of the collected data to get the desired output and then finally determining the relation among the different climatic parameters and analyzing them carefully. An estimation of the health of forest vegetation was conducted by time series analysis using NDVI images (Fig. 2).

**Table 4** Name of data and software used and their sources

| Sr. No. | Description | Sources |
|---|---|---|
| 1 | Rainfall data | www.maharain.gov.in |
| 2 | Temperature data | www.globalweather.com |
| 3 | Arc-GIS | ESRI |
| 4 | LC/LU MAP | Arc-GIS |

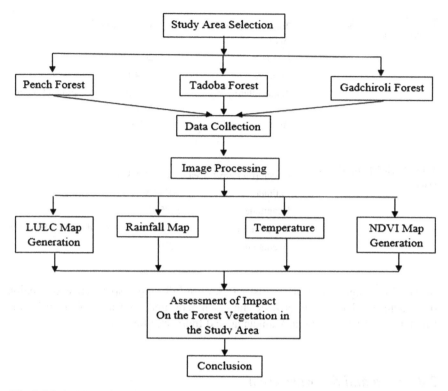

**Fig. 2** Methodology flowchart

## 3.1 Preliminary Data Generation

It consists of acquiring satellite images, rainfall data, temperature data, boundary of study area, etc. Data generation process consists of generation of various maps needed for the study and also data needed for statistical analysis.

### 3.1.1 Collection of Satellite Images

Earth explorer has been used to observe and download the satellite images of the study areas Pench, Tadoba, and Gadchiroli, for different years and months.

For which, the procedure used is as follows:

- Set time of which the image is required.
- Locate the study area and mark it. Zoom it to the required resolution.
- Place four pins at corners and download the satellite images (Table 5).

**Table 5** Datasets used in the present study and their source

| Sr. No. | Data type | Sensor | Year | Resolution | Source |
|---|---|---|---|---|---|
| 1 | LANDSAT 5 | TM | 2000 | 30 m | USGS |
| 2 | LANDSAT 7 | ETM+ | 2010 | 30 m | USGS |
| 3 | LANDSAT 8 | OLI | 2015 | 30 m | USGS |
| 4 | SENTINEL 2 | MSI | 2020 | 10 m | USGS |

### 3.1.2 Projection System and Digitization of the Images

a. Open Arc Map > Table of Contents > Right click on Layers > select Properties.
b. Window named "Data frame properties" will open. Select coordinate system of "WGS 1984 UTM Zone 43N" > Apply > OK.

### 3.1.3 Rainfall data

Rainfall data were obtained from the Maharashtra State Department of Agriculture (http://maharain.gov.in/) Rainfall Recording and Review website. Average annual average rainfall data is obtained for the districts of Nagpur, Chandrapur, and Gadchiroli for the period 2000–2020. Then, around each rain gauge spot, Thiessen polygons were formed. Then Pench and Tadoba and Gadchiroli were found to have mean annual rainfall using the formula given below.

$$\overline{P} = \frac{P_1 A_1 + P_2 A_2 + P_3 A_3 + \ldots P_n A_n}{A_1 + A_2 + A_3 + \ldots A_n} = \frac{\sum_{i=1}^{n} P_i A_i}{\sum_{i=1}^{n} A_i}$$

$\overline{P}$    Mean precipitation
$P_i$    Rainfall observed at the $i$th station inside or outside the basin
$A_i$    In-region portion of the area of the polygon surrounding the $i$th station
$N$    The number of areas
pagebreak

### 3.1.4 Temperature data

The Temperature data has been collected from the Global weather website for the study area for a duration of 21 years from 2000 to 2020. The data annual highest and lowest temperature has been calculated for the study. The annual average temperature has been calculated using the following formula,

$$Tavg = \frac{Tmax + Tmin}{2}$$

Tavg    annual average temperature

Tmax  maximum annual average temperature
Tmin  minimum annual average temperature.

## 3.2 Land Use Land Cover Data

Using ArcMap software, the land use map for the years 2000 and 2020 was developed by raster-based supervised classification. The Land Use Land Cover satellite images were downloaded from United States Geological Survey (URL: https://earthexplorer.usgs.gov). Before supervised classification, images were projected. Then, the area was classified as:

- Built-up (Urban and Rural)
- Agriculture
- Forest
- Mining
- Water-body
- Wasteland.

## 3.3 NDVI Data Processing

The NDVI satellite data has been downloaded from USGS for a duration of 21 years from 2000 to 2020. The following processing is done on these images.

a. The downloaded image for a particular time is opened in ArcMap.
b. Now mosaic satellite data and from the processed image the desired area is clipped.
c. Then go to the Windows>image analysis>image analysis option>select band R band; NIR band>NDVI>OK.
d. The image is then classified for different break values (right click the NDVI image>go to symbology>click on classified>select number of classes> classify> ok.

$$\text{NDVI} = \frac{\text{NIR} - \text{RED}}{\text{NIR} + \text{RED}}$$

NIR  Reflectance in Near-Infrared band for a cell
RED  Reflectance in Red band for the cell (Table 6).

## 3.4 Analysis

Data obtained from the NDVI time series analysis in the form of NDVI value and area tables were then used to calculate the mean NDVI value of each vegetation class

**Table 6** NDVI range for different features

| NDVI range | Feature |
|---|---|
| −1 to 0 | Water, snow, cloud |
| 0 to 0.2 | Built-up, barren land, rock |
| 0.2 to 1 | Vegetation |

and the coverage area of that class, respectively. For the period 2000–2020, plots were made for the command areas.

The trends were identified and studied. Relation of climatic parameters upon the vegetation with the forest vegetation has been carried out on the basis of spatial and temporal variation

## 4 Results and Discussions

The LULC maps of Nagpur, Chandrapur, and Gadchiroli have been first studied in a chronological manner within the study time, i.e., 2000 and 2020. In the second part, the major focus of the study was on the forest area of Pench, Tadoba, and Gadchiroli and how the urban development has affected the climate and NDVI of the mentioned forests. The meteorological parameters are also related to the LULC change of the affecting urban areas; in this process, we can predict the effect of LULC changes of the urban areas on the NDVI of the study areas. Finally, the conclusion would be drawn on the effect of urban developments on the vegetation changes in nearby forest (Fig. 3, 4, 5 and 6).

The above study can help to understand the nature of effects that the anthropogenic activities in these cities can have on the forest ecosystem of the above-said forests. The study proved to be successful and some positive inferences were obtained about the anthropogenic activities and their effects on forest ecosystem. The land use and land cover map of the cities were studied in a time span and the probable meteorological effects were predicted that could occur in the close proximity of these cities which include the forest areas under study (Fig. 7, 8, 9, 10, 11, 12, 13 and 14).

The trend graphs provide a mathematical approach toward the understanding of the NDVI trends in the study area. The images were enough to understand the coarse patterns of NDVI. But, to understand the fine changes the graphs are necessary.

While from the NDVI graph it is evident that the vegetation has been decreasing through the years. All three forest areas including Pench, Tadoba, and Gadchiroli show similar trends of decreasing NDVI i.e., the density of the forest. But as compared to Pench and Tadoba, Gadchiroli show less decrement in vegetation.

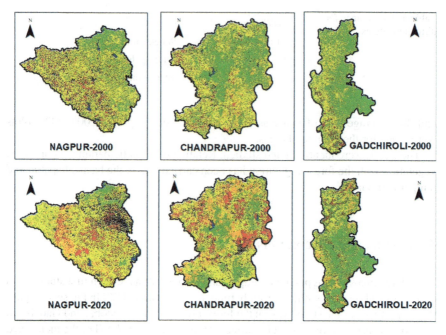

**Fig. 3** LU/LC change in study area for year 2000 and 2020

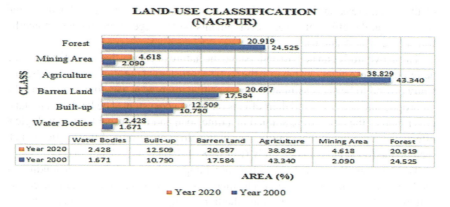

**Fig. 4** Changes in LU/LC in Nagpur in 2000 and 2020

## 5 Conclusion

- In the last few years, there is a rapid rise in anthropogenic practices that have detrimental effects on the environment over time, so we have to consider the impact on the forest environment of human activities.

Anthropogenic Impacts on Forest Ecosystems ...

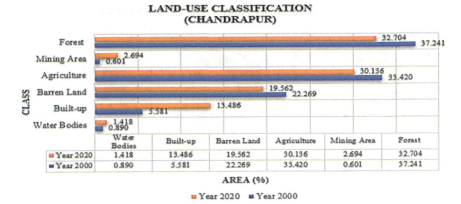

**Fig. 5** Changes in LU/LC in Chandrapur in 2000 and 20220

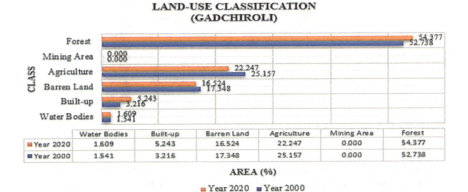

**Fig. 6** Changes in LU/LC in Gadchiroli in 2000 and 2020

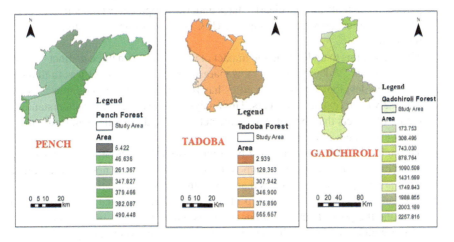

**Fig. 7** Area of influence under each rain gauge station in the study area

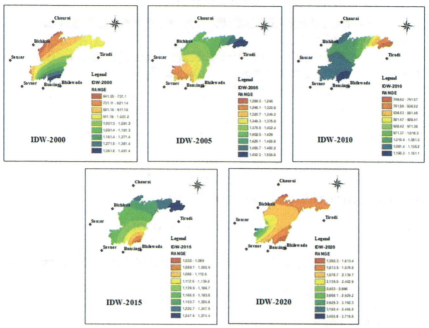

Fig. 8 Rainfall map of Pench forest using IDW method

- The Pench forest area has been decreased by 1.60% in the process of urbanization. The Tadoba forest area has been decreased by 4.53% in the process of urbanization. A drastic decrease of 4.53% has been monitored through the LULC maps. In the last twenty-one years, the Gadchiroli forest cover rises by 1.63%, because a small percentage in built-up area clearly states underdevelopment in Gadchiroli district. The positive scenario in the change of LULC of Gadchiroli is the increase in water bodies by 1.23%.
- This study aims to understand the impact of urbanization on three distinct Central Indian forests (Vidharba region of Maharashtra state). The target was to find out that if the development in these cities has been affecting the natural ecosystem adversely in last twenty-one years. This analysis can also clarify the understanding nature of effects that the anthropogenic activities in these cities can have on the forest ecosystem of near future.
- Through this study, it could have been understood that the protective measures taken by state and central government have been effective or not, and if not, then what are the probable measures that can be implemented.
- The land use and land cover map of the cities were analyzed over a period of time and the possible meteorological effects that could arise in the near vicinity of these cities, including the forest areas under study. Then, along with NDVI, the meteorological parameters (rainfall and temperature) were analyzed in these

## RAINFALL MAP OF TADOBA FOREST

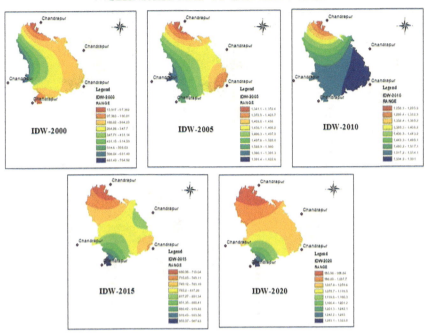

**Fig. 9** Rainfall map of Tadoba forest using IDW method

forests to explain the patterns in the last twenty-one years and the patterns are then contrasted with the estimates of the transition from the previously collected causative cities on land use and land cover maps.
- In order to validate the predictions and inferences mathematically, a thorough analysis can be performed on the same, and this method can be pursued anywhere in other areas to figure out the impact of anthropogenic behavior on forest ecosystems.

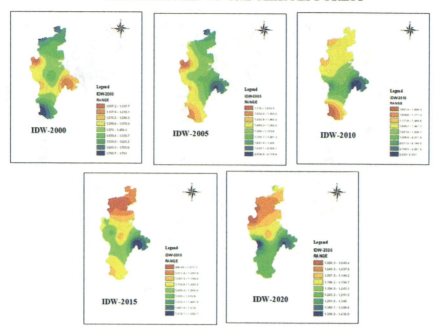

**Fig. 10** Rainfall map of Gadchiroli forest using IDW method

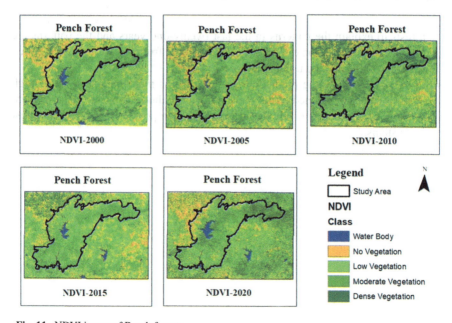

**Fig. 11** NDVI image of Pench forest

Anthropogenic Impacts on Forest Ecosystems ... 495

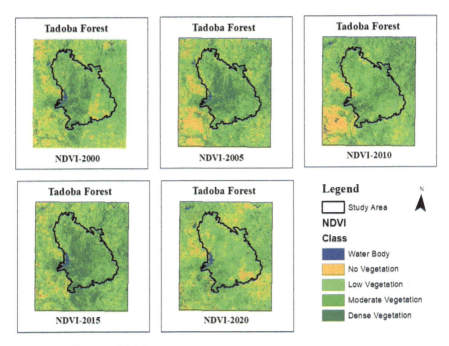

**Fig. 12** NDVI image of Tadoba forest

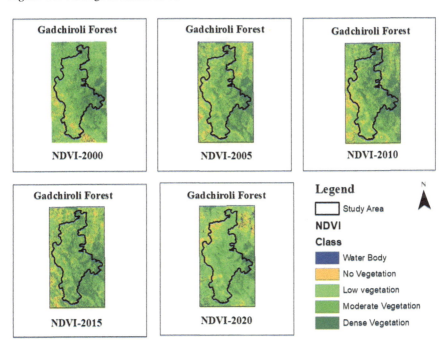

**Fig. 13** NDVI image of Gadchiroli forest

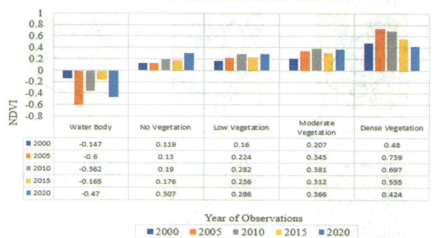

**Fig. 14** Trend in NDVI variation in Pench forest

# Seasonal and Lockdown Effects on Air Quality in Metro Cities in India

**K. Krishna Raj and S. Shrihari**

**Abstract** Air pollution is one of the worst avoidable threats in developing nations across the world. India has undergone a substantial number of infrastructure changes during recent years due to the ever-increasing population. This and the consequent industrialization, the air quality of Indian cities became worsened. The changes in climatic conditions across various cities in India also contribute to air pollution. To control the air pollution within the acceptable limit several control measures have been imposed in India, despite these efforts the air pollution level has not decreased considerably. In India, the first COVID-19 case has reported on 30th January 2020 in the state of Kerala. To control the quick spread of COVID-19 in India, the central government executed a three-week nationwide lockdown from 24th March 2020, and further, it has extended into several phases. It was the first time in India a long-term shutting down of all the sectors happening and which resulted in positively on the environment. This study is dealing with the lockdown effect on air quality in metro cities in India and is compared with the pre-existing conditions. Also, the seasonal variations in air quality in the course of the past two years are compared. The data of pollutants $PM_{10}$, $PM_{2.5}$, $SO_2$, $NO_2$, $O_3$, CO, and $NH_3$ from metro cities were collected and by adopting the National Air Quality Index to depict the variations in overall air quality. During the lockdown period, most of the cities experience a considerable improvement in overall air quality and $PM_{10}$, $NO_2$, $PM_{2.5}$, and CO concentrations. Whereas, the Ozone shows some increasing trend in a few cities might be due to the increment in the temperature caused by the exposure of sun during the summer season.

**Keywords** Lockdown · Metro cities · India · Air quality · COVID-19

---

K. Krishna Raj (✉) · S. Shrihari
Department of Civil Engineering, NITK Surathkal, Mangaluru, Karnataka, India
e-mail: krishnarajk1621@gmail.com

© The Author(s), under exclusive license to Springer Nature Singapore Pte Ltd. 2023
M. S. Ranadive et al. (eds.), *Recent Trends in Construction Technology and Management*, Lecture Notes in Civil Engineering 260,
https://doi.org/10.1007/978-981-19-2145-2_38

## 1 Introduction

Air pollution is one of the serious issues that developing nations like India face nowadays. The impacts of the abnormal population boost, urban growth, industrialization, and increment in vehicular count have created serious environment degradation, particularly, the deterioration of air quality in Indian cities. Along with the anthropogenic activities, the seasonal changes, differing climatology, and geography across the various regions also affect air pollution. According to Indian Meteorological Department (IMD), the winter season starts from December to February, summer season from March to May, monsoon season from June to September, and autumn season from October to November are the four major seasons in India [1]. Thus, air pollution causes serious health-associated difficulties related to organs like lung, heart, liver, etc. Therefore, effective air quality assessment at the city level could be a chief method for knowing the air pollution levels and might help to select appropriate schemes for opposing the condition. Identification of sources, technique adopted by the governing officials to reduce the emission, and checking the advancement in the technique adopted frequently are the chief elements of effective air quality [2].

Air Quality Index (AQI) is used to express the magnitude of air pollution of a region. In India, the monitoring of air quality was started in 1984 to quantify integrated air quality and in 2015 the CPCB revised the National Air Quality Index (NAQI) based on the maximum operator perspective [3]. Few of the Indian cities exceeding the air quality standards recommended by the World Health Organization (WHO) and Central Pollution Control Board (CPCB) for the past few years and thus falling into the list of most polluted cities of the world [4]. Non-local sources contribute around 70% of particulate matter in the North Indian region [5] and the activities like transportation, industrial, power generation and domestic fuel combustion, etc. helps to originate pollutants such as $NO_X$, CO, and $SO_2$ [6].

The Ministry of Environment, Forest, and Climate Change (MoEFCC) runs as the focal organization in collaboration with the UN Environment Programme for controlling air pollution and thus to protect the nature. The CPCB runs as a statutory body that obliged under the Air Act 1981. The Air Act (1981), Environment Act (1986), The Motor Vehicles Act (1988), The Public Liability Insurance Act (1991), National Environmental Tribunal Act (1995), National Environment Policy (2006) are also helping to protect the environment [3]. The biofuels scheme aimed to transform a minimum of 20% conventional fuel engines into biofuel engines by 2017 and the National Clean Air Programme introduced in 2016 focused on reducing particulate matter by 30% nationwide by the end of 2021 [7]. In addition to these, strategies like odd and even system in Delhi, introduced Metros to improve public transport, etc., were tried to reduce the air pollution in various cities. Despite these regulation measures, the air quality is not much improved across India.

From 24th of March 2020 to a period of three weeks a nationwide lockdown was forced in India due to the spread of pandemic COVID-19. During this period, activities related to transportation, industries, constructions, commercial, institutional, administrative, etc., have been restricted nationwide. As a result, the emission of chief pollutants associated with these activities gets reduced in metro cities and which ultimately helped to improve the overall air quality of the cities during the lockdown as per the data from the CPCB. The current work is intended to know the degree of air quality change due the lockdown and also the variations in air quality during different seasons for the past two years at metro cities in India. Also, it is to evaluate the functionality of the lockdown as an approach for the reduction of air pollution in metro cities. The current work has focused on Delhi, Noida, Gurugram, Lucknow, Kolkata, Jaipur, Mumbai, Hyderabad, Bangalore, and Chennai. The objective of the study is to collate the chief atmospheric pollutant concentrations in metro cities from 24th of March to 14th of April in recent past years, to know the seasonal variations in air quality in metro cities, to estimate the general air quality due to the restrictions made during lockdown period and to assess the usefulness of the lockdown as an approach for reduction of air pollution in metro cities.

## 2 Materials and Methodology

The data From CPCB portal (https://app.cpcbccr.com/ccr/#/caaqm-dashboard-all/caaqm-landing) for air quality data dissemination, the concentrations of pollutants were obtained [8]. The pollutants selected for the analysis were $PM_{2.5}$, $PM_{10}$, $SO_2$, $NO_2$, $CO$, $O_3$, and $NH_3$. To compare the overall air quality of cities NAQI was used. The sub-indices for pollutants at a monitoring station were computed centered on 24-h average for $PM_{2.5}$, $SO_2$, $PM_{10}$, $NO_2$, and $NH_3$ and eight hours for $O_3$ and $CO$. Also, it is not possible to monitor all seven pollutants at all the locations concurrently. In order to measure NAQI, data of minimum of three pollutants concentration is necessary and it include either one among PM10 or PM2.5 is requisite within those. Else, the data are inadequate for the calculation of NAQI. Similarly, for the estimation of subindex of a pollutant, a minimum of 16 h data is necessary and the air quality of a pollutant is the subindex value calculated for that pollutant.

## 3 Results and Discussion

### 3.1 Changes in NAQI During Lockdown

The metro cities have witnessed a substantial reduction of the pollutants after the declaration of the lockdown measures. During the three-week lockdown in the year 2020, the sectors like industries, mass transportation, construction, commercial, and institutional establishments were remains closed. Thus, the lockdown helped the cities to reduce the emission rate of air pollutants. As a result, the average NAQI of metro cities had shown a considerable reduction during the lockdown period (Fig. 1). The North Indian cities had shown a maximum depletion in air pollution

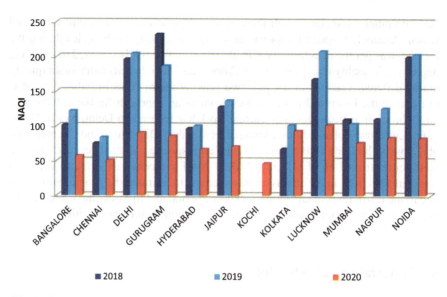

**Fig. 1** Changes in average NAQI during 24 March–14 April in recent past years at metro cities

during the study period. The overall air quality of cities Noida and Gurugram had been improved by about 59.12% during the lockdown in comparison with the average of previous years. Similarly, the cities Delhi, Jaipur, and Lucknow had undergone 54.98%, 46.85%, and 46.12% improvements in NAQI, respectively. Whereas, metro cities located other than North India also had considerable enhancement in overall air quality. During the lockdown period, the average NAQI values of all cities except Lucknow fall below 100. The cities Chennai, Hyderabad and Nagpur NAQI was falls under the category of satisfactory or good throughout the lockdown period. This improvement in air quality is primarily due to the depletion in the concentration of $PM_{10}$, $NO_2$, $PM_{2.5}$, and CO.

## 3.2 Changes in Pollutant Concentration During Lockdown

The closure of all sectors during the lockdown period helped to reduce the emission of Particulate Matters (PM) drastically. The average concentration of significant pollutants at different metro cities during the period starting from 24th of March–14th of April during the past years is indicated in Table 1. Table 1 outlines the slight increment in the concentration of pollutants in the year 2019 than that of 2018. Whereas, during the lockdown, the $PM_{10}$ concentration reduced by more than 60% in cities Delhi and Noida in comparison to 2018 and 2019. Thus, the cities located in the northern part of India have undergone a maximum declination in pollutant concentration during the lockdown period in comparison with the previous years. It

**Table 1** Variations of average concentration of dominant pollutants during 24 March–14 April in recent past years at metro cities

| Span | 24 March–14 April | | | | | | | | | | | |
|---|---|---|---|---|---|---|---|---|---|---|---|---|
| | 2018 | | | | 2019 | | | | 2020 | | | |
| Cities | Average pollutants concentration ($\mu g/m^3$) | | | | | | | | | | | |
| | $PM_{10}$ | $PM_{2.5}$ | $NO_2$ | $O_3$ | $PM_{10}$ | $PM_{2.5}$ | $NO_2$ | $O_3$ | $PM_{10}$ | $PM_{2.5}$ | $NO_2$ | $O_3$ |
| Bangalore | 92.7 | 30.5 | 26.0 | 46.2 | 113.1 | 51.3 | 35.7 | 55.1 | 46.8 | 25.9 | 11.3 | 43.2 |
| Chennai | – | 39.2 | 18.3 | 33.1 | 32.8 | 38.5 | 13.7 | 25.6 | 145.4 | 22.6 | 8.5 | 35.4 |
| Delhi | 219.0 | 87.6 | 46.7 | 51.2 | 225.9 | 86.2 | 34.9 | 44.7 | 84.8 | 40.7 | 20.0 | 39.5 |
| Gurugram | 173.0 | 70.5 | 24.8 | 25.9 | 213.7 | 82.5 | 9.0 | 32.6 | 73.8 | 35.3 | 14.7 | 55.8 |
| Hyderabad | 99.3 | 45.1 | 41.1 | 44.2 | 109.0 | 43.6 | 36.7 | 38.3 | 64.3 | 33.8 | 19.5 | 32.1 |
| Jaipur | 132.4 | 50.9 | 34.3 | 74.1 | 154.4 | 49.4 | 33.0 | 61.4 | 59.1 | 23.0 | 13.4 | 48.2 |
| Kochi | – | – | – | – | – | – | – | – | 23.2 | 39.3 | 2.5 | 1.32 |
| Kolkata | 77.7 | 46.0 | 12.5 | 49.3 | 94.2 | 48.5 | 42.4 | 31.0 | 65.3 | 37.3 | 12.6 | 56.9 |
| Lucknow | – | 85.5 | 42.1 | 33.1 | – | 102.4 | 47.7 | 39.2 | 49.9 | 34.5 | 13.3 | 32.8 |
| Mumbai | – | – | – | – | 95.7 | 25.8 | 18.5 | 26.4 | 74.0 | 25.4 | 8.8 | 27.7 |
| Nagpur | 103.3 | 48.9 | 38.9 | 46.4 | 105.2 | 52.5 | 46.7 | 115.3 | 45.7 | 22.5 | 22.9 | 57.7 |
| Noida | 271.6 | 68.4 | 27.5 | 43.3 | 237.0 | 90.1 | 35.2 | 28.0 | 77.0 | 34.7 | 14.3 | 41.5 |

is primarily because of the restriction given on transportation and industrial activities during the span. Due to the prevailing local conditions in the course of the lockdown span in Chennai, the $PM_{10}$ concentration has increased more than fourfold upon comparing with the average of the same period in 2019. Similar to the trend of $PM_{10}$, the pollutant $PM_{2.5}$ has also declined drastically in North Indian cities and moderate to slight in other cities during the lockdown period. About 53.12% reduction in $PM_{2.5}$ has been observed in Delhi in the course of the lockdown period upon comparing with the average concentration for same period in 2018–2019. Table 2 shows the area-wise concentration of particulate matter in typical metro cities. In metro cities, industrial and transportation sectors play a vital role in the increment of pollution rate. In Delhi, the pollutant concentrations from industrial and transportation sector has reduced more than 60% in case of $PM_{10}$ and more than 50% in the case of $PM_{2.5}$. Thus, this data opens our eyes to the need for restriction implementation on anthropogenic activities. During the lockdown period, vehicle counts on the road decreased, the manufacturing processes paused in industries helped to reduce the dust resuspension in metro cities.

In metro cities, the source for $NO_2$ and CO is from vehicles on road, combustion practice, manufacturing industry, and power plants. The lockdown also reduced the concentration of pollutants $NO_2$ by more than 60% in cities like Lucknow, Bangalore, and Jaipur upon comparing with the average value of 2018–2019 for the same period. Gurugram was the only city which shows increment in $NO_2$ during the lockdown period by about 63.3% during the lockdown in comparison with the average value of 2019. Figure 2 shows the trends of variations of CO in metro cities from

Table 2 Area wise variation of concentration of chief pollutants during 24 March–14 April in recent past years at typical metro cities

| Cities | pollutants | Span from 24 March–14 April ||||  Average concentration during lockdown 2020 |||| Overall variations ||
| | | Average concentration of 2018–2019 |||| | | | | | |
| | | Avg | Industrial area | Transport area | Residential and other areas | Avg | Industrial area | Transport area | Residential and other areas | Net | % |
|---|---|---|---|---|---|---|---|---|---|---|---|
| Delhi | $PM_{10}$ | 222.5 | 246.5 | 244.8 | 203.7 | 84.8 | 96.1 | 90.4 | 79.7 | −137.7 | −61.9 |
| | $PM_{2.5}$ | 86.9 | 94.3 | 93.2 | 81.84 | 40.7 | 45.9 | 43.4 | 39.0 | −46.2 | −53.2 |
| Hyderabad | $PM_{10}$ | 104.2 | 113.9 | 110.0 | 99.6 | 64.3 | 60.9 | 80.6 | 61.7 | −39.9 | −38.3 |
| | $PM_{2.5}$ | 44.4 | 47.3 | 51.7 | 44.2 | 33.8 | 31.1 | 47.0 | 33.5 | −10.6 | −23.9 |
| Bangalore | $PM_{10}$ | 102.9 | – | 138.4 | 94.9 | 46.8 | – | 39.5 | 48.5 | −56.1 | −54.5 |
| | $PM_{2.5}$ | 40.9 | – | 50.5 | 39.9 | 25.9 | – | 30.6 | 24.9 | −15.0 | −36.7 |
| Lucknow | $PM_{2.5}$ | 94.0 | 112.5 | 100.9 | 88.3 | 34.5 | 72.6 | 65.2 | 33.6 | −59.5 | −63.3 |

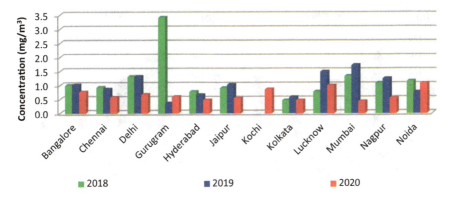

**Fig. 2** Average variation of concentration of CO at different metro cities during 24 March–14 April in recent past years

24th of March–14th of April in recent past years. The mean concentration of CO in cities Mumbai and Gurugram has declined by about 72.37% and 68.62% respectively during the lockdown than that of the averaged value of 2018–2019 for the same duration. Whereas, Noida is the only city which shows increment in concentration of CO by about 12.63% in comparison with the averaged value of 2018–2019. Significant declining trends in $NO_2$ and CO were observed in the first two weeks of lockdown period, afterward, it shows some increase in the concentration of these pollutants. The reason for the increase might include the open burning of nonbiodegradable materials at households and intermittent checking of parked vehicles such as buses and heavy motors.

On the other hand, the pollutant Ozone shows a slight rise in the course of the lockdown span in majority of cities. Whereas, few cities like Bangalore, Delhi, Hyderabad, etc., show decreasing trends too. The reason for the changes in $O_3$ concentration might be because of the usual rise or fall of exposure of sunlight and temperature during the summer period in different cities. The pollutants $SO_2$ and $NH_3$ experience a slight fall in the mean concentration in the course of the lockdown. The major emission source for $SO_2$ in metro cities includes fossil fuel combustion at power plants and industries, burning of materials that contain sulfur, metals processing, and smelting facilities, and vehicles. Whereas, sources of $NH_3$ include industrial processes, vehicular emissions, and uses of $NH_3$-based fertilizers. The lockdown helps to reduce the pollutants in metro cities due to the closure of these sectors during the period.

## *3.3 Seasonal Variations in NAQI*

The differing climates in metro cities have a dominant part in making the air quality of a city poor. The majority of cities exceed the permissible levels of NAQI during the

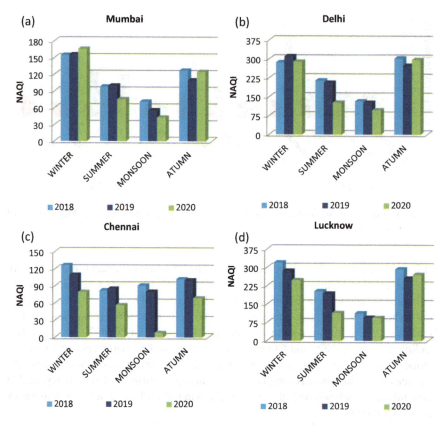

**Fig. 3** Seasonal changes in average NAQI in typical metro cities for the recent past years

autumn and winter seasons (Fig. 3). It is primarily because of the practice of open-burning biodegradable and nonbiodegradable materials. The direction of wind plays a crucial role in transporting pollutants from one place to another. Thus, along with the anthropogenic activities, the natural phenomenon also had a prominent role in deciding the air quality of a city. During the autumn season, farmers burn agricultural wastes like straw stubble that makes the air quality very poor in North Indian cities. Similarly, the burning of agricultural straw stubble happens during the end of the winter season in Southern India. Whereas in monsoon season, the continuous rain and cyclonic winds help to improve the air quality. The overall air quality became improved exceptionally than that of other seasons because of the flowing of pollutants with rainwater in the monsoon season. The average NAQI during the monsoon season in the year 2020 is nearly equal to or more improved than that of the lockdown period in selected cities. The reason for the extra enhancement in air quality in the course of the monsoon season in 2020 is due to the extension of restrictions on certain activities into several phases. Similarly, the phased unlock procedure also helped to improve the air quality during this season. On the other hand, the average NAQI during the summer season has much above the monsoon season but falls below the average values of the autumn and winter season. The North Indian cities had poor air quality in all seasons than that of the other locations might be due to the increased number of vehicular sources, manufacturing industries, and power plants located nearby cities. There are not many changes observed in the NAQI in 2018 and 2019 in all seasons. Similarly, during the winter and autumn season in the year 2020, the air quality becomes worsened equal to or more than that in the same seasons in previous years. Thus, it may be indicating the need for a strategy to improve the overall air quality levels in metro cities during the winter and autumn seasons.

## 3.4 Seasonal Variations in Dominant Pollutants

The concentrations of air pollutants have a drastic increment during the autumn and winter season in metro cities due to various reasons (Fig. 4). The prevailing local condition in different cities during the autumn and winter season plays a vital role in the variation of air pollutants. As a result, the particulate matter, $NO_2$, and $O_3$ levels have shown high variation than those in the monsoon and summer seasons. Open burning of biomass during the autumn and winter season increasing the concentration of $PM_{10}$ and $PM_{2.5}$ far above the acceptable limits. In North Indian cities like Delhi, Noida, Gurugram, etc., particulate matter has been deteriorating the air quality during the autumn and winter season. The yearly average concentration of $PM_{10}$ for the city Delhi was 244.3, 188.9, and 160.9 $\mu g/m^3$ for 2018, 2019, and 2020, respectively. Whereas, the average seasonal concentration of $PM_{10}$ for the winter and autumn lies far above the annual average values for the past years from 2018 to 2020 in Delhi. Similarly, in the case of $PM_{2.5}$, the trends of variation were identical as that of $PM_{10}$. Other cities also had more or less identical behavior that of the city Delhi in the case of particulate matters. On the other hand, the concentration of CO, $NO_2$, and $O_3$ was almost having same value during the autumn and winter seasons in all cities. Upon comparing with other seasons, the concentration of these two pollutants were high in the winter and autumn season. Along with anthropogenic activities, the natural phenomenon like wind direction, wind velocity, moisture content, etc., has a significant role in maintaining the air quality very poor. Thus, during the autumn and winter season, it is a must to introduce some strategies to control the pollutants within the acceptable limits.

Similar to the trend of NAQI, the air pollutants concentration has also shown declining trends during the monsoon season. Upon comparing with other seasons, during monsoon, all the pollutants have shown declining trends in cities. The average monsoon season concentration for the year 2020 for most of the selected cities falls below the three-week average value during the lockdown period. It is because of the continued number of lockdown phases followed by a number of unlocking phases. Even in the years 2018 and 2019, the average monsoon season concentration of air pollutants were almost equal to or slightly more than the corresponding average concentration of pollutants in the lockdown period. Thus, during monsoon season,

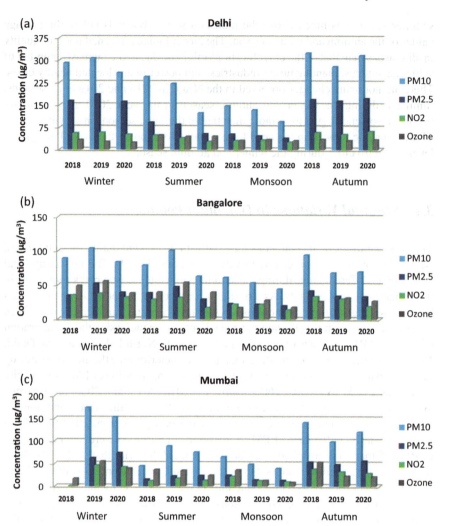

**Fig. 4** Seasonal changes in average concentration of chief pollutants at typical metro cities **a** Delhi **b** Bangalore **c** Mumbai

every city has shown a minimum level of concentration of pollutants. Due to the decline in the natural exposure of sunlight and temperature during monsoon season, the concentration of $O_3$ has shown decreasing trend than that of the other seasons. Similarly, other pollutants also had reducing trend during the monsoon season.

Whereas in summer, the concentration of pollutants except $O_3$ has a value greater than the concentration of corresponding pollutants in monsoon and less than that of the autumn and winter. The pollutant Ozone has shown maximum concentration during the summer season. It might be due to the increment in the natural exposure of sunlight and temperature during the summer season. Due to the dust resuspension and open burning the $PM_{10}$ and $PM_{2.5}$ remains high value in the course of summer also. Upon comparing the North Indian cities with other cities, the cities located in North India have high concentration of particulate matter. Whereas, in certain cities located in Southern India, low concentration of PM might be due to the intermittent summer rain and cyclonic winds.

## 3.5 Annual Average Variation of Particulate Matter

Particulate matter (PM) decides the overall air quality in terms of NAQI in a city most of the time. Table 3 has shown the variation in annual mean of $PM_{10}$ and $PM_{2.5}$

Table 3 Annual average concentration of particulate matter in recent past years at metro cities

| Cities | Pollutants | Average concentration ($\mu g/m^3$) ||| Variation between 2020 and 2019 || Variation between 2020 and 2018 ||
|---|---|---|---|---|---|---|---|---|
| | | 2018 | 2019 | 2020 | Net | Percentage | Net | Percentage |
| Bangalore | $PM_{10}$ | 79.6 | 77.3 | 63.6 | −13.7 | −17.7 | −16.0 | −20.1 |
| | $PM_{2.5}$ | 34.4 | 35.8 | 29.2 | −6.6 | −18.4 | −5.2 | −15.11 |
| Chennai | $PM_{10}$ | – | 50.1 | 80.6 | +30.5 | +60.87 | – | – |
| | $PM_{2.5}$ | 53.3 | 42.7 | 30.7 | −12.0 | −28.1 | −22.6 | −42.4 |
| Delhi | $PM_{10}$ | 244.3 | 217.2 | 160.9 | −56.3 | −25.9 | −83.4 | −34.1 |
| | $PM_{2.5}$ | 114.7 | 108.2 | 81.4 | −26.8 | −24.8 | −33.3 | −29.0 |
| Gurugram | $PM_{10}$ | 233.0 | 188.9 | 148.5 | −40.4 | −21.4 | −84.5 | −36.3 |
| | $PM_{2.5}$ | 111.5 | 93.2 | 81.9 | −11.3 | −12.1 | −29.6 | −26.6 |
| Hyderabad | $PM_{10}$ | 95.6 | 91.7 | 77.6 | −14.1 | −15.4 | −18.0 | −18.8 |
| | $PM_{2.5}$ | 43.2 | 41.5 | 35.8 | −5.7 | −13.7 | −7.4 | −17.1 |
| Jaipur | $PM_{10}$ | 141.5 | 114.6 | 102.7 | −11.9 | −10.4 | −38.8 | −27.4 |
| | $PM_{2.5}$ | 62.1 | 49.1 | 45.1 | −4.0 | −8.2 | −17 | −27.4 |
| Kochi | $PM_{10}$ | – | – | 59.4 | – | – | – | – |
| | $PM_{2.5}$ | – | – | 32.8 | – | – | – | – |
| Kolkata | $PM_{10}$ | 104.6 | 125.2 | 88.6 | −36.6 | −29.2 | −16.0 | −15.3 |
| | $PM_{2.5}$ | 61.3 | 68.4 | 48.6 | −19.8 | −29.0 | −12.7 | −20.7 |
| Lucknow | $PM_{10}$ | – | – | 210.3 | – | – | – | – |
| | $PM_{2.5}$ | 116.4 | 97.4 | 89.3 | −8.1 | −8.3 | −27.1 | −23.3 |
| Mumbai | $PM_{10}$ | 88.4 | 96.5 | 90.8 | −5.7 | −5.9 | +2.4 | +2.7 |
| | $PM_{2.5}$ | 33.1 | 34.8 | 37.5 | +2.7 | +7.8 | +4.4 | +13.3 |
| Nagpur | $PM_{10}$ | 88.4 | 79.2 | 56.1 | −23.1 | −29.2 | −32.3 | −36.5 |
| | $PM_{2.5}$ | 44.5 | 42.4 | 29.3 | −13.1 | −30.9 | −15.2 | −34.2 |
| Noida | $PM_{10}$ | 252.7 | 225.1 | 184.8 | −40.3 | −17.9 | −67.9 | −26.9 |
| | $PM_{2.5}$ | 116.8 | 110.6 | 97.2 | −13.4 | −12.1 | −19.6 | −16.8 |

in metro cities. In general, the annual average concentration of particulate matters in most of the cities had a decreasing trend from 2018 to 2020. Upon comparing the annual average concentration of $PM_{10}$ in 2020, Chennai is the only city which shows an increase of about 60.87% than that of 2019. Similarly, in Mumbai, the annual mean of $PM_{2.5}$ in 2020 has increased beyond 7% than that of 2019. Whereas, in 2020, about 2.7% and 13.3% increment has been observed in $PM_{10}$ and $PM_{2.5}$, respectively, in Mumbai than that of the year 2018. Mumbai is the only city which has shown increasing trend in the concentration of $PM_{2.5}$ both in 2019 and 2020 than that of the annual average concentration in the year 2018. All other cities have shown declining trend in the level of particulate matter in the year 2020. The decrease in $PM_{10}$ and $PM_{2.5}$ was primarily due to the lockdown measures implemented to control the rapid spread of COVID-19. Also, the extended lockdown measures and phased unlock of lockdown during the monsoon season drastically reduced the particulate matters.

The cities located in North India have undergone a maximum reduction in concentration of particulate matter in the year 2020. Upon comparing the average concentration of $PM_{10}$ in 2018–2019 with 2020, about 30.27%, 29.60%, 15.89%, and 22.65% reduction has observed in cities Delhi, Gurugram, Jaipur, and Noida respectively. Similarly, in the case of $PM_{2.5}$ about 26.96%, 19.98%, 18.88%, and 14.51% reduction has observed in cities Delhi, Gurugram, Jaipur, and Noida respectively.

## 4 Conclusions

The study was conducted to determine the seasonal and lockdown effect on air quality in metro cities in India. The result reveals that the metro cities undergo significant improvement in overall air quality in terms of NAQI during the lockdown duration. The reason for the improvement in NAQI is due to the reduction in concentration of chief pollutants such as $PM_{10}$, $PM_{2.5}$, $NO_2$, and CO. This improvement in overall air quality directly reveals the role of anthropogenic activities on air pollution. Similarly, the overall air quality in terms of NAQI becomes worsened during the autumn and winter season. Therefore, this study is a helpful addendum to the governing authorities to plan a new measure like short duration lockdown (say up to 7 days) in an aim to improve the air quality, especially, during the autumn and winter season.

## References

1. Roy SS (2019) Spatial patterns of trends in seasonal extreme temperatures in India during 1980–2010. Weather Climate Extremes 24
2. Brauer M, Guttikunda SK, Nishad KA, Deye S, Tripathig SN, Weagleh C, Martin RV (2019) Examination of monitoring approaches for ambient air pollution: a case study for India. Atmos Environ 216:116940

3. Mahato S, Pal S, Ghosh KG (2020) Effect of lockdown amid COVID-19 pandemic on air quality of the megacity Delhi, India. Sci Total Environ 730:139086
4. Garaga R, Sahu SK, Kota SH (2018) A review of air quality modeling studies in India: local and regional scale. Current Pollution Reports
5. Guo H, Kota SH, Sahu SK, Zhang H (2019) Contributions of local and regional sources to PM2.5 and its health effects in north India. Atmos Environ 214:116867
6. Sunil Gulia SM, Nagendra S, Khare M, Khanna I (2015) Urban air quality management-A review. Atmos Pollut Res 6:286–304
7. Sharma S, Zhang M, Anshika JG, Zhang H, Kota SH (2020) Effect of restricted emissions during COVID-19 on air quality in India. Sci Total Environ 728:138878
8. Central Control Room for Air Quality Management Delhi NCR, CAAQMS Information, https://app.cpcbccr.com/ccr/#/caaqm-dashboard-all/caaqm-landing, Accessed date: June 2020 to Dec 2020

# Inter-Basin Pipeline Water Grid for Maharashtra

Raibbhann Sarnobbat, Pritam Bhadane, Vaibhav Markad, and R. K. Suryawanshi

**Abstract** Maharashtra is facing water crises of unprecedented nature. There are floods during monsoon but in the same area, there is no water during the summer season. Some regions conically face years of droughts and the rivers' current have webbed. Some reservoirs spill during monsoon whereas some do not get filled. On the other hand, water demands are continuously increasing. To overcome these issues, in this paper an attempt is made to develop an inter-basin water transfer pipeline grid for the river basins in the state of Maharashtra with the objective of transfer of water from surplus to deficits for both spatial and temporal scales. Major storages in all the river basins in the state are located at the georeferenced locations on the Digital Elevation Model (DEM) of Maharashtra. Using the DEM to contour map and FRL and DSL of each of the reservoirs, pipeline water grid is delineated. The pipeline grid is filled during monsoon so that surplus waters can be transferred to deficit reservoirs. This also can be done during non-monsoon to serve the scarcity regions. Since the pipeline grid is designed for gravity flow and no major land acquisition issues are involved, the inter-basin pipeline water grid can be an effective solution to mitigate the water problems in the state.

**Keywords** Water crises · Pipeline water grid · Gravity flows · ArcGIS

## 1 Introduction

State of Maharashtra in India faces varied rainfall conditions. The north–south small strip of Konkan region experiences very high rainfall whereas the widely spread Marathwada region receives very less rainfall compared to the national average. Rainfall is restricted to monsoon months of June to September only. Rest of the period is practically dry. Thus, the spatial and temporal distribution of surface water resources is highly skewed [1, 2]. Major portion of the state includes river basins of Krishna, Godavari, and Tapi and west-flowing rivers and a small portion in north falls

R. Sarnobbat · P. Bhadane · V. Markad · R. K. Suryawanshi (✉)
Emeritus Prof COEP, Pune, India
e-mail: rksurya2000@yahoo.com

© The Author(s), under exclusive license to Springer Nature Singapore Pte Ltd. 2023
M. S. Ranadive et al. (eds.), *Recent Trends in Construction Technology and Management*, Lecture Notes in Civil Engineering 260,
https://doi.org/10.1007/978-981-19-2145-2_39

in th Narmada basin. The major rivers flowing through the state include Godavari, Krishna, Tapi, Indravati, Wardha, Manjara, Penganga, Purna, and the west-flowing rivers of Konkan strip. Many times, there is excess water in some basins whereas other basins face high water scarcity. Water demands are also varying spatially as well as temporally in view of increasing population, rapid urbanization, and industrialization. Thus, it has become highly essential to have long-term planning for water resources management to meet the rapidly changing supply and demand scenario [3–5]. The inter-basin water transfer is an alternative to balance the non-uniform temporal and spatial distribution of availability of water resources and water demands [6, 7].

## 2 The Need

Generally, inter-basin water transfer is attempted by linking two or more rivers by creating a network of canals to transfer the water from surplus river basin to the deficit river basin which can mitigate the droughts and floods situations to some extent. In this study, an attempt is made to analyze the feasibility of the inter-basin water transfer in Maharashtra using pipeline grid with gravity flow. Accordingly, the main four river basins of Godavari, Krishna, Tapi, and west-flowing rivers are considered for transfer of water from basin to basin. Some of the major reservoirs in these basins are interconnected so that the excess water in the reservoir can be transferred to the deficit reservoir under gravity flow using a pipeline. The tools used include ArcGIS with Google Pro [1, 8].

## 3 Study Area

The objective is to study the feasibility of establishing a water transfer grid under gravity flow using pipes instead of conventional canal grid for the state of Maharashtra. The basins selected include Krishna, Godavari, Tapi, west-flowing rivers as depicted in Fig. 1. The major reservoirs in these basins are selected for the establishment of pipeline links.

The Krishna basin with catchment area of about 258,948 km$^2$ has average annual rainfall of 921 mm. Total of 11 major storages are selected from this basin for the pipeline grid. The Godavari basin has a catchment area of about 312,811 km$^2$ with average annual rainfall of 947 mm. Total of 15 reservoirs are selected from this basin. The Tapi basin has a catchment area of about 51,504 sq. km having average annual rainfall of 709 mm and total 5 dams are selected from this basin. West flowing river basins have a catchment area of about 33,017 sq. km with average annual rainfall of 3145 mm and 4 reservoirs are selected from this basin for pipeline grid. The basin wise reservoirs selected and their relevant characteristics such as FRL, MDDL, storage capacity are shown in Tables 1, 2, 3, and 4.

Inter-Basin Pipeline Water Grid for Maharashtra

**Fig. 1** River basins in Maharashtra

**Table 1** Krishna basin reservoirs

| Sr No | Reservoir | FRL (m) | MDDL (m) | Gross storage (MCM) | Height of dam (m) |
|---|---|---|---|---|---|
| 1 | Dhom | 747.70 | 725.0 | 382.97 | 50.00 |
| 2 | Dimbhe | 719.15 | 711.9 | 380.00 | 67.21 |
| 3 | Pimpalgaon Joge | 686.80 | 678.4 | 235.53 | 28.60 |
| 4 | Koyna | 657.90 | 609.5 | 2797.40 | 103.00 |
| 5 | Varasgaon | 639.50 | 628.7 | 374.00 | 63.40 |
| 6 | Panshet | 635.81 | 609.8 | 303.00 | 63.56 |
| 7 | Kasarsai | 626.60 | 610.6 | 17.38 | 17.27 |
| 8 | Pawana | 613.26 | 602.0 | 305.00 | 42.37 |
| 9 | Radhanagari | 590.00 | 566.1 | 236.79 | 42.68 |
| 10 | Khadakwasla | 582.87 | 578.6 | 86.00 | 32.90 |
| 11 | Ujjani | 497.58 | 491.0 | 332.00 | 56.40 |

**Table 2** Godavari basin reservoirs

| Sr. No | Reservoir | FRL (m) | MDDL (m) | Gross storage (MCM) | Height of dam (m) |
|---|---|---|---|---|---|
| 1 | Bhandardara | 744.74 | 720.7 | 312.4 | 82.35 |
| 2 | Karanjvan | 661.00 | 553.0 | 248.0 | 39.61 |
| 3 | Nilwande | 660.00 | 640.0 | 236.0 | 73.91 |
| 4 | Manjara | 642.37 | 633.4 | 250.7 | 25.00 |
| 5 | Gangapur | 612.50 | 606.0 | 215.0 | 36.57 |
| 6 | Darna | 593.00 | 570.0 | 226.9 | 28.00 |
| 7 | Mula | 552.30 | 534.2 | 738.9 | 48.17 |
| 8 | Nandur-Madhyameshwar | 533.53 | 529.3 | 29.9 | 10.90 |
| 9 | Totladoh | 490.00 | 464.0 | 1241.0 | 74.50 |
| 10 | Jayakwadi | 483.91 | 441.6 | 2909.0 | 41.30 |
| 11 | Yeldari | 461.67 | 455.6 | 934.0 | 51.20 |
| 12 | Isapur | 441.00 | 418.7 | 1241.5 | 57.00 |
| 13 | Upper wardha | 342.00 | 332.5 | 786.5 | 46.20 |
| 14 | Gosekhurd | 245.00 | 241.3 | 1146.1 | 22.40 |
| 15 | Erai | 207 | 200.5 | 144.69 | 30.0 |

**Table 3** Tapi basin reservoir

| Sr No | Reservoir | FRL (m) | MDDL (m) | Gross storage (MCM) | Height of dam (m) |
|---|---|---|---|---|---|
| 1 | Chankapur | 673.52 | 658.1 | 79.7 | 39.00 |
| 2 | Girna | 398.07 | 382.8 | 608.9 | 54.56 |
| 3 | Akkalpada | 390.00 | 370.0 | 369.0 | 33.00 |
| 4 | Wagur | 234.10 | 223.6 | 325.0 | 13.60 |
| 5 | Hatnur | 214.00 | 207.7 | 388.0 | 25.50 |

**Table 4** West flowing rivers basin reservoirs

| Sr No | Reservoir | FRL (m) | MDDL (m) | Gross storage (MCM) | Height of dam (m) |
|---|---|---|---|---|---|
| 1 | Upper vaitarna | 663.5 | 580.0 | 353.96 | 41.0 |
| 2 | Tillari | 462.0 | 437.0 | 462.20 | 73.0 |
| 3 | Bhatsa | 142.7 | 79.2 | 976.10 | 88.5 |
| 4 | Talamba | 142.4 | 100.0 | 308.75 | 57.4 |

Table 1 shows the reservoirs selected in the Krishna basin. Among the storages considered in this study, it can be observed that from the point of view of feasibility of water transfer with gravity pipe flow, the highest FRL elevation in Krishna basin is for Dhom reservoir, and lowest FRL is for Ujjani reservoir. The largest storage capacity is of Koyna reservoir with gross storage of 2797.4 MCM.

Table 2 shows the reservoirs selected in the Godavari basin. Among the storages considered in this study, the reservoir with the highest elevation in the Godavari basin is Bhandardara and at lowest being Gosekhurd. The largest storage capacity is of Jayakwadi reservoir with gross storage of 2909 MCM.

Table 3 shows reservoirs selected in Tapi basin. Among the storages selected it is seen that the storage with highest FRL elevation in Tapi basin is Chanakapur and at lowest being Hatnur. The largest storage is Girna with a gross storage capacity of 608.9MCM.

Table 4 shows reservoirs selected in west-flowing river basins. Among the storages selected, the highest FRL elevation in west-flowing rivers basin is of Upper Vaitarna reservoir and the lowest is for Talamba. The largest storage is of Bhatsa having gross storage capacity of 976.1 MCM.

## 4 Methodology

The study is conducted as per the study flow chart as depicted in Fig. 2. For establishment of pipe water grid in the state of Maharashtra, total of 35 major reservoirs in four river basins are selected representing all the regions of the stat.

Methodology followed mainly includes the following assumptions:

- This water grid is based on purely technical aspects related to the flow of water in the pipeline grid. Interstate water agreements for sharing of water, etc., are beyond the scope of this study [7].
- Water is taken out at MDDL of the upper reservoir and is to be emptied at FRL of the connection lower reservoir considering that the reservoir whose MDDL

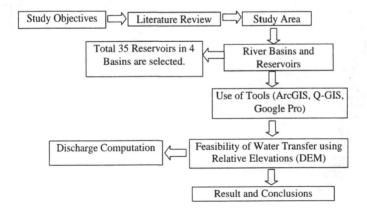

**Fig. 2** Methodology flow chart

is higher than FRL of the lower reservoir so that water can be transferred under gravity flow.
- To save land acquisition cost, pipelines are proposed along either side of state and national highways.
- Wherever local height is greater than MDDL of the upper reservoir, such lengths can be under cuttings or tunnels which can be identified and computed.
- The maximum diameter of the pipeline is 1.8 m to restrict the water transfer from basin to basin.
- Only selected major projects are considered to represent all the basins in the state.
- The idea is to transfer water from water surplus to water deficit reservoirs at any point of time especially during monsoon months when reservoirs are spilling or to mitigate the drought situation. with respect to time and regions [5].

## 5 Analysis and Results

Tools used include mainly ArcGIS with images from Google Pro. SRTM 90 m DEM (Fig. 3) is used for the entire state of Maharashtra. Using ArcGIS morphological analysis is carried out and the drainage basins are delineated and the locations of the selected major projects in each basin are marked (Fig. 4) The reservoir-to-reservoir

**Fig. 3** DEM image of Maharashtra

Inter-Basin Pipeline Water Grid for Maharashtra 517

**Fig. 4** Location of selected reservoirs

pipeline grid within the basins and inter basins are digitized using the alignments of State and National Highways using google images. The results are as per Figs. 5, 6, 7, 8 and 9 for all respective intra basins grids as well as for inter-basin grid. The maximum possible discharges in each of the links are computed using the Darcy–Weisbach equation.

To calculate discharge through the pipeline using head loss.

$$H_l = f \frac{L}{D} \frac{v^2}{2g} \qquad (1)$$

where

$H_l$ = *Head Loss*,

$f$ = *friction factor* = $64/Re$, $Re$ = $1500$ *for laminar flow condition*,

$L$ = *Length of Link*,

$D$ = *Diameter of pipe*,

$v$ = *Velocity of flow* = *Discharge(Q)/area(A)* = $Q/A$,

**Fig. 5** Krishna basin links

$g = $ *acceleration due to gravity* $= 9.81$ m/s$^2$,

$$H_l = f \frac{L}{D} \frac{Q^2}{2gA^2} \qquad (2)$$

Using the Eq. (2), the Discharge (Q) in each of the links is computed.

The links selected include intra basin as well as inter-basin. The results of the link and the respective maximum discharges for Krishna basin are as per Table 5. In Krishna Basin, the selected 11 reservoirs are linked depending on the relative elevations of MDDL and FRL. The ten links having maximum discharge of 1.93 cumecs between Dhom and Khadakwasla with many links having lengths more than 100 km.

In Godavari basin, selected 15 reservoirs with total storage capacity of 43,662 MCM are linked and depending on the relative elevations of MDDL and FRL the links are studied and results are depicted in Table 6. It is seen that maximum discharge of 2.51 cumecs is observed in the link between Darna to Mula reservoirs. Some of the links show lengths over 200 km with minimum lengths of about 50 km.

In the Tapi basin, selected five major reservoirs are linked depending on the relative elevations of MDDL and FRL. The suggested four links with their respective lengths

**Fig.6** Godavari basin links

and maximum discharges possible are depicted in Table 7. It can be seen that the maximum length is 142 km whereas the maximum possible discharge is 2.61 cumecs in Girna to Waghur link.

In West Flowing River basins, selected four reservoirs are linked depending on the relative elevations of MDDL and FRL, The three links possible are depicted in Table 8 It can be seen that the maximum discharge possible is 3.90 cumecs between Upper Vaitarna to Bhatsa and the maximum length observed is of 297 km.

For Inter Basin Water Transfer many of the above reservoirs are considered. Depending upon the relative elevations of MDDL and FRL 18 possible links as depicted in Table 9 are analyzed. It can be seen that most of the links have discharge of more than 1 cumec with maximum discharge of 3.85 cumecs. The maximum length observed is 379 km between Koyana to Jayakwadi.

# 6 Conclusion

The scope of this study is to analyze the feasibility of a pipeline water grid for the state of Maharashtra using modern tools of RS-GIS for terrain analysis. Using the basin-wise existing data of various reservoirs, modem geospatial tools available the

**Fig. 7** Tapi basin links

analysis is conducted. From the results obtained, the following conclusions can be drawn.

- Considering the reservoir levels in the intra or inter-basins, it is seen that waters can be generally transferred from higher reservoir to lower reservoirs using pipelines placed along the sides of the existing road network. The transfer of water can be mostly under gravity though in a few cases pumping is essential.
- It is to indicate that using modern technological tools of terrain analysis, depending upon the relative elevations of the storages, the existing reservoirs can be linked by pipeline networks in order to transfer water from surplus to deficit during floods or drought situations.
- The links analyzed in this study are purely based on topographical/terrain analysis in the entire state of Maharashtra. There can be many more links which can be more feasible which have not been studied here. The links considered are representative only.

**Fig. 8** West flowing rivers basin links

- Scope of this study is macro analysis. There are several other major/ medium reservoirs in the basins which have not been included in this study. Thus, there is further scope for consideration of these reservoirs for such links.
- Considering the relative elevations, it can be seen that there is scope for transfer of water from West flowing rivers and Krishna basin to deficit regions of Godavari and Tapi Basins
- From the analysis it can be seen that the pipeline network interconnecting large number of possible existing reservoirs along the existing roads is feasible option for uniform supply of water for all regions which can eliminate the land acquisition cost, reduce the evaporation and theft of precious water resources with faster implementation and less R&M expenditure.

**Fig. 9** Inter basin links

**Table 5** Krishna basin links

| No | From | To | Length (m) | Head Loss (m) | Pipe Diameter(m) | Discharge (m$^3$/s) |
|---|---|---|---|---|---|---|
| 1 | Dimbhe | Khadakwasla | 97,804.29 | 129.03 | 1.8 | 1.67 |
| 2 | Dimbhe | Ujjani | 234,015.84 | 214.32 | 1.8 | 1.39 |
| 3 | Dhom | Khadakwasla | 80,491.44 | 142.13 | 1.8 | 1.93 |
| 4 | Dhom | Koyana | 104,256.68 | 67.10 | 1.8 | 1.17 |
| 5 | Koyana | Radhanagari | 160,867.19 | 19.50 | 1.8 | 0.51 |
| 6 | Dhom | Ujjani | 145,088.23 | 227.42 | 1.8 | 1.82 |
| 7 | Koyana | Ujjani | 186,320.33 | 111.92 | 1.8 | 1.13 |
| 8 | Pimpalgaon Joge | Kasarsai | 112,227.12 | 52.40 | 1.8 | 0.99 |
| 9 | Varasgaon | Pawana | 82,929.28 | 15.44 | 1.8 | 0.63 |
| 10 | Dimbhe | Panshet | 116,837.46 | 76.09 | 1.8 | 1.17 |

**Table 6** Godavari Basin links

| No | From | To | Length (m) | Head Loss (m) | Pipe Diameter (m) | Discharge (m³/s) |
|---|---|---|---|---|---|---|
| 1 | Mula | Jayakwadi | 97,677.57 | 50.3 | 1.8 | 1.04 |
| 2 | Yeldari | Isapur | 76,757.13 | 21.0 | 1.8 | 0.76 |
| 3 | Upper Wardha | Gosekhurd | 183,636.19 | 87.5 | 1.8 | 1.00 |
| 4 | Gosekhurd | Erai | 117,619.38 | 34.3 | 1.8 | 0.78 |
| 5 | Isapur | Erai | 231,224.15 | 211.7 | 1.8 | 1.39 |
| 6 | Manjara | Jayakwadi | 108,012.87 | 149.5 | 1.8 | 1.71 |
| 7 | Gangapur | Nandur-madhyameshwar | 51,528.45 | 72.5 | 1.8 | 1.72 |
| 8 | Darna | Mula | 125,101.71 | 22.7 | 1.8 | 0.62 |
| 9 | Bhandardara | Nalwande | 20,305.84 | 60.7 | 1.8 | 2.51 |
| 10 | Nalwande | Mula | 103,945.86 | 87.7 | 1.8 | 1.33 |
| 11 | Karanjwan | Gangapur | 37,934.31 | 47.5 | 1.8 | 1.63 |
| 12 | Gangapur | Darna | 44,419.35 | 14.0 | 1.8 | 0.82 |
| 13 | Manjara | Yeldari | 205,678.33 | 148.4 | 1.8 | 1.23 |
| 14 | Totaladoh | Upper Wardha | 146,664.14 | 122.0 | 1.8 | 1.32 |

**Table 7** Tapi Basin Links

| No | From | To | Length (m) | Head Loss(m) | Pipe Diameter(m) | Discharge (m³/s) |
|---|---|---|---|---|---|---|
| 1 | Girna | Waghur | 46,052.26 | 148.7 | 1.8 | 2.61 |
| 2 | Waghur | Hatnur | 40,548.04 | 9.6 | 1.8 | 0.71 |
| 3 | Chankapur | Akkalpada | 99,564.04 | 315 | 1.8 | 2.58 |
| 4 | Akkalpada | Waghur | 142,103.1 | 135.9 | 1.8 | 1.42 |

**Table 8** West flowing rivers links

| No | From | To | Length (m) | Head Loss(m) | Pipe Diameter(m) | Discharge (m³/s) |
|---|---|---|---|---|---|---|
| 1 | Upper Vaitrana | Bhatsa | 60,509.24 | 437.3 | 1.8 | 3.90 |
| 2 | Tillari | Talamba | 70,721.67 | 297 | 1.8 | 2.98 |
| 3 | Upper Vaitrana | Natuwadi | 297,530.7 | 460 | 1.8 | 1.81 |

**Table 9** Inter-basin links

| No | From | To | Length (m) | Head Loss(m) | Pipe Dia (m) | Discharge (m³/s) |
|---|---|---|---|---|---|---|
| 1 | Dhom | Mula | 234,224.2 | 172.7 | 1.8 | 1.25 |
| 2 | Dimbhe | Mula | 116,405.7 | 159.6 | 1.8 | 1.70 |
| 3 | Dimbhe | Nandur-Madh'war | 166,427.1 | 178.4 | 1.8 | 1.50 |
| 4 | Jaikwadi | Girna | 200,168.8 | 43.5 | 1.8 | 0.68 |
| 5 | Yeldari | Waghur | 230,344.7 | 227.9 | 1.8 | 1.44 |
| 6 | Upper Vaitrana | Nandr- Madh'war | 86,299.27 | 130.0 | 1.8 | 1.78 |
| 7 | Nandr-Madh'war | Akkalpada | 140,361.9 | 125.3 | 1.8 | 1.37 |
| 8 | Pimpalgaon Joge | Nalwande | 34,098.57 | 18.4 | 1.8 | 1.07 |
| 9 | Dhom | Manjara | 276,328.9 | 82.6 | 1.8 | 0.79 |
| 10 | Bhandardara | Chankyapur | 122,583.4 | 15.7 | 1.8 | 0.52 |
| 11 | Ujjani | Jayakwadi | 207,652.9 | 7.1 | 1.8 | 0.27 |
| 12 | Radhanagari | Talamba | 81,856.03 | 426.1 | 1.8 | 3.31 |
| 13 | koyana | Natuwadi | 69,717.31 | 489.5 | 1.8 | 3.85 |
| 14 | koyana | Jayakwadi | 379,187.8 | 125.6 | 1.8 | 0.84 |
| 15 | Totaladoh | Upper Wardha | 146,664.14 | 122.0 | 1.8 | 1.32 |
| 16 | Yeldari | Isapur | 76,757.13 | 21.0 | 1.8 | 0.76 |
| 17 | Upper Wardha | Gosekhurd | 183,636.19 | 87.5 | 1.8 | 1.00 |
| 18 | Gosekhurd | Erai | 117,619.38 | 34.3 | 1.8 | 0.78 |
| 19 | Isapur | Erai | 231,224.15 | 211.7 | 1.8 | 1.39 |

# References

1. Suryawanshi RK, Gavhale SH (2019) River rejuvenation by adopting continuous stream storage approach in mula-mutha basin using geospatial and simulation tools. In: Proceedings of Sixth India Water Week, 24–28 Sept 2019, Vigyan Bhavan New Delhi
2. Jain SK, Reddy NSRK, Chaube UC (2009) Analysis of a large inter-basin water transfer system in India. Hydrol Sci J
3. Tian J and others (2019) Impacts of inter-basin water transfer projects on optimal water resources allocation in the Hanjiang River Basin, China. State Key Laboratory of Water Resources and Hydropower Engineering Science. Wuhan University,Wuhan (MDPI J)
4. Yano S, Okazumi T, Iwasaki Y, Yamaguchi M (2018) How inter- basin transfer of water alters basin water stress used for water footprint characterization. MDPI J
5. Strategic analyses of inter-linking of rivers in drought- prone tehsils of Jalgaon district—Maharashtra (2018) Asian J Sci Technol
6. Karamouz M, Ali Mojahedi S, Ahmadi A (2010) Inter-basin water transfer: economic water quality-based model. J Irrig Drainage Eng

7. Anupriya, Chamat LR (2018) Study of interlinking of rivers by using geographic information system (GIS) With Quantum-GIS. Int J Eng Res Technol (IJERT)
8. Prakash I, Patel AK, Singh NP Planning of river inter- linking canal system between Shetrunji River and Dhatarwadi River, Saurashtra, India, using Remote Sensing and GIS

# Removal of Heavy Metals from Water Using Low-cost Bioadsorbent: A Review

**Praveda Paranjape and Parag Sadgir**

**Abstract** Current review paper presents the performance of natural low-cost (economical) bioadsorbent, i.e., plant biomass and agricultural waste used for the removal or extraction of copper and other heavy metals. The current study focuses on use of various leaves, seed, peel, root, bark, husk, shell, fiber, and other bioadsorbents in their raw, pretreated as well as modified forms. This review paper objects to evaluate the applicability of said biomass for single and multimetal removal from water/wastewater. Researchers' variations in different experimental parameters such as contact time, initial concentration, adsorbent mass, speed of agitation, temperature and pH, etc., are considered along with isotherm, kinetics, thermodynamic, and desorption studies. Researchers using various instruments such as UV vis. spectrophotometer, transmitted electron microscope (TEM), atomic absorption spectroscopy (AAS), molecular absorption spectrophotometer, SEM–EDX analyses, XDR spectroscopy and FTIR analysis, etc., to obtain the characterization of bioadsorbent. In this review, an extensive summary of previous literature on plant biomass, agricultural waste, and other biomass were compiled to provide information on experimental and equilibrium conditions and their potential as a low-cost sorbent.

**Keywords** Copper removal · Plant-based bioadsorbents · Agricultural waste · Adsorption mechanism

## 1 Introduction

In India, nowadays, the quality of environment is deteriorating due to extensive industrial processes or operations which particularly include pharmaceuticals, mining, processing of metals, tanneries, pesticides, organic chemicals manufacturing, processing of plastics and rubber industries, wood products manufacturing, agricultural processes, and disposal of industrial waste materials, etc. The wastewater and its effluents which are released from most of the industries contain toxic materials

---

P. Paranjape (✉) · P. Sadgir
Department of Civil Engineering, College of Engineering Pune, Pune 411005, India
e-mail: pnp20.civil@coep.ac.in

and heavy metals into rivers without appropriate treatment hence heavy metals are major pollutants in marine water, groundwater as well as treated wastewater. Heavy metals in industrial wastewater comprise Cd, Pb, Fe, Cr, Cu, Hg, Ag, Zn, Ni, etc. These metals are very responsive even at very less concentrations and can accumulate in the food web, causing adverse effects on community health. Adverse effects on heavy metals on human health have been regularly looked over by international bodies like the World Health Organization (WHO) and problems arise due to metal toxicity are anemia, arthritis/rheumatoid arthritis, hypertension, nausea/vomiting, schizophrenia, insomnia, heart problem, cystic/pulmonary fibrosis, inflammation and enlargement of liver, thyroid dysfunction, genetic defects, muscle weakness, skin, lung, and bladder cancer, etc., and many more. Various chemical, physical, and biological treatments are area unit accessible for the removal of harmful toxic metals like chemical oxidization, reduction, chemical precipitation, ultrafiltration, reverse diffusion, electrochemical analysis, and adsorption. Most of those standard ways suffer from some limitations like low potency, high operational price, generation of huge quantity of sludge and their disposal issues, etc. Among these standard techniques, adsorption looks low price operating, efficient, straightforward in style, and effective at lower concentrations. In adsorption, especially bioadsorption (usually uses inactive waste biomass) is one among the in effect method for removal of noxious or toxic heavy metal ions from water or waste material because it offers the most potency even at lower concentrations. The purpose of the current paper is to review info supported, varied plant biomass and agricultural waste as bioadsorbents that are employed in recent studies for copper and alternative serious metals removal from completely dissimilar media. Current paper conjointly purposes to review the pertinence of the plant biomass and agricultural waste bioadsorbents, factors poignant the bioadsorption method, metal removal potency of adsorbents in numerous experimental conditions, etc.

## 2 Plant Biomass Used for Heavy Metal Removal

Plant biomass is made up of logs and agricultural waste and is available in leaf, seed, root, bark, peel, and other biomass formats. Lignocellulosic constituents are the structural fundamentals of wood and other plant materials that are existing in abundant amounts in the biosphere, are reasonable and have little to low cost-effective value, and these are very promising for the future.

### 2.1 Leaves Biomass

Leaves are in the group of abundantly available natural resources, often non-toxic and renewable. In adverse situations, most plants shelter their leaves and these leaves can be used as bioadsorbents in the form of raw powder or with some modifications. The adsorption potential of leaf biomass, in the bioadsorption of metal ions is compiled in Table 1.

Work carried out on the separation of Cu (II) ions from aqueous solution by means of spent-tea-leaves improved with Ca (OH)$_2$ by Ghosh et al. [21]. Mathematical models were used to investigate the simulation and optimal adsorption conditions for adsorbent dose, pH, and reaction time and concluded that RSM gives superior predictions than ANN. Also, the author discovered that adsorption equilibrium statistics were close-fitting with Langmuir isotherm and follows the pseudo-second-order kinetics. Maximum removal efficiency of Cu (II) was observed to be 96.12% using modified spent tea leaves. For the separation of cadmium, copper, zinc, and lead from aqueous solution, activated carbon prepared from olive branches (H$_3$PO$_4$ treated) were used as raw material [15]. They found that the process of bioadsorption depends on the pH and increases with rise in temperature and the initial concentration of metal ions. Maximum bioadsorption for Pb (II), Cu (II), and Cd (II) was obtained at pH 5 and at pH 3 for Zn (II). The overall adsorption potential in mg/g for Pb (II) was 41.32, 34.97 for Zn (II), 43.10 for Cu (II) and for Cd (II), respectively. The author also set up that for all metal ions, adsorption equilibrium statistics were best described with Freundlich isotherm model, and progress of bioadsorption was observed to be non-spontaneous.

## 2.2 Seed Biomass

Researchers have analyzed seed biomass as resourceful bioadsorbent for extraction of many heavy metals from water and wastewater and summarized it as follows.

Oboh et al. [35] observed the elimination of Pb, Cu, Zn, and Ni using Luffa ylindrical (sponge and seeds) as a bioadsorbent from aqueous solution. The FTIR and SEM analysis used for characterization of bioadsorbents and found that the carboxylic and amine groups were accountable for metal complex formation in the adsorption of all metal ions. The maximum bioadsorption was reported 98.2% for lead, 95.2% for zinc, 87.6% for copper, and 43.5% for nickel at pH 5 with 1 g dose of adsorbent. Within 120 min of time of contact, the adsorption equilibrium was reached and obeys the pseudo-second-order kinetics model for all metals. Garba et al. [17] studied prime conditions using defatted papaya seeds (DPS) as a bioadsorbent for the removal of copper (II) and lead (II) from the solution. The optimal bioadsorption conditions for adsorbent dose, shaking speed, and initial metal concentration were examined by applying central composite design (CCD) which is a subgroup of response surface methodology (RSM) and 97.55% removal of Cu (II) and 99.96% removal of Pb (II) were acquired at bioadsorbent dose of 0.30 gm, shaking speed of 180irpm, and primary metal concentration of 150img/L. Langmuir isotherm model provides the best result with monolayer adsorption capacities measured in mg/g was 17.29 for copper (II) and 53.02 for lead (II).

Table 1 Summary of heavy metals removal using leaf biomass

| Leaf biomass | Metal | pH | Adsorbent dose (g/l) | Contact time (minutes) | Adsorption capacity (mg/g) | Isotherm model | KINETICSMODEL | References |
|---|---|---|---|---|---|---|---|---|
| Caesalpinia bonducella | Cu (II) | 5 | 0.7 | 120 | 76.92 | Langmuir | Pseudo-second-order | [54] |
| Syzygium cumini leaves | Cu (II) | 4 | 15 | 12 (hr) | 15.87 | Langmuir | Pseudo-second-order | [27] |
|  | Ni (II) | 6 |  |  | 15.60 |  |  |  |
|  | Cr (VI) | 2 |  |  | 16.07 |  |  |  |
| Populus deltoides leaves | Cu (II) | 4 | 15 | 12 (hr) | 17.76 |  |  |  |
|  | Ni (II) | 6 |  |  | 16.45 |  |  |  |
|  | Cr (VI) | 2 |  |  | 18.31 |  |  |  |
| Lawsonia Inermis Leaves | Cu (II) | 5 | 0.5 | 90 | 6.06 | Langmuir | Pseudo-second-order | [7] |
| Dry cabbage leaves | Cu (II) | 6 | 10 | 120 | 5.74 | Langmuir | Pseudo-second-order | [26] |
|  | Cd (II) |  |  |  | 5.06 |  |  |  |

## 2.3 Peel Biomass

The feasibility of peel biomass (peel is considered waste material) as bioadsorbents in the removal of metals has been demonstrated in recent studies. Table 2 summarizes some of the adsorption parameters of peel biomass in the bioadsorption of heavy metal ions.

Chowdhury et al. [11] investigated the use of skins of garlic and onion for elimination of Cu (II) from aqueous solution and observed that onion skin has maximum adsorption capacity (76.9 mg/g) as w.r.t. garlic skin (66.7 mg/g) at optimum pH of 5.3 at 303 K. Endothermic and spontaneous cycles of bioadsorption of Cu (II) on both onion and garlic skin was recorded. The adsorption equilibrium data was close-fitting with the Langmuir isotherm model and follows the pseudo-second-order kinetics for both bioadsorbents. Later on, Lagenaria siceraria (LS) peel has been explored for the elimination of Fe, Hg, Cr, Ag, Co, Zn, and Cu from industrial wastewater [2]. The untreated and Formaldehyde treated Lagenaria siceraria-LS biomass were used for elimination of the said metal ions and concluded that said biomass has notable affinity toward copper ions (99% removal efficiency) followed by iron and silver (95% removal efficiency). The removal efficiency of metallic ions at pH 4–5 is in order mercury < zinc < silver < chromium < cobalt < iron < copper. The adsorption equilibrium of all metal ions follows Langmuir model except Iron (follows Freundlich model). Extreme adsorption capacity in mg/g was observed to be 7.34 for Cu, 8.54 for Co, 5.03 for Zn, 8.92 for Cr, 6.05 for Hg, 4.34 for Ag, and 11.36 for Fe.

## 2.4 Root Biomass

The use of water hyacinth roots as an economical bioadsorbent for deduction of Cu (II) from the water-based solution was scrutinized [55] and concluded that the bioadsorption of Cu (II) on water hyacinth roots was both a chemisorption method and a quick and endothermic one. The adsorption potential in mg/g of Cu (II) was found to be 22.7 and equilibrium is best fitted with Langmuir isotherm model. From FTIR analysis they have concluded that the –OH and C = O groups have more affinity towards Cu (II) ions. Sorghum root (SR) powder has been explored as bioadsorbent [10] for bioadsorption of copper (II) and chromium (VI) from dissolvable water solution. The author found that optimum pH 2 with a bioadsorbent dosage of 0.45 mg/100 ml for 50 mg/l of initial concentration of metal ions has maximum separation efficiencies of 93% for copper (II) and 91.98% for chromium (VI) within a duration of 60 min at $20^0C$. The highest adsorption capacities for Cu (II) and Cr (VI) were observed to be 18.6 and 18.39, respectively, measured in mg/g. The adsorption process follows the pseudo-second-order kinetic model with Langmuir isotherm model for both said metal ions. The bioadsorption of copper (II) and chromium (VI) on SR powder was possible, exothermic, and spontaneous.

Table 2 Summary of heavy metals removal using peel biomass

| Peel biomass | Metal | pH | Adsorbent dose (g/l) | Contact time (minutes) | Adsorption capacity (mg/g) | Isotherm model | Kinetics model | References |
|---|---|---|---|---|---|---|---|---|
| Zalacca edulis peel | Cu (II) | 5 | 0.1 | 24 (hr) | 27.03 | Langmuir and Freundlich | Pseudo-second-order | [47] |
| Orange peel | Cd (II) | 5 | 4 | 30 | 13.7 | Langmuir and Freundlich | Pseudo-second-order | [29] |
| | Cu (II) | | | | 15.27 | | | |
| | Pb (II) | | | | 73.53 | | | |
| | Pb | | | | 116.2 | | | |
| Sweet lime peel | Cu (II) | 5 | 1 | 30 | 37.45 | Langmuir | Pseudo-second-order | [37] |
| Pomegranate peel | Cu | 5.8 | 3 | 120 | 30.12 | Langmuir | Pseudo-second-order | [6] |

## 2.5 Bark Biomass

Bark is a by-product of the timber industry and abundantly available at a very cost-effective value. The use of bark in water/wastewater treatment converts the agricultural waste byproduct into a cost-effective adsorbent and reduces the cost of residue disposal. Bark biomasses have been suggested for its effectiveness as bioadsorbents for metals removal as follows.

Chakravarty et al. [8] carried out a study on the separation of Cu (II) ions from aqueous solution utilizing heartwood powder of Areca catechu as a bioadsorbent and concluded that optimum pH is 5.5 with bioadsorbent dose of 0.5 g/l having maximum removal efficiency within a duration of 30 min. The FT-IR and SEM characterization of the bioadsorbent concluded that N–H, C–O, and O–H groups were primary binding groups for Cu (II) ions. The bioadsorption method was physical in nature and balance is ideally well-matched with the Langmuir isotherm model having extreme bioadsorption potential of 9.578 in mg/g and follows the kinetic model of the pseudo-second-order. Cutillas-Barreiro et al. [12] observed the removal of Cd, Cu, Ni, Pb, and Zn utilizing Pine bark powder as a bioadsorbent and effectively removed Pb (98–99%) and Cu (83–84%) followed by Cd (78–84%), Zn (77–83%) and Ni (70–75%) for 1–48 h of the time of the contact. The bioadsorption equilibrium statistics was close-fitted with Langmuir isotherm model for all said metals. The maximum uptake was 39.4, 65.9, 39.7, 87.9, and 38.1 mmol/kg for Cd, Cu, Ni, Pb, and Zn, respectively.

## 3 Agricultural Waste Used for Heavy Metals Removal

Agricultural products and by-products like husk, stalk, fiber, and shell are the abundantly available waste materials and their proper disposal is necessary. When disposed of by burning, they produce carbon dioxide and other sources of emissions. As a consequence, it is important to transform agricultural products and by-products into useful and value-added products. As described below, the usage of different agricultural products and byproducts for the separation of heavy metal from the solution was discovered by the number of researchers and proved to be economical [19].

## 3.1 Husk/ Hull

Husk is the dried form of outer shell or coting of grains generated from various agricultural processes, which contains cellulose, hemicelluloses, lignin, mineral ash, and silica. It can be chemically improved to increase adsorption capacity also coconut waste contains hydroxyl and carboxyl group which also helps in bioadsorption of heavy metals. The summary of removal of heavy metals by husk/hull reported in the literature is compiled in Table 3.

Marula seed husk was utilized as bioadsorbent for extraction of Pb (II) and Cu (II) ions from water-based solution [32]. The Langmuir model defined the equilibrium data well and it was found that the maximum bioadsorption abilities in mg/g were 20 and 10.20 for lead (II) and copper (II), respectively. In current studies, adsorption kinetics follows the pseudo-second-order model. The optimal pH for the removal of lead (II) is 5 and for copper (II), it is 6. The equilibrium was accomplished within 60 min of contact time for Pb (II) ions while 180 min was required for Cu (II) ions. Bioadsorption of both the ions on marula seed husk was endothermic, spontaneous, and feasible. A proportional studies on removal of zinc (Zn) and copper (Cu) from aqueous solution by using sunflower hull/husk (SFH) and durian leaves (DL) as a bioadsorbent was carried out by Abood et al. [1]. The ideal pH required for the removal of both metals was 8 for sunflower husk and 9 for durian leaves. The equilibrium was reached within 120 min of contact time in case of SFH and within 60 min in case of DL for both the metals and close-fitted with Freundlich isotherm model. The removal efficiency of copper and zinc was found to be 83% and 93% correspondingly using SFH while using DL it was 70% and 78%, respectively at 10 mg/l of initial metal concentration, agitated at 150 rpm at 25 °C. The maximum bioadsorption capacities of copper was found to be 0.39 mg/g (SFH), 0.06 mg/g (DL) and for zinc 0.27 mg/g (SFH), 0.15 mg/g (DL).

## 3.2 Shell

Shells are byproducts of various agricultural activities which mostly contain Lignocellulosic material. These materials are highly porous, with more surface area and thus easily used to extract heavy metals (because of the attraction of water molecules to cell wall components) as summarized as follows.

Utilization of cashew nut shells as an operative bioadsorbent for removal of Cu (II) ions was studied by SenthilKumar et al. [46]. The maximum removal efficiency obtained was 82.11% with best suitable pH of 5 having dosage of 3 g/l in 20 mg/l copper (II) solution. Adsorption equilibrium was reached at 120 rpm agitation speed in 30 min of contact time and the ability of monolayer adsorption in mg/g was observed to be 20. The bioadsorption equilibrium statistics was upright provided by the Langmuir and Freundlich isothermal models at 30 °C. Mechanism of bioadsorption was feasible, exothermic, and spontaneous and follows the model of kinetics of the pseudo-second-order. Hossain et al. [24] explored the use of palm oil fruit covering as bioadsorbent for the removal of copper ions from water. They have concluded that an extreme removal efficiency obtained is 98.5% with the best suitable pH of 6.5 and adsorption capacity is between 28 and 60 mg/g at room temperature. The adsorption kinetics obeys pseudo-second-order model. Further, [22] investigated the use of ultrasound in bioadsorption procedure for the removal of Cu (II) ions using watermelon shells make activate with citric acid (AWS) and calcium hydroxide

**Table 3** Summary of heavy metals removal using Husk/Hull biomass

| Husk biomass | Metal | pH | Adsorbent dose (g/l) | Contact time (minutes) | Adsorption capacity (mg/g) | Isotherm model | Kinetics model | References |
|---|---|---|---|---|---|---|---|---|
| watermelon seed hulls | Cu (II) | 5 | 2 | 60 | 33.90 | Langmuir | Pseudo-first order | [3] |
|  | Pb (II) | 5 | 2 | 80 | 42.19 |  |  |  |
| Cicer arietinum husk | Cu (II) | 5 | 20 | 120 | 9.7 | Langmuir | Pseudo-second order | [38] |
|  | Cd (II) | 7 |  |  | 8.58 |  |  |  |
| Moringa oleifera seed husk | Cd | 6 | 1 | 60 | 13.1 | Langmuir | Pseudo-second order | [18] |
|  | Cu |  |  |  |  |  |  |  |
| Pigen pea hull | Cu (II) | 5 | 0.4 | 60 | 20.83 | Langmuir | Pseudo-second order | [48] |
|  | Cd (II) |  |  |  | 19.81 |  |  |  |
| Pigen pea hull AC | Cu (II) | 6 | 0.6 |  | 37.39 |  |  |  |
|  | Cd (II) |  |  |  | 35.85 |  |  |  |
| Sesame husk | Cu (II) | 6 | 10 | 30 | 10.83 | Langmuir | Pseudo-second order | [14] |
| Maize hull | Cu (II) | 6 | 5 | 90 | 19.6 | Langmuir | Pseudo-second order | [20] |

(SWS). The author observed that in the existence of ultrasound the adsorption rate increases and equilibrium was reached early as compared to conventional process. At pH 5, 90 W ultrasonic power, and contact time of 20 min and 30 °C temperature, the maximum removal efficiency was attained. The maximum potential for adsorption was 27.027 mg/g for ASW and 31.25 mg/g for SWS. For both adsorbents, adsorption system obeys the kinetic model of pseudo-second-order and is ideally equipped with the isotherm model of Langmuir.

## 4 Other Biomass

Other than plant biomass and agricultural waste biomass like cone biomass, flower, and fruit waste can also be utilized as a bioadsorbent for the removal of heavy metals as shown in Table 4.

Pretreated black tea waste has been used to extract Cui(II) from water-based solution as a bioadsorbent [50]. The NaOH treated black tea waste proved to be more efficient than other pretreatments. The maximum removal efficiency was nearly 90% for a duration of 10 min with optimum pH of 4.4 at a temperature of $26^0C$. FT-IR research reveals that the responsible groups of black tea waste for copper adsorption were the –CH, –OH, and –C = C groups and adsorption governs diffusion and reaction follows physisorption. The adsorption of copper on NaOH treated black tea waste was spontaneous and endothermic. The Langmuir isotherm was well represented in the adsorption equilibrium data and follows pseudo-second-order kinetics model with the maximum bioadsorption potential in mg/g of 43.18. (Yusoff et al., 2014) investigated the removal of Cu (II), Pb (II), and Zn (II) ions from aqueous solution by exploiting diverse types of agricultural waste materials like sawdust of Durian tree (DTS), coconut coir (CC) and oil palm empty fruit bunch (EFB) as economical bioadsorbents. The utmost adsorption was obtained at pH 5 and the bioadsorption of all metal ions to all adsorbents was favorable. The maximum bioadsorption capacities in mg/g was obtained to be 18.42 (DTS),18.38 (CC), and 26.95 (EFB) for Cui(II), 20.37 (DTS), 37.04 (CC), and 37.59 (EFB) for Pb (II) while 22.78 (DTS), 24.39 (CC), and 21.19 (EFB) for Zn (II) calculated by Langmuir isotherm model.

## 5 Mechanism of Bioadsorption

The elimination of elements from solution by biological materials is well defined as biosorption/bioadsorption (Insoluble or soluble and gaseous inorganic or organic forms can be such substances). Bioadsorption is a physicochemical system that can involve processes such as adsorption, ion exchange, precipitation, absorption, and texture of the surface [16]. The mechanism of bioadsorption is very complex and, under some circumstances, a number of mechanisms can be operative. Because of the concentration gradient and diffusion across cell walls and membranes, primarily

Table 4 Summary of heavy metals removal using different biomass

| Other biomass | Metal | pH | Adsorbent dose (g/l) | Contact time (min) | Adsorption capacity (mg/g) | Isotherm model | Kinetics model | References |
|---|---|---|---|---|---|---|---|---|
| Pine cone | Cu (II) | 5 | 0.3 | 120 | 19.27 | Langmuir | Pseudo-second-order | [13] |
|  | Cd (II) | 7 |  |  | 26.11 |  |  |  |
| Olive stone waste (AC) | Fe (II) | 5 | 0.3 | 180 | 57.47 | Langmuir | Pseudo-second-order | [4] |
|  | Pb (II) |  |  |  | 22.37 |  |  |  |
|  | Cu (II) |  |  |  | 17.83 |  |  |  |
|  | Cu (II) |  |  |  | 124.21 |  |  |  |
| Tomato waste | Cu (II) | 8 | 4 | 60 |  | Langmuir | Pseudo-second-order | [52] |
| Moringa pods | Cu | 1 |  | 40 | 6.07 | Freundlich and Temkin | Pseudo-second-order | [31] |
|  | Ni |  |  | 30 | 5.53 | Temkin and D–R |  |  |
|  | Cr |  |  | 40 | 5.49 | Langmuir |  |  |

ion exchange, adsorption by physical forces, chelation, and trapping in inter and intrafibrillary capillaries and spaces of the structural polysaccharide network occurs. The variety of structural components found in biomass contributes to the numerous functional groups that can interact with metal types and improve their affinity, e.g., carboxyl, phosphate, hydroxyl, amino, amido and thiol, etc. (for strong or weak bond with ions). The existence of such functional groups does not guarantee bioadsorption [43]. X-Ray Diffraction (XRD), X-Ray Photo Electron Spectroscopy (XPS), Fourier Transform Infrared spectroscopy (FTIR), Scanning Electron Microscopy (SEM), Elemental analysis and Energy Dispersive, X-ray fluorescence spectroscopy (EDX) can also help to determine the process of bioadsorption [5].

## 6 Bioadsorption Isotherm Modeling

Isothermal adsorption is the process by which the amount of material adsorbed is calculated as a function of concentration (at a constant temperature). Adsorption isotherms are created by exposing a fixed amount of adsorbate to varying amounts of adsorbent in a fixed volume of liquid. The amount of adsorbate that can be taken up by an adsorbent depends on both the temperature and the adsorbate's characteristics and concentration. The isotherms do not provide any information about the adsorption process, but they do aid in determining adsorption ability (Metcalf and Eddy, Inc., 2003). The following equation is used to develop adsorption isotherm.

$$q_e = \frac{(Co - Ce)V}{m}$$

where, $q_e$ = adsorbent phase concentration after equilibrium in mg/g.

Co = initial concentration of adsorbate in mg/l.

Ce = final equilibrium concentration after absorption has occurred in mg/l.

$V$ = Volume of liquid in the reactor in lit.

$m$ = mass of adsorbent in g.

Langmuir, Fruendlich, Braunaure-Emmett Telleri (BET), Temkin, and Dubinin-Rudushkevich (D-R) models can explain the equilibrium isotherm. Among them, Langmuir and Fruendlich are the most ordinarily used isotherm equilibrium models. Table 5 shows the isotherm equations and assumptions for commonly used isotherms.

**Table 5** Isotherm equations and Assumptions

| Sr. No | Model | Equation | Remark/Assumptions |
|---|---|---|---|
| 1 | Freundlich isotherm | $q_e = KC^\beta$ or $\log q_e = \log K + \beta \log C$ | 1. Defines adsorption on Surfaces which are heterogeneous, i.e., surfaces with different affinities having adsorption sites<br>2. Only at a constant pH value they can be implemented |
| 2 | Langmuir isotherm | $q_e = Q^0 KC/1 + KC$ | 1. The adsorbate has an equal affinity for all binding sites<br>2. Adsorption is restricted to monolayer formation<br>3. The number of surface sites does not exceed the total number of adsorption species<br>4. Only at a constant pH value they can be implemented |

# 7 Factors Affecting Bioadsorption

## 7.1 Effect of pH

The metal ions uptake in adsorption is intensely dependent on pH and as pH increases metal ion adsorption increases because of competition between metal cation and proton at low pH. Along with this positive surface charge is decreases as pH increases which leads to the dropping of Coulombic repulsion of metal ions [49]. Yargiç et al. [52] observed the impact of pH on extraction of Cu (II) ions using tomato waste and found that adsorption capacity changes from 9.34 to 46.04 mg/g as pH increases from 2 to 8 at 293 K. Pandey [38] explored the use of Cicer arietinum husk as bioadsorbent for separation of copper (II) and cadmium (II) ions and results indicates that adsorption to be influenced by the pH of the sorbent media. The experimental investigation indicates that at lower pH, bioadsorbent surface is positively charged due to preferential adsorption of $H^+$ ions but on increasing pH concentration of $H^+$ ions decrease and metal uptake increases. The best pH for removing Cu (II) and Cd (II) was determined to be 5 and 7, respectively. Kamar et al. [26] found that the ideal pH for removal of Cadmium (II) and Copper (II) was 6 using dry cabbage leaves as a bioadsorbent. The trials were carried out between pH 3 and 8 and observed that removal increases as pH rises up to 6 and then decreases due to decrease in the $H^+$ ions in the solution which form some facilities with metal ions and precipitate as metals hydroxide.

## 7.2 Effect of Dose of Adsorbent

A significant parameter for studying the quantitative absorption of contaminants/ metal ions is the adsorbent dose i.e. pollutant or metal ions retention was observed in relation to the adsorbent dose. The percentage of removal increases as the adsorbent dose increases, after which it remains constant as the number of active sites rises as the dose bioadsorbent of increases [4]. The impact of adsorbent dose on adsorption was studied [42] and observed that percentage adsorption of Cu (II) on Stalk powder of Jerusalem artichoke (Helianthus tuberosus L.) rises with rise in adsorbent dose, then decreases and after that becomes constant at 5 g/l of dose. Değirmen et al. [13] performed a number of experiments to determine the effect of bioadsorption dose on the percentage removal of Cu (II) and Cd (II) from pine cones. The bioadsorbent concentration varies between 1 and 10 g/l keeping temperature and pH constant and found that as bioadsorbent dose increases yield increases because more active sites will be available but adsorption capacity decreases because of unsaturated adsorption sites or might be due to particle interaction (i.e., aggregation, resulting in a reduction in surface area).

## 7.3 Effect of Initial Metal Concentration

The bioadsorption efficiency would be reliant on the initial metal concentration. The percentage of bioadsorption tends to decrease as the initial metal concentration increases, but the ability to adsorb increases. At lower initial concentrations, the surface area available for initial moles of metal ions was limited, and fractional adsorption was independent of initial metal concentration [44]. The increase in the potential for adsorption is proportional to the enhancement of the initial concentration of metals. The concentration gradient's driving forces are stronger at higher initial concentrations, which aids in increasing the initial adsorption rate [37, 51]. The effect of initial metal concentration on the removal of Cu (II), Pb (II), and Zn (II) using durian tree sawdust was investigated, and it was discovered that as the initial metal concentration accumulated upto 200 mg/L, the adsorption potential increased from 2.09 to 16.62 mg/g for Copper (II), 2.37 to 20.01 mg/g for Lead (II), and 1.57 to 15.65 mg/g for (II) (Yusoff et al., 2014).

## 7.4 Effect of Contact Time

Metal ion bioadsorption rates are initially higher being the accessibility or availability of more surface area adsorbents, but they gradually decrease over time until they reach equilibrium (equilibrium state happens when adsorbent is saturated with the adsorbate i.e. metal ions). As the number of active sites decreases over time/exhaustion, the rate of uptake is governed by the rate of adsorbent transport from the outside to the inside site. In monolayer adsorption, each active adsorbent site can only adsorb one metal ion as it has a fixed number of sites, so surface metal absorption is initially rapid and subsequently decreases, resulting in metal ions remaining in the media [46]. Pavan et al. [40] observed that adsorption of nickel (II), cobalt (II), and copper (II) from solution by utilizing waste ponkan mandarin peel was increased with time and reach equilibrium within 5 min of contact time in case of Ni (II) (Also, 5 min of contact time to reach equilibrium was reported by Ofomaja et al. [36] utilizing pine cone powder for removal of Cu (II), this time was very low as related to the other research conclusions), while it takes 10 min for Co(II) and 15 min for Cu (II). Paulino et al. [39] observed that elimination of Cu (II), Zn (II), Cd (II), and Pb (II) by using citric acid-treated natural cotton fibers from aqueous solution was very fast at early stage and gradually increases with time and within 10 min of contact, equilibrium was achieved.

## 7.5 Effect of Temperature

The thermodynamic factors such as variation in Gibb's free energy ($\Delta G°$), variation in enthalpy ($\Delta H°$) and variation in entropy ($\Delta S°$) should be considered in order to identify the essence of the adsorption procedure. Both entropy and energy variables are considered to determine the spontaneity of the operation [13]. The negative value of $\Delta G°$ specifies the viability and random nature of the adsorption, while the negative value of ($\Delta H°$) indicates the exothermic nature. (positive value specifies adsorption process is endothermic) where as positive value of $\Delta S^0$ specifies the increasing uncertainty at the solid/liquid boundary [30, 33]. At high temperatures, the adsorption is faster and more beneficial because temperature changes will travel along the driving force of diffusion through the external interface layer and increase the diffusion rate inside the pores, making it easier for metal ions to reach the inner pores of bioadsorbents [51]. Most of the process of bioadsorption shows the exothermic nature, but some research shows that, for example, adsorption of copper on chemically treated rice husk is mostly physical adsorption and absorbs energy, i.e., endothermic [25]. The adsorption of copper and nickel ions decreases with rise in temperature indicates exothermic nature while adsorption of chromium ions increases with increase in temperature specifies endothermic nature using Moringa pods as a bioadsorbent [31]. Hossain et al. [24] observed that $\Delta G°$ decreases as temperature rises from 30 to 70 °C and positive value of $\Delta H°$ presented that the adsorption of Cu (II) ions on banana peels was endothermic.

## 7.6 Effect of Size of Particles

In the adsorption procedure, size of particles plays a significant role. Yuvaraja et al. [54] studied the impact of particle size (50, 75, 100, and 125 μm) of Caesalpinia bonducella leaf powder on the bioadsorption of Cu (II) and found that as particle size increases removal efficiency decreases due to decrease in surface area and binding probability between metal ions and bioadsorbent functional groups. The variation of particle size from 75 to 125 μm gives decrease in efficiency from 85.22, to 72.46, to 32.23%. Hossain et al. [24] performed the batch experiment to study the effect of particle size (six particle sizes varies between 600 and <75 μm) on removal of copper using banana peel and observed that as particle size reduce from 600 to <75 μm the removal efficiency increased from 74 to 96%. Also [6] reported that as particle size of pomegranate peel powder reduced from 1.6 mm to 630 μm, the removal efficacy of Cu (II) ions increases from 57.39 to 89.99%.

## 7.7 Effect of Agitation Speed/rate

The speed of agitation develops the potential for adsorption. Choudhary et al. [10] was discovered that as agitation speed increased from 80 to 120 rpm, copper and chromium adsorption ability increased from 17.55 to 18.60 mg/g and 16.75 to 18.39 mg/g, respectively. The author also reported that increasing the agitation speed promoted bulk quantity transfer of metal ions from the bulk solution to the adsorbent surface and shortened the adsorption equilibrium time. Similar trend was observed by Banerjee [28] in adsorption of Cu (II) on watermelon shells and [34] during utilization of biomass of Thuja orientails for bioadsorption of Cu (II) from water-based media.

# 8 Desorption

Desorption of loaded biomass, regardless of the fact that the desorbing mediator does not significantly harm or destroy biomass, encourages biomass reuse and the recovery and/or containment of the adsorbed material. Desorption therapies may result in a loss of biomass efficiency in some cases, but they may also result in increased sorption capability in others. Acids, alkalis, and other complexing agents are the most commonly used desorbents, depending on the material sorbed, procedure conditions, and economic factors; selective desorption (e.g., for certain metals) is also possible. The desorption processes in continuous flow system occur without significant interruption as, columns in parallel allow easy sorption and desorption [16]. Cinnamomum camphora leaves powder (CLP) was used [9] to extract copper ions from solution. They inferred from EDS and FTIR study that the basic removal mechanism was ion exchange and a certain amount of surface complexation mechanism also coexisted. At pH 4 and 30–60 °C temperature, the bioadsorption power of Cu (II) ions was found to be 16.76–17.87 mg/g. Further, [37] utilized sweet lime peel (SLP) as the bioadsorbent for extraction of Cu (II) from water-based solution and concluded that max removal is obtained at pH 5 and $20^0$C temperature. The adsorption balance is best equipped with monolayer adsorption Langmuir isotherm with an extreme potential of 37.45 mg/g at 293 K and kinetics obey the model of pseudo-second-order. The maximum desorption efficiency of Cui(II) is observed to be 90% with 0.1 N HCL.

# 9 Conclusion

The bioadsorption process is economical and more efficient within a short contact time but limitation is that bioadsorbents are not available in readily usable form and pretreatment or modification (it is required to increase bioadsorption properties by alkali or acidic solutions) may increase the overall cost. Although researching various bioadsorbents, much of the research performed has limited batch experimentation, but there is still limited application to real wastewater on a large scale. Adsorption equilibrium data for copper removal is ideally suited to the isotherms of both Langmuir and Freundlich, which showed single and multilayer bioadsorption. Bioadsorption is firmly influenced by physiochemical constraints (parameters) such as solution pH, size of adsorbent and dosage, initial concentration of metals, surface chemistry, etc. The desorption studies and successful recovery and reuse of bioadsorbent is still challenge.

# References

1. Abood MM, Istiaque F, Azhari NN (2019) Remediation of heavy metals using selected agricultural waste: sunflower husk and durian leaves. IOP Conf Ser Mater Sci Eng 557. https://doi.org/10.1088/1757-899X/557/1/012080
2. Ahmed D, Abid H, Riaz A (2018) Lagenaria siceraria peel biomass as a potential biosorbent for the removal of toxic metals from industrial wastewaters. Int J Environ Stud 75:763–773. https://doi.org/10.1080/00207233.2018.1457285
3. Akkaya G, Güzel F (2013) Bioremoval and recovery of Cu(II) and Pb(II) from aqueous solution by a novel biosorbent watermelon (Citrullus lanatus) seed hulls: kinetic study, equilibrium isotherm, SEM and FTIR analysis. Desalin Water Treat 51:7311–7322. https://doi.org/10.1080/19443994.2013.815685
4. Alslaibi TM, Abustan I, Ahmad MA, Foul AA (2014) Kinetics and equilibrium adsorption of iron (II), lead (II), and copper (II) onto activated carbon prepared from olive stone waste. Desalin Water Treat 52:7887–7897. https://doi.org/10.1080/19443994.2013.833875
5. Arief VO, Trilestari K, Sunarso J, Indraswati N, Ismadji S (2008) Recent progress on biosorption of heavy metals from liquids using low cost biosorbents: characterization, biosorption parameters and mechanism studies. Clean - Soil, Air, Water 36:937–962. https://doi.org/10.1002/clen.200800167

6. Ben-Ali S, Jaouali I, Souissi-Najar S, Ouederni A (2017) Characterization and adsorption capacity of raw pomegranate peel biosorbent for copper removal. J Clean Prod 142:3809–3821. https://doi.org/10.1016/j.jclepro.2016.10.081
7. Bhatia, Amarpreet Kour FK (2015) Biosorptive removal of copper (II) ion from Aqueous solution. J Environ Earth Sci 5:21–30
8. Chakravarty P, Deka DC, Sarma NS, Sarma HP (2012) Removal of copper (II) from wastewater by heartwood powder of Areca catechu: kinetic and equilibrium studies. Desalin Water Treat 40:194–203. https://doi.org/10.1080/19443994.2012.671167
9. Chen H, Dai G, Zhao J, Zhong A, Wu J, Yan H (2010) Removal of copper(II) ions by a biosorbent-Cinnamomum camphora leaves powder. J Hazard Mater 177:228–236. https://doi.org/10.1016/j.jhazmat.2009.12.022
10. Choudhary S, Goyal V, Singh S (2015) Removal of copper(II) and chromium(VI) from aqueous solution using sorghum roots (S. bicolor): a kinetic and thermodynamic study. Clean Technol Environ Policy 17:1039–1051. https://doi.org/10.1007/s10098-014-0860-2
11. Chowdhury A, Bhowal A, Datta S (2012) Equilibrium, thermodynamic and kinetic studies for removal of copper (II) from aqueous solution by onion and garlic skin. Water 4:37–51. https://doi.org/10.14294/WATER.2012.4
12. Cutillas-Barreiro L, Ansias-Manso L, Fernández-Calviño D, Arias-Estévez M, Nóvoa-Muñoz JC, Fernández-Sanjurjo MJ, Álvarez-Rodríguez E, Núñez-Delgado A (2014) Pine bark as bioadsorbent for Cd, Cu, Ni, Pb and Zn: Batch-type and stirred flow chamber experiments. J Environ Manage 144:258–264. https://doi.org/10.1016/j.jenvman.2014.06.008
13. Değirmen G, Kiliç M, Çepelioğullar Ö, Pütün AE (2012) Removal of copper(II) and cadmium(II) ions from aqueous solutions by biosorption onto pine cone. Water Sci Technol 66:564–572. https://doi.org/10.2166/wst.2012.210
14. El-Araby HA, Ibrahim AMMA, Mangood AH, Abdel-Rahman AA-H (2017) Sesame husk as adsorbent for copper(II) ions removal from aqueous solution. J Geosci Environ Prot 05:109–152. https://doi.org/10.4236/gep.2017.57011
15. Elsherif KM, Alkherraz AM, Ali AK (2020) Removal of Pb(II), Zn(II), Cu(II) and Cd(II) from aqueous solutions by adsorptiononto olive branches activated carbon: equilibrium and thermodynamic studies. Chem Int 6:11–20
16. Gadd GM (2009) Biosorption: Critical review of scientific rationale, environmental importance and significance for pollution treatment. J Chem Technol Biotechnol 84:13–28. https://doi.org/10.1002/jctb.1999
17. Garba ZN, Bello I, Galadima A, Lawal AY (2016) Optimization of adsorption conditions using central composite design for the removal of copper (II) and lead (II) by defatted papaya seed. Karbala Int J Mod Sci 2:20–28. https://doi.org/10.1016/j.kijoms.2015.12.002
18. Garcia-Fayos B, Arnal JM, Piris J, Sancho M (2016) Valorization of Moringa oleifera seed husk as biosorbent: isotherm and kinetics studies to remove cadmium and copper from aqueous solutions. Desalin Water Treat 57:23382–23396. https://doi.org/10.1080/19443994.2016.1180473
19. Ghaedi M, Mosallanejad N (2013) Removal of heavy metal ions from polluted waters by using of low cost adsorbents: review. J Chem Heal Risks 3:7–22
20. Ghasemi SM, Mohseni-Bandpei A, Ghaderpoori M, Fakhri Y, Keramati H, Taghavi M, Moradi B, Karimyan K (2017) Application of modified maize hull for removal of cu(II)ions from aqueous solutions. Environ Prot Eng 43:93–103. https://doi.org/10.5277/epe170408
21. Ghosh A, Das P, Sinha K (2015) Modeling of biosorption of Cu(II) by alkali-modified spent tea leaves using response surface methodology (RSM) and artificial neural network (ANN). Appl Water Sci 5:191–199. https://doi.org/10.1007/s13201-014-0180-z
22. Gupta H, Gogate PR (2016) Intensified removal of copper from waste water using activated watermelon based biosorbent in the presence of ultrasound. Ultrason Sonochem 30:113–122. https://doi.org/10.1016/j.ultsonch.2015.11.016
23. Hossain A, Ngo H, Guo W, Nguyen V (2012) Biosorption of Cu(II) from water by banana peel based biosorbent: experiments and models of adsorption and desorption 2:87–104. https://doi.org/10.11912/jws.2.1.87-104

24. Hossain MA, Ngo HH, Guo WS, Nguyen TV (2012) Palm oil fruit shells as biosorbent for copper removal from water and wastewater: experiments and sorption models. Bioresour Technol 113:97–101. https://doi.org/10.1016/j.biortech.2011.11.111
25. Jaman H, Chakraborty D, Saha P (2009) A study of the thermodynamics and kinetics of copper adsorption using chemically modified rice husk. Clean—Soil, Air, Water 37:704–711. https://doi.org/10.1002/clen.200900138
26. Kamar FH, Nechifor AC, Nechifor G, Sallomi MH, Jasem AD (2016) Study of the single and binary batch systems to remove copper and cadmium ions from aqueous solutions using dry cabbage leaves as biosorbent material. Rev Chim 67:1–7
27. Kaur R, Singh J, Khare R, Cameotra SS, Ali A (2013) Batch sorption dynamics, kinetics and equilibrium studies of Cr(VI), Ni(II) and Cu(II) from aqueous phase using agricultural residues. Appl Water Sci 3:207–218. https://doi.org/10.1007/s13201-012-0073-y
28. Banerjee K et al (2012) A novel agricultural waste adsorbent, watermelon shell for the removal of copper from aqueous solutions Iran. J Energy Environ 3:143–156. https://doi.org/10.5829/idosi.ijee.2012.03.02.0396
29. Lasheen MR, Ammar NS, Ibrahim HS (2012) Adsorption/desorption of Cd(II), Cu(II) and Pb(II) using chemically modified orange peel: equilibrium and kinetic studies. Solid State Sci 14:202–210. https://doi.org/10.1016/j.solidstatesciences.2011.11.029
30. Farhan AM, Salem NM, Ahmad AL, Awwad AM (2013) Kinetic, equilibrium and thermodynamic studies of the biosorption of heay metals by *Ceratonia Siliqua* Bark. Am J Chem 2:335–342.https://doi.org/10.5923/j.chemistry.20120206.07
31. Matouq M, Jildeh N, Qtaishat M, Hindiyeh M, Al Syouf MQ (2015) The adsorption kinetics and modeling for heavy metals removal from wastewater by Moringa pods. J Environ Chem Eng 3:775–784. https://doi.org/10.1016/j.jece.2015.03.027
32. Moyo M, Guyo U, Mawenyiyo G, Zinyama NP, Nyamunda BC (2015) Marula seed husk (Sclerocarya birrea) biomass as a low cost biosorbent for removal of Pb(II) and Cu(II) from aqueous solution. J Ind Eng Chem 27:126–132. https://doi.org/10.1016/j.jiec.2014.12.026
33. Munagapati VS, Yarramuthi V, Nadavala SK, Alla SR, Abburi K (2010) Biosorption of Cu(II), Cd(II) and Pb(II) by Acacia leucocephala bark powder: kinetics, equilibrium and thermodynamics. Chem Eng J 157:357–365. https://doi.org/10.1016/j.cej.2009.11.015
34. Nuhoglu Y, Oguz E (2003) Removal of copper(II) from aqueous solutions by biosorption on the cone biomass of Thuja orientalis. Process Biochem 38:1627–1631. https://doi.org/10.1016/S0032-9592(03)00055-4
35. Oboh IO, Aluyor EO, Audu TO (2011) Application of luffa cylindrica in natural form as biosorbent to removal of divalent metals from aqueous solutions—kinetic and equilibrium study. Waste Water—Treat Reutil 546.https://doi.org/10.5772/16150
36. Ofomaja AE, Naidoo EB, Modise SJ (2009) Removal of copper(II) from aqueous solution by pine and base modified pine cone powder as biosorbent. J Hazard Mater 168:909–917. https://doi.org/10.1016/j.jhazmat.2009.02.106
37. Panadare DC, Lade VG, Rathod VK (2014) Adsorptive removal of copper(II) from aqueous solution onto the waste sweet lime peels (SLP): equilibrium, kinetics and thermodynamics studies. Desalin Water Treat 52:7822–7837. https://doi.org/10.1080/19443994.2013.831789
38. Pandey G (2016) Removal of Cd(II) and Cu(II) from aqueous solution using Bengal gram husk as a biosorbent. Desalin Water Treat 57:7270–7279. https://doi.org/10.1080/19443994.2015.1026280
39. Paulino ÁG, da Cunha AJ, da Sliva Alfaya RV, da Sliva Alfaya AA (2014) Chemically modified natural cotton fiber: a low-cost biosorbent for the removal of the Cu(II), Zn(II), Cd(II), and Pb(II) from natural water. Desalin Water Treat 52:4223–4233.https://doi.org/10.1080/19443994.2013.804451
40. Pavan FA, Lima IS, Lima ÉC, Airoldi C, Gushikem Y (2006) Use of Ponkan mandarin peels as biosorbent for toxic metals uptake from aqueous solutions. J Hazard Mater 137:527–533. https://doi.org/10.1016/j.jhazmat.2006.02.025
41. Prasanna Kumar Y, King P, Prasad VSRK (2006) Equilibrium and kinetic studies for the biosorption system of copper(II) ion from aqueous solution using Tectona grandis L.f. leaves powder. J Hazard Mater 137:1211–1217. https://doi.org/10.1016/j.jhazmat.2006.04.006

42. Prokopov T (2015) Removal of copper (ii) from aqueous solution by biosorption onto powder of jerusalem artichoke. Ecol Eng Environ Prot 1:24–32
43. Rao LN, Prabhakar G (2011) Removal of heavy metals by biosorption–an overall review. J Eng Res II:17–22
44. Rathnakumar S, Sheeja RY, Murugesan T (2009) Removal of copper (II) from aqueous solutions using teak (Tectona grandis L. f) Leaves. World Acad Sci Eng Technol 3:880–884
45. Romero-Cano LA, García-Rosero H, Gonzalez-Gutierrez LV, Baldenegro-Pérez LA, Carrasco-Marín F (2017) Functionalized adsorbents prepared from fruit peels: equilibrium, kinetic and thermodynamic studies for copper adsorption in aqueous solution. J Clean Prod 162:195–204. https://doi.org/10.1016/j.jclepro.2017.06.032
46. SenthilKumar P, Ramalingam S, Sathyaselvabala V, Kirupha SD, Sivanesan S (2011) Removal of copper(II) ions from aqueous solution by adsorption using cashew nut shell. Desalination 266:63–71. https://doi.org/10.1016/j.desal.2010.08.003
47. Sirilamduan C, Umpuch C, Kaewsarn P (2011) Removal of copper from aqueous solutions by adsorption using modify Zalacca edulis peel modify. Songklanakarin J Sci Technol 33:725–732
48. Venkata Ramana DK, Min K (2016) Activated carbon produced from pigeon peas hulls waste as a low-cost agro-waste adsorbent for Cu(II) and Cd(II) removal. Desalin Water Treat 57:6967–6980. https://doi.org/10.1080/19443994.2015.1013509
49. Villaescusa I, Martínez M, Miralles N (2000) Heavy metal uptake from aqueous solution by cork and yohimbe bark wastes. J Chem Technol Biotechnol 75:812–816. https://doi.org/10.1002/1097-4660(200009)75:9%3c812::AID-JCTB284%3e3.0.CO;2-B
50. Weng CH, Lin YT, Hong DY, Sharma YC, Chen SC, Tripathi K (2014) Effective removal of copper ions from aqueous solution using base treated black tea waste. Ecol Eng 67:127–133. https://doi.org/10.1016/j.ecoleng.2014.03.053
51. Weng CH, Wu YC (2012) Potential low-cost biosorbent for copper removal: Pineapple leaf powder. J Environ Eng (United States) 138:286–292. https://doi.org/10.1061/(ASCE)EE.1943-7870.0000424
52. Yargiç AS, Yarbay Şahin RZ, Özbay N, Önal E (2015) Assessment of toxic copper(II) biosorption from aqueous solution by chemically-treated tomato waste. J Clean Prod 88:152–159. https://doi.org/10.1016/j.jclepro.2014.05.087
53. Yusoff SNM, Kamari A, Putra WP, Ishak CF, Mohamed A, Hashim N, Isa IM (2014) Removal of Cu(II), Pb(II) and Zn(II) Ions from aqueous solutions using selected agricultural wastes: adsorption and characterisation studies. J Environ Prot (Irvine,. Calif) 05:289–300. https://doi.org/10.4236/jep.2014.54032
54. Yuvaraja G, Subbaiah MV, Krishnaiah A (2012) Caesalpinia bonducella leaf powder as biosorbent for Cu(II) removal from aqueous environment: Kinetics and isotherms. Ind Eng Chem Res 51:11218–11225. https://doi.org/10.1021/ie203039m
55. Zheng JC, Feng HM, Lam MHW, Lam PKS, Ding YW, Yu HQ (2009) Removal of Cu(II) in aqueous media by biosorption using water hyacinth roots as a biosorbent material. J Hazard Mater 171:780–785. https://doi.org/10.1016/j.jhazmat.2009.06.078

# Adsorptive Removal of Acridine Orange Dye from Industrial Wastewater Using the Hybrid Material

**Vibha Agrawal and M. U. Khobragade**

**Abstract** In recent years, a lot of development and industrialization has led to the environmental pollution to a greater extent. Water pollution resulting due to toxic and non-biodegradable dyes present in the effluent of many industries related to textile, rubber, plastics, paper, cosmetics, leather is frequently seen nowadays. This research study uses hybrid material, combination of Surfactant Modified Alumina (SMA) and Surfactant modified Silica (SMS) for the removal of acridine orange dye (ACO) using the concept of adsolubilization. ACO dye is an organic dye capable of binding to the nucleic acid of various organisms. The surfactant which is used to modify the surface of alumina is Sodium Dodecyl Sulphate (SDS) which is anionic in nature whereas cationic surfactant Cetyl Trimethyl Ammonium Bromide (CTAB) is used to modify the surface of silica. The mode of analysis was batch study and the various parameters affecting the batch study of adsorption were studied. The batch studies indicated that as the dye concentration of ACO increased, the sorption capacities increased while percentage removal efficiency decreased and achieved equilibrium at 120 min contact time with 120 rpm agitation at $29 \pm 2$ °C. The removal efficiency of hybrid adsorbent material was around 92.5% and 81% for synthetic and real wastewater sample respectively with adsorption capacity of 46.25 mg/gm for synthetic sample and 40.5 mg/gm for the sample of real textile industry wastewater sample at adsorbent dose of 2 g/l. EDX and SEM analysis was carried out to determine the elemental composition of the adsorbent as well as to check the removal of the dye. It was concluded that the removal took place through pseudo-second-order kinetic mechanism ($R^2 > 0.99$) and data was a good fit to the Freundlich isotherm model ($R^2 > 0.998$).

**Keywords** Surfactant modified alumina · Surfactant modified silica · Acridine orange dye · Adsorption · Batch study

V. Agrawal (✉) · M. U. Khobragade
Department of Civil Engineering, College of Engineering Pune, Pune 411005, India
e-mail: vibhagrwl071@gmail.com

## 1 Introduction

In recent years, a lot of development and industrialization has led to the environmental pollution to a greater extent. Water pollution resulting due to the dyes present in the effluent of many industries related to textile, rubber, plastics, paper, cosmetics, leather is frequently seen nowadays. The ecosystem poses threats due to these dyes as many dyes are toxic and non-biodegradable [4]. Some dyes are also mutagenic, toxic, and carcinogenic [10]. Variety of treatment methods such as membrane separation technologies, electrochemical method, ion exchange, photocatalytic degradation, adsorption are explored by various scientists [13]. Adsorption technique is the most efficient one to remove various dyes. The most commonly used adsorbent is activated carbon but it requires high energy and cost in its synthesis and regeneration [3]. Thus, researchers are finding alternative adsorbent materials.

Acridine Orange Dye (ACO) is a basic cationic dye used in textile industry for coloring textiles. The dye also had wide applications in various fields. The biological affects of ACO including mutagenic actions are another wide area of research [12].

Nanomaterials have special physical and chemical properties thus it has received impetus for adsorption in research work. But the hydroxyl group on their surface causes them to agglomerate and leads to poor dispersion capacity [6]. Thus, using surfactant-modified metal oxides as adsorbent for adsorption has been in trend in recent years [11]. Depending on the surfactant concentration, the single layer or double layer structure of ionic surfactants forms on the surface of charged metal oxide [9]. These surfactant layers formed known as micelles have the ability to remove the organic molecules and solubilize them within its structure and this concept is called adsolubilization [2]. There are not much studies on engineering application of surfactant modified metal oxides for treatment of wastewater. The preliminary study has shown less efficiency of individual adsorbents SMA and SMS for the removal of dye. Since this study is out of the scope of this research paper it is not mentioned here.

In this research, we have worked towards the removal of acridine orange dye (ACO) from textile industry wastewater using adsorption process. The adsorbents used are hybrid materials—combination of surfactant modified alumina (SMA) and surfactant modified silica (SMS). The batch study of adsorption is used in this work. The various entities affecting the batch study such as adsorbent dose, time of contact, temperature, agitation speed as well as concentration of dye are studied and analyzed. The isotherm studies as well as the kinetic studies are carried to check the pollutants uptake rate and also to study the removal mechanism. The Scanning Electron Microscopy (SEM) and Electron Dispersive Xray (EDX) analysis are also carried out to determine the modification of the adsorbent material and removal of the contaminants.

## 2 Materials, Instrumentation, and Methodology Adopted

### 2.1 Materials

#### 2.1.1 Chemicals Used

Acridine orange dye ($C_{17}H_{19}N_3$, molecular weight—265.360 g·mol$^{-1}$), cationic surfactant cethyltrimethylammoniumbromide (CTAB—C19H42BrN)), anionic surfactant Sodium Dodecyl Sulphate (SDS—NaC12H25SO4), Alumina ($Al_2O_3$—molecular weight 101.96 gm), Silica gel and all other chemicals of high purity were obtained from a chemist in Pune and used as received.

#### 2.1.2 Instrumentation

High-end precision instruments such as electrical balance for measurement, digital pH meter, oven for drying, orbital incubator shaker was used. UV spectrophotometer was used to measure the concentration of ACO dye in the solution by simple principle of measuring light absorbed. Glassware used was made up of borosil. The distilled water was used to wash all the glassware thoroughly several times and later on dried in oven.

#### 2.1.3 Preparation of Synthetic Adsorbate Solution of ACO Dye

Working solution of 100 mg/l of ACO dye was prepared from the stock solution of 500 mg/l. The wavelength of 463 nm was used for testing the absorbance by acridine orange dye in the UV spectrophotometer.

#### 2.1.4 Preparation of Adsorbent

**Surfactant Modified Alumina (SMA)**—The positive charge is present on the surface of the alumina so the surfactant Sodium Dodecyl Sulphate (SDS) which carries negative charge on surface is used to modify its surface. The anionic surfactant forms micelles on its surface which solubilize the organic molecules. The alumina of 70–290 mesh ASTM and Zero Point of charge of 9.15 is used. Alumina of weight 200 g was shaken with SDS concentration of 20,000 mg/l for 24 h. After shaking for 24 h in orbital shaker, the supernatant obtained was discarded and the alumina was washed with distilled water and dried at 60 °C for 24 h. The final product obtained was the surfactant modified alumina (SMA) and it was further used for adsorption process [1].

**Surfactant Modified Silica (SMS)**—The surface of the negatively charged silica gel was modified using the cationic surfactant cetyltrimethylammoniumbromide

(CTAB). 180 g silica gel with the dose of 30 g/l was shaken with the CTAB solution of 6 L for 24 h which had 7500 mg/l concentration. The supernatant was again discarded and the silica gel was washed thoroughly with the distilled water and later on dried at 60 °C to obtain Surfactant Modified Silica (SMS) [2].

The SMA and SMS were taken in 1:1 proportion to form hybrid adsorbent material for the removal of ACO dye.

## 2.2 Methodology

### 2.2.1 Batch Study of Adsorption

The batch study of adsorption was used in this research work. The various factors affecting the batch study were analyzed. The adsorbent dose was varied from 0.1 g to 0.5 g to find the optimum dose for removal of dye. The variation of the contact time was done in the range from 30 to 180 min and agitation speed varied from 90 to 180 RPM. The inlet concentration was varied from 20 mg/l to 100 mg/l and even variation of temperature was studied.

### 2.2.2 Removal Efficiency and Adsorption Capacity

The UV–Vis Spectrophotometer was used to study the removal of the ACO dye. The adsorbent efficiency of the removal and the capacity of adsorption of the hybrid adsorbent material (mg/gm) were calculated using Eqs. (1) and (2), respectively [12, 14].

$$\% \text{ Removal} = \frac{C_0 - C_e}{C_0} * 100 \tag{1}$$

$$\text{Adsorbent Capacity} = \frac{C_0 - C_e}{M} \tag{2}$$

where,

$C_0$ represents the initial concentration (mg/l) of ACO dye
$C_e$ represents the final concentration (mg/l) of ACO dye
$M$ is the mass (g/l) of SMA adsorbent.

### 2.2.3 Adsorption Isotherms

Isotherms for adsorption process are plotted to describe the mechanisms of adsorption process. They describe the relationship between the adsorbate and adsorbent. The

Langmuir isotherm and Freundlich isotherm models are the two most used isotherm models.

**Langmuir adsorption Isotherms**—The assumption of the Langmuir isotherm is that the adsorbent surface has active sites with uniform monolayer formation. Here it is assumed that at every point on adsorbent there is same affinity for the adsorbate. Equation (3) below is the non-linearized form of Langmuir isotherm [7, 14].

$$\frac{1}{q_e} = \left[\frac{1}{K_L * q_e} * \frac{1}{C_e}\right] + \frac{1}{q_m} \quad (3)$$

where,

$q_e$ is equilibrium capacity in mg/gm
$K_L$ is Langmuir constant (l/mg)
$C_e$ is concentration at equilibrium (l/mg)
$q_m$ is the maximum concentration at equilibrium (mg/gm).

**Freundlich adsorption isotherms**—The Freundlich isotherm is valid for multi-layer adsorption based on the assumption that adsorption of ions occur on a heterogeneous adsorbent surface. It can be represented in linear form by Eq. (4) and its modified form by Eq. (5). The value of $n$ and $K_f$ is found by plotting the graph between $\ln C_e$ and $\ln q_e$.

$$q_e = K_f + \left(C_e^{1/n}\right) \quad (4)$$

$$\ln q_e = \log K_f + 1/n \, \ln C_e \quad (5)$$

where,

$q_e$ is equilibrium capacity in mg/gm
$K_f$ is freundlich constant (l/mg)
$C_e$ is concentration at equilibrium (mg/l)
$n$ indicates the affinity towards the adsorbent.

### 2.2.4 Adsorption Reaction Kinetics

Reactions that are time-dependent are called as kinetic reactions. Adsorption kinetics are used to test linearity between time and adsorption capacity. The rate of adsorption of ions on adsorbent material can be analyzed using two kinetic models.

**Pseudo first-order kinetic model**—This model suggested by Lagergren is given below in its integrated form (Eq. 6) [12].

$$\log(q_e - q_t) = \log(q_e - K_{s_1}) \quad (6)$$

where,

$q_e$ is equilibrium capacity in mg/gm
$q_t$ is equilibrium capacity at any time t
$K_{s_1}$ is rate constant in l/min = 2.303 K.

**Pseudo-second-order kinetic model**—The pseudo-second-order kinetic model based on the sorption equilibrium capacity is suggested in 1999. Following Eq. (7) gives the integrated linearized form of the model [5].

$$\frac{t}{q_t} = \frac{1}{K_{s2}q_e^2} + \frac{1}{q_e}t \qquad (7)$$

where,

$K_{s_2}$ is pseudo-second-order rate constant in gm/mg-min.

## 3 Analysis, Results, and Discussion

### 3.1 SEM and EDX Analysis

Scanning Electron Microscopy (SEM) rasters a focused electron beam toward the surface and detect backscattered electrons providing detailed high-resolution images of the sample. An Energy Dispersive X-Ray (EDX or EDA) gives identification of element and information about composition of the sample. The SEM image of hybrid adsorbent material after removal of ACO dye (Fig. 2) is darker than the previous image before removal (Fig. 1) which shows the adsorption of dye on surface of hybrid material [8].

SEM indicated the large surface area and rough surface texture of adsorbent before adsorption of dye, which provides suitable binding sites for ACO molecules. The change in the surface texture of the adsorbent material was found before and after the adsorption process. After adsorption, dye molecules led to coverage of most of the pores to become saturated.

The EDX analysis showed the chemical composition of dye adsorbed on hybrid material -mixed modified alumina and silica. It was observed that the amount of oxygen, bromium, carbon, nitrogen, aluminium, silicon has increased significantly which indicates the adsorption of dye on the surface of hybrid material and its removal from the wastewater. The results clearly indicated the increase in atomic percentage of carbon and nitrogen (main chemical composition of ACO dye).

**Fig. 1** SEM image of hybrid adsorbent material before treatment of dye

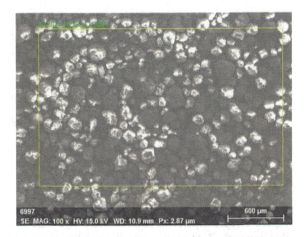

**Fig. 2** SEM image of hybrid adsorbent material after treatment of dye

## 3.2 Effect of Various Parameters

### 3.2.1 Effect of Dose of Adsorbent

The dose of the adsorbent is an important parameter which affects the process of adsorption. The removal efficiency increases as the active adsorbent site increases after increasing the dose of adsorbent. The adsorbent dose effect on the efficiency is illustrated below in Fig. 3. The adsorbent dose was varied from 1 g/l to 5 g/l and it was observed that at dose of 2 g/l, the percentage removal of ACO dye was 91%. It was also seen that after 2 g/l dose, the efficiency was not increasing significantly. The other operating conditions were as mentioned, pH was fixed at 5.3 at 29 °C with agitation of 120 RPM for 120 min for 100 mg/l concentration of dye.

**Fig. 3** Effect of adsorbent dose on the removal of ACO dye

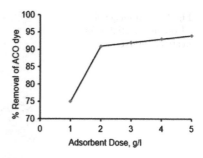

### 3.2.2 Effect of Time of Contact and Initial Concentration

Figure 4 represents adsorption of ACO dye onto hybrid material. It was realized that removal of ACO dye onto the combination of SMA and SMS reached to the equilibrium in 120 min. It was observed that at the initial stage of adsorption of ACO dyes, the removal is less than 30 min and was about 64.46% for combination of SMA and SMS. When the contact time of adsorption was increased, there was gradual increase in the removal of ACO dye and was reached upto 82.56% at contact time 120 min. The further increase in the contact time between the adsorbent and the dye showed that there was a slight increase in the percentage removal of ACO. Thus, 120 min is treated or used as equilibrium time for further experiments.

The other operating conditions were as mentioned, pH was fixed at 5.3 at 29 °C with agitation of 120 RPM for dose of adsorbent of 2 g/l for 100 mg/l concentration of ACO dye.

The concentration of ACO dye was varied from 20 mg/l to 100 mg/l under the controlled operating conditions (Speed of agitation = 120 rpm at 29 °C for 120 min contact time). It was noticed that the percentage removal efficiency decreased with the increase in the dye concentration. The decrease observed was in the range of 99.8–92.4% for concentration of 20–100 mg/l of ACO dye.

**Fig. 4** Effect of time of contact on the efficiency of ACO dye

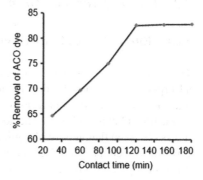

**Fig. 5** Effect of concentration on the removal of ACO dye

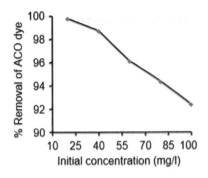

### 3.2.3 Effect of Speed of Agitation and Temperature

Figure 6 shows the effect of speed of agitation on efficiency of removal of ACO dye. The range of agitation speed was 90RPM-180RPM and the efficiency of ACO dye removal was studied in this range. The highest removal efficiency of 92.41% was observed for 120 RPM agitation speed. This shows that after a particular speed of agitation the adsorbed contaminant dye is removed from the surface of the adsorbent due to excess speed.

Figure 7 shows the effect of temperature on removal efficiency of ACO dye. Temperature changes the capacity of adsorption so it is another important physiochemical parameter. The temperature varied was in the range of 293, 302, and

**Fig. 6** Effect of speed of agitation on the removal of ACO dye

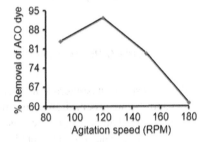

**Fig. 7** Effect of temperature on the removal of ACO dye

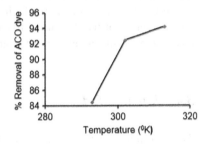

**Fig. 8** Langmuir adsorption isotherm

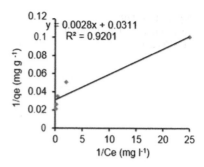

313 °K. The other parameters used are agitation speed (120 rpm) with concentration of 100 mg/l at 120 min time of contact and adsorbent dose (2 g/l). The 302 °K (29 degrees) was the optimum temperature with the maximum removal efficiency of around 95%. The increase in removal was not much on increasing the temperature beyond 302 °K.

#### 3.2.4 Real Wastewater Analysis

When the real wastewater was analyzed by the above-optimized parameters of adsorbent dose of 2 g/l at agitation speed of 120 RPM at temperature of 29 degrees for 120 min contact time in 100 mg/l dye concentration, the removal efficiency observed was 81% with adsorption capacity of 36.5 mg/gm.

### 3.3 Adsorption Isotherms

We plot adsorption isotherms in order to establish the relationship between the adsorbate and the adsorbent. Langmuir and Freundlich adsorption isotherms are plotted in this study. The best fit of these isotherms was checked using the data of varying concentration. The plots indicated that the value of correlation coefficient $R^2$ was more for (Fig. 9) Freundlich isotherm (0.993) as compared to the (Fig. 8) Langmuir isotherm (0.9201). Thus, Freundlich isotherm was a good fit to the data indicating the adsorption of the dye in multilayers on the adsorbent. It also displays that the adsorption has taken place in heterogenous manner.

### 3.4 Kinetic Model

The adsorption kinetic model represents the pollution uptake rate. The pseudo-first order and pseud-second-order kinetic model are analyzed here for the study. The

**Fig. 9** Freundlich adsorption isotherm

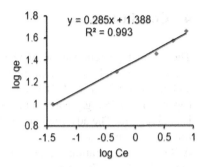

$R^2$ value of first-order kinetic model (Fig. 10) is less than the $R^2$ value of second-order model (Fig. 11). From this we concluded that adsorption is a chemisorption controlled mechanism. The contaminant dye particles are attached to the surface of the adsorbent by means of chemical forces, which are harder to desorb as compared to ones attached through physical forces.

**Fig. 10** Pseudo first order kinetic model

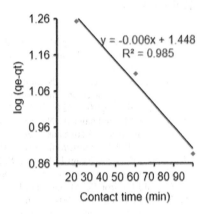

**Fig. 11** Pseudo second order kinetic model

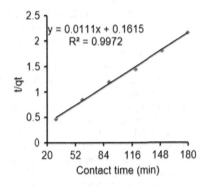

## 4 Conclusions

The percentage removal of ACO dye molecules from synthetic as well as real wastewater sample using hybrid adsorbent material made of combination of SMA and SMS is 92.5% and 81% respectively. For synthetic sample, the adsorption capacity is 46.25 mg/gm as well for real wastewater sample the adsorption capacity is 40.5 mg/gm. The adsorbent dose of 2 g/l of hybrid material along with the contact time of 120 min at speed of agitation of 120 RPM at 29 °C for the 100 mg/l ACO dye concentration is reported to be the optimum experimental condition. The removal followed Freundlich isotherm model and pollution uptake was controlled by pseudo-second-order mechanism. It was seen that the combination of SMA and SMS could provide the effective and efficient way to remove the ACO dye. Thus, it is suggested to explore the concept of hybrid adsorbent material for the removal of other contaminants and dyes from wastewater.

## References

1. Adak A, Bandyopadhyay M, Pal A (2005) Removal of crystal violet dye from wastewater by surfactant-modified alumina. Sep Purif Technol 44(2):139–144. https://doi.org/10.1016/j.seppur.2005.01.002
2. Adak A, Koner S (2016) Adsolubilization of organic dye through surfactant modified silica gel waste from aquatic environment batch and fixed bed studies wastewater treatment. 26684561:1–5. https://doi.org/10.15224/978-1-63248-114-6-12
3. Alnajjar M, Hethnawi A, Nafie G, Hassan A, Vitale G, Nassar NN (2019) Silica-alumina composite as an effective adsorbent for the removal of metformin from water. J Environ Chem Eng 7(3):102994. https://doi.org/10.1016/j.jece.2019.102994
4. Gupta VK, Suhas Tyagi I, Agarwal S, Singh R, Chaudhary M, Harit A, Kushwaha S (2016) Column operation studies for the removal of dyes and phenols using a low cost adsorbent. Global J Environ Sci Manage2(1):1–10. https://doi.org/10.7508/gjesm.2016.01.001
5. Kasprzyk-Hordern B (2004) Chemistry of alumina, reactions in aqueous solution and its application in water treatment. Adv Coll Interface Sci 110(1–2):19–48. https://doi.org/10.1016/j.cis.2004.02.002
6. Khan AM, Shafiq F, Khan SA, Ali S, Ismail B, Hakeem AS, Rahdar A, Nazar MF, Sayed M, Khan AR (2019) Surface modification of colloidal silica particles using cationic surfactant and the resulting adsorption of dyes. J Mol Liq 274:673–680. https://doi.org/10.1016/j.molliq.2018.11.039
7. Khobragade MU, Pal A (2016) Adsorptive removal of Mn(II) from water and wastewater by surfactant-modified alumina. Desalin Water Treat 57(6):2775–2786. https://doi.org/10.1080/19443994.2014.982195
8. Khobragade MU, Pal A (2016) Fixed-bed column study on removal of Mn(II), Ni(II) and Cu(II) from aqueous solution by surfactant bilayer supported alumina. Separat Sci Technol (Philadelphia) 51(8):1287–1298. https://doi.org/10.1080/01496395.2016.1156698
9. Koner S, Pal A, Adak A (2012) Use of surface modified silica gel factory waste for removal of 2,4-D pesticide from agricultural wastewater: a case study. Int J Environ Res 6(4):995–1006. https://doi.org/10.22059/ijer.2012.570
10. Koner S, Pal A, Adak A (2010) Cationic surfactant adsorption on silica gel and its application for wastewater treatment. Desalin Water Treat 22(1–3):1–8. https://doi.org/10.5004/dwt.2010.1465

11. Nguyen TMT, Do TPT, Hoang TS, Nguyen NV, Pham HD, Nguyen TD, Pham TNM, Le TS, Pham TD (2018) Adsorption of anionic surfactants onto alumina: characteristics, mechanisms, and application for heavy metal removal. Int J Polym Sci 2018:5–9. https://doi.org/10.1155/2018/2830286
12. Qadri S, Ganoe A, Haik Y (2009) Removal and recovery of acridine orange from solutions by use of magnetic nanoparticles. J Hazard Mater 169(1–3):318–323. https://doi.org/10.1016/j.jhazmat.2009.03.103
13. Wang H, Zhao W, Chen Y, Li Y (2020) Nickel aluminum layered double oxides modified magnetic biochar from waste corncob for efficient removal of acridine orange. Biores Technol 315(May):123834. https://doi.org/10.1016/j.biortech.2020.123834
14. Yadav S, Asthana A, Chakraborty R, Jain B (2020) Cationic dye removal using novel magnetic/activated charcoal/β—cyclodextrin/alginate polymer nanocomposite 1–20

# Basin Delineation and Land Use Classification for a Storm Water Drainage Network Model Using GIS

Kunal Chandale and K. A. Patil

**Abstract** Due to rapid urbanization and the recent trends in climate change there is a need for an efficient and sustainable storm water drainage network. In order to design a storm water drainage network there are some preliminary studies to be taken into consideration. This paper aims at using ArcMap 10.3 for GIS based basin delineation, supervised classification and nodal elevation extraction. The study area covers about 13 km$^2$ which spans over villages Handewadi, Wadachiwadi, and Pisoli in the South-Eastern region of the Pune city. Digital Elevation Models (DEM) were used for basin delineation which helped in identification of ridge line for the study area. To identify the percentage of impervious area supervised classification was performed using the Landsat 8 images, upon analysis which concluded in 43.37% of area being impervious, i.e., (buildings and roads). 660 plus nodes were digitized for an area using MIKE Urban + which were then exported, and then imported in ArcMap 10.3. Elevation values were extracted for each node using the DEMs.

**Keywords** Basin delineation · Land classification · GIS

## 1 Introduction

Storm water modelling plays a significant role in checking issues such as urban flooding, water logging and urban water-quality problems [1]. Before designing a storm water drainage network there are some prerequisite tasks which are mandatory and help design a sustainable drainage network. Rapid construction and land development causes an increase in the imperviousness of land which leads to increased surface runoff causing risk of urban flooding. To mitigate risks of urban flooding, an efficient storm water drainage system is mandatory. In order to administer and

---

K. Chandale (✉) · K. A. Patil
Department of Civil Engineering, College of Engineering Pune, Pune 411005, India
e-mail: chandaleka19.civil@coep.ac.in

K. A. Patil
e-mail: kap.civil@coep.ac.in

© The Author(s), under exclusive license to Springer Nature Singapore Pte Ltd. 2023
M. S. Ranadive et al. (eds.), *Recent Trends in Construction Technology and Management*, Lecture Notes in Civil Engineering 260,
https://doi.org/10.1007/978-981-19-2145-2_42

understand the landscape modifications, the quantitative assessment of land use and land-cover and the transformations in it must be performed [2]. The various factors operating either on local, regional, or global scale define the land use changes which are basically dynamic in nature [3]. The land use pattern is dynamically being affected by the swift and unrestrained population growth, along with rapid industrialization and the continuously growing economy [4].

Remote Sensing (RS) and Geographic Information System (GIS) are the most acknowledged tools being used for estimating and quantifying these changes in the land use. In order to gather, store, recover, analyze, influence or exploit, and display any georeferenced or spatial data for a particular development-oriented program, GIS proves to be an efficient tool capable of performing the required tasks [5]. Recent advances in remote sensing technology allow to acquire data at spatial and temporal resolutions beyond the human capability, at any point in space and time. Remote sensing data along with geographic information systems has made more-compendious analysis of spatial information feasible [6]. Among various Land use land cover classification techniques, supervised and unsupervised classifications are used the most [7].

## 2 Study Area

Pune located in state of Maharashtra is one of the eight most populous city of India. Known as the "Oxford of the East," it is an education hub since decades and in the recent times has become a booming IT hub. It is situated 560 m above mean sea level on the Deccan Plateau on the banks of Mutha River. Given the scenario, there is a massive hike in the population of the city in the recent years. According to the Pune Municipal Corporations (PMC) development plan the population of the city in the next 20 years is projected to be thrice of what it was in the year 2011 (PMC City Development Plan 2012). The villages Handewadi (18°45' N, 73°93'E), Pisoli (18°44'N, 73°90'E), and Wadachiwadi (18°43'N,73°91'E) situated in the South-Eastern region of the city of Pune together forming an area of 13 km$^2$, is considered in this study.

## 3 Data Collection

ArcMap 10.3 software was used for basin delineation and land use classification studies. Cartoset-1 Digital Elevation Model (CartoDEM) was downloaded from the Indian Space Research Organization's (ISRO) bhuvan.nrsc.gov.in website. Landsat 8 OLI (Operational Land Imager) and TIRS (Thermal Infrared Sensor) with 15–30 m multispectral data images downloaded from United States Geological Survey (USGS) website.

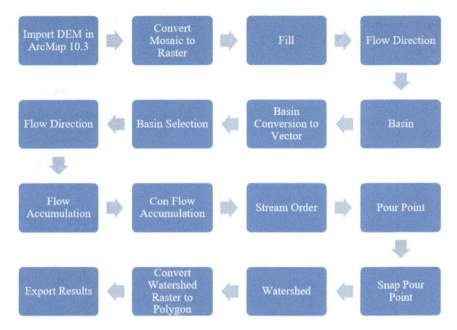

**Fig. 1** Flowchart for basin delineation process

## 4 Methodology

### 4.1 Basin Delineation

To design a storm water drainage network, there is a need to first identify the boundary or ridge line of that particular region which shows the area that will be contributing the runoff to the drainage network. Spatial Analysis Tools and Data Management Tools of the ArcToolbox were primarily used for basin delineation. The following flowchart shows the commands used in order to obtain the desirable results (Fig. 1).

### 4.2 Land Use Classification

Urban floods are caused due to an increase in the recent trends of population density, haphazard development of urban infrastructure without paying due consideration to drainage aspects and increase in paved surfaces. Paved surfaces are relatively more impervious than unpaved surfaces which restrain surface infiltration and cause increased surface runoff. To identify the percentage of paved or impervious land use classification is necessary. Supervised classification of Landsat eight images was done using ArcMap 10.3. The remote sensing images are displayed in three primary

**Fig. 2** Flowchart for supervised classification process

colors (red, green, and blue) known as color composite images. It means associating each spectral band to primary color results in a color composite image [8]. False Color Composite (FCC) scheme was used to detect various classes.

Red = Short Wave Infrared (SWIR) band (SPOT4 band 4, Landsat TM band 5)
Green = Near Infrared (NIR) band (SPOT4 band 3, Landsat TM band 4)
Blue = Red band (SPOT4 band 2, Landsat TM band 3)

Sample training was done for five classes, i.e., buildings, roads, barren land, water bodies, and vegetation. Interactive supervised classification was carried out accordingly. The following flowchart shows the process for supervised classification (Fig. 2).

## 4.3 Nodal Elevations

For regular inspection, cleaning, and maintenance of the drainage line, there has to be some sort of access available. This access is provided by constructing orifices along the drainage line which are called Manholes. At every change of orientation, slope of land, change in diameter of pipe, at the origin of all pipes and branches, and at every intersection of two or more pipes, a manhole must be provided mandatorily [9]. In the software, these manholes are represented as nodes. 660 nodes were digitized along the road network for the study area using MIKE Urban+ software. MIKE Urban+

**Fig.3** Fill clip raster

features complete high performing MIKE 1-D multicore engine for modeling large stormwater and sewerage systems with fast execution, extracting model data directly from a high-performing database. The node shapefile was imported to ArcMap 10.3 and elevation values for each node were extracted using extraction tool of the ArcToolbox from the DEM.

## 5 Results and Discussion

### 5.1 Delineated Basin

The resulted delineated basin will form the boundary for the area contributing to surface runoff and eventually contributing to the stormwater drainage network. The basin's area is 29.49 km$^2$ and the perimeter of the basin is 32.43 km. The resultant shapefiles of each and every command generated shapefiles in ArcMap are as follows (Figs. 3, 4, 5, 6 and 7).

### 5.2 Supervised Classification

In the composite stacked image, after using the 4-3-2 FCC, water bodies resembled black color. The buildings showed pink color and the roads showed dark pink color. The barren land was seen in shades of blue. The aim of the classification was to get percentage of paved area, i.e., roads and buildings. The composite image shows bifurcation of landmass with respect to different classes in ArcMap (Figs. 8, 9 and Table 1).

The classification results above in the classified image states that approximately 45% of study area comprises of barren land and vegetation. Around 43.37% of study

**Fig. 4** Flow direction raster

**Fig. 5** Vector polygon

**Fig. 6** Stream order

area comprises of impervious land and the remaining consists of water body. Visual inspection was done for verification of these classification results which was in favor of the results (Table 2).

**Fig. 7** Delineated basin

**Fig. 8** Composite image

## 5.3 Node Elevation

Post importing the node shape file from MIKE Urban+, Extract Values to Points command from the Arc Toolbox was used which is quite user-friendly. The elevation values efficiency majorly depend on the accuracy of the DEMs used to extract these values. Figure 10 shows node shapefile in ArcMap whereas Fig. 11 shows imported node shapefile in MIKE Urban+ with elevation values.

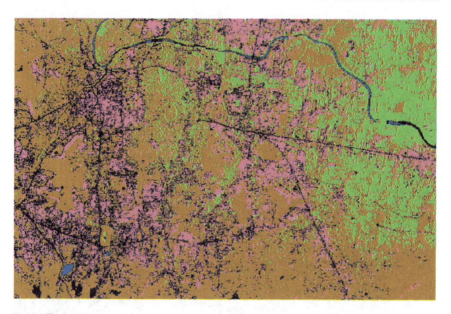

**Fig. 9** Classified image

**Table 1** Class name and value

| Sr. No. | Value | Class name |
|---|---|---|
| 1 | 27 | Vegetation |
| 2 | 1 | Water bodies |
| 3 | 7 | Barren land |
| 4 | 41 | Buildings |
| 5 | 40 | Roads |

**Table 2** Classification results

| Sr. No. | Class value | % of area | Class name |
|---|---|---|---|
| 1 | 27 | 1.815175 | Vegetation |
| 2 | 1 | 0 | Water bodies |
| 3 | 7 | 54.80489 | Barren land |
| 4 | 41 | 23.58139 | Buildings |
| 5 | 40 | 19.79855 | Roads |

## 6 Conclusions

Based on the study carried out different conclusions drawn are as given below.

**Fig. 10** Node shapefile in ArcMap 10.3

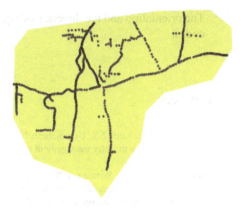

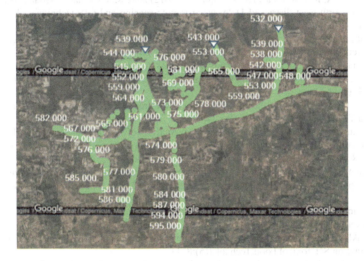

**Fig. 11** Nodes with elevation values

1. The resultant shape file of basin delineation is useful as a foundation layer for study area and to understand the topography.
2. Resultant shape file is useful to demarcate the study area, digitize nodes, land use classification, delineate catchments, and countoring.
3. The land use classification helps to find the magnitude of the imperviousness of the land. Percentage imperviousness helps to calculate the groundwater abstractions from the surface runoff which gives the net runoff contributing to the drainage system.
4. The percentage area of vegetation, barren land, buildings, and roads in the study area found to be 1.82%, 54.80%, 23.58%, and 19.80%, respectively.
5. The flow in a collection system or drainage system is always Freudian, i.e., under the influence of gravity. The nodes and its elevations define the pipe network.

6. The orientation and the direction of pipe flow are governed by the elevations of nodes.

## References

1. Rangari VA, Prashanth SS, Umamahesh NV, Patel AK (2018) Simulation of urban drainage system using a storm water management model (SWMM). Asian J Eng Appl Technol 7(S1):7–10
2. Singh SK, Mustak S, Srivastava PK, Szabó S, Islam T (2015) Predicting spatial and decadal LULC changes through cellular automata Markov chain models using earth observation datasets and geo-information. Environ Processes 2:61–78
3. Rahman A, Kumar S, Fazal S, Siddiqui MA (2012) Assessment of land use/land cover change in the north-west district of Delhi using remote sensing and GIS techniques. J Indian Soc Remote Sens 40:689–697
4. Dutta D, Rahman A, Paul SK, Kundu A (2019) Changing pattern of urban landscape and its effect on land surface temperature in and around Delhi. Environ Monit Assess 191:551
5. Mallupattu PK, Sreenivasula Reddy JR (2013) Analysis of land use/land cover changes using remote sensing data and GIS at an urban area, Tirupati, India. The ScientificWorld J 2013(268623)
6. Quinn NWT, Kumar S, Imen S (2019) Overview of remote sensing and GIS uses in watershed and TMDL analyses. J Hydrol Eng 24(4)
7. Naikoo MW, Rihan M, Ishtiaque M (2020) Analysis of land use land cover (LULC) change and built-up expansion in the suburb of a metropolitan city: spatio-temporal analysis of Delhi NCR using landsat datasets. J Urban Manage 09:347–359
8. Imam E (2019) Remote sensing and GIS module: colour composite images and visual image interpretation. University Grand Commission (UGC), MHRD, Government of India
9. Manual on Storm Water Drainage Systems, Central Public Health and Environmental Engineering Organisation (CPHEEO), Ministry of Housing and Urban Affairs (MoHUA), Government of India
10. Shukla H, Watershed management: it's role in environmental planning and management. IOSR J Environ Sci Toxicol Food Technol 1(5):8–11
11. Pune Municipal Corportation (2013) Draft city development plan 2012/2013. Pune Municipal Corporation, Pune

# Prediction of BOD from Wastewater Characteristics and Their Interactions Using Regression Neural Network: A Case Study of Naidu Wastewater Treatment Plant, Pune, India

Sanket Gunjal, Moni Khobragade, and Chirag Chaware

**Abstract** Analysis of variance (ANOVA) results were used to analyze the behavior of various water quality parameters like Total Suspended Solids, Biochemical Oxygen Demand, and Turbidity. The turbidity is not a recommended water testing parameter for wastewater testing, still it was measured on experimental basis. The datasets used in this work were derived from a detailed experimental investigation of inlet and outlet water parameters of Naidu wastewater treatment plant, a major conventional treatment plant in Pune City, India. It has an average flow rate of 115 MLD. The samples were collected daily for over three months. Statistical data analysis methods such as correlation coefficient, regression analysis, and graphical representation of the data were used to find the interrelations between selected parameters. The regression neural network model obtained from the analysis of data was used to predict BOD parameter considering standard error. The $p$-value obtained by ANOVA analysis was observed to be significant at $p < 0.05$. The ANN model predicted the results at maximum accuracy of 96% and average accuracy of 94%. Good interrelation between the selected parameters was observed and it was observed that BOD can be predicted using suspended solids and turbidity of water sample.

**Keywords** BOD prediction · Turbidity · TSS · ANN

## 1 Introduction

The wastewater systems are very complex systems and difficult to model in any statistical relations. The time consumption of the environmental experiments greatly

---

S. Gunjal · M. Khobragade · C. Chaware (✉)
Department of Civil Engineering, College of Engineering Pune, Pune 411005, India
e-mail: chawarecy19.civil@coep.ac.in

S. Gunjal
e-mail: sanketdg1996@gmail.com

M. Khobragade
e-mail: muk.civil@coep.ac.in

© The Author(s), under exclusive license to Springer Nature Singapore Pte Ltd. 2023
M. S. Ranadive et al. (eds.), *Recent Trends in Construction Technology and Management*, Lecture Notes in Civil Engineering 260,
https://doi.org/10.1007/978-981-19-2145-2_43

needs to be reduced to make fast decisions and increase the wastewater treatment efficiencies. The legitimate operation and regulation of wastewater treatment plants (WWTPs) are increasing due to concerns regarding environmental problems. Inappropriate activity of a WWTP may achieve genuine ecological and general well-being issues, as its gushing to an accepting water body can spread different diseases to individuals. A good control of a wastewater treatment plant could be accomplished by building up a powerful numerical tool, for predicting plant execution, based on previous interpretations of such critical boundaries. "The measurement of complex physical, biological and chemical processes engaged with wastewater treatment displays non-linear behaviour and is hard to depict by direct numerical models" [1]. Be that as it may, demonstrating a WWTP is a troublesome errand because of the multifaceted nature of the treatment measures. The artificial neural network is tasted to be best to model various non-uniform, and complex systems. This study focuses on prediction of water quality parameters like BOD which are time and resource consuming. This paper presents prescient models dependent on the idea of artificial neural network (ANN), a generally utilized use of fake insight that has demonstrated a serious guarantee in an assortment of utilizations in designing, design acknowledgment, and monetary market investigation. The experimental work was conducted for more than three months to get more than 90 data sets from November 2020 to February 2021. The created models are appeared to perform reliably well in the face of changing precision and size of information. Though turbidity is not a recognized parameter for wastewater quality testing, it was used for analysis on experimental basis. The methods like ANN are being used progressively in water and wastewater area. As it has been observed to be great while handling nonlinear and non-uniform datasets.

## 2 Material and Methods

The standard procedures for measurement of the desired parameters BOD, TSS, and turbidity were used. The APHA manual was referred for the experimental procedures. Analytical grade chemicals were used for the experimental work. The chemicals were purchased from the Aishwarya chemicals. The standard procedure for measuring the parameters like BOD and TSS takes huge time. The prediction method that we are targeting to develop will be helpful in getting the Bod value in just few minutes once we get the TSS and turbidity values ready for fresh analysis. The main disadvantage of this system is that it needs good quality data or else the system will end up predicting very random value of the targeted parameter, in our case the BOD. The huge amount of data sets doesn't seem to be required in this kind of analysis, as the readings of a particular plant mostly lie in a particular range. The seasonal variations are unavoidable and hence should be tackled by collecting the datasets all over the year and creating seasonal models will help solving this issue. Although this problem has not solved in our work. Validation of the results will be done by using 10% of the data selected randomly from the collected data sets. The unique thing about this work is that we have tried establishing the interrelation between TSS and Turbidity with respect to BOD for the selected wastewater treatment plant for a case study. There are various limitations regarding sample collection and constraints to perform the study or carry out the experimentations which are tried to be eliminated with current work. The prediction of the BOD from less time-consuming parameters, TSS and turbidity will save most of the resources such as chemicals, manpower, machines and the most valuable and important time. Conventionally the parameters are analyzed by the lab experimental procedures of the respective parameters. The conventional methods are pretty time-consuming and cannot be used on the field. The final artificial neural network that was ready after training on the data that was generated by us gave the average accuracy of 95% and maximum prediction accuracy of 96%.

## 2.1 Case Study

The created ANN models were used for Naidu wastewater treatment plant. This plant is an important WWTP on the bank of the Mula stream, India. It is being developed in three phases to serve an extreme populace of 1 million (115 MLD). It is anon nitrifying actuated overflow type plant. It incorporates fine coarse and fine screening and grit removal, sedimentation, diffused aeration, last explanation, and chlorination offices. The records taken over the time of 3 months collected for BOD concentration and TSS at the inlet station, and at the stream from the last stage.

### 2.1.1 Model Development

The steps shown in Fig. 1 were carried out for ANN model development process.

Collection of Data and Preprocessing of Data

Data collection was done for over 90 days. The fundamental target of the information preprocessing is to filter out the data which will help improving the predicted results. For excluding the outlier datasets, we used regression line and buffer zone of 20% around the line. Generally, the preprocessing also includes the filling of missing data in the datasets, but in our case, there was no any missing data so that we skipped the step of filling the missing values. Dependent and independent variables were

**Fig. 1** ANN model development

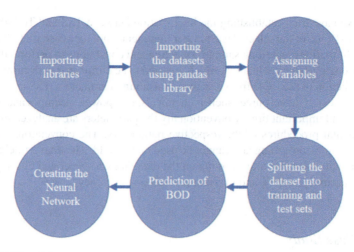

**Fig. 2** ANN model steps

decided logically as the BOD is a parameter dependent on the solids present in the water so that TSS and Turbidity were chosen as independent parameters and BOD as a dependent parameter. The dissolved solids were not considered as the dissolved impurities are mostly minerals and they contribute very less in the BOD and that's why can be neglected for our current project work (Fig. 2).

Model Design and Network Training

Multivariate linear regression (MLR)

It has multiple independent variables contributing to the value of dependent variable and has multiple coefficients and complex in nature. In our case turbidity and TSS are considered to be independent variables and BOD is considered to be a dependent variable.

Libraries such as Numpy, Sklearn, Pandas, and Tensorflow were used for the pre-processing ad actually building the neural network.

NumPy is a Python library used for working with arrays, domain of linear algebra, Fourier transform, and matrices. Travis Oliphant created This open-source library in 2005. NumPy stands for Numerical Python.

The scikit-learn (sklearn) library is used to deal with machine learning and statistical problems including clustering of data, classification of data, regression, and dimensionality reduction of the data. Scikit-learn is an open-source library.

Pandas is also an open-source library created for the Python programming language and it is used for data analysis and data manipulation. The data for our work was imported and exported in the model using pandas library.

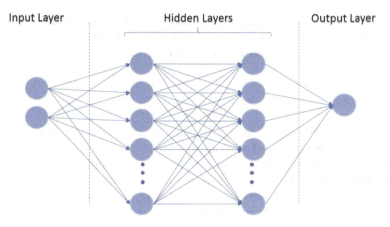

**Fig. 3** Neural network

TensorFlow: TensorFlow is an open-source library developed by Google. This beautiful library is primarily developed for deep learning applications. It also supports traditional machine learning.

The ANNs are tools of mathematical modeling that are especially helpful in the field of prediction and in dynamic environments forecasting. The ANN accomplishes this by a vast number of strongly integrated computing components (neurons) working together to solve particular challenges, such as predicting and pattern detection (Fig. 3).

Each neuron has different coefficients or weights related to some of its neighbors, which reflect the relative effect of the various neuron inputs to other neurons. The Python coding language was used for he coding the neural network. The model engineering, preparing strategy, and preparing rates were resolved utilizing an experimentation approach. The BOD, TSS and turbidity values (100 sets for every parameter) were used for training and initial testing of model. 90% (90 entries) of total datasets were used to train the model and remaining 10% datasets (10 entries) were used for validation of constructed ANN model. Feedforward-error back-propagation type neural network was constructed with the help of TensorFlow library design by google. The loss function used for the network was "Root Mean Squared Error" and various optimization functions like Adagrad, Adadelta, Adam, Nadam were used while constructing the neural network and were observed to obtained the best results. Various neuron activation functions like sigmoid, Relu, selu, leaky-relu were used and observed till getting the best results. To arrive at the reasonable organization engineering, a few preliminaries for every gathering were led till the appropriate learning rate, number of hidden layers, and number of neurons per each layer were reached. Reasonable arrangements are those that have delivered an insignificant blunder term in the training and testing of data. The reasonable design for BOD prediction was resolved to comprise an input layer with two neurons (for input TSS and Turbidity). The hidden layer has five hundred neurons. The output layer has one neuron (for anticipated BOD).To arrive at this network structure, various network structures were tested and were assessed. Network structures with four layers containing five hundred neurons were observed to be perform great.

Model Testing

The observed ANN model average accuracy for the Naidu wastewater treatment plant was 94% and It reached a maximum value of 97% for inlet properties and 85–89% for outlet water quality parameters. The activation functions, loss functions, and optimizer functions along with a number of epochs affect the accuracy of model greatly. We should also observe carefully while testing the model that mistakenly we have not got the local minima of the error surface otherwise the readings will appear to be very random and diverted instead of occurring in certain range (e.g., range of 4–6% from the estimated values) of BOD. Graphs of testing are as shown (include graphs of testing errors, etc. Also include descriptions of the graphs in this paragraph). Various combinations of number of neurons and number of layers were checked and the behavior of the model was noted to optimize the number of neurons and number of layers by trial and error method. The model was behaving weird at lower neuron count but higher epochs, so that number of neurons were set to five hundred to get stable outcomes.

## 3 Results and Discussion

The ANN model constructed for the prediction of inlet water quality parameters performed really well. The observed values (Red color line) and predicted values (Blue color line) by ANN model for inlet water parameter are as shown in Fig. 4. The accuracy of the model was observed in between 94 and 97%. The ANN model constructed for prediction of outlet water quality parameters didn't performed well as our selected parameters were not sufficient to predict the outlet BOD. The model's accuracy was observed to be in between 85 and 89%. The observed values (Orange

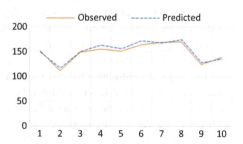

**Fig. 4** Plotted values of BOD observed and predicted for inlet water

**Fig. 5** Plotted values of BOD observed and predicted for outlet water

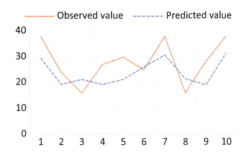

color line) and predicted values (Blue color line) by ANN model for inlet water parameter are as shown in Fig. 5.

The *p*-value for the ANOVA analysis at $p < 0.05$ was observed to be $p < 0.00001$ and the F-ratio was observed to be 1347.95699. These values signify that the data that we collected doesn't have much randomness. *P*-values for all the intermediate interrelationships of all three parameters taken two at a time was observed to be less than 0.00001. The mean value of Turbidity, TSS, and BOD for inlet water are 40.79, 214.375, and 149.67, respectively.

## 4 Conclusion

The ANN-dependent model was created to predict the influent and effluent-BOD concentration for a WWTP in Pune. The neural network model gave great prediction to the BOD values. The datasets we used for the prediction covered a wide range of data for training and testing the model. The error variations were very less and observed to be smooth over the range of data. The limitations in data, in any case, ought to be highlighted. On the off chance that more data sets were gathered, if the information were less noisy, it would have brought about an improved prediction capacity of the model. In any case, the ANN is observed to be a great tool for the prediction of BOD.

## Reference

1. Hamed MM, Khalafallah MG, Hassanien EA (2004) Prediction of wastewater treatment plant performance using artificial neural networks. Environ Model Softw 19:919–928

# Floodplain Mapping of Pawana River Using HEC-RAS

**Tejas R. Bhagwat and Aruna D. Thube**

**Abstract** A flood is catastrophic natural high stage at which the stream overflows its banks and inundates the surrounding area of river banks. It is a natural hazard and threat to life and property. In this study, flood inundation mapping is done for the downstream of Pawana dam located on Pawana river in Upper Bhima sub-basin of Krishna basin in Pune district. Flood mapping has been done for entire river stretch of 58 km till its confluence with the Mula river. Mapping involves locating blue line, orange line, and red line which are the extent of the inundated area for the discharges of 25-year, 50-year, and 100-year return period flood and identification of the most affected villages as they come under prohibitive and restrictive regions. The study utilizes the Digital Elevation Model (DEM) data for the procurement of cross-section and HEC-RAS software developed by USACE for the steady flow analysis. which provides the water surface profiles and water depths obtained are further utilized to create the floodplain map. For model calibration, Manning's roughness coefficient is varied between 0.015 to 0.045 for channel and 0.010–0.050 for floodplain. The Coefficient of determination ($R^2$) and Nash Sutcliffe Coefficient (NSE) values is found to be above 0.60 and 0.75 for both calibration and validation. The locations coming under the flood lines are identified by importing the flooding map to Google Earth as a layer. The significant findings found from the study was that under prohibitive zone, up to river station 29,966, i.e., Godumbre village, there is not much urbanization present. But there are urbanized locations downstream of river station 27,965, i.e., Gahunje village. From this river station 27,965, there are residential sectors, buildings, and small slums located on the banks of river which are getting flooded even for 25-year return period flood and fall under prohibitive zone. Such type of construction is strictly prohibited in this zone, but structure exists, so government agencies should monitor existing structures and should be well prepared in advanced with Emergency Action Plan (EPA) and Disaster Management Plan if the flooding situation arises.

---

T. R. Bhagwat (✉)
Environmental and Water Resources Engineering, College of Engineering Pune, Pune, India
e-mail: bhagwattr19.civil@coep.ac.in

A. D. Thube
College of Engineering Pune, Pune, India

**Keywords** HEC-RAS · DEM · Return period · Pawana river · Floodplain mapping

## 1 Introduction

India is one of the most disaster-prone countries in the world due to its geoclimatic conditions. Out of 329 million hectare of total geographical area, 40 mha of area is flood-prone. Every year on an average 75 lakh hectares of land is getting affected. Nowadays, floods have also been occurring in areas which was not considered to be flood prone earlier. Flooding is a condition of partial or complete inundation by water. It is temporary covering of land by water due to heavy rainfall when capacity of water bodies to hold water exceeds. Among all the disasters, floods are the most common and cause damages to infrastructure apart from the loss of lives. It also affects the economy due to reconstruction costs, shortage of food which causes increase in prices, etc. The flood disaster cannot be controlled completely but many methods have been practiced to mitigate the flood. Flooding is one of the world's most severe threats and accounts for 40% of all-natural disaster casualties, with many flood events occurring in developing and tropical regions. According to Abhas et al. (2012), 178 million people were affected by floods in 2010 alone and total financial losses in the exceptional years such as 1998 and 2010 exceeded $40 billion. There are four types of floods, specifically river floods, flash floods, tidal floods, and storm surge floods, which cause tremendous loss to property and lives [1]. Flood zone recognition has been used extensively in river management studies by defining high-risk flooding areas, set mandatory limits in high-risk areas, avoid flood risk, coordinate, and optimize rivers by deciding flood zones with different return periods (Forgani et al. 2016). Floodplain mapping and river cross-sectional boundaries play a major role in planning and optimizing the utilization of the floodplain areas on the banks of river in order to reduce flood damage. Due to the complexity of rivers during floods, modeling using software is one of the cost-effective tools for studying and simulating the behavior of rivers during floods (Parsa et al. 2016). According to [2]. through floodplain maps, areas that are vulnerable to flooding hazards can be identified and floodplain analysis indicated that more than 400% area is likely to be inundated as compared to the normal flow of the river, hence there is a need of policymakers and planners for the development of flood mitigation measures.

India is facing the same global challenges of droughts and floods resulting from extreme climate change. In India, the majority of floods occur due to excessive rainfall that leads to the bursting of river banks, causing flood and the consequent high discharge of rivers damages crops and infrastructure. They also result in siltation of reservoirs and hence limit the capacity of existing dams to control floods. There are several factors which are responsible for occurrence of floods and droughts which including climatic changes, global warming. urban development and population density in exposed areas. In the recent past, Maharashtra too have witnessed many flood situations. The recurring flood losses hinder the state's economic growth. The population living in the region where flooding frequently occurs is constantly threatened. Unplanned growth and increased floodplain invasions have contributed to the intensity of river floods. A call for better flood preparedness to ensure that adequate and efficient flood emergency preparedness steps are taken in order to prevent the loss of life, property, and the environment. In order to reduce vulnerability, it is crucial to have an effectual flood prevention plan and mitigation strategy, in addition to effective disaster response management plan. For planning and relief operations, flood inundation mapping plays a vital role for conveying the available flood risk information to decision-makers, evacuation teams, and to the general public.

Main objective of this study is to perform steady flow analysis and to develop floodplain inundation maps for the segment of the Pawana river and to develop flood line marking for 25-year, 50-year, and 100-year return period flood and determining critical locations affected due to flooding.

## 1.1 Flood Lines

Flood lines are lines indicating the maximum water level likely to be reached and delimits the area that would get inundated by floodwaters once every 25, 50, or 100 years.

### 1.1.1 Blue Line

For the discharge of 25-year return period flood or 1.5 times river capacity whichever is more. The area which gets flooded on either side of river bank comes under this line and zone is called prohibitive zone. Under this zone, any type of construction is prohibited and should be left open and can be used for gardens, playgrounds, or light crops.

### 1.1.2 Orange Line

For the discharge of 50-year return period flood, the area which gets flooded on either side of the banks of river comes under this line.

### 1.1.3 Red Line

For the discharge of 100-year return period flood or the design flood for dam, the area which gets flooded on either side of the banks of river comes under this line and the zone is called restrictive zone. The area under restrictive zone can be used for constructions having their plinth at safe levels from which evacuation can be facilitated and the construction can be such that it will not get demolished in floods.

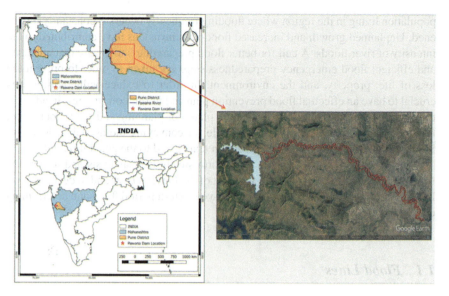

**Fig. 1** Study area map of Pawana river

## 2 Materials and Procedures

### 2.1 Selection of Study Area

Pawana river originates in south of Lonavala in Pune district and is located on the western side of Maharashtra, India. Pawana river confluences with the Mula river at Pimpale Gurav and further flows downstream as the Mula-Mutha river, which further meets Bhima River around 54 km east of the city. The Pawana dam an Earth fill gravity dam has been built on the Pawana river in Pune district of Maharashtra state. The latitude of dam is 18° 21′ 30″ and longitude is 73° 40′ 30″. Dam has a height of 38.10 m and length of 1329 m. Water of this dam is used for hydroelectricity generation and for water supply to the area of Pimpri-Chinchwad city. The study area of Pawana river is covered between 18° 40′ 55″ N and 18° 34′ 18″ N latitudes and 73° 29′ 56″E to 73° 49′ 58″E longitudes. The location of the study area of Pawana river in India, has been shown in Fig. 1.

### 2.2 Data Collected and Processing

The necessary data needed for this study was obtained from different sources. Among the collected data, type of data, sources of data, and location has been listed in Table 1.

# Floodplain Mapping of Pawana River Using HEC-RAS

**Table 1** Data collected list

| Type of data | Source of data | Location |
|---|---|---|
| Discharge data | Irrigation Division, Pune | Pawana river |
| Gauge water level | Hydrology Project, Nashik | Pimpale-Gurav |
| Return flood period data | Irrigation Division, Pune | Pawana river |
| DEM (30 m * 30 m) | Bhuvan site | Pune, Maharashtra |

The major inputs to HEC-RAS model are Digital Elevation Model (DEM) of 30 m × 30 m resolution, Manning's roughness coefficient, and spillway discharge for calibration and validation of model.

## 2.3 Analysis Process for HEC-RAS

HEC-RAS uses multiple number of input parameters for hydraulic analysis of river channel geometry and water flow. With the help of these parameters cross-section are created along the stream center line. At each cross-section, stream bank lines are identified so as to divide the cross-section into section of left floodplain channel, and right floodplain channel as visualized in Fig. 2.

For open channel flow, the total energy per unit weight has three components: elevation head, pressure head, and velocity head as shown in Fig. 3.

$$H = Z + Y + \frac{\alpha V^2}{2g} \qquad (1)$$

where:

$H$ = Total head or Energy head in m.
$Z$ = Elevation head or potential head in m.

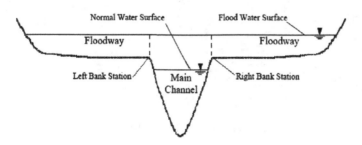

**Fig. 2** Schematic view of stream cross-section [3]

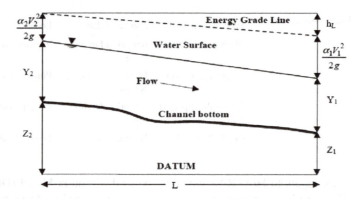

**Fig. 3** Representation of energy equation parameters [3]

$Y$ = Depth of water above channel bed in m.
$\alpha$ = Energy correction factor = 1 for Gradually varied flow.

Analysis has been done to get water depths, velocity, and water encroachment over the bank lines for different flood discharges which will be used further for purpose of mapping. In Arc-GIS, each attribute, i.e., cross-sections, river banks lines, etc., are created and saved in different features called as RAS layer, these layers are imported in HEC-RAS to make geometric files for HEC-RAS. In HEC-RAS software input the geometric data, steady flow data, boundary condition, etc. The schematic representation of geometry created and imported in HEC-RAS has shown in Fig. 4.

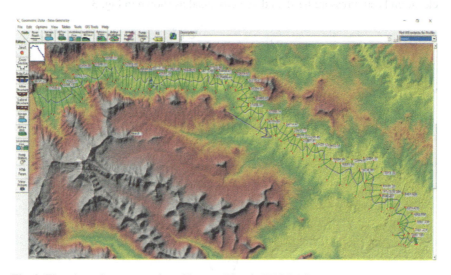

**Fig. 4** The schematic representation of Pawana River in HEC-RAS

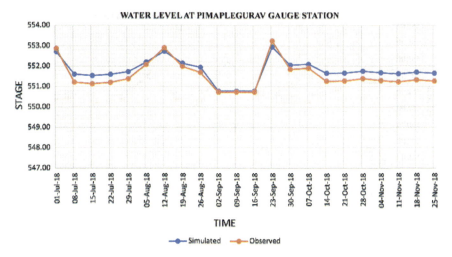

**Fig. 5** Model calibration observed and simulated hydrograph at Pimpale Gurav

## 2.4 Model Calibration for Steady Flow

For HEC-RAS model calibration the daily spillway discharge of four months from July to November of 2018 has been used for this study. The model has been calibrated using measured water level data at Pimpalegurav gauging stations and effort has been made to calibrate Manning's roughness coefficient to a single value. Manning's roughness coefficient was varied between 0.015 and 0.045 for the channel and 0.010 and 0.050 for floodplain in order to justify the adequacy of simulation of flow in the Pawana River. Finally, "$n$" value as 0.030 for main channel and 0.037 for flood plain has been calibrated. After calibration, the comparison of observed and simulated stage hydrograph at Pimpale Gurav gauging stations has shown in Fig. 5. Coefficient of determination ($R^2$) and Nash–Sutcliffe Efficiency (NSE) have been found as 0.63 and 0.81, respectively, for the steady flow calibration.

## 2.5 Model Validation for Steady Flow

For validation of model the daily spillway discharge of four months from July to October of 2017 has been used. After validation, the comparison of observed and simulated stage hydrograph at Pimpale Gurav gauging stations is in close accord with each other and have been shown in Fig. 6. Coefficient of determination ($R^2$) and Nash–Sutcliffe Efficiency (NSE) has been found as 0.61 and 0.76 respectively for the steady flow validation.

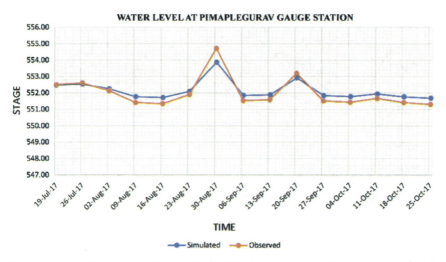

**Fig. 6** Model validation observed and simulated hydrograph at Pimpale Gurav

## 2.6 Running Steady-State Simulation

After the model has been calibrated and validated, then the model is simulated for 25-year, 50-year, and 100-year return flood period data. For the Blue line, 25-year return period flow or 1.5 times river capacity whichever is higher is chosen for simulation. The 1.5 times river capacity is 811 $m^3/s$ and 25-year return period flood is 929 $m^3/s$ as per provided by Irrigation Division, Pune, therefore, flow values for Blue lines are taken as 929 $m^3/s$. For the orange line, discharge value is 1061 $m^3/s$ and for red line discharge value is 1261 $m^3/s$ by taking these values as input model was run to perform steady flow analysis. Normal depth for boundary condition of bed slope is specified as 0.001. Once the steady flow data and boundary conditions are completed the model is simulated.

## 2.7 Visualize the Results

The model will take a few minutes for simulation and then visualization of the results, the cross-section, 3-D perspective plots, depth of water, water surface level, velocity, summary output table, and floodplain map can be viewed in each respective tab. The detailed flood inundation map can be exported to Google-Earth to view the areas which are getting affected. The detailed methodology in the form of flowchart in Fig. 7.

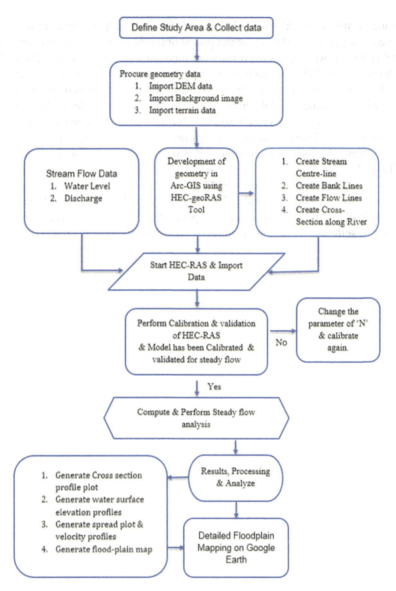

**Fig. 7** General method for modeling floodplains using Arc-GIS and HEC-RAS

## 3 Results and Discussion

Floodplain inundation mapping is a valuable task to be performed by government agencies, planners, and engineers, and with the help of the map we can understand the extent of flooding caused and floodplain inundated, so in the best way, we can allocate resources to prepare for emergency action plans and be well prepared in advance if the situation arises. The analysis for the floodplain areas inundated for 25-year, 50-year, and 100-year return period floods has been performed using HEC-RAS software. River cross-sections were inputted at around 500 m distance from each other. Total 115 cross-sections were plotted for 58 km long Pawana river, out of which 31 important cross-sections which are nearer to villages, settlements, etc., are selected. The coefficient of determination ($R^2$) and Nash Sutcliffe Coefficient (NSE) values are found to be above 0.60 and 0.75 for both calibration and validation. Then the results of maximum water depth, maximum velocity, and water spread on left and right of banks on each cross-section of Pawana river has been tabulated and shown in Table 2.

The floodplain map created using HEC-RAS has been imported to Google Earth and blue line, orange line, and red line for 25-year, 50-year and 100-year return period flood are marked. Once the floodplain map has been generated the affected locations, settlements, etc., are identified (Figs. 8, 9, 10, 11 and Table 3).

**Table 2** Maximum water depth and maximum velocity for affected locations on the banks of the river for 25-year, 50-year, and 100-year return period flood

| River station | Affected location on left of bank | Affected location on right of bank | 25-year return period flood Max water depth (m) | 25-year return period flood Max velocity (m/s) | 50-year return period flood Max water depth(m) | 50-year return period flood Max velocity (m/s) | 100-year return period flood Max water depth (m) | 100-year return period flood Max Velocity (m/s) |
|---|---|---|---|---|---|---|---|---|
| 57,437 | Pawananagar | – | 2.98 | 0.48 | 3.18 | 0.49 | 3.47 | 0.52 |
| 52,898 | Kadadhe village | – | 2.19 | 1.21 | 2.4 | 1.22 | 2.68 | 1.24 |
| 42,233 | – | Pimpalkhunte | 5.84 | 1.81 | 6.06 | 1.91 | 6.38 | 2.02 |
| 37,536 | – | Bebadohal | 3.9 | 1.25 | 4.15 | 1.33 | 4.5 | 1.5 |
| 34,856 | Parandwadi | Dhamaane | 6.99 | 0.99 | 7.22 | 1.05 | 7.55 | 1.13 |
| 31,516 | Shirgaon | | 7.25 | 0.95 | 7.48 | 1.01 | 7.8 | 1.08 |
| 29,966 | | Godumbre | 4.57 | 1.1 | 4.8 | 1.16 | 5.1 | 1.28 |
| 27,965 | Gahunje | – | 5.81 | 1.04 | 6.01 | 1.13 | 6.29 | 1.15 |
| 25,487 | Lodha-towers | – | 2.92 | 1.16 | 3.1 | 1.23 | 3.31 | 1.33 |
| 22,917 | Dattawadi | – | 0.86 | 0.32 | 1.02 | 0.42 | 1.23 | 0.53 |
| 19,997 | Ravetgaon | BuddhaVihar | 6.9 | 0.65 | 7.06 | 0.73 | 7.28 | 0.83 |
| 19,431 | Sector 32 A | Vishnudev nagar | 6.59 | 1.14 | 6.74 | 1.24 | 6.95 | 1.38 |
| 18,300 | Walhekarwadi | Newale wasti | 2.82 | 0.55 | 2.96 | 0.61 | 3.17 | 0.69 |
| 15,612 | Keshvanagar | Bijlinagar | 5.23 | 2.93 | 5.45 | 3.16 | 5.74 | 3.48 |
| 14,772 | Deowada | | 5.14 | 1.71 | 5.41 | 1.7 | 5.8 | 1.67 |
| 14,035 | Moryarajpark | – | 6 | 1.46 | 6.27 | 1.52 | 6.66 | 1.64 |
| 12,386 | Manik colony | Pawananagar | 7.26 | 1.03 | 7.52 | 1.1 | 7.88 | 1.19 |
| 11,900 | Yashopram housing soc | Shanti colony | 6.98 | 1.31 | 7.23 | 1.4 | 7.58 | 1.52 |
| 11,304 | Ambedkar colony | Nirmalnagar | 4.26 | 2.02 | 4.48 | 2.16 | 4.8 | 2.36 |
| 10,129 | Mali Ali | Shraddha colony | 4.75 | 1.4 | 4.91 | 1.49 | 5.13 | 1.63 |
| 9002 | Pimprigaon | Rahatnigaon | 5.29 | 1.27 | 5.44 | 1.37 | 5.63 | 1.51 |
| 7053 | – | Krantinagar | 8.16 | 1.17 | 8.46 | 1.21 | 8.88 | 1.29 |
| 6442 | Kasarwadi | – | 8.77 | 1.68 | 9.07 | 1.8 | 9.46 | 2 |
| 4589 | | Kadamwada | 5.54 | 2.59 | 5.78 | 2.75 | 6.13 | 2.93 |
| 4228 | Sevanagar | Agarsenagar | 7.49 | 1.44 | 7.74 | 1.53 | 8.1 | 1.64 |
| 3785 | Devkarnagar | Dapodi | 9.03 | 1.1 | 9.27 | 1.19 | 9.63 | 1.31 |
| 3399 | Devkarnagar | Ganeshnagar | 7.57 | 0.9 | 7.81 | 0.97 | 8.17 | 1.03 |

(continued)

**Table 2** (continued)

| River station | Affected location on left of bank | Affected location on right of bank | 25-year return period flood | | 50-year return period flood | | 100-year return period flood | |
|---|---|---|---|---|---|---|---|---|
| | | | Max water depth (m) | Max velocity (m/s) | Max water depth(m) | Max velocity (m/s) | Max water depth (m) | Max Velocity (m/s) |
| 3097 | Pimpalegurav road | Gangotri nagar | 2.25 | 1.63 | 2.49 | 1.67 | 2.85 | 1.73 |
| 2754 | – | Shivramnagar | 3.53 | 5.07 | 3.76 | 5.27 | 4.08 | 5.54 |
| 2269 | – | Samarthnagar | 6.39 | 1.64 | 6.64 | 1.76 | 7 | 1.89 |
| 826 | Budhvihar | Mamtanagar | 6.65 | 1.08 | 6.84 | 1.18 | 7.1 | 1.32 |

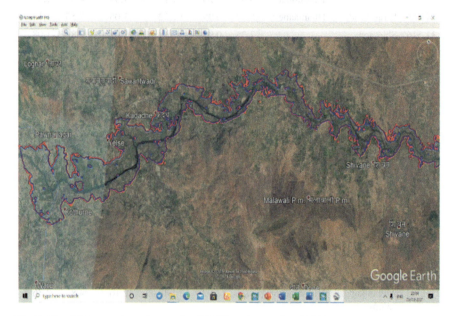

**Fig. 8** Floodplain map from river station 57,437 to river station 42,233

Floodplain Mapping of Pawana River Using HEC-RAS 591

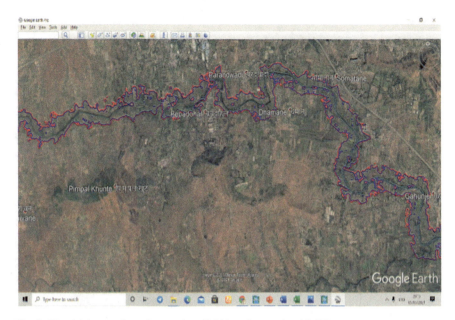

**Fig. 9** Floodplain map from river station 42,233 to river station 25,487

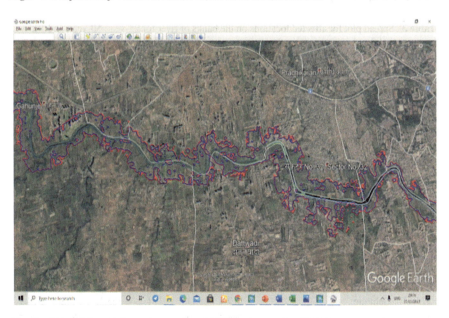

**Fig. 10** Floodplain map from river station 25,487 to river station 11,900

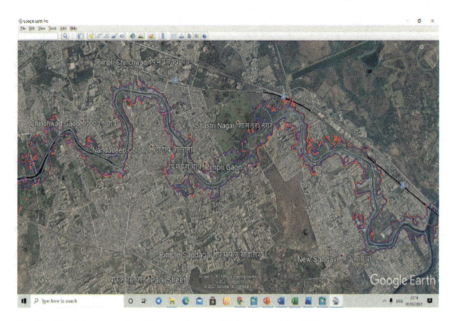

**Fig. 11** Floodplain map from river station 11,900 to river station 826

**Table 3** Spread plot for affected locations on the banks of river for 25-year, 50-year, and 100-year return period flood

| River station | Affected location on left of bank | Affected location on right of bank | 25-year return period flood | | 50-year return period flood | | 100-year return period flood | |
|---|---|---|---|---|---|---|---|---|
| | | | Water spread on LOB (m) | Water spread on ROB (m) | Water spread on LOB (m) | Water spread on ROB (m) | Water spread on LOB (m) | Water spread on ROB (m) |
| 57,437 | Pawananagar | – | 578 | 679 | 578.8 | 691.7 | 600 | 696 |
| 52,898 | Kadadhe village | – | 557 | 46 | 561.2 | 48.7 | 580 | 53.1 |
| 42,233 | – | Pimpalkhunte | 218 | 47 | 231.8 | 55.5 | 238 | 61 |
| 37,536 | – | Bebadohal | 113 | 153 | 120.5 | 160.7 | 125 | 185 |
| 34,856 | Parandwadi | Dhamaane | 70 | 334 | 81.2 | 336.1 | 83 | 342 |
| 31,516 | Shirgaon | | 264.1 | 257.1 | 270.7 | 255.9 | 282.1 | 262.2 |
| 29,966 | | Godumbre | 371.4 | 133.7 | 382.5 | 137.8 | 392.9 | 140.2 |
| 27,965 | Gahunje | – | 119.4 | 353 | 131.1 | 363.1 | 142.9 | 374.9 |
| 25,487 | Lodha-towers | – | 286.5 | 200.5 | 289.1 | 214.5 | 294.6 | 220.7 |
| 22,917 | Dattawadi | – | 286.7 | 406.9 | 311.5 | 407.1 | 357 | 410.3 |

(continued)

Table 3 (continued)

| River station | Affected location on left of bank | Affected location on right of bank | 25-year return period flood Water spread on LOB (m) | Water spread on ROB (m) | 50-year return period flood Water spread on LOB (m) | Water spread on ROB (m) | 100-year return period flood Water spread on LOB (m) | Water spread on ROB (m) |
|---|---|---|---|---|---|---|---|---|
| 19,997 | Ravetgaon | BuddhaVihar | 482.5 | 71.3 | 494.9 | 72.6 | 533.1 | 74.4 |
| 19,431 | Sector 32 A | Vishnudev nagar | 66.9 | 426.8 | 67.4 | 432.1 | 70.1 | 436.8 |
| 18,300 | Walhekarwadi | Newale wasti | 421.7 | 340.1 | 427.8 | 348 | 433.9 | 351.6 |
| 15,612 | Keshvanagar | Bijlinagar | 100.8 | 18.8 | 76.9 | 20.1 | 101.6 | 69.9 |
| 14,772 | Deowada | | 210.3 | 194.1 | 211.1 | 195 | 211.7 | 195.3 |
| 14,035 | Moryarajpark | – | 222.7 | 224.2 | 227.7 | 227.6 | 232.8 | 231.1 |
| 12,386 | Manik colony | Pawananagar | 139 | 185.1 | 143.1 | 190.3 | 145.2 | 197 |
| 11,900 | Yashopram housing soc | Shanti colony | 201.5 | 57.8 | 205.1 | 60.8 | 208.4 | 71.2 |
| 11,304 | Ambedkar colony | Nirmalnagar | 36.3 | 78 | 38.3 | 79.7 | 42.4 | 86.3 |
| 10,129 | Mali Ali | Shraddha colony | 233.1 | 156.4 | 241.7 | 160.8 | 256.6 | 165.6 |
| 9002 | Pimprigaon | Rahatnigaon | 246.7 | 194.2 | 253.3 | 197.7 | 261.7 | 202.6 |
| 7053 | – | Krantinagar | 378.1 | 86.8 | 383.4 | 88.5 | 420.3 | 99.9 |
| 6442 | Kasarwadi | – | 101.8 | 182.4 | 120.4 | 192.9 | 193 | 201.1 |
| 4589 | | Kadamwada | 91.6 | 60.7 | 97.1 | 63.4 | 203.9 | 65.8 |
| 4228 | Sevanagar | Agarsenagar | 113.1 | 175 | 116.7 | 176.8 | 121.5 | 182.7 |
| 3785 | Devkarnagar | Dapodi | 117.4 | 168.1 | 120.3 | 177.7 | 124.9 | 188.5 |
| 3399 | Devkarnagar | Ganeshnagar | 137.9 | 355.7 | 176 | 363.1 | 189.3 | 375.2 |
| 3097 | Pimpalegurav road | Gangotri nagar | 219.6 | 47.6 | 223 | 53.3 | 225.4 | 58.2 |
| 2754 | – | Shivramnagar | 3.4 | 32 | 5.1 | 33.7 | 7 | 36.2 |
| 2269 | – | Samarthnagar | 97 | 35 | 106.7 | 37 | 116 | 40.8 |
| 826 | Budhvihar | Mamtanagar | 194 | 19 | 195.2 | 20.8 | 200 | 22 |

## 4 Conclusion

1. The floodplain mapping for 25-year, 50-year, and 100-year return period floods has been done, by calibrating and validating the model on HEC-RAS with the real-time observed data to get accuracy of the flooding caused.
2. The extent of the flooding caused has shown with the help of the flood lines, i.e., blue line, orange line, and red line marking have been done for the Pawana river at the downstream of Pawana dam.
3. It is observed that there are many critical locations that are affected due to this flooding. Under prohibitive zone, up to river station 29,966, i.e., Godumbre village, there is not much urbanization present. Hence under prohibitive zone there are structures mainly temples, small industrial, and few temporary structures.
4. The urbanized locations at downstream of river station 27,965, i.e., Gahunje village. From this river station 27,965, there are residential sectors, buildings, and small slums located on banks of river which are getting flooded even for 25-year return period flood and fall under prohibitive zone.
5. Flood control measures such as construction of dykes and levees, structures which can absorb water or divert the floodwater away from banks, repair, and maintenance of the embankments are suggested. Also, government agencies should monitor existing structures and should be well prepared in advanced with Emergency Action Plan (EPA) and Disaster Management Plan if a flooding situation arises.
6. It is recommended to do floodplain mapping and investigate flood risk management for dams and rivers, well in advance using suitable measures to tackle the situation in an effective and efficient way.

## References

1. Rahman MM, Ali MM (2016) Floodplain mapping of the Jamuna River using HEC-RAS and HEC-GeoRAS. J. Presidency University, Part: B 3(2):24–32. ISSN: 2224–7610
2. Shahzad Khattak M, Anwar F, Saeed TU, Sharif M, Sheraz K, Anwaar A (2016) Floodplain mapping using hec-ras and ArcGIS: a case study of Kabul River. Arab J Sci Eng 41:1375–1390
3. Tate E, Maidment DR (1999) Floodplain mapping using HEC-RAS and ArcView GIS. Master's Thesis, Department of Civil Engineering, the University of Texas at Austin
4. Forghani MR, Irandoust M, Jalalkamali N (2016) Flood plain analysis and flow simulation of river using HEC-RAS model. Sci Arena Publ Spec J Agri Sci 2(2):27–36
5. ShahiriParsa A, Qalo N, Heydari M, bt Mohd Amin NF (2013) Introduction to floodplain zoning simulation models through dimensional approach. Int J Adv Civil Struct Environ Eng 1(1):20–23
6. Tate EC, Maidment DR, Olivera F, Anderson DJ (2002) Creating a terrain model for floodplain mapping. J Hydrol Eng 7:100–108

7. USACE, River Analysis System HEC-RAS, Hydraulic Reference Manual version 5.0.7, https://www.hec.usace.army.mil/software/hec-ras/
8. USACE, User's Manual of HEC-RAS version 5.0.7

# Performance Evaluation of Varying OLR and HRT on Two Stage Anaerobic Digestion Process of Hybrid Reactor (HUASB) for Blended Industrial Wastewater as Substrate

Rajani Saranadagoudar, Shashikant R. Mise, and B. B. Kori

**Abstract** The study was carried out to evaluate the performance of a two-stage pilot-scale hybrid up-flow anaerobic sludge blanket reactor (HUASB) of 9.5-L capacity treating blended Industrial Wastewater as substrate under varying organic and hydraulic loading conditions under ambient laboratory conditions. The reactor was operated for around 6 months with four different hydraulic retention time of 48, 36, 24 and 12 h. Imposed organic loading rates (OLR) ranged from 2.4, 3.2, 4.8, and 9.6 (kgCOD/m$^3$·d). The steady-state performance of HUASB reactor showed a tremendous performance on the treatment of blended industrial wastewater, with an average COD and BOD5 removal efficiency of 94.4% and 92.7%, respectively, with a maximum biogas production of 12600 cc/d. The results show that the HUASB reactor could serve as a good alternative for anaerobic digestion of Blended industrial wastewater and methane production.

**Keywords** Anaerobic · COD · HUASB reactor · Blended industrial wastewater · Biogas

## 1 Introduction

The dairy industry is a food processing industry which is one of the major polluting industries, both in terms of quality and quantity of waste generated. It is estimated that about 120 million tons of the milk and approximately 275 million tons of wastewater will be produced every year from the Indian dairy industries by the year 2020. A series of operations involved in dairy industry are receiving, Pasteurizing, bottling, condensing, dry milk manufacturing, cheese production, butter, and Casein making.

---

R. Saranadagoudar (✉) · S. R. Mise
Department of Civil Engineering, Poojya Dodappa Appa College of Engineering, Kalaburagi, Karnataka 585102, India
e-mail: rajanigndecb@gmail.com

B. B. Kori
Department of Civil Engineering, Guru Nanak Dev Engineering College, Bidar, Karnataka 585403, India

Wastewater is mainly generated from the following actions such as cleaning of equipment, tankers, and washing of containers, floor washing, water softening unit, boiler house, and refrigeration plant waste [1]

Distillery industries are included under the category of most polluting Industries in regard with water pollution. The total volume of wastewater may vary from 70–120 L per liter of alcohol produced. Distillery using molasses for the production of alcohol by fermentation an distillation processes produces a very high organic and inorganic effluent. The manufacturing processes are feed preparation, yeast propagation, and continuous fermentation, multipurpose distillation with isolated spent-wash evaporator, stillage processing, molecular sieve dehydration for fuel alcohol, and packaging unit. The waste streams comprise of spent wash, which in turn is one of the major source of wastewater. Spent wash, contains dissolved salts and has persistent dark brown color. It also contains acids and alkalies in traces which are produced from cleaning and sanitizing.

Aerobic processes, which are widely used for the treatment of wastewater have two distinct disadvantages like high energy requirement and excess sludge production, which require handling, treatment, and disposal [2]. In contrast, anaerobic processes generate energy in the form of biogas, and produce sludge in significantly lower amounts than those resulting from aerobic systems [3].

## 2 Experimental Section

### 2.1 Hybrid UASB Reactor Setup

- A laboratory-scale Hybrid UASB reactor is fabricated using a 5 mm thick acrylic pipe with external diameter of 34 cm. The reactor has a working volume of 9.5 L. Height of the hybrid reactor is 251 cm which includes GLSS separator. To create electrolysis process in the reactor, two electrodes (zinc rods) are provided with a diameter of about 10 mm and with the height of about 150 mm, One rod is connected to anode and other rod to cathode, these rods are connected to DC power supply with the volt of about 4.5, resulting in continuous supply of oxygen and hydrogen. At high inter-electrode distances, removal efficiency decreases due to reduced due to reduced rate of mass transfer and due to elevated ohmic drop which reduces the anodic oxidation [4].
- The settler consists of baffles, which help, in guiding the gas bubbles to enter into the separator, to capture the evolved gas, and to allow the settling of suspended solids.
- Bedding media made up of PVC material in the form of Circular Rings (Diameter—1 cm, Thickness—0.1 cm, Height—0.2 cm) was added up to 30 cm height into the reactor to enhance granulation process by increasing the surface area. Gas collected was measured by water displacement method. The arrangement

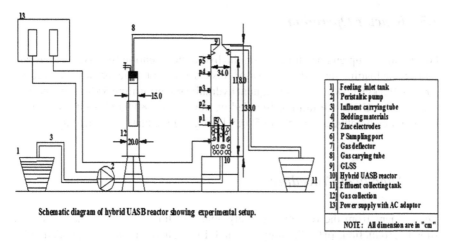

Fig. 1 Schematic representation of hybrid UASB reactor

consists of a 15 cm diameter internal cylinder, 20 cm diameter external cylinder with capacity of 20 L.
- The reactor was mounted on a mild steel-framed structure. The Schematic representation of Hybrid UASB reactor is as shown in Fig. 1.

## 2.2 Wastewater and Seed Sludge

The study was carried out under ambient environmental conditions. The dairy wastewater is collected from Karnataka Milk Federation (KMF) milk processing plant, Kalaburagi, Karnataka. The distillery industry wastewater is collected from Sanjeevani Sahakari Sakharkharkhana limited, Shingnapur, At: Kolpewadi, Tal-Kopergaon, Dist. Ahmednagar, Maharashtra. The general characteristics of blended industrial wastewater (dairy and distillery) and seed sludge are given in Table 1 [2, 5].

Table 1 Physico-characteristics of blended industrial wastewater and seed sludge

| S.No. | Parameters | Unit | Seed sludge (distillery) | Blended industrial wastewater |
|---|---|---|---|---|
| 1 | COD | mg/L | 1,33,200 | 7200 |
| 2 | $BOD_5$ | mg/L | 82,947 | 4380 |
| 3 | pH |  | 4.2 | 7.10 |
| 4 | DO | mg/L | Nil | Nil |
| 5 | Alkalinity | mg/L | 8658 | 2600 |
| 6 | Total solids | mg/L | 60,040 | 16,624 |
| 7 | Dissolved solids | mg/L | 44,377 | 13,182 |
| 8 | Suspended solids | mg/L | 15,662 | 3442 |

## 2.3 Reactor Operation

For initial startup and acclimatization the hydraulic retention time (HRT) was 48 h for organic loading rate 2.4 (kgCOD/m$^3$·d) subsequent organic loading rates are 3.2, 4.8, and 9.6 kgCOD/m$^3$·d at subsequent hydraulic retention time 36 h, 24 h, and 12 h, respectively. COD:N:P ratio of Blended wastewater is 480:9.6:1, COD:N:P should be maintained between 250:5:1 to 500:5:1 for good methanogenic activity.

## 2.4 Analytical Procedure

The effluent from the reactor was daily analyzed for various parameters such as Biogas production, pH, alkalinity, TS and TVS, alkalinity, and suspended solids (TS) analysis was carried out in accordance with Standard Methods (APHA 2017).

## 3 Results and Discussion

The reactor was operated continuously for 131 days. Initially, reactor was operated at a loading rate of 2.4 (kgCOD/m$^3$·d) and Hydraulic-Retention-Time of 48 h. The reactor was operated for varying OLR of 2.4, 3.2, 4.8, and 9.6 (kgCOD/m$^3$·d) by changing HRTs 48 h to 36 h, 24 h, and 12 h, respectively. During the operation, COD removal efficiencies were observed by performing tests. Due to the presence of large granules, accumulations of hydrolytic fermentative bacteria around the granules were more in number. Due to this complex, organic matter is converted into simpler dissolved form. After achieving the successful granulation the COD removal efficiency increased. During the operation, optimum COD removal efficiency recorded was 94.5%.

BOD$_5$ removal efficiency is the major parameter to study the performance of a reactor. The BOD$_5$ removal efficiency in the beginning was 73.06% later it increased up to 92.6%. During the operation pH was maintained within the range (6.7–8.0) which resulted in the optimum BOD removal efficiency and also due to successful granulation, BOD removal efficiency was increased. The Solid removal is the most important parameter to study the performance of the reactor, as hybrid reactor is provided with bedding media to improve the sludge settling characteristics. Initially, the total solids removal was 49%, Dissolved solids removal was 50.7% and suspended solids removal was 43.3% for aOLRof 2.2 kgCOD/m$^3$·d. Later, it increased to 95.4%, 97.3%, and 87.8% for a OLRof 9.6 kgCOD/m$^3$·d.

**Fig. 2** Variation of biogas with OLR

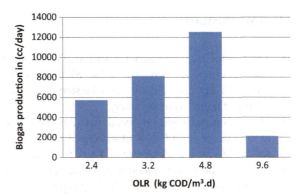

Figure 2 illustrates variation of Biogas with OLR. At an OLR of 2.4 (kgCOD/m$^3$·d), biogas production of 5800 cc/d. At an increased loading of 3.2 (kgCOD/m$^3$·d), it was 8200 cc/d, at OLR 4.8 (kgCOD/m$^3$·d) biogas production was maximum 12600 cc/d, as the loading was increased to 9.6 (kgCOD/m$^3$·d) biogas production was reduced to 2200 cc/d. An increase in OLR had shown increase in the biogas production. When the OLR was increased to 9.6 (kgCOD/m$^3$·d) biogas production decreased to 2200 cc/d and the reactor became instable, this indicates that the HRT of the reactor was not to be reduced to 12 h or less than 12 h.

The COD concentration of 7200 mg/L was maintained in the reactor at an HRT of 48 h and later HRT were reduced to 36 h, 24 h, and 12 h respectively. The COD concentration was maintained constant throughout until the COD removal efficiency has reached steady-state conditions for varying organic-loading rates 2.4, 3.2, 4.8, and 9.6 (kgCOD/m$^3$·d). Figure 3 shows the variation of organic loading rate along with HRT.

**Fig. 3** Variation of HRT with OLR

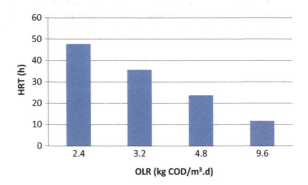

## 4 Conclusions

- The reactor was operated for various OLRs and HRTs. The maximum COD removal efficiency was 76%, 84%, 87.8%, and 94.5% after 50th, 75th, 100th, and 120th day, respectively, from the start up at a loading rate of 2.4, 3.2, 4.8, and 9.6 (kgCOD/m$^3$·d) respectively.
- The maximum biogas produced was 12,600 mL after 120th day from the start-up of the reactor at an O.L.R of 4.8 (kgCOD/m$^3$·d). The BOD$_5$ removal efficiency was 92.6% for a loading rate of 4.8 (kgCOD/m$^3$·d).
- Hence, It can be concluded that, as the daily feed ratio was increased from 2.4, 3.2, 4.8, and 9.6 (kgCOD/m$^3$·d), the gas production and percentage COD removal increased.
- However, as the O.L.R was increased to 15.36 (kgCOD/m$^3$·d), reactor became instable with an increase in VFA/Alkalinity ratio of 0.44–0.89.

## References

1. Patyal V, Lallotra B (2015) Study of USAB method of treatment of distillery wastewater. IOSR J Mech Civil Eng (IOSR-JMCE) 27-32, e-ISSN: 2278-1684, p-ISSN: 2320-334X
2. Bhatti ZA (2014) UASB reactor startup for the treatment of municipal wastewater followed by advanced oxidation process. Brazil J Chem Eng 31(03):715–726, ISSN 0104-6632
3. Venkatesh KR, Rajendran M, Murugappan A (2013) Start-up of an upflow anaerobic sludge blanket reactor treating low-strength wastewater inoculated with non-granular sludge. Int Ref J Eng Sci (IRJET) 2(5):46–53
4. Fayad N (2017) The application of electrocoagulation process for wastewater treatment and for the separation and purification of biological media. Chem Process Eng. Université Clermont Auvergne, 2017. English. ffNNT: 2017CLFAC024ff. fftel-01719756f
5. Shirule PA (2013) Treatment of dairy waste water using UASB reactor and generation of energy. Pratibha: Int J Sci Spirit Bus Technol (IJSSBT) 2(1):2277–7261

# Analysis of Morphometric Parameters of Watershed Using GIS

**Bhairavi Pawar and K. A. Patil**

**Abstract** Watershed management is the study for sustainable development of water bodies as well as land in the watershed region. Morphometric parameters of watershed act as a yardstick in analyzing, planning, and maintaining of watersheds. In this paper, data acquired from LANDSAT and NASA satellites was interpolated, supervized, and classified in order to obtain DEM and delineate watershed boundary in Arc GIS 10.3. Pour point was taken on dam wall of Palkhed dam to delineate watershed on the upstream region of the dam. Area of 811.15 sq. km of watershed was obtained. Various topographic, geomorphometric, and geological parameters of the watershed were extracted and analysed using GIS as a tool.

**Keywords** Watershed · GIS · Arc map · Morphometric parameters

## 1 Introduction

Watershed is a hydrological unit and is an exemplary unit for management of resources like water and land. According to National Water Mission (NWM), Govt. of India, the estimated usable surface water is 690BCM. Most of this water precipitated is lost in the form of excessive surface flow due to the absence of management of water. Watershed management aims at storing this excess water and utilizing it for various onsite purposes. Watershed regions have a huge extent of area and thus the study of various morphological parameters becomes tedious. Morphological parameters are obtained by using satellite imagery data and by processing this obtained data using GIS and remote sensing software like Arc-GIS.

In this era of digitization, most of the manual and traditional systems of analysis of data collection, software, and tools like remote sensing and GIS replace data analysis

---

B. Pawar (✉) · K. A. Patil
Department of Civil Engineering, College of Engineering Pune, Pune 411005, India
e-mail: pawarbv19.civil@coep.ac.in

K. A. Patil
e-mail: kap.civil@coep.ac.in

© The Author(s), under exclusive license to Springer Nature Singapore Pte Ltd. 2023
M. S. Ranadive et al. (eds.), *Recent Trends in Construction Technology and Management*, Lecture Notes in Civil Engineering 260,
https://doi.org/10.1007/978-981-19-2145-2_46

**Fig. 1** Location of study area

and data interpretation. Remote sensing tool has made it possible to gain detailed information without even having physical contact with that place.

## 2 Study Area

Dindori is a town and taluka located in Nashik District of Maharashtra, which is also known as the "Wine capital of India" because of its grape farming. Palkhed dam is a significant water resource project that falls in this region. It is an earth-fill dam on Kadwa river with gross storage capacity of 230,100.00 cu km. This study considers watershed region of 811.15 sq km on upstream side of Palkhed dam (20.1911225°N 73.8834431°E) as the study area. This region marks the presence of River Kadwa which is the tributary of Godavari river. The significance of this region is the purpose of the dam, which is solely for irrigation, and it meets the needs of almost 1.77 lakh population receding in the watershed region. Figure 1 shows the study area location.

## 3 Data Collection

DEM (30 m resolution) from ASTER DEM of the study region downloaded from the earth explorer website of USGS. Hydrological soil group data from ORNL DDAC Sponsored by NASA downloaded from daac.ornl.gov website. Landsat 8 OLI, i.e., Operational Land Imager and TIRS, i.e., Thermal Infrared Sensor with 30 m multispectral data images downloaded from USGS website, Bhukosh portal of Geological survey of India and toposheet from Survey of India Website.

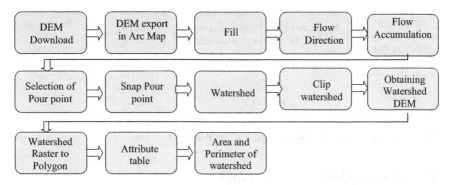

**Fig. 2** Flowchart for watershed delineation process

## 4 Methodology

### 4.1 Watershed Delineation

The delineation of watershed includes the use of Digital Elevation Models (DEMs) obtained from earth explorer web portal of U.S. geological survey website. The DEMs obtained are of 30 m resolution and act as a source of stored data that help us to obtain spatial information related to the selected study area. Prerequisite for spatial study of watershed is to delineate that watershed from the given DEM. Arc Map10.3 version facilitates the user to clip the watershed raster from the DEM. It has Spatial Analysis Tool (SAT) and Data Management Tool (DMT) that are used for watershed delineation. Following are the steps (Fig. 2).

### 4.2 Morphometric Parameters

Morphometric analysis is the practice of geomorphology that aims at quantitative description and analysis of landforms of drainage basins. One such basic parameter under morphometry is the stream order. According to the method put forward by Strahler (1964), stream orders of the given basin are further classified into five stream orders. Most of the streams are stream order 1 present in the basin. All other morphometric parameters are further obtained using stream order data. Following are the steps to obtain stream order. Figure 3 shows a flow chart of morphometric parameter analysis.

**Fig. 3** Flowchart for morphometric parameters analysis

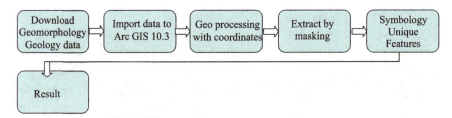

**Fig. 4** Flowchart for geomorphology and geological analysis

## 4.3 Geomorphology and Geology

Geomorphology is the scientific study of effects of physical, chemical, or biological processes on the topographic and bathymetric features of the region. Geomorphology data and geology data of India were obtained from "Bhookosh" portal of Geological Survey of India website. The data is imported into Arc Map 10.3 and terrain features are extracted using masking tool and assigning unique values and unique color codes to each feature. Raster file is geo-referenced with WGS 1984 UTM Zone 43 N coordinate system. Figure 4 shows the flow chart of for geomorphology and geological analysis.

## 4.4 TIN Model Generation

**TIN** an acronym for **Triangulated Irregular Network** is a portrayal of a continuous surface consisting completely of triangular facets, used mainly as Discrete Global Grid in primary elevation modeling. TIN model gives a 3D feel of real features on grounds. It is generated using Arc Map and can be simulated in Arc Scene. TIN Model also provides with Slope and Aspect of the region, which is essential in formulating discharge accumulation sites and depressions and elevations in the given region. Figure 5 shows steps to formulate TIN model.

**Fig. 5** Flowchart for TIN analysis

# 5 Results and Discussion

## 5.1 Watershed and Watershed Parameters

The Watershed obtained is having an area of 811.15 sq. km and a perimeter of 131.34 km. The length of the longest stream is 46.26 km. Figure 6 shows DEM clip of watershed. Figures 7, 8, and 9 show flow direction raster, drainage density, and longest stream, respectively. Figure 10 shows stream orders the values of which are ranging between 1 and 5 and Fig. 11 shows TIN model of the watershed.

**Fig. 6** DEM clip watershed

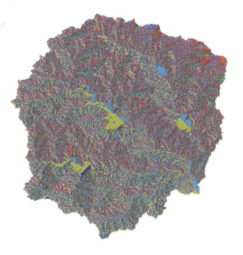

**Fig. 7** Flow direction raster

**Fig. 8** Drainage density

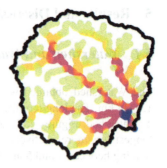

**Fig. 9** Longest stream

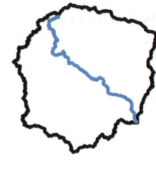

**Fig. 10** Stream order

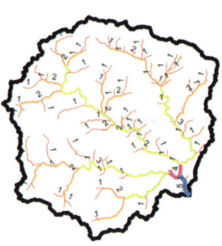

**Fig. 11** TIN model

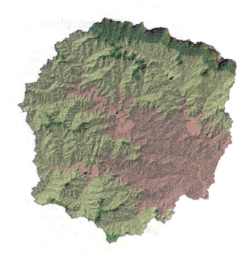

## 5.2 *Geology and Geomorphology Results*

Figure 12 shows geology of the watershed and Fig. 13 shows geomorphology of the watershed. From these figures, it is seen that the region has a late cretaceous—Palaeocene, Deccan trap type of geological features where the rock is Deccan Trap Basalt.

It is seen in the geomorphology of the watershed that the watershed has dissected plateau region. There are four reservoirs in the catchment.

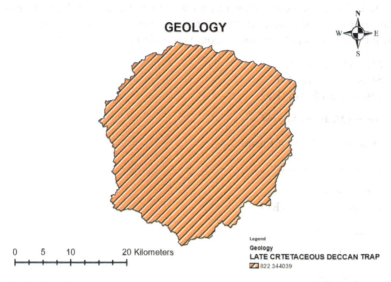

**Fig. 12** Geology of the watershed

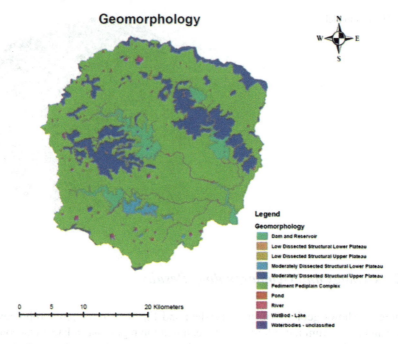

**Fig. 13** Geomorphology of the watershed

## 5.3 Morphometric Parameters Analysis

A table is formulated containing various watershed parameters like stream order, stream length, stream ratio, bifurcation ratio, stream frequency, length of overland flow, form factor, circulatory ratio, texture ratio, drainage density, etc.

1. **Stream Order: u**
2. **Stream Length: Lu**
3. **Elongation Ratio (Re)**

    **Re = (2/Lb) * (A/$\pi$)$^{0.5}$**

    **Where A** = Area of Basin

    **Lb** = Length of Basin

    **P** = Perimeter
4. **Circularity Ratio (Rc)**

    **Rc = 4 $\pi$ A/P$^2$.**

    Where A = Area of Basin

    P = Perimeter
5. **Texture Ratio (T)**

Table1 Morphometric parameters

| Sr No. | Parameters | Stream orders | | | | |
|---|---|---|---|---|---|---|
| | | I | II | III | IV | V |
| 1 | Stream order (U) | 5050 | 2633 | 2342 | 132 | 102 |
| 2 | Stream length (LU) | 173.00 | 90.49 | 80.99 | 4.4 | 3.68 |
| 4 | Stream length ratio (RL) | 1.0058 | 1.0112 | 1.0125 | 1.2941 | 1.3731 |
| 5 | Bifurcation ratio (Rb) | | 1.918 | 1.124 | 17.742 | 1.294 |
| 7 | Perimeter (P) (in km) | 141.34 | | | | |
| 8 | Basin length (Lb) (km) | 46.26 | | | | |
| 9 | Basin area (sqkm) | 811.15 | | | | |
| 12 | Elongation ratio (Re) | 0.69 | | | | |
| 13 | Length of over land flow (Lg) (km) | 4.5977 | | | | |
| 14 | Drainage density (D) (km/km$^2$) | 0.43 | | | | |
| 15 | Stream frequency (Fs) (per sq. km) | 12.64 | | | | |
| 16 | Texture ratio (Rt) | 72.5 | | | | |
| 17 | Form factor (Rf) | 0.37 | | | | |
| 18 | Circulatory ratio (Rc) | 0.51 | | | | |

$T = Nu/P$

Where P = Perimeter

Nu = Total number of stream segments of order u

6. **Drainage Density (Dd)**

$Dd = Lu/A$

Where Lu = Total stream length of order u

A = Area of Basin

7. **Bifurcation Ratio (Rb)**

$Rb = Nu/Nu\ 0 + 1$

Where Nu = Total number of stream segment of order u (Table 1).

# 6 Conclusions

Following conclusions based on analysis of morphometry:

1. Most of the streams in the study area are found to be of order 1.
2. The elongation ratio of watershed is below 0.7, thus the basin is elongated.
3. Drainage density of watershed is less than two, so the basin is coarse in nature.

4. Length of overland flow is greater than 0.4, this indicates a gentle slope and a long flow path.
5. The elongated watershed has high relief and young geomorphic stage, thus the watershed is prone to erosion.
6. The erosion causes silting of the dam and affects its storage capacity.
7. The rock type is—Palaeocene, Deccan trap type and hence it has low permeability.
8. The stream order is inversely proportional to the number of obtained streams.

## References

1. Bisht BS, Kothyari BP (2001) Land-cover change analysis of garur ganga watershed using GIS/remote sensing technique. J Indian Soc Remote Sens 29(3):2001
2. Ramakrishnan D, Durga Rao KHV, Tiwari KC (2001) Delineation of potential sites for water harvesting structures through remote sensing and GIS techniques: a case study of Kali watershed, Gujarat, India. Geocarto Int 23(2):95–108
3. Singh JP, Singh D, Litoria PK (2008) Selection of suitable sites for water harvesting structures in soankhad watershed, Punjab using remote sensing and geographical information system (RS&GIS) approach—a case study. Springer, J Indian Soc Remote Sens (March 2009) 37:21–35
4. Al Marsumi KJ, Al Shamma AM (2017) Selection of suitable sites for water harvesting structures in a flood prone area using remote sensing and GIS—case study. J Environ Earth Sci 7(4). www.iiste.org ISSN 2224–3216(Paper) ISSN 2225–0948 (Online)
5. Khan MA, Gupta VP, Moharana PC (2001) Watershed prioritization using remote sensing and geographical information system: a case study from Guhiya, India. J Arid Environ 49:465–475. https://doi.org/10.1006/jare.2001.0797, available online at http://www.idealibrary.com
6. Quinn NWT, Kumar S, Imen S (2019) Overview of remote sensing and GIS uses in watershed and TMDL analyses. J Hydrol Eng 24(4)

# Mapping Ground Water Potential Zone of Fractured Layers by Integrating Electric Resistivity Method and GIS Techniques

R. Chandramohan and B. Kesava Rao

**Abstract** This research summarizes the results of integrating GIS with electric resistivity methods to draw awareness to the significance, understanding and improving groundwater governance and management. Schlumberger Vertical Electrical Sounding (VES), the electric resistivity method was selected to examine earth's sub-layers in the selected zone. Later, the collected resistivity data was explicated by IPI2WIN software to establish Earth's subsurface zones of different regions such as top soil, weathered rocks, first and second fractured layers. The results obtained from geophysical method were exported to GIS software for identifying the most favourable and favourable groundwater-potential areas. In Palani taluk, most of the groundwater potential zones falls under hornblende-biotite gneiss. In Palani taluk 25.51 $Km^2$ area was covered with the most favourable groundwater zones and 31.32 $km^2$ was covered with favourable groundwater potential zone. The result showed that the integration of GIS and the method of electrical resistivity can be effectively used for categorizing potential groundwater areas. By using proper water recharge arrangements, the areas of the most favourable and favourable zone can be made more capable of accumulating more groundwater. This research can be useful to identify either to build or avoid constructing recharge structure in selected area. These results are useful to develop, monitor and manage groundwater resources in different hydro-geological environments.

**Keywords** GIS · VES survey · Hydrology · Fractured zone · Groundwater potential areas

R. Chandramohan · B. Kesava Rao (✉)
Department of Civil Engineering, RVR & JC College of Engineering (A), Chowdavaram, Guntur 522019, India
e-mail: kesava.battena@gmail.com

# 1 Introduction

Water is one of the earth's natural resources required for the survival of man, animals and plants. Water is distributed unevenly on the earth's surface and the earth's subsurface. Among the few foremost resources for drinking water, groundwater plays a vital role. Without groundwater, humanity cannot survive in this world. It is a foremost resource for three significant requirements of day-to-day life such as the cultivation of crops, manufacturing industries and human consumption. Insufficient groundwater may result in setbacks to the nation in the key profitable activities. Along with identification suitable groundwater potential areas, nowadays it has become essential to conserve and monitor this important resource. Overlay GIS analysis is used to locate potential groundwater area [1–3]. When compared to all geophysical method, the Schlumberger VES resistivity ground survey is an economical, less time-consuming and suitable technique for arid and semi-arid areas. Understanding the earth's various subsurface layers, especially in a fractured zone in semi-arid and arid areas, is still a very difficult task. The above-mentioned techniques are attempted by many authors [4–6]. Most of the hydrologists and geologists [7, 8] used GIS tool to identify, monitor and conserve the groundwater. The present effort involved a comprehensive electric resistivity study of the study region. Later the geophysical results were taken into GIS. In GIS, multiple thematic maps were produced such as the first fractured thickness, first fracture resistivity, second fracture thickness, second fractured resistivity and geological map. All thematic maps were overlaid to identify the apt groundwater-potential region. Overlaying all thematic maps, the most favourable, favourable and unfavourable groundwater potential zones are identified. By integrating GIS and resistivity techniques any location within the study area can be easily recognized whether it is having good or poor groundwater potential zone as well as it is possible to construct an appropriate artificial recharge structure in the study area.

# 2 Methodology

The review area lies between 10°20'2" N and 10°38'24" N and longitudes and 77°18'6" E to 77°35'41" E. Study area covers 766.83 km$^2$, of which 116.85 km$^2$ are occupied by hilly landforms. The study region falls in Tamil Nadu state, Dindigul District. Recharge of groundwater here depends on monsoon rainfall. The average rainfall of a Palani Taluk is 690 mm (1980–2013) [9, 10]. The review area is mostly covered by Archean crystalline rocks and surrounded by hillocks. Groundwater is found mostly in the fractured zones easy.

Palani taluk base map was collected from the Statistical Department of Dindigul district, and the map was geo-referenced with SOI (Survey of India) toposheet number 58f of 1:50,000 scale. Geo-referencing was performed in GIS software. The geological map was obtained from Bhuvan website, traced and digitized and then imported to

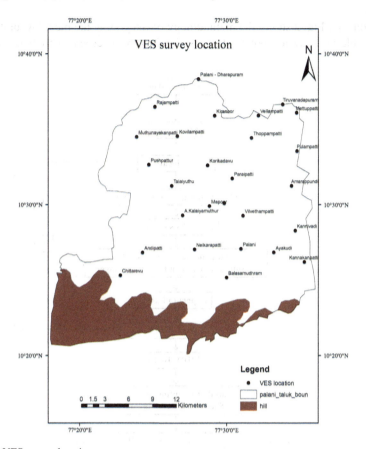

**Fig. 1** VES survey location

a GIS platform. Schlumberger VES (Vertical Electrical Sounding) ground survey was conducted at 27 locations in Palani Taluk as shown in Fig. 1. The map coordinates of the locations were confirmed with a GPS device. The geophysical Schlumberger VES ground survey was performed at 250 m maximum electrode spacing. The potential electrode and current electrode spacing vary from 0.2 m to 15 m and 0.5 m to 125 m. The collected data was exported to IPI2WIN software to evaluate the different layers of the earth's subsurface by its resistivity and thickness. The results obtained from IPI2WIN software were imported into attribute table. The ground survey coordinates and attribute data were exported in GIS software. By IDW (Inverse Distance Weight) techniques in GIS software, the spatial interpolation map was drawn for thickness and resistivity values of all earth sub-layers. This technique was used for the fractured layers to prepare the corresponding thickness and resistivity map. Later the above map was overlaid with the geological map. The suitable zones for groundwater were identified by low resistivity and greater thickness. Figure 1 shows the VES Survey location, and Fig. 2 shows the methodology of integrating the most favourable and favourable-groundwater-potential zones identified by electric resistivity method.

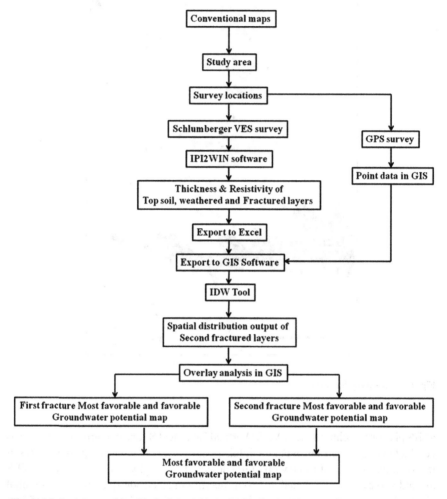

**Fig. 2** Methodology to identify the favourable and most favourable groundwater potential zone by electric resistivity

## 3 Results and Discussions

### 3.1 Groundwater-Potential Zone Using Geophysical Survey

The field survey results of VES Schlumberger are shown in Fig. 3 (thickness) and Fig. 4 (resistivity) furnish the topsoil; weathered, fractured zones (first and second). The obtained geophysical results are added to the GIS environment as a point feature for further analysis. The maximum resistivity value in Paraipatti, Mappoor, etc.

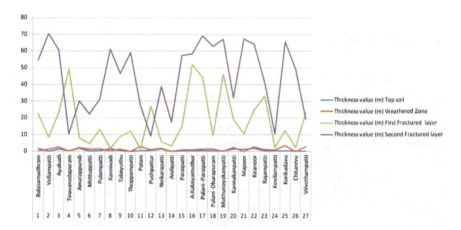

**Fig. 3** Thickness values for various VES survey location

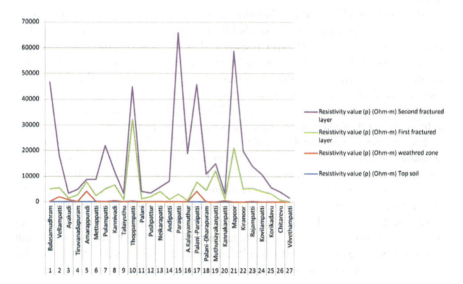

**Fig. 4** Resistivity values for various VES survey location

confirms compact rock formation. It was also evident in the survey that the borehole drilled closer to this area doesn't have a good quantity of groundwater. Low resistivity values in Amarapundi, Neikarapatti, etc. show the availability of good quantity of groundwater. The areas falling in high thickness with low resistivity value will yield a good quantity of groundwater. In the study area, high thickness, i.e. 52.14 m with low resistivity 507 Ωm value falls in the first fracture zone in a village named A. Kalaiyamuthur. By conducting the field check in A. Kalaiyamuthur, the area has good groundwater quantity. Therefore, the high thickness zone indicates a good sum of groundwater storage.

## 3.2 GIS Analysis

All VES surveyed locations' point coordinate values were captured by the GPS device. Later, the coordinate points are added as a layer into GIS software as point data and the resulting VES survey values were imported and related to the appropriate locations to generate various thematic maps. The fracture zone of the first layer resistivity, shown in Fig. 5a, was superimposed over the fracture zone of the first layer thickness, as shown in Fig. 5b. The first fractured groundwater potential zone map was derived and shown in Fig. 5c. Figure 5c gives the 15 combinations value; it was shown in Fig. 5g. Among these combinations, the values are indicated by abbreviations as follows: LT means Low Thickness, MT means Medium Thickness, HT means High Thickness, VHT means Very High Thickness, LR means Low Resistivity, MR means Medium Resistivity, HR means High Resistivity and VLR means Very Low resistivity.

Among these, HT & LR combination value covers a huge area of 97 km$^2$. Due to its shallow depth, this area is suitable for dug well construction. Similarly, The fracture zone of the second layer resistivity Fig. 5d was superimposed over the fracture zone of the second layer thickness map, shown in Fig. 5e. The Second fracture zone resistivity and thickness were derived and shown in Fig. 5f. Results show 13 number of the combination, represented in Fig. 5h. Among these MT & LR combination value covers a large area of 97.16 km$^2$. This combination shows water to be at greater depth and boring tube wells needs to be considered.

The groundwater potential zone of first fractured (Fig. 5c) was superimposed with groundwater potential zone of second fractured (Fig. 5f), which was overlaid with geological map and the resultant map was shown in Fig. 6 and it represented 107 combinations value. Of these, a few combinations value are shown in Fig. 7. The first and second fracture most favourable and favourable-groundwater-potential zone were identified; it was shown in the corresponding Figs. 8 and 9. Table 1 provides the combination of values of favourable and most favourable groundwater-potential zones by electric resistivity method for Palani Taluk. The results show that the combinations of most favourable and favourable locations in groundwater potential zone

# Mapping Ground Water Potential Zone of Fractured Layers ...

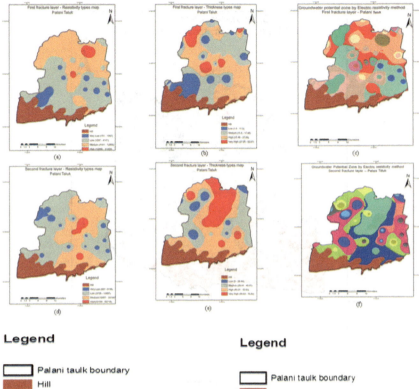

**Fig. 5** First and second fracture layer thickness, resistivity, groundwater potential zone and its legends

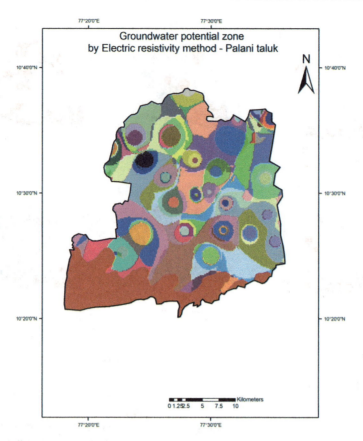

**Fig. 6** Integrated groundwater potential map of first and second fractured layer

fall on geology type of hornblende-biotite gneiss, because most of the study area comes under this type. High Thickness and Low Resistivity, Medium Thickness and Low Resistivity and hornblende-biotite gneiss cover 27.02 km$^2$; these are high groundwater potential zones. This was also verified in the field. The most favourable-groundwater-potential zone covers 25.51 km$^2$ and favourable groundwater potential zone covers 31.32 km$^2$. Figure 10 shows the integration of the most favourable and favourable groundwater potential zones of the first and second fractured layers of study area. By integrating GIS and electric resistivity method government, farmers, etc. can easily identify the location where the groundwater potential is more and where the groundwater potential is very poor. After identifying the most favourable and favourable groundwater potential zone government appropriate artificial recharge structure can be constructed in the selected locations to increase groundwater level.

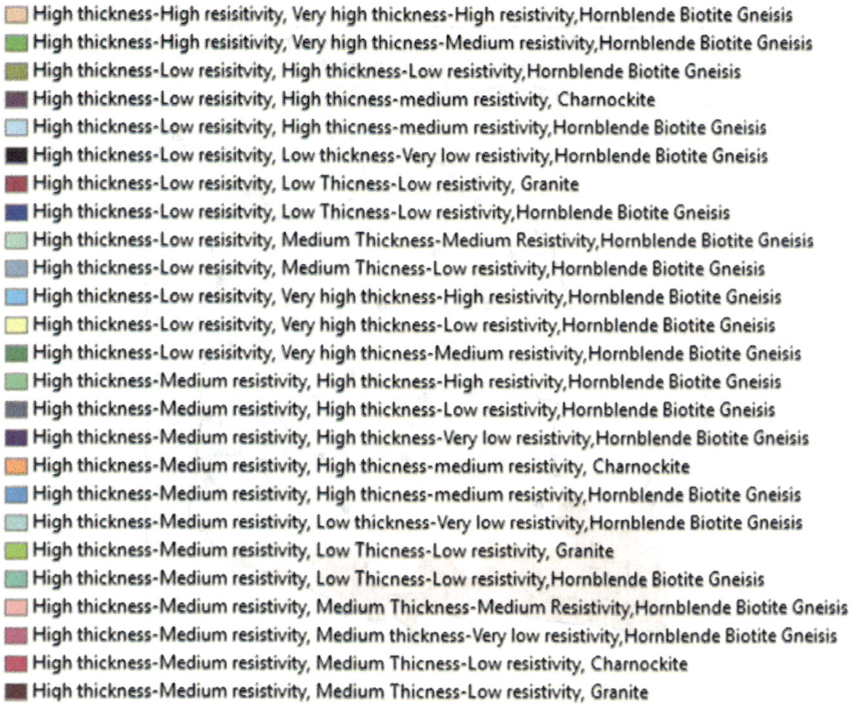

- High thickness-High resisitivity, Very high thickness-High resistivity, Hornblende Biotite Gneisis
- High thickness-High resisitivity, Very high thicness-Medium resistivity, Hornblende Biotite Gneisis
- High thickness-Low resisitvity, High thickness-Low resistivity, Hornblende Biotite Gneisis
- High thickness-Low resisitvity, High thicness-medium resistivity, Charnockite
- High thickness-Low resisitvity, High thicness-medium resistivity, Hornblende Biotite Gneisis
- High thickness-Low resisitvity, Low thickness-Very low resistivity, Hornblende Biotite Gneisis
- High thickness-Low resisitvity, Low Thicness-Low resistivity, Granite
- High thickness-Low resisitvity, Low Thicness-Low resistivity, Hornblende Biotite Gneisis
- High thickness-Low resisitvity, Medium Thickness-Medium Resistivity, Hornblende Biotite Gneisis
- High thickness-Low resisitvity, Medium Thicness-Low resistivity, Hornblende Biotite Gneisis
- High thickness-Low resisitvity, Very high thickness-High resistivity, Hornblende Biotite Gneisis
- High thickness-Low resisitvity, Very high thickness-Low resistivity, Hornblende Biotite Gneisis
- High thickness-Low resisitvity, Very high thicness-Medium resistivity, Hornblende Biotite Gneisis
- High thickness-Medium resistivity, High thickness-High resistivity, Hornblende Biotite Gneisis
- High thickness-Medium resistivity, High thickness-Low resistivity, Hornblende Biotite Gneisis
- High thickness-Medium resistivity, High thickness-Very low resistivity, Hornblende Biotite Gneisis
- High thickness-Medium resistivity, High thicness-medium resistivity, Charnockite
- High thickness-Medium resistivity, High thicness-medium resistivity, Hornblende Biotite Gneisis
- High thickness-Medium resistivity, Low thickness-Very low resistivity, Hornblende Biotite Gneisis
- High thickness-Medium resistivity, Low Thicness-Low resistivity, Granite
- High thickness-Medium resistivity, Low Thicness-Low resistivity, Hornblende Biotite Gneisis
- High thickness-Medium resistivity, Medium Thickness-Medium Resistivity, Hornblende Biotite Gneisis
- High thickness-Medium resistivity, Medium thickness-Very low resistivity, Hornblende Biotite Gneisis
- High thickness-Medium resistivity, Medium Thicness-Low resistivity, Charnockite
- High thickness-Medium resistivity, Medium Thicness-Low resistivity, Granite

**Fig. 7** Few combination values of potential groundwater zone by electric resistivity method

In recent years, most of the researchers performed VES survey to identify groundwater potential zone; this may help to identify potential groundwater region of a specific sample point location, etc. But very few researchers such as [11, 12] imported the result of geophysical method into GIS environment and applied interpolation techniques. In GIS IDW (Inverse Distance Weighted) interpolation techniques are used to estimate pixel values by averaging the values of VES survey sample points. These techniques may be useful especially in arid and semi-arid areas to identify suitable groundwater potential zone.

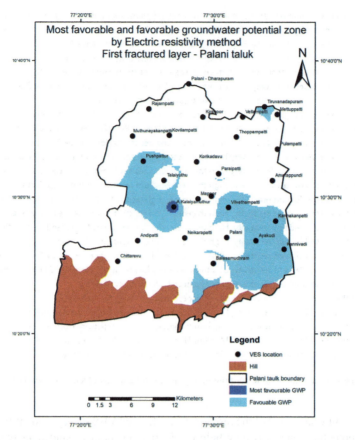

**Fig. 8** First fractured layers most favourable and favourable groundwater potential

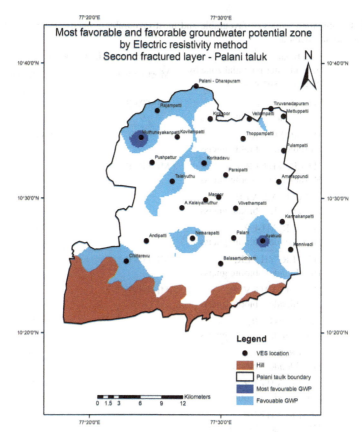

**Fig. 9** Second fractured layers most favourable and favourable groundwater potential

**Table 1** Combination values of favourable and most favourable-groundwater-potential zone by electric resistivity method

| S. No. | Combination values of groundwater potential zone by electric resistivity | Electric resistivity GWP type ||
| --- | --- | --- | --- |
| | | Most favourable GWP | Favourable GWP |
| 1 | HT-LR, HT-LR, Hornblende—biotite gneiss | MF-GWP | |
| 2 | HT-LR, MT-LR, Hornblende—biotite gneiss | | F-GWP |
| 3 | HT-LR, VHT-LR, Hornblende—biotite gneiss | MF-GWP | |
| 4 | HT-VLR, MT-VLR, Hornblende—biotite gneiss | | F-GWP |
| 5 | HT-VLR, MT-LR, Hornblende—biotite gneiss | | F-GWP |
| 6 | HT-VLR, VHT-LR, Hornblende—biotite gneiss | MF-GWP | |
| 7 | HT-VLR, VHT-VLR, Hornblende—biotite gneiss | MF-GWP | |
| 8 | MT-LR, HT-LR, Hornblende—biotite gneiss | | F-GWP |
| 9 | MT-LR, MT-LR, Hornblende—biotite gneiss | | F-GWP |
| 10 | MT-LR, VHT-LR, Hornblende—biotite gneiss | | F-GWP |
| 11 | MT-LR, VHT-LR, Hornblende—biotite gneiss | | F-GWP |

a. GWP—Groundwater potential zone
b. MF-GWP—Most favourable-groundwater-potential zone
c. F-GWP—Favourable-groundwater-potential zone

# Mapping Ground Water Potential Zone of Fractured Layers ... 625

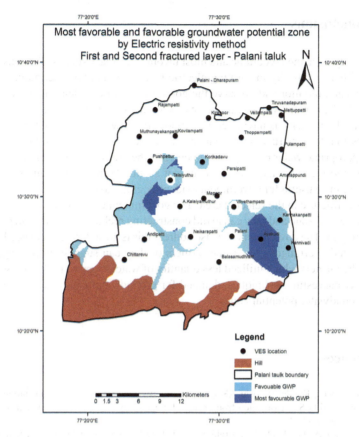

**Fig. 10** Integrated most favourable and favourable-groundwater-potential zone of first and second fracture layer

## 4 Conclusions

In the study area, the favourable and most favourable area is suitable for constructing dug and bore wells. By integrating electric resistivity and GIS techniques, we can easily locate, monitor and preserve the good groundwater potential zones. By integrating GIS and resistivity technique any location within the study area can be easily recognized whether it is having good or poor groundwater potential zone as well as it is possible to construct appropriate artificial recharge structure in the study area. With appropriate water recharge arrangements, the areas of high thickness and very low resistivity can be made more capable of accumulating more groundwater. This research can be useful for government to identify either to build or avoid constructing recharge structure in the selected area. The government can also focus poor groundwater potential zone and they can avoid constructing recharge structures in particular in that region. The government can also address this issue to farmers and need to carry out alternative arrangements to the farmers or they can suggest crops, vegetables, etc. to the farmers which utilized less quantity of water to grow. By building various rainwater harvesting techniques, demand for groundwater can reduce especially in poor groundwater potential zone.

## References

1. Rokade VM, Kundal P, Joshi AK (2007) Groundwater potential modelling through remote sensing and GIS: a case study from Rajura Taluka, Chandrapur District, Maharastra. J Geol Soc India 69:943–948
2. Adeyeye OA, Ikpokonte EA, Arabi SA (2019) GIS-based groundwater potential mapping within Dengi area, North Central Nigeria. Egyptian J Remote Sens Space Sci 22(2):175–181
3. Allafta H, Opp C, Partra S (2020) Identification of groundwater potential zones using remote Sensing and GIS techniques: a case study of the Shatt Al-Arab-Basin. Remote Sens 13
4. Gurugnanam B, Prabhakharan N, Suvetha M, Vasudevan S, Gobu B (2008) Geographic information technologies for hydrogeomorphological mapping in parts of VellarBasin, Central Tamilnadu, South India. J Geol Soc India 72:471–478
5. Ravindran A, Ramanujam A, Somasundaram P (2012) Wenner array resistivity and sp logging for groundwater exploration in Sawerpuram Teri deposits, Thoothukudi District, Tamil Nadu, India. ARPN J Earth Sci 1(1):1–5
6. Gopalan CV (2011) A comparative study of the groundwater potential in hard rock areas of Rajapuram and Balal, Kasaragod, Kerala. J Indian. Geophys Union 15(3):179–186
7. Venkateswaran S, Prabhu MV, Karuppannan S (2014) Delineation of groundwater potential zones using geophysical and GIS techniques in the Sarabanga Sub Basin, Cauvery River, Tamil Nadu, India. Int J Curr Res Academ Rev 2:58–75
8. Omolaiye GE, Oladapo IM, Ayolabi AE (2020) Integration of remote sensing, GIS and 2D resistivity methods in groundwater development. Appl Water Sci 10(129)
9. Chandramohan R, Kanchanabhan TE, Siva VN, Krishnamoorthy R (2017) Groundwater fluctuation in Palani Taluk, Dindigul District, Tamilnadu, India. Int J Civil Eng Technol 8(7):1041–1049

10. Chandramohan R, Kanchanabhan TE, Siva VN (2018) Identification of artificial recharges structures using remote sensing and gis for arid and semi-arid areas. Nat Environ Pollut Technol 18(1):183–189
11. Mohamaden MII, El-Sayed HM, Hamouda AZ (2016) Combined application of electrical resistivity and GIS for subsurface mapping and groundwater exploration at El-Themed, Southeast Sinai, Egypt. Egyptian J Aquatic Res 42(4):417–426
12. Byamungu M, Ngenzebuhoro C, Kiro K, Shungu L (2018) The application of vertical electrical sounding (ves) for the hydrogeological and geophysical investigations in kibumba area. Int J Innov Appl Stud 24(1):9–16

# Sustainable Development in Circular Economy: A Review

**Mohnish Waikar and Parag Sadgir**

**Abstract** Circular Economy (CE) is the new sustainable policy by which we can reduce waste generation by reusing and recycling the products to form a loop. The new resources are utilized less in this policy and the waste generation is minimized. This reuse and recycling will influence the business models to emerge and contribute to the country and nature. Based on the existing investigations, the current work builds up an orderly financial model in agreement with the attributes, destinations and standards of CE. Also, government mediation is basic to building a waste administration division in its beginning phase. This paper suggests the model for CE research in Pune Metropolitan Regional Development Authority (PMRDA) region for the holistic development of the industry and the betterment of society.

**Keywords** Circular economy · Water · Wastewater

## 1 Introduction

Circular economy (CE) is the sustainable use of resources while making less waste. It's all about making growth sustainable. It involves making the most use of our natural resources and constructing our goods so that processed raw materials are reused as much as possible. They ought not to wind up in a dump, however, in another item.,we create a closed loop by reusing and recycling the resources. By doing this, we can save a lot on the use of primary resources and stop generating waste, pollution and carbon emissions. The transition to CE involves action to be taken at all stages of the life cycle, beginning with product design, through raw material procurement, refining, manufacturing, use, collection of waste and ending with its management. If waste is already produced in the CE approach, it should be treated as secondary raw materials and used for re-production. This is to be achieved through mechanisms implemented at earlier stages of the life cycle. The CE approach is inextricably linked to the growth of innovation, the emergence of new business models and the growing

---

M. Waikar (✉) · P. Sadgir
Department of Civil Engineering, College of Engineering Pune, Pune 411005, India
e-mail: waikarmm20.civil@coep.ac.in

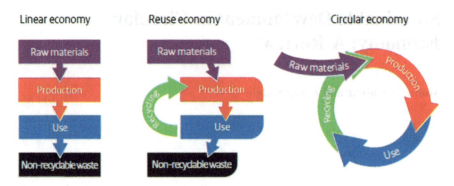

**Fig. 1** Difference between linear economy and circular economy

environmental consciousness of society. Thus, CE differs from the linear economy model, based on the "take-produce-use-throw away" concept. As the government of Netherlands has mentioned in their circular "A Circular Economy in the Netherlands by 2050" the biggest challenge of the twenty-first century is the raw materials, as the population is increasing day by day the extraction is increasing so, aside from the increased environmental impact it will likewise include expanding harm to and fatigue of common capital a deficiency of biodiversity, a danger of depleting the inventory of raw materials, and climate change. The ministry of environment and food of Denmark has also launched a strategy for CE as "Danish Strategy for Circular Economy" in which they have stated that "It is the objective of the government to promote circular economy, including better use and recycling of resources and the prevention of waste." Geissdoerfer et al. states that there are various objectives related to the CE and supportability in the writing. While it appears clear to most authors that the CE is focusing on a closed loop, taking out all asset information sources and waste and outflow spillages of the framework, the objectives of manageability are open-ended what's more, various authors address a significant huge number of objectives, which additionally move contingent upon the thought about specialists and their interests (Fig. 1).

## 2 Studying Different Models Used for CE

### 2.1 France

The government of France has announced a law, i.e. "anti-waste law for a circular economy" which proposes a more calm day by day life wherein producers are more responsible, local specialists are soothed and buyers are better educated to turn out to be more engaged with the change during their buys. There is a period when manufacturing for the purpose of destroying is no longer appropriate.

The law provides for a ban on all single-use plastics by 2040. They have implemented five-year plans for a progressive and reasoned method for phasing out which is, plastic reduction, reuse and recycling targets. Their goal is to reduce disposable tableware in the fast-food sector, by using reusable tableware, they can save up to 180,000 tonnes of packing products. The government issued a ban on polystyrene boxes, plastic confetti, etc. while promoting the use of biodegradable plastics.

Disposal, in other words, the landfilling and incineration, of unsold non-food products will be prohibited. From now on, companies will have to donate or recycle their unsold products. These include items such as daily hygiene products, clothing, electronics, shoes, books and household appliances, among others. This is according to the publication "A French act of law against waste and for a circular economy" published by The Ministry of the Ecological Transition in February 2020.

## 2.2 Netherlands

The government aims to be fully practicing CE by the end of 2050 as per their strategy "A Circular Economy in the Netherlands by 2050" published by The Ministry of Infrastructure and the Environment and the Ministry of Economic Affairs, also on behalf of the Ministry of Foreign Affairs and the Ministry of the Interior and Kingdom Relations in September 2016. In practical terms, this ensures that by 2050, raw materials can be easily used and reused, with no harmful emissions into the atmosphere. If new raw materials are needed, they will be obtained in a sustainable manner, preventing further harm to the social and physical living environments as well as public health.

They deduced that, if the CE is involved, then the extra turnover of 7.3 billion can be generated, and the use of raw materials can be reduced by 100,000 kilo tonnes, i.e. one-fourth of their total annual import for raw materials. The use of bioplastics should be increased to reduce the non-degradable plastic. With the help of students from Amsterdam and some entrepreneurs, they have started manufacturing skateboards from disposed plastic bottles and caps and are making carpets out of discarded fishing nets.

## 2.3 Denmark

Ministry of Environment and Food and Ministry of Industry, Business and Financial Affairs of Denmark, in 2018, published with state of green the publication "Strategy for Circular Economy". They have taken the following steps to achieve CE.

Empower organizations to go about as impetuses for the CE. Organizations can improve their primary concern by planning items and parts with long valuable life and the capacity to be effectively fixed, reused, and reused. This is to give SMEs induction to accurately that data and those resources that can help them convert their endeavour to the use of a more circular plan of action. Creating a typical mark of correspondence with government offices for firms that utilize circular plans of action. For organizations with roundabout plans of action, the public authority will make a solitary resource with specialists so that they can find a quicker solution on whether another innovation, plan of action, or creation strategy can be utilized under current guidelines. Increasing the inventory of financing for roundabout plans of action, The Danish Green Pension Fund will actually want to have responsibilities because of the public authority's endeavours. Thus, the asset will want to subsidize a more extensive assortment of activities. This will give new green endeavours, including circular organizations, extra decisions for finding the imperative financing for the creation, headway and creating of their game plan. The public authority is attempting to guarantee that products are planned so that they add to a circular economy indeed while guaranteeing an elevated requirement of ecological and general wellbeing security while growing Danish presence in European circular standards work. Circular procurement is promoted. Denmark's government will ensure that the country is a leader in renewable and circular public procurement. This can be achieved, among other items, by widening and improving the Relationship between Green Public Procurement and the Forum on Fair Procurement, all of which have recently been granted a joint Secretariat for Procurement to ensure co-thinking and cooperation. The government needs to see a more uniform waste collection in order to ensure improved environmental waste management while also releasing industry economic benefits from a more effective waste collection into a wider and more efficiently operating sector. For creating a fair playing field for waste and recycled raw materials on the market, the government needs to ensure that companies operate on an equal basis regardless of where they are based. Simultaneously, the government would foster a better-functioning market for waste and recycled raw materials without jeopardizing efficiency, public health, or the environment.

## *2.4 Spain*

In order to address this situation, the Spanish Strategy for the Circular Economy, called Espada Circular 2030, has been launched in June 2020. Espada Circular 2030 establishes the bases to advance another creation and utilization model in which the estimation of items, materials and assets are kept up inside the economy for as far as might be feasible, with negligible waste and reusing however much as could be expected the waste that can't be stayed away from.

The overall standards which are the foundation of this Strategy, motivated by the European and Spanish legitimate systems, are accompanying:

(i) Waste progressive system: Successful implementation of the waste order law, progressing waste anticipation, reusing and demonizing waste that cannot be reused for power production or other purposes.

(ii) Production quality: Adopting suggestions to improve worldwide proficiency and imagination of effective cycles; this should be possible by the utilization of computerized administrations and foundations, just as the presentation of natural administration plans, with the general objective of encouraging business intensity and long-haul development.
(iii) Sustainable usage: Promoting inventive models for cognizant and practical utilization, including things and organizations, similar to the use of modernized establishments and organizations.
(iv) Collaboration and mindfulness: Promoting the value of moving towards a circular economy, as well as encouraging and motivating effective networks for cooperation among organizations and data sharing between public bodies, financial and social partners, as well as creative and conventional academics, to build mutual energies that can fuel this transition.

Before 2030, they intend to keep greenhouse gas emissions under 10 million tonnes $CO_2$eq.

## 2.5 Finland

The government has a goal of implementing the CE by 2025 as mentioned in their Sitra journal, "Sitra studies 121" published in 2016. The following steps need to be taken:

(i) Rather than maximizing the amount of timber, the national forest policy should concentrate on the total benefit of Finnish forest-based goods and services.
(ii) Encourage the utilization of wood-based and different items produced using renewables in open acquisitions.
(iii) Create a business opportunity for natural reused supplements
(iv) Minimize food waste by removing obstacles and providing rewards.
(v) Promote the use of biogas systems and other green energy technologies in agriculture to minimize the use of fossil fuels.
(vi) Develop motivating forces and strategy instruments to hasten the transition to a more service-based transportation structure.
(vii) Create tax and other measures to encourage the phase-out of fossil fuels in private vehicles by 2040 and the use of renewable biofuels.

## 2.6 Germany

The publication of the government of Germany "Federal Ministry for the Environment, Nature Conservation, Building and Nuclear Safety" has been studied for the CE work. They have been in this system for a long time from about 2012 and their goal of taking resources efficiently has been on the radar. They implemented the law for procuring raw materials sustainably in 2016 and looking for the supply, fortifying natural, social and straightforward principles globally and making supply chains more supportable. The government has not drafted a combined report or policy for the whole country, whereas divided it into the city level so that optimization and increase in efficiency are possible for the cities and towns.

Different laws and goals are provided for the respective city.

## 2.7 Greece

The Ministry of environment and energy of Greece has published "National strategy for Circular Economy" in December 2018 which is referred for the CE work in Greece. Due to the variety of possibilities and potential for using the country's capital, the expertise and specialization of young Greek professionals, and the recent developments in our country's economy and development, in general, as well as the waste management industry, in particular, the circular economy concept can be easily adapted to the Greek economy. The emergency our nation has been encountering as of late, joblessness—and youth joblessness specifically—and underdevelopment set out more open doors for Circular Economy. The lack of funds to procure raw materials, the versatility of SMEs and social businesses, the need to provide jobs for young professionals, as well as environmental legislation's commitments, all lead to recycling and reuse programmes. It has the ability to generate new jobs, help small and medium-sized enterprises, grow new trades and improve the social economy.

Implementation actions taken by them:

(i) Completion of the waste management regulatory process.
(ii) Convincing execution of prioritization of waste organization, propelling the contravention of making waste and engaging in reuse and reusing are significant.
(iii) Processing proposals for reducing food loss.
(iv) A strong difference between waste and goods, making for a seamless transition to secondary raw materials.
(v) Using recycled water and sludge from wastewater treatment plants
(vi) Developing a methodology for calculating and monitoring food waste.
(vii) Promoting the utilization of waste as optional fuel in the industry.

## 2.8 Poland

The Polish government has published "Road Map towards the Transition to Circular Economy" in September 2019. Their aim for transitioning to CE is:

(i) Development of guidelines for Zero Waste Coal Power (ZWCP) aimed at minimizing the environmental impact associated with coal mining and electricity and heat generation from coal combustion.
(ii) Feasibility study on creating a dedicated platform for secondary raw materials.
(iii) Monitoring the effectiveness and efficiency of the current regulations and developing recommendations for adapting and amending national regulations on municipal waste.
(iv) Preparation of proposals for regulations concerning hazardous waste.
(v) A feasibility report on the creation and advancement of neighbourhood biorefineries.
(vi) A public awareness programme targeted at farmers, with the intention of educating them and steering their efforts towards CE.
(vii) Drawing up standards and guidelines for specific classifications of items produced using biomass.

## 2.9 Ireland

"A Waste Action Plan for a Circular Economy" prepared by the Department of Communications, Climate Action and Environment in September 2020 has been referred. They have drafted Ireland's National Waste Policy 2020–2025 in which they have included:

1. For Household and Business

    (a) Recycling targets should be given to waste collectors so that they will work efficiently
    (b) Standardized bin colours should be used across the state: green for recycling, black for residual and brown for organic waste.
    (c) A body to regulate waste control and customer protection.
    (d) A waste segregation education and awareness drive.

2. Plastic, Packaging and Single-Use Plastic (SUP)

    (a) A scheme for depositing the plastic and aluminium cans should be developed for the encouragement of people.
    (b) Single-use items will be banned from July 2021.
    (c) Significantly reduce sups being available in the market by 2026 so that the waste generation can be reduced.
    (d) All packaging should be reusable or recyclable by 2030 for the minimization of waste.

## 3 Methodology

By referring to the mentioned documents, we can carry out the CE planning and implementation work for PMRDA. In terms of the developing community, this is important for moving forward, continuing with the framework and setting targets for waste minimization, product reuse, recycling and compliance with the CE. We should primarily concentrate on issues such as agriculture, manufacturing and the Air Quality Index as tools for assessing whether the process is being applied or not. Waste management, such as electronic waste, biomedical waste, industrial waste, urban waste and so on, is also an important step in ensuring CE compliance.

### 3.1 Data Collection and Analysis

The data used for the PMRDA work is referred from the annual Environmental Status Report (ESR) 2018–2019. As PMRDA majorly consists of Pune and Pimpri Chinchwad, the ESR from Pune Municipal Corporation (PMC) and Pimpri-Chinchwad Municipal Corporation (PCMC) of 2018–2019 is used for reference.

The waste generated from the city is categorized as given below. By seeing the data, we can definitely say that the waste that is going into the landfill is more, i.e. 1100 MT. So, if we develop a method where the waste generation is minimized and the reuse of products is increased, then the landfill mass will be less. If goods are to be reused, careful segregation is necessary. A policy for the consideration of waste and what constitutes a secondary product should be given. The biodegradable waste should also be used to generate energy using biogas plants, etc. The wet waste can also be used for making compost and used as fertilizers for farming. Sludge from the wastewater treatment can also be used for the same. Treated water can also be used for gardening, washing, in industry for cooling towers and so on.

As we can see, the daily production of waste around 2000–2200 MT is generated, and from that, around 50% goes to landfills. If we reduce the amount of waste generated first-hand at the primary stage, then we can save on the land of the landfill, we can also save the vehicles needed to transport the waste to the site, thus helping nature in multiple ways. The larger part of the daily waste generated is that of the household waste in comparison to others (Figs. 2 and 3).

The utilization of inexhaustible resources like sun-based energy, wind energy ought to likewise be focus on as the essential resources for other energy creation for example petroleum derivatives, etc. will be minimized and there will be reduction of the extraction of resources from nature. Using rainwater harvesting, the extraction of water can be minimized and load on the water distribution system can be shared. The data shows that the usage of rainwater harvesting, vermicomposting and solar heaters have increased in the present year as people are becoming more aware of

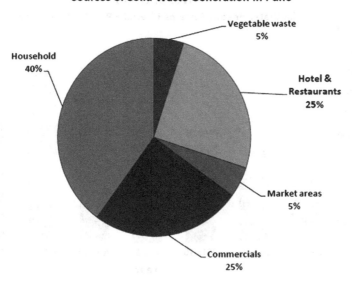

**Fig. 2** Generation of solid waste in Pune (*Source* ESR 2018–2019)

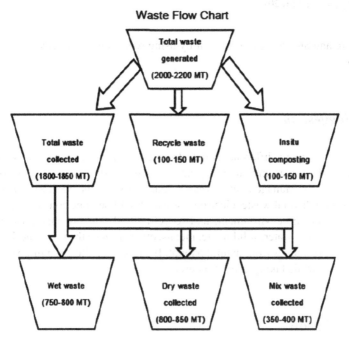

**Fig. 3** Collected waste categorized (*Source* ESR 2018–2019)

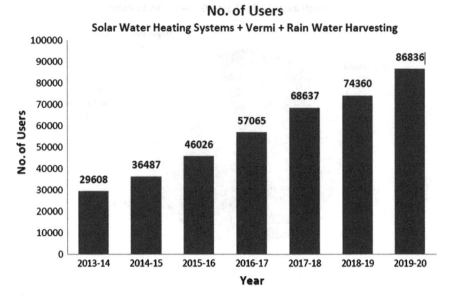

**Fig. 4** Number of users using solar water heating systems, vermicomposting and rainwater harvesting (*Source* ESR 2018–2019)

the nature and are trying to protect it by using renewable resources as their sources (Fig. 4).

## 3.2 Suggestion

As we can compare different models from the European countries, we can also achieve CE if we start implementing the norms and start focusing on how to manage our city properly. Management of water resources (water and wastewater), municipal solid waste, industrial waste, electronic waste should be done considering the effect of products on the environment, and recycling and reuse of materials should be increased, such as water, solid waste, metals, etc. to reduce the raw material load and depletion of resources from natural habitat. If succeeded, the CE will give a boost to the economy while taking care of nature.

## 4 Conclusion

This paper has talked about the Circular Economy idea which is presently pervasive in the arrangement and business advancement banter on the economic advancement of modern creation. CE is seen by strategy improvement offices and business relationships as a significant component to advance economical creation and is seen as a potential or potential future paradigmatic move, which will, therefore, bring about mechanical changes. Besides, the paper tended to the subsequent examination question—How is the Circular Economy thoughtfully identified with sustainability? We found that the Circular Economy is seen as a condition for sustainability, an advantageous connection, or a compromise in writing. This can be separated into eight distinct connections. In light of the examined writing, this paper contends that the subset relationship is by all accounts proper to keep up variety while, correspondingly, revealing insight into the wide scope of correlative techniques that managers and policymakers can adopt. In the current conditions, a Circular Economy can reinforce business and advancement, along with a significant level of natural preservation. Putting resources into circular economy, energy productivity and confronting environmental change may turn into a switch for changing the profitable model, in this way, turning around the pervasive patterns of de-venture, while likewise advancing new speculation and making new openings all through the inventory network of industrial items.

# Biogas Generation Through Anaerobic Digestion of Organic Waste: A Review

**Vaishali D. Jaysingpure and Moni U. Khobragade**

**Abstract** The main aim of the present study is to focus on the generation and optimization of methane gas from organic waste by using anaerobic digestion (AD) technique. Anaerobic digestion is a traditional method that is in use for decades. It is the biggest source of renewable energy. It gives an optimal solution to the treatment of organic waste by giving useful byproducts. The byproducts of AD, mainly biogas, are directly used as an energy resource. It can be transformed into any form of energy. The efficiency of the AD process and the amount of gas production fully depend on parameters such as pH, temperature, holding time and recirculation of leachate (Gopal in J King Saud Univ Sci 33, 2021). Recently, a concept of co-digestion came into focus in which two organic wastes are digested together to enhance the efficiency of the digestion process. The results obtained by different researchers proved that co-digestion of two substrates produces a large quantity of biogas than a single substrate. Results obtained by experimental analysis can be optimized by using different types of software like MATLAB, RSM, i.e. Response Surface Methodology and ANN, i.e. Artificial Neural Network.

**Keywords** Anaerobic digestion (AD) · Co-digestion · Biogas

## 1 Introduction

Nowadays, the whole world is facing the most threatening issue, i.e. Solid waste disposal. Approximately 980 million tonnes of waste generation is observed annually in India, which includes waste generated from households, commercial, industries and agriculture waste. Anaerobic Digestion (AD) is a traditional method used for organic solid waste treatment in an efficient manner. It is a technology in which

---

V. D. Jaysingpure (✉) · M. U. Khobragade
Department of Civil Engineering, College of Engineering Pune, Pune 411005, India
e-mail: vaishalij18.civil@coep.ac.in

M. U. Khobragade
e-mail: muk.civil@coep.ac.in

© The Author(s), under exclusive license to Springer Nature Singapore Pte Ltd. 2023
M. S. Ranadive et al. (eds.), *Recent Trends in Construction Technology and Management*, Lecture Notes in Civil Engineering 260,
https://doi.org/10.1007/978-981-19-2145-2_49

waste is stored in an air-tight tank for some specific duration. Therefore, this technology has been successfully used to stabilize the organic sludge produced from municipal wastewater treatment. Recently, a number of researchers are using this technology for treating many organic solid wastes such as wastes from energy crops. Goapl et al. [1] studied the AD for flower waste. Wang et al. [2] used AD concept for waste-activated sludge digestion. Chandra et al. [3] studied methane production from anaerobic digestion of oil cakes of jatropha and pongamia. Nowadays, mixed anaerobic digestion, i.e. mixing of one organic waste with other organic wastes (2 or 3 wastes) appears to be an efficient technique for enhancement of biogas production. Biogas, which is one of the most useful byproducts of the anaerobic digestion process, is a type of biofuel resulting from the disintegration of organic matter. It is a mixture of various gases like $CO_2$, $H_2S$ and Methane. Biogas is considered as the largest energy source. It is a key source of green energy in the terms of electricity and heat for the local grid. Korai et al. [4] studied the mono-, co- and tri-digestion of fruits, vegetables and yard waste. He recorded the data for three different seasons. He reported that tri-digestion of fruits, vegetables and yard waste with an equal mixing ratio gives the highest methane production. Cheng et al. [5], in his study, added food waste with rusted iron shavings to improve the process of digestion. He recorded a significant increase in daily methane yield and rate of methane production after addition of rusted iron shavings to the food waste. Co-digestion of food waste with vegetable waste is performed in a leach bed reactor [6].

Nguyen et al. [7] tested anaerobic digestion (AD) of various organic wastes, such as wastes from livestock farms, a cattle slaughterhouse waste and waste streams from the agricultural zone, to check its ability to produce methane. The studies are carried out under thermophilic conditions. He studied the highest methane yield of 110.83 $mLCH_4$/g VS added from animal carcasses on the 4th day. The experimental results proved the model organic wastes to have the good potential to produce biomethane. To overcome the deficiency of nitrogen in rice straw, it is digested anaerobically with co-substrate which is rich in nitrogen named, hydrilla verticillata [7].

## 2 Materials and Methods

### 2.1 Feedstocks and Inoculums

Various types of organic wastes called as feedstocks, also called as substrates are used for biogas production by anaerobic digestion. Different researchers used a variety of organic feedstocks which includes agricultural waste, kitchen waste, garden waste, dairy waste, vegetable and fruit waste, industrial sludges, etc. [8]. The biogas potential of each feedstock is different. Kainthola et al. [9] carried out the anaerobic co-digestion of hydrilla verticillata with rice straw. Ranjithav [10] explained the results of the anaerobic co-digestion of vegetables and flower wastes.

The process of addition of a suitable number of active microbes into the feedstock during anaerobic digestion to convert the organic material into biogas is called inoculation. Therefore, the success of digestion process mostly depends on the proper selection of inoculums. The best inoculum consists of a large count of active microbes. The factors such as rate of degradation, composition of biogas, time taken for complete digestion and stability of reactor are directly influenced by the proper selection of inoculums. Commonly used inoculums include industrial sludges or sludge from wastewater treatment plants and cow dung.

Nguyen et al. [7] collected cow dung from Amingaon, Assam, India, from the outlet of operating biogas plant through a fixed dome and used it as an inoculum. He observed that the cow dung consisted of abundant anaerobic microorganisms and analyzed it for various parameters like VS% of 32.63%, sCOD of 8.81 g/L and a pH of 7.75.

## 2.2 Pretreatment of Feedstock

Treatments given to the feedstock prior to feeding in the digester are called as pretreatments. Pretreatment improves the AD treatment of organic waste. It improves biogas production. The main advantage of pretreatment is that it breaks the lignin layer which protects the cellulose and hemicelluloses, which, in turn, make the organic material easily accessible to the microorganisms and thus improves the digestion process. Pretreatment also helps to decrease the crystalline nature of cellulose which increases its porosity.

Main pretreatments given to the substrates include physical, chemical and biological treatments [1]. Physical pretreatment includes mechanical and thermal or microwave pretreatments. Mechanical pretreatment includes particle size reduction of the substrate which increases the speed of the digestion process by making a large surface area available for the microorganisms to feed on the organic matter. Thermal pretreatment maintains the required temperature of the digester by using different techniques. In Chemical pretreatment, some chemicals such as acids, alkalis and oxidizers are added to the feedstock. Whereas in biological pretreatment, few microorganisms are introduced into the substrate. The most commonly used pretreatments include Physical and chemical methods, whereas biological treatment methods are used rarely. These pretreatments can be used in combination with Physical-chemical and thermo-chemical methods. Wang et al. [2] carried out combined alkaline/acid pretreatments to waste-activated sludge using bipolar membrane electrodialysis to produce equal concentrations of acid and alkali.

The co-digestion of wheat straw and chicken manure was carried out with the waste-activated sludge. Initially, it was thermally pretreated before being sent into the digester. It was analyzed for different four C/N ratios: 35:1, 30:1, 25:1 and 20:1. After experiments, he studied that the composition having C/N value of 20:1 as the optimum value [11].

## 2.3 Substrate to Inoculum Ratio (S/I Ratio)

The substrate to inoculum ratio can be expressed as the amount of volatile solids present in the substrate per the amount of volatile solids or amount of volatile suspended solids (VSS) present in the inoculum. The rate of the anaerobic digestion process is highly influenced by the S/I ratio. It is considered one of the most important parameters in biomethane potential testing. Sri Bala Kameswari et al. [12] studied the optimization of I/S ratio for co-digestion of fleshings with the sludge from the tannery. He carried out batch studies and maintained 6 different reactors with 6 different I/S ratios: 0.25, 0.50, 0.67, 1.00, 1.50 and 2.20. Biogas generation was monitored daily. It was noted that methanogenic activity slow down as the I/S ratio decreased, which results in a decrease in the generation of biogas. But with an increased I/S ratio from 0.25 to 2.30, 1846 ml of increase in biogas generation was observed. Gopal et al. [1] studied the AD of flower waste by running 3 experimental setups with 3 different S/I ratios 1:2, 1:1 and 2:1, respectively.

Korai et al. [4], while studying co-digestion and combined digestion of 3 substrates including fruit (FW), Vegetables (VW) and yard (YW) waste, prepared the samples by using different ratios on a weight basis such as FW:VW (1:1), FW:YW (1:1) and VW:YW (1:1). And for combined digestion of 3 substrates, the samples were prepared by using different ratios on a weight basis such as FW: VW: YW (0.5:1:1), Fr: V: Y (1:0.5:1), FW: VW: YW (1:1:0.5) and FW: VW: YW (1:1:1).

Co-digestion of kitchen waste and black water in the ratio of 2:1 gives a larger amount of biogas than single digestion of kitchen waste. Also, by the addition of granular activated carbon by maintaining the same ratio, biomethane potential (BMP) is enhanced [13]. Food waste is digested with raw sludge (co-digestion) but with a higher concentration of food waste gives higher methane production. Okoro-Shekwaga et al. [14] in his study used two different food waste samples with less particle sizes of 1 mm and 2 mm, using two different inoculum to substrate ratios for testing biomethane potential. He studied that the two inoculums to substrate ratios, i.e. 3:1 and 4:1 were helpful to stabilize experiment reactors, respectively. For the co-digestion of vegetable and flower waste, pig manure is used as an inoculum [8].

## 3 Experimental Analysis

The biological process, with zero presence of oxygen, in which bacterial breakdown of organic materials occurs resulting in the generation of biogas, which is composed of different gases including methane, carbon dioxide and hydrogen sulphide, etc. is called as anaerobic digestion. Feedstocks used to produce gas include organic solid wastes from different sources, manure resulting from live stocks and wet organic materials. Anaerobic digestion has two stages, namely: fermentation and methanogenesis. Fermentation includes hydrolysis, acidogenesis and acetogenesis. In fermentation, 10 to 20 % of the energy of the substrate is transformed into gaseous products such as hydrogen and carbon dioxide. Whereas, as the name suggests, in methanogenesis, 80 to 90 % energy of the substrate gets transformed into gaseous products, i.e. methane. The complete anaerobic digestion process takes approximately 21 days [15]. On the basis of stages of operation, anaerobic digestion is carried out in two ways, batch and continuous process.

## 3.1 Batch Mode Technique of Anaerobic Digestion

In a batch mode anaerobic digestion process, the reactor vessel is fed with raw feedstock and inoculum in proper ratio. Then it is sealed to maintain the anaerobic condition and kept undisturbed until complete digestion is over. Biomethane potential called as BMP test is conducted to check the biodegradability of feedstocks and their capacity to produce methane gas. It is carried out in small bottles called as BMP bottles varying from size 125 ml to 2000 ml. Wang et al. [2] studied biomethane potential for sludge treatment. Acid and alkaline pretreatment were given to the waste and the experiments were conducted in 250 ml serum bottles. To mix the substrate and inoculums thoroughly, the shaker was operated at a shaking speed of 150 rpm maintaining a temperature of 30 °C.

Korai et al. [4] studied the biomethane potential of wastes arising from fruits, vegetables and yards with combinations of one, two and three substrates. Anaerobic digestion is carried out in fifteen reactors, each having a capacity of 500 ml. The tests are carried out under mesophilic conditions by maintaining the temperature up to 38 ± 1 °C or under thermophilic conditions by maintaining the temperature up to 50–58 °C. The digesters are expected to be agitated manually or automatically 3 to 4 times per day for the proper bacterial contact with the waste and to avoid the sedimentation of solid particles. According to studies for statistical analysis, BMP should be carried out in triplicates [16]. BMP of inoculums is measured separately to identify the BMP of substrate individually. Accumulated gas collection can be done daily by using the liquid displacement method. The percentage of methane, carbon dioxide and hydrogen sulphide content are estimated by using Gas chromatography.

## 3.2 Continuous Mode Anaerobic Digestion

In a continuous digestion process, the reactor is continuously fed with feedstock and the sludge, i.e. digested material is removed continuously at a specific interval. Continuous mode anaerobic digestion process is carried out on a magnifying scale. The main difference between batch and continuous methods is that in the batch process, the reactor is fed only once at the beginning and operated under required controlled conditions, whereas, in the continuous process, it is fed daily. A batch process is technically simple and less capital intensive. Still, the land required for the batch process is comparatively larger. Besides, as waste generation is a continuous process, the batch system is not practical. Also, another major concern in the batch process is its hectic loading and unloading. Having these short comings, the batch process has got some restrictions on the users for their study.

## 3.3 Characterization of Waste

The biodegradability of organic matter completely depends on its composition. Hence, it is essential to know the exact characteristics of a substrate and inoculum before carrying out a BMP test of the substrate [8]. As per [7], food/microorganism (F/M) ratio, carbon/nitrogen (C/N) ratio and pH are the main environmental parameters that have a large influence on biogas generation includes.

Generally, the feedstocks and inoculums are analyzed for total solids (TS), Suspended solids (SS), moisture content (MC), Volatile solids (VS), pH, total alkalinity and COD. Elemental analysis is carried out for the presence of C, H. N, O and S. Surface morphology of the substrate is studied by using SEM, i.e. Scanning electron microscopy.

Korai et al. [4] analyzed feedstocks, inoculums and material for TS, VS, pH, total alkalinity, volatile fatty acids and ammonia nitrogen ($NH_3$–N) according to the standard methods defined in APHA, 2005. Elemental analysis for C, H, O, N, was carried out as per the method prescribed in BBOT23122013. Wang et al. [2] studied the surface morphology of pretreated sludge and E. coli cells by scanning electron microscopy, SEM USA make. Nguyen et al. [7] has studied elemental composition (C, H, N, O) of each by an elemental analyzer Italy make, which is used to study the elemental composition C, H, N, O of substrates.

## 3.4 Measurement of Biogas

Techniques involved in gas measurements evaluates gas either by using manometers maintaining the volume constant and increase in the pressure indicates the volume of gas. Another method measures gas by volume maintaining constant pressure conditions and measuring the volume of gas [8]. Low permeability gas sampling bags can be the alternative biogas collection method that nullifies the absorption problem during long periods of contact with a barrier solution.

Measurement of biogas generated from lab scale models is done either by water displacement method or by using lubricated syringes. Liquid displacement is the most commonly used method for the collection of biogas [17]. Other instruments such as manometers, pressure transducers and low pressure flow meters are also used for the measurement of biogas. For large-scale models, it can be measured by using automatic gas flow meters. The water displacement method includes the arrangements of three tanks placed in sequence connected to each other. The first one is a digestion tank in which feeding material is added for a particular period. The second is the water displacement tank. This tank is filled with water and is connected to the digester through a small pipe. The third unit is the collection tank which is again connected to the displacement tank through a pipe. By this method, gas generated in the digestion tank is measured in terms of ml of water displaced by the displacement tank. The water displacement method is used to measure biomethane production. Instead of water, 1.5 ml of normal sodium hydroxide solution is used in the displacement tank, and an indicator, named thymol blue is used. The displaced sodium hydroxide volume is measured by measuring cylinder which gives the per day quantity of biomethane generated [9].

In the syringe method, normally a 1000 ml syringe is used to measure biogas. A gas tight syringe is used in each digester at regular intervals of time and measures the volume of biogas production. By injecting the needle through the butyl bung, the syringe is connected to the reactors. Pressure in the headspace is dropped to ambient pressure by drawing the plunger out and the amount of biogas is collected, and thus the volume of gas in the syringe gives the measurement of the gas produced. After collection, a gas chromatograph is used to analyze the biogas for its contents. A gas chromatograph is equipped with a capillary column having split ratio 3:1 and dimensions: 30 m length, inner diameter 0.53 mm, 40 $\mu$m film and having a thermal conductivity detector to measure the methane concentration [6, 8].

## 3.5 Optimization of Biogas

Optimization is the process of enhancing system capabilities and integration of all the parameters to the extent that all components operate with maximum efficiency. It is done by using different types of software. These software provide standardized, efficient, accurate and comprehensible 2D and 3D graphs which gives clarity to the results obtained from experiments. Nowadays different types of software are used to optimize biochemical methane potential (BMP) of various substrates including MATLAB, RSM (Response Surface Methodology), ANN (Artificial Neural Network), etc.

Nguyen et al. [7] used RSM, a central composite design (CCD) for optimization of the design and performed twenty trials for this experimental design. The three important variables selected for the optimization and modelling include $X1 = C/N$ ratio, $X2 = F/M$ ratio and $X3 = pH$.

Sharma et al. [17] used response surface methodology for optimization of various combinations of flower waste with cow dung. For defining the experimental design in an anaerobic co-digestion of rice straw with hydrilla verticillata, a central composite design, response surface methodology was used efficiently [18]. The stability of the digestion, potential of biomethane production and rate of biomethane production for waste from farms of livestock, slaughterhouse waste and streams of waste generated from agricultural lands were assessed followed by the experimental analysis of results using four kinetic models [7].

## 4 Discussion and Concluding Remarks

The current research primarily focuses on the processes and methods of the anaerobic digestions through various well-known researchers' work into consideration. This study has focused on detailed methods available for anaerobic digestion, its significance and limitations. An effort is paid to find the optimum conditions required for an efficient digestion process. Rigorous insights into the study yield the following conclusions:

- Anaerobic digestion is an efficient technology which has remarkable advantages compared to other waste stabilization techniques. It converts the organic waste into useful byproducts like biogas which is the cheapest source of energy. Digestate resulting from the process have the highest nutritive value and is used as a fertilizer.
- It is studied that anaerobic digestion of multi substrates gives higher methane production as compared to anaerobic digestion of a single substrate. Rather the productivity of biogas reactors can be improved further by a mix-digestion strategy, as compared to mono-digestion and co-digestion of organic waste.
- Temperature, pH and Mixing rate of waste/agitation, and Hydraulic retention time are the major controlling parameters of the anaerobic digestion process.
- Proper inoculum to substrate ratio (I/S ratio) has a significant influence on biogas generation in the anaerobic digestion process.
- After a detailed review of the work of researchers, it was observed that the pretreatment of substrates gives more favourable results than that of untreated waste. Also, it is noted that thermal alkaline pretreatment enhances biogas production significantly.
- Optimization of parameters influencing the anaerobic digestion process gives accurate predictions, thus improving the efficiency of the digestion process to a large extent.

## References

1. Gopal LC et al (2021) Optimization strategies for improved biogas production by recycling of waste through response surface methodology and artificial neural network: sustainable energy perspective research. J King Saud Univ Sci 33(1). https://doi.org/10.1016/j.jksus.2020.101241
2. Wang S et al (2019) Development of an alkaline/acid pre-treatment and anaerobic digestion (APAD) process for methane generation from waste activated sludge. Sci Total Environ 134564. https://doi.org/10.1016/j.scitotenv.2019.134564
3. Chandra R, Vijay VK, Subbarao PMV, Khura TK (2012) Production of methane from anaerobic digestion of jatropha and pongamia oil cakes. Appl Energy 93:148–159. https://doi.org/10.1016/j.apenergy.2010.10.049
4. Korai MS, Mahar RB, Uqaili MA (2018) The seasonal evolution of fruit, vegetable and yard wastes by mono, co and tri-digestion at Hyderabad, Sindh Pakistan. Waste Manag 71:461–473. https://doi.org/10.1016/j.wasman.2017.09.038

5. Cheng J, Zhu C, Zhu J, Jing X, Kong F, Zhang C (2020) Effects of waste rusted iron shavings on enhancing anaerobic digestion of food wastes and municipal sludge. J Clean Prod 242:118195. https://doi.org/10.1016/j.jclepro.2019.118195
6. Chakraborty D, Venkata Mohan S (2019) Efficient resource valorization by co-digestion of food and vegetable waste using three stage integrated bioprocess. Bioresour Technol 284(January):373–380. https://doi.org/10.1016/j.biortech.2019.03.133
7. Nguyen DD et al (2019) Thermophilic anaerobic digestion of model organic wastes: evaluation of biomethane production and multiple kinetic models analysis. Bioresour Technol 280(February):269–276. https://doi.org/10.1016/j.biortech.2019.02.033
8. Raposo F, De La Rubia MA, Fernández-Cegrí V, Borja R (2012) Anaerobic digestion of solid organic substrates in batch mode: an overview relating to methane yields and experimental procedures. Renew Sustain Energy Rev 16(1):861–877. https://doi.org/10.1016/j.rser.2011.09.008
9. Kainthola J, Kalamdhad AS, Goud VV (2019) Optimization of methane production during anaerobic co-digestion of rice straw and hydrilla verticillata using response surface methodology. Fuel 235. https://doi.org/10.1016/j.fuel.2018.07.094
10. Ranjithav J (2014) Production production of bio-gas from flowers and vegetable wastes using anaerobic digestionof bio-gas from flowers and vegetable wastes using anaerobic digestion. Int J Res Eng Technol 03(08):279–283. https://doi.org/10.15623/ijret.2014.0308044
11. Hassan M, Ding W, Shi Z, Zhao S (2016) Methane enhancement through co-digestion of chicken manure and thermo-oxidative cleaved wheat straw with waste activated sludge: a C/N optimization case. Bioresour Technol 211:534–541. https://doi.org/10.1016/j.biortech.2016.03.148
12. Sri Bala Kameswari K, Kalyanaraman C, Porselvam S, Thanasekaran K (2012) Optimization of inoculum to substrate ratio for bio-energy generation in co-digestion of tannery solid wastes. Clean Technol Environ Policy 14(2):241–250. https://doi.org/10.1007/s10098-011-0391-z
13. Zhang Q, Li R, Guo B, Zhang L, Liu Y (2021) Thermophilic co-digestion of blackwater and organic kitchen waste: impacts of granular activated carbon and different mixing ratios. Waste Manag 131(June):453–461. https://doi.org/10.1016/j.wasman.2021.06.024
14. Okoro-Shekwaga CK, Turnell Suruagy MV, Ross A, Camargo-Valero MA (2019) Particle size, inoculum-to-substrate ratio and nutrient media effects on biomethane yield from food waste. Renew Energy 151:311–321. https://doi.org/10.1016/j.renene.2019.11.028
15. Cremiato R, Mastellone ML, Tagliaferri C, Zaccariello L, Lettieri P (2018) Environmental impact of municipal solid waste management using Life cycle assessment: the effect of anaerobic digestion, materials recovery and secondary fuels production. Renew Energy 124:180–188. https://doi.org/10.1016/j.renene.2017.06.033
16. Filer J, Ding HH, Chang S (2019) Biochemical methane potential (BMP) assay method for anaerobic digestion research. Water (Switzerland) 11(5), MDPI AG. https://doi.org/10.3390/w11050921
17. Sharma D, Yadav KD, Kumar S (2018) Biotransformation of flower waste composting: optimization of waste combinations using response surface methodology. https://doi.org/10.1016/j.biortech.2018.09.036
18. Kainthola J, Kalamdhad AS, Goud VV (2019) Optimization of methane production during anaerobic co-digestion of rice straw and hydrilla verticillata using response surface methodology. Fuel 235:92–99. https://doi.org/10.1016/j.fuel.2018.07.094